METHODS IN MOLECULAR BIOLOGY

Series Editor
John M. Walker
School of Life and Medical Sciences
University of Hertfordshire
Hatfield, Hertfordshire, AL10 9AB, UK

For further volumes:
http://www.springer.com/series/7651

Mitochondrial Medicine

Volume I, Probing Mitochondrial Function

Edited by

Volkmar Weissig

Midwestern University, Glendale, AZ, USA

Marvin Edeas

ISANH, Paris, France

 Humana Press

Editors
Volkmar Weissig
Midwestern University
Glendale, AZ, USA

Marvin Edeas
ISANH
Paris, France

ISSN 1064-3745   ISSN 1940-6029 (electronic)
ISBN 978-1-4939-2256-7   ISBN 978-1-4939-2257-4 (eBook)
DOI 10.1007/978-1-4939-2257-4
Springer New York Heidelberg Dordrecht London

Library of Congress Control Number: 2014960347

Cover illustration: From Figure 1 of Chapter 25 (Prigione)

Printed on acid-free paper

Humana Press is a brand of Springer
Springer is part of Springer Science+Business Media (www.springer.com)

Preface

Mitochondrial Medicine is an interdisciplinary and rapidly growing new area of biomedical research comprising genetic, biochemical, pathological, and clinical studies aimed at the diagnosis and therapy of human diseases which are either caused by or associated with mitochondrial dysfunction. The term "Mitochondrial Medicine" was probably used for the first time by Rolf Luft [1] who is widely accepted as the father of Mitochondrial Medicine. Over 50 years ago, it was he who described for the very first time a patient with clinical symptoms caused by malfunctioning mitochondria [2].

The beginning of mitochondria-related research dates back to the end of the nineteenth century. During the 1890s, early cytological studies revealed the existence of bacteria-resembling subcellular particles in the cytosol of mammalian cells. Robert Altman termed them bioblasts, and he hypothesized that these particles were the basic unit of cellular activity. The name mitochondrion, which means thread-like particles, was coined in 1898 by Carl Benda. During the 1940s, progress was made in the development of cell fractionation techniques which ultimately allowed the isolation of intact mitochondria from cell homogenates, thereby making them more accessible to biochemical studies. Subsequently, by the end of the 1940s, activities of a variety of enzymes needed for fatty acid oxidation, the Krebs cycle, and other metabolic pathways were found to be associated with mitochondrial fractions.

Human mitochondrial DNA was discovered in 1963 [3], and Mitchell's disputed chemiosmotic theory [4] of ATP synthesis became generally accepted in the early 1970s. In 1972, Harman proposed the Mitochondrial Theory of Aging, according to which aging is the result of the cumulative effects of mitochondrial DNA damage caused by free radicals [5, 6]. In 1986, Miquel and Fleming published their hypothesis about the involvement of mitochondria-originated free radicals in the process of ageing [7]. By 1981, mitochondrial DNA was completely sequenced [8], and, 5 years later, its entire genetic content had been described [9, 10]. Obviously, research on and with mitochondria has been conducted for over 120 years continuously and with steady success. Nevertheless, the last decade of the twentieth century saw another significant boost of interest in studying mitochondrial functions. First, in 1988, two papers, one published in *Science* and the other in *Nature* [11, 12], revealed for the very first time deletions and point mutations of mitochondria DNA to be the cause for human diseases. Second, by around 1995, mitochondria well known as the "powerhouse of the cell" have also been accepted as the "motor of cell death" [13] reflecting the organelle's key role in apoptosis. It is nowadays recognized that mitochondrial dysfunction is either the cause of or at least associated with a large number and variety of human disorders, ranging from neurodegenerative and neuromuscular diseases, obesity, cardiovascular disorders, migraine, liver and kidney disease to ischemia-reperfusion injury and cancer. Subsequently, increased pharmacological and pharmaceutical efforts have led to the emergence of mitochondrial medicine as a new field of biomedical research [1, 14]. Future developments of techniques for probing and manipulating mitochondrial functions will eventually lead to the treatment and prevention of a wide variety of pathologies and chronic diseases, "the future of medicine will come through mitochondria" [15].

Our book is dedicated to showcasing the tremendous efforts and the progress that has been made over the last decades in developing techniques and protocols for probing, imaging, and manipulating mitochondrial functions. All chapters were written by leading experts in their particular fields. The book is divided into two volumes. Volume I (*Probing Mitochondrial Function*) is focused on methods being used for the assessment of mitochondrial function under physiological conditions as well as in healthy isolated mitochondria. Volume II (*Manipulating Mitochondrial Function*) describes techniques developed for manipulating and assessing mitochondrial function under general pathological conditions and specific disease states.

Volume I

Stefan Lehr and coworkers critically evaluate in a review chapter a commonly used isolation procedure for mitochondria utilizing differential (gradient) centrifugation and depict major challenges to achieve "functional" mitochondria as basis for comprehensive physiological studies. The same authors provide in a protocol chapter an isopycnic density gradient centrifugation strategy for the isolation of mitochondria with a special focus on quality control of prepared intact, functional mitochondria. The isolation of interorganellar membrane contact sites is described by Alessandra d'Azzo and colleagues. They outline a protocol tailored for the isolation of mitochondria, mitochondria-associated ER membranes, and glycosphingolipid-enriched microdomains from the adult mouse brain, primary neurospheres, and murine embryonic fibroblasts. The analysis of single mitochondria helps uncovering a new level of biological heterogeneity and holds promises for a better understanding of mitochondria-related diseases. Peter Burke and colleagues describe a nanoscale approach for trapping single mitochondria in fluidic channels for fluorescence microscopy. Their method reduces background fluorescence, enhances focus, and allows simple experimental buffer exchanges. Stephane Arbault and colleagues describe the preparation and use of microwell arrays for the entrapment and fluorescence microscopy of single isolated mitochondria. Measuring variations of NADH of each mitochondrion in the array, this method allows the analysis of the metabolic status of the single organelle at different energetic-respiratory stages.

Deep resequencing allows the detection and quantification of low-level variants in mitochondrial DNA (mtDNA). This massively parallel ("next-generation") sequencing is characterized by great depth and breadth of coverage. Brendan Payne and colleagues describe a method for whole mtDNA genome deep sequencing as well as short amplicon deep sequencing. In another chapter, the same group provides a method for characterizing mtDNA within single skeletal muscle fibers. This approach allows the detection of somatic mtDNA mutations existing within individual cells which may be missed by techniques applied to the whole tissue DNA extract. The authors also apply single-cell mtDNA sequencing for analyzing differential segregation of mtDNA during embryogenesis. They demonstrate how to study this phenomenon by single-cell analysis of embryonic primordial germ cells. Next-generation sequencing (NGS) as an effective method for mitochondrial genome sequencing is also the subject of Shale Dames' chapter. He and his group describe an mtDNA enrichment method including library preparation and sequencing on "Illumina NGS platforms" and provide also a short command line alignment script for downloading via FTP. Conventional methods for mitochondrial DNA (mtDNA) extraction do not yield the level of mtDNA enrichment needed for direct sequencing, and the necessary subsequent long-range PCR amplification may introduce bias into the sequence results. Alexander Maslov and colleagues provide a protocol involving a paramagnetic bead-based purification step for the preparation of mtDNA-enriched samples ready for direct sequencing. Lars Eide and coworkers give a detailed protocol for the use of real-time qPCR to analyze the integrity of mitochondrial DNA and RNA quantitatively. Their method has low material requirement, is low cost, and can detect modifications with high resolution.

Mitochondria in species ranging from yeast to human have been found to import a small number of nucleus-encoded RNAs. With the advent of high-throughput RNA sequencing, additional nucleus-encoded mitochondrial RNAs are being identified. Michael Teitell and his group describe both an in vitro and in vivo import system for studying mitochondrial RNA import, processing, and functions.

In the last decade an increasing number of studies have been conducted aimed at quantifying acquired changes in the concentration of circulating mitochondrial DNA (mtDNA) as an indicator of mitochondrial function. Afshan Malik and colleagues provide a protocol for accurately measuring the amount of human mtDNA in peripheral blood samples which is based on the use of real-time quantitative PCR (qPCR) to quantify the amount of mtDNA relative to nuclear DNA. Their protocol is suitable for high-throughput use and can be modified for application to other body fluids, human cells, and tissues. The characterization of mtDNA processing at the single-cell level is poorly defined. Laurent Chatre and Miria Ricchetti describe a mitochondrial transcription and replication imaging protocol which is based on modified fluorescence in situ hybridization and which allows the detection of qualitative and quantitative alterations of the dynamics of mtDNA processing in human cells undergoing physiological changes.

William Sivitz and colleagues describe a highly sensitive and specific nuclear magnetic resonance-based assay which allows the simultaneous quantification of ATP and reactive oxygen species using small amounts of mitochondrial isolates or permeabilized cells. Their novel assay also avoids the problem of changing mitochondrial membrane potential while ADP is converted to ATP, as occurs in conventional assays. Accurate detection of mitochondrial superoxide especially in living cells remains a difficult task. Werner Koopman and coworkers describe a live-cell microscopy-based method for detecting superoxide in both mitochondria and the entire cell using dihydroethidium. Boronate-based probes were developed over the last decade for detection of hydrogen peroxide and peroxynitrite in biological systems. However, most boronates lack specificity needed to distinguish between hydrogen peroxide and peroxynitrite within a complex biological system. Jacek Zielonka and colleagues describe how a newly developed mitochondria-targeted phenylboronic acid can be used to detect and differentiate peroxynitrite-dependent and independent probe oxidation. Time-resolved fluorescence spectrometry can be used to detect and characterize mitochondrial metabolic oxidative changes by means of endogenous fluorescence. Alzbeta Marcek Chorvatova and coworkers describe the detection and measurement of endogenous mitochondrial NAD(P)H fluorescence in living cells in vitro using fluorescence lifetime spectrometry imaging after excitation with a 405 nm picoseconds laser. Quantifying the mitochondrial membrane potential is essential for understanding mitochondrial function. Most of the current methodologies are based on the accumulation of cation indicators. Roger Springett describes a new methodology which allows calculating the membrane potential from the measured oxidation states of the b-hemes. To better understand the impact of oxygen on cellular function, James Hynes and Conn Carey outline the procedure for measuring in situ oxygenation of cells in 2D and 3D cultures. These authors also illustrate how the impact of drug treatment on cell oxygenation can be assessed and how the link between oxygenation and glycolytic metabolism can be examined. Egbert Mik and Floor Harms have developed a method called Protoporphyrin IX—Triplet State Lifetime Technique as a potential tool for noninvasive monitoring of mitochondrial function in the clinic. In their chapter they describe the application of mitochondrial respirometry for monitoring mitochondrial oxygen tension and mitochondrial oxygen consumption in the skin of experimental animals. The selective monitoring of mitochondria-produced hydrogen peroxide inside living systems can be challenging. Alexander Lippert and colleagues describe the synthesis of the small molecular probe MitoPY1 and its application for measuring hydrogen peroxide in vitro and in live cells. The authors also provide an example procedure for measuring mitochondrial hydrogen peroxide in a cell culture model of Parkinson's disease. Erich Gnaiger and colleagues describe how the Amplex Red assay can be used to detect hydrogen peroxide production in combination with the simultaneous assessment of mitochondrial bioenergetics by high-resolution respirometry. They have optimized instrumental and methodological parameters to analyze the effects of various substrate, uncoupler, and inhibitor titrations (SUIT) on respiration versus hydrogen peroxide production. The authors also show an application example using isolated mouse brain mitochondria as an experimental model for the simultaneous measurement of mitochondrial respiration and hydrogen peroxide production in SUIT protocols. Andrey Abramov and Fernando Bartolome describe a strategy for assessing NADH/NAD(P)H and FAD autofluorescence in a time course-dependent manner. Their method provides information about NADH and FAD redox indexes both reflecting the activity of the mitochondrial electron transport chain. Their analysis of NADH autofluorescence after induction of maximal respiration can also offer information about the pentose phosphate pathway activity where glucose can be alternatively oxidized instead of pyruvate. Coenzyme Q10 (CoQ10) is an essential part of the mitochondrial respiratory chain. Outi Itkonen and Ursula Turpeinen describe an accurate and sensitive liquid chromatography tandem mass spectrometry method for the determination of mitochondrial CoQ10 in isolated mitochondria.

Assessing bioenergetic parameters of human pluripotent stem cells (hPSCs), including embryonic stem cells (ESCs) and induced pluripotent stem cells (iPSCs), provides considerable insight into their mitochondrial functions and cellular properties, which allows exposing potential energetic defects caused by mitochondrial diseases. Alessandro Prigione and Vanessa Pfiffer describe a method that facilitates the assessment of the bioenergetic profiles of hPSCs in a noninvasive fashion, while requiring only small sample sizes and allowing for several replicates.

Due to the complexity of the interactions involved at the different levels of integration in organ physiology, current molecular analyses of pathologies should be combined with integrative approaches of whole organ function. By combining the principles of control analysis with noninvasive ^{31}P NMR measurement of the energetic intermediates and simultaneous measurement of heart contractile activity, Philippe Diolez and colleagues have developed MoCA (Modular Control and Regulation Analysis), which is an integrative approach designed to study in situ control and regulation of cardiac energetics during contraction in intact beating perfused isolated heart. In their review chapter the authors present selected examples of the applications of MoCA to isolated intact beating heart, and they also discuss wider application to cardiac energetics under clinical conditions with the direct study of heart pathologies.

Mitochondrial proteins encoded on the cytosolic ribosomes carry specific patterns in the precursor sequence needed for mitochondrial import. Rita Casadio and colleagues discuss the feasibility of utilizing computational methods for detecting such mitochondrial targeting peptides in polypeptide sequences. These authors also introduce their newly implemented web server and demonstrate its application to the whole human proteome for detecting mitochondrial targeting peptides. Fabiana Perocchi and Yiming Cheng describe evolutionary biology approaches for studying mitochondrial physiology. One strategy, which they refer to as "comparative physiology," allows the de novo identification of mitochondrial proteins involved in a physiological function. Another approach known as "phylogenetic profiling" allows predicting the function of uncharacterized proteins as well as functional interactions by comparing phylogenetic profiles of uncharacterized and known components. Besides DNA mutations, faulty posttranslational modifications can also cause malfunction of mitochondrial proteins. Suresh Mishra and colleagues describe procedures for the isolation of mitochondria from cells and for separating the mitochondrial proteins by two-dimensional gel electrophoresis. The employment of antibodies specific to each posttranslational modification allows them to assess posttranslational modifications of mitochondrial proteins. Posttranslational protein glutathionylation regulates protein function in response to cellular redox changes and is involved in carbon monoxide-induced cellular pathways. Helena Viera and Ana S. Almeida describe a technique for the assessment of mitochondrial protein glutathionylation in response to carbon monoxide exposure.

High-resolution melting (HRM) allows detecting homozygous or heterozygous point sequence variants and small deletions within specific PCR products. Marketa Tesarova and colleagues provide an updated HRM-based protocol for routine variant screening of nuclear genes encoding assembly factors and structural subunits of cytochrome c oxidase (COX). Their general recommendations given for HRM analysis are applicable for examining any genetic region of interest. Anton Vila-Sanjurjo and colleagues have designed a computational approach named Heterologous Inferential Analysis or HIA for making predictions on the disruptive potential of a large subset of mt-rRNA variants. The authors demonstrate that in the case of certain mitochondrial variants for which sufficient information regarding their genetic and pathological manifestation is available, HIA data alone can be used to predict their pathogenicity.

Mitochondria play a key role in apoptosis. Vladimir Gogvadze and coworkers describe how to evaluate the release of intermembrane space proteins during apoptosis, alterations in the mitochondrial membrane potential, and oxygen consumption in apoptotic cells. Fluorescent lifetime imaging microscopy-Förster resonant energy transfer (FLIM-FRET) is a high-resolution technique for the detection of protein interactions in live cells. David Andrews and colleagues provide a detailed protocol for applying this technique to assess the interaction between BclXL and Bad at the mitochondrial outer membrane in live MCF7 breast cancer cells. Mitochondrial Ca^{2+} uptake is essential for regulating mitochondrial function. Markus Waldeck-Weiermair and colleagues analyze the benefits and drawbacks of various established old and new techniques to assess dynamic changes of mitochondrial Ca^{2+} concentrations in a wide range of applications.

Untargeted lipidomics profiling by liquid chromatography-mass spectrometry (LC-MS) allows the examination of lipids without any bias towards specific classes of lipids. Bruce Kristal and group describe a workflow including the isolation of mitochondria from liver tissue, followed by mitochondrial lipid extraction and the LC-MS conditions used for data acquisition. The authors also highlight how, in this method, all ion fragmentation can be used for the identification of species of lower abundances, which are often missed by data-dependent fragmentation techniques.

Mitochondrial dynamics, i.e., mitochondrial location, number, and morphology, has an essential function in numerous physiological and pathophysiological phenomena in the developing and adult human heart. Elizabeth Lipke and colleagues describe the application of a computer-based tool (MATLAB, MQM) to quantify mitochondrial changes, in particular number, area, and location of mitochondria, during human pluripotent stem cell differentiation into spontaneously contracting cardiomyocytes. Helena Bros and coworkers present an ex vivo method for monitoring the movement of mitochondria within myelinated sensory and motor axons from spinal nerve roots.

Volume II

The development of mitochondria-targeted pharmaceutical nanocarriers began at the end of the 1990s with an accidental discovery of the vesicle-forming capacity of dequalinium chloride. Volkmar Weissig describes a detailed protocol for the preparation, characterization, and application of dequalinium-based nano vesicles called DQAsomes. Whether small molecule xenobiotics (biocides, drugs, probes, toxins) will target mitochondria in living cells without the assistance of any mitochondria-targeted delivery system can be predicted using an algorithm derived from QSAR modeling and is described in detail by Richard Horobin. Small molecules can be physicochemically targeted to mitochondria via conjugation to mitochondriotropic triphenylphosphonium cations. Utilizing this strategy, Richard Hartley describes the preparation of MitoB and MitoP as exomarkers of mitochondrial hydrogen peroxide. Gerard D'Souza and his group describe the use of triphenylphosphonium cations for the preparation of phospholipid conjugates which in turn are the basis for preparing mitochondria-targeted liposomes. Triphenylphosphonium cations are used by Jung-Joon Min and Dong-Yeon Kim for the synthesis of 18F-labeled fluoroalkyl triphenylphosphonium conjugates as mitochondrial voltage sensors for PET myocardial imaging. Fernanda Borges and her group describe the utilization of triphenylphosphonium cations for the development and application of a new antioxidant based on dietary cinnamic acid. Tamer Elbayoumi and colleagues have utilized the intrinsic mitochondriotropism of Genistein to design mitochondria-targeted cationic lipid-based nanocarrier systems including micelles and nanoemulsions. Since Genistein, a major soy isoflavone, exhibits extensive proapoptotic anticancer effects which are mediated predominantly via induction of mitochondrial damage, this delivery system is potentially suited to enhance anticancer efficacy of different coformulated chemotherapeutic agents. Shanta Dhara and her group outline in one chapter the synthesis and characterization of a functionalized polymer for building mitochondria-targeted nanoparticles (NPs), and in a second chapter she describes the application of such mitochondria-specific nanoparticles for the delivery of a photosensitizer to mitochondria for photodynamic therapy. Hideyoshi Harashima and Yuma Yamada describe the construction and application of a mitochondria-targeted dual function liposome-based nanocarrier termed DF-MITO-Porter.

The utilization of α-aminophosphonates as external probes in combination with [31P-NMR] allows the simultaneous pH measurements of cytosolic and acidic compartments in normal and stressed cultured cells. Sylvia Pietria's group has developed this strategy further by using triphenylphosphonium derivatives of aminophosphonates as mitochondria-targeted pH probes. The authors describe the synthesis and [31P]-NMR pH titrating properties of such mitochondria-targeted pH probes as well as their application in green alga cultures. The formation of reactive oxygen species (ROS) in the inner mitochondrial membrane can cause mitochondrial dysfunction eventually followed by induction of apoptosis. Antioxidants conjugated with mitochondria-targeted, membrane-penetrating cations can be used to scavenge ROS inside mitochondria. Vladimir Skulachev and coworkers describe some essential methodological aspects of the application of mitochondria-targeted cations belonging to the MitoQ and SkQ groups which have shown promise for treating oxidative stress-related pathologies. Lucia Biasutto and colleagues describe a step-by-step procedure for synthesizing mitochondria-targeted derivatives of resveratrol and quercetin, two plant polyphenols exhibiting potential health-promoting properties, as well as a method for assessing their mitochondrial accumulation.

The mitochondrial respiratory chain is stress-responsive and responds to mitochondrially targeted anticancer agent by destabilization and induction of massive ROS production eventually leading to apoptosis. Jiri Neuzil's group has developed mitochondrially targeted anticancer agents epitomized by the mitochondrially targeted analogue of the redox-silent compound vitamin E succinate, which belongs to the group of agents that kill cancer cells via their mitochondria-destabilizing activity. The authors describe the use of native blue gel electrophoresis and clear native electrophoresis coupled with in-gel activity assays as methods of choice for trying to understand the molecular mechanism of the effect of such mitochondria-destabilizing agents. Many low-molecular-weight agents that may be of potential clinical relevance act by targeting mitochondria, where they may suppress mitochondrial respiration. Jiri Neuzil and coworkers describe the methodology for assessing respiration in cultured cells as well as in tumor tissue exposed to mitochondria-targeted anticancer agents.

Nina Entelis, Ivan Tarassov, and colleagues have developed mitochondria-targeting RNA vectors for the delivery of therapeutic oligoribonucleotides into human mitochondria. Their group provides a detailed protocol for the transfection of cultured human cells with small recombinant RNA molecules as well as methods for characterizing the mitochondrial transfection efficiency. Genetic transformation of mitochondria in multicellular eukaryotes is of fundamental importance for basic investigations and for applications to gene therapy or biotechnology. Andre Dietrich's group has developed a strategy to target nuclear transgene-encoded RNAs into mitochondria in plants. In their chapter they give a detailed protocol for mitochondrial targeting of trans-cleaving ribozymes destined to knockdown organelle RNAs for regulation studies, inverse genetics, and biotechnological purposes.

Allotopic expression (AE) of mitochondrial proteins, i.e., nuclear localization and transcription of mtDNA genes followed by cytoplasmic translation and transport into mitochondria, has been suggested as a strategy for gene replacement therapy in patients harboring mitochondrial DNA mutations. Carl Pinkert and David Dunn describe the use of AE for transgenic mouse modeling of the pathogenic human T8993G mutation in mtATP6 as a case study for designing AE animal models.

There is increasing evidence that exposure to air pollutants is associated with human disease and may act through epigenetic modification of the nuclear genome, but there have been few publications describing their impact upon the mitochondrial genome. Hyang-Min Byun and Timothy M. Barrow describe a protocol for the isolation of mitochondrial DNA from peripheral blood samples and the analysis of 5-methlycytosine content by bisulfite-pyrosequencing. Stanislaw Pshenichnyuk and Alberto Modelli describe the application of two complementary experimental techniques, Electron Transmission Spectroscopy (ETS) and Dissociative Electron Attachment Spectroscopy (DEAS), for studying the transfer of electrons unto xenobiotics in the intermembrane space of mitochondria. Additional support of experimental procedures by suitable quantum-chemical calculations is described in detail and illustrated by an example of ETS/DEAS study of rhodamine which shows rich fragmentation under gas-phase resonance electron attachment.

The link between mitochondrial dynamics and human pathologies has spawned significant interest in developing methods for screening proteins involved in mitochondrial dynamics as well as small molecules that modulate mitochondrial dynamics. Antonio Zorzano and Juan Pablo Munoz describe in their chapter functional screening protocols for the in vitro examination of mitochondrial parameters such as mitochondrial morphology, reactive oxygen species (ROS) levels, mitochondrial calcium, and oxygen consumption rate. Dysfunctional mitochondria communicate via retrograde signaling with the nucleus leading to cell stress adaptation by changes in nuclear gene expression. Mitochondria to nucleus signaling pathways have been widely studied in *Saccharomyces cerevisiae*, where retrograde-target gene expression is regulated by RTG genes. Sergio Giannattasio and coworkers describe a method for the assessment of the mitochondrial retrograde pathway activation in yeast cells based on monitoring the mRNA levels of a variety of RTG-target genes. Adaptations to energy stress or altered physiological condition can be assessed by measuring changes of multiple bioenergetic parameters. Dmitri Papkovsky and Alexander Zhdanov describe a simple methodology for high-throughput multiparametric assessment of cell bioenergetics, called Cell Energy Budget (CEB) platform, and demonstrate its practical use with cell models.

Viable disease models for mitochondrial DNA diseases are much needed for elucidating genotype/phenotype relationships and for improving disease management. Alessandro Prigione discusses the potential advantages and critical challenges for the utilization of induced pluripotent stem cells (iPSCs) from patients affected by mtDNA disorders for modeling debilitating mtDNA diseases.

Heteroplasmic mice can be used for studying the segregation of different mtDNA haplotypes in vivo against a defined nuclear background. Thomas Kolbe and colleagues describe two methods involving either the transfer of ooplasm or the fusion of two blastomeres for the creation of such mice models.

H. van der Spek and coworkers describe a robust and efficient method for visualizing and quantifying mitochondrial morphology in *Caenorhabditis elegans*, which is a preferred model for studying mitochondrial deficiencies caused by disease or drug toxicity. Their method allows for a comprehensive analysis of mitochondrial morphology. Mitochondrial DNA (mtDNA) is a useful and reliable biomarker of UV-induced genetic damage in both animal and human skin. Mark Birch-Machin and Amy Bowman describe in their protocol chapter the assessment of UV-induced mtDNA damage, including the extraction of cellular DNA, qPCR to determine the relative amount of mtDNA, and qPCR to determine UV-induced damage within a long strand of mtDNA.

Mitochondrial dysfunction is associated with the pathogenesis of septic disorders, eventually leading to a decline in energy supply. Matthias Hecker and colleagues give a protocol for assessing the influence of short- and medium-chain fatty acids on mitochondrial respiration using high-resolution respirometry under inflammatory and baseline conditions. Hun-Kuk Park and Gi-Ja Lee describe the application of Atomic Force Microscopy (AFM)-based shape analysis for the characterization of nanostructural changes of mitochondria. The authors use AFM to study mitochondrial swelling in heart mitochondria during myocardial ischemia-reperfusion injury employing a rat model. In some tissues such as the heart, abnormal mitochondrial fusion and fission can go along with mitochondrial apoptosis, but its contribution as cause vs. a consequence remains to be defined. Catherine Brenner and her group give a protocol for the isolation of fresh mitochondria from rat heart by a procedure adapted to the myofibrillar structure of the tissue, and they describe several miniaturized enzymatic assays for probing mitochondria-mediated apoptosis. The pathogenesis of Parkinson's disease (PD) is poorly understood and under intensive investigation. Mitochondrial dysfunction has been linked to the sporadic form of PD. Daniella Arduino's group describes a method for the generation of cytoplasmic hybrid cells as a cellular model of sporadic PD which is based upon the fusion of platelets harboring mtDNA from PD patients with cells in which the endogenous mtDNA has been depleted. JC-1, a commercially available fluorescence dye, is widely used for measuring changes in the mitochondrial membrane potential. Dorit Ben-Shachar and coworkers show that JC-1 can also be used to follow alterations in mitochondrial distribution and mitochondrial network connectivity. The authors describe various applications of JC-1 staining to study mitochondrial abnormalities in different cell types derived from schizophrenia patients and healthy subjects.

We are extremely grateful to all authors for having spent significant parts of their valuable time to contribute to this book. It is our hope that together we have succeeded in providing an essential source of know-how and a source of inspiration to all researchers who are as fascinated as we are about this tiny organelle which so much seems to control life and death of a single cell and the whole organism alike. Last but not least we would like to thank John Walker, the series editor of *Methods in Molecular Biology*, for having accepted our book proposal, which originated from our efforts in organizing a series of annual conferences on Targeting Mitochondria, the fifth one of which has taken place in October 2014 in Berlin, Germany (www.targeting-mitochondria.com). We are also grateful to John Walker for his unlimited guidance and help throughout the whole process.

I (V.W.) would like to thank my wife, Angelina Lynn Weikel, for her understanding and strong support throughout the duration of this project.

Glendale, AZ, USA   *Volkmar Weissig*
Paris, France   *Marvin Edeas*

References

1. Luft R (1994) The development of mitochondrial medicine. Proc Natl Acad Sci U S A 91:8731–8738

2. Luft R, Ikkos D, Palmieri G, Ernster L, Afzelius B (1962) A case of severe hypermetabolism of nonthyroid origin with a defect in the maintenance of mitochondrial respiratory control: a correlated clinical, biochemical, and morphological study. J Clin Invest 41:1776–1804

3. Nass S, Nass MM (1963) Intramitochondrial fibers with DNA characteristics. II. Enzymatic and other hydrolytic treatments. J Cell Biol 19:613–629

4. Mitchell P (1961) Coupling of phosphorylation to electron and hydrogen transfer by a chemi-osmotic type of mechanism. Nature 191:144–148

5. Harman D (1972) Free radical theory of aging: dietary implications. Am J Clin Nutr 25:839–843

6. Harman D (1972) The biologic clock: the mitochondria? J Am Geriatr Soc 20:145–147

7. Miquel J, Fleming J (1986) Oxygen radical-mitochondrial injury hypothesis of ageing. In: Johnson JE (ed) Free radicals, ageing and degenerative diseases. Alan R. Liss, New York, NY, pp 51–74

8. Anderson S, Bankier AT, Barrell BG et al. (1981) Sequence and organization of the human mitochondrial genome. Nature 290:457–465

9. Chomyn A, Cleeter MW, Ragan CI, Riley M, Doolittle RF, Attardi G (1986) URF6, last unidentified reading frame of human mtDNA, codes for an NADH dehydrogenase subunit. Science 234:614–618

10. Chomyn A, Mariottini P, Cleeter MW et al. (1985) Six unidentified reading frames of human mitochondrial DNA encode components of the respiratory-chain NADH dehydrogenase. Nature 314:592–597

11. Holt IJ, Harding AE, Morgan-Hughes JA (1988) Deletions of muscle mitochondrial DNA in patients with mitochondrial myopathies. Nature 331:717–719

12. Wallace DC, Singh G, Lott MT et al. (1988) Mitochondrial DNA mutation associated with Leber's hereditary optic neuropathy. Science 242:1427–1430

13. Brown GC, Nicholls DG, Cooper CE (1999) Mitochondria and cell death. Princeton University Press, Princeton, NJ, pp vii–viii

14. Weissig V (2003) Mitochondrial-targeted drug and DNA delivery. Crit Rev Ther Drug Carrier Syst 20:1–62

15. Edeas M, Weissig V (2013) Targeting mitochondria: strategies, innovations and challenges: the future of medicine will come through mitochondria. Mitochondrion 13:389–390

Contents

Contributors

ANDREY Y. ABRAMOV • *Neurosciences group, Instituto de Investigacion Hospital 12 de Octubre (i+12), Av Cordoba, Madrid, Spain*

JAN ADAMUS • *Institute of Applied Radiation Chemistry, Lodz University of Technology, Lodz, Poland*

SAIMA AJAZ • *Diabetes Research Group, School of Medicine, King's College London, London, UK*

ANA S. ALMEIDA • *Chronic Diseases Research Center (CEDOC), Faculdade de Ciências Médicas, Universidade Nova de Lisboa, Lisbon, Portugal; Instituto de Biologia Experimental e Tecnológica (IBET), Oeiras, Portugal; Instituto de Tecnologia Química e Biológica (ITQB), Universidade Nova de Lisboa, Oeiras, Portugal*

SUDHARSANA R. ANDE • *Department of Internal Medicine, University of Manitoba, Winnipeg, MB, Canada*

DAVID W. ANDREWS • *Department of Biochemistry, University of Toronto, Toronto, ON, Canada; Odette Cancer Research Program, Sunnybrook Research Institute, Toronto, ON, Canada*

IDA ANNUNZIATA • *Department of Genetics, St. Jude Children Research Hospital, Memphis, TN, USA*

STÉPHANE ARBAULT • *ISM, CNRS, University of Bordeaux, Pessac, France*

LAURENT ARSAC • *INSERM U1045, Centre de Recherche Cardio-Thoracique de Bordeaux & LIRYC, Institut de Rythmologie et Modélisation Cardiaque, Université de Bordeaux, Lormont, France*

FERNANDO BARTOLOMÉ • *Neurosciences group, Instituto de Investigacion Hospital 12 de Octubre (i+12), Av Cordoba, Madrid, Spain*

SUSAN S. BIRD • *Thermo Fisher Scientific, Cambridge, MA, USA*

ALEXANDER I. BONDARENKO • *Institute of Molecular Biology and Biochemistry, Center of Molecular Medicine, Medical University of Graz, Graz, Austria*

HELENA BROS • *NeuroCure Clinical Research Center, Institute for Medical Immunology, Charité-Universitätsmedizin Berlin, Berlin, Germany; Experimental and Clinical Research Center, a joint cooperation between the Charité-Universitätsmedizin Berlin and the Max-Delbrück Center for Molecular Medicine, Berlin, Germany*

PETER BURKE • *Integrated Nanosystem Research Facility, Electrical Engineering and Computer Science, University of California Irvine, Irvine, CA, USA*

BLAKELY S. BUSSIE • *Department of Chemical Engineering, Auburn University, Auburn, AL, USA*

GUILLAUME CALMETTES • *Department of Medicine (Cardiology), David Geffen School of Medicine, University of California, Los Angeles, CA, USA*

CONN CAREY • *Luxcel Biosciences Ltd., BioInnovation Centre, UCC, Cork, Ireland*

RITA CASADIO • *Biocomputing Group, CIRI Health Sciences & Technologies (HST), University of Bologna, Bologna, Italy*

LAURENT CHATRE • *Team "Stability of nuclear and mitochondrial DNA", Unité de Génétique Moléculaire des Levures, CNRS UMR3525, Institut Pasteur, Paris, France*

YIMING CHENG • *Gene Center, Ludwig-Maximilians-Universität, Munich, Germany; Institute of Human Genetics, Helmholtz Zentrum Munich, Munich, Germany*

PATRICK F. CHINNERY • *Mitochondrial Research Group, Institute of Genetic Medicine, Newcastle University, Newcastle-upon-Tyne, UK*

DUSAN CHORVAT JR. • *Department of Biophotonics, International Laser Center, Bratislava, Slovakia*

ALZBETA MARCEK CHORVATOVA • *Department of Biophotonics, International Laser Center, Bratislava, Slovakia*

JONATHAN COXHEAD • *Mitochondrial Research Group, Institute of Genetic Medicine, Newcastle University, Newcastle-upon-Tyne, UK*

LYNSEY CREE • *School of Medicine, University of Auckland, Auckland, New Zealand*

ANNA CZAJKA • *Diabetes Research Group, School of Medicine, King's College London, London, UK*

ALESSANDRA D'AZZO • *Department of Genetics, St. Jude Children Research Hospital, Memphis, TN, USA*

SHALE DAMES • *ARUP Laboratories, ARUP Institute for Clinical and Experimental Pathology, Salt Lake City, UT, USA; Department of Biomedical Informatics, University of Utah, Salt Lake City, UT, USA*

ANDRÁS T. DEAK • *Institute of Molecular Biology and Biochemistry, Center of Molecular Medicine, Medical University of Graz, Graz, Austria*

VÉRONIQUE DESCHODT-ARSAC • *INSERM U1045, Centre de Recherche Cardio-Thoracique de Bordeaux & LIRYC, Institut de Rythmologie et Modélisation Cardiaque, Université de Bordeaux, Lormont, France*

KATHERINE M. DESIMONE • *Department of Chemical Engineering, Auburn University, Auburn, AL, USA*

BRYAN C. DICKINSON • *Department of Chemistry, The University of Chicago, IL, Chicago, USA*

PHILIPPE DIOLEZ • *INSERM U1045, Centre de Recherche Cardio-Thoracique de Bordeaux & LIRYC, Institut de Rythmologie et Modélisation Cardiaque, Université de Bordeaux, Lormont, France*

DAVID A. DUNN • *Department of Chemical Engineering, Auburn University, Auburn, AL, USA*

LARS EIDE • *Department of Medical Biochemistry, Oslo University Hospital, University of Oslo, Oslo, Norway*

KAREN EILBECK • *Department of Biomedical Informatics, University of Utah, Salt Lake City, UT, USA*

JOANNA L. ELSON • *Institute of Genetic Medicine, Newcastle University, Newcastle upon Tyne, UK*

YING ESBENSEN • *Institute of Clinical Epidemiology and Molecular Biology, Akershus University Hospital, Lørenskog, Norway*

PIERO FARISELLI • *Biocomputing Group, CIRI Health Sciences & Technologies (HST), University of Bologna, Bologna, Italy; DISI Department of Computer Science and Engineering, University of Bologna, Bologna, Italy*

MARIO FASCHING • *OROBOROS INSTRUMENTS, Innsbruck, Austria*

BRIAN D. FINK • *Department of Internal Medicine/Endocrinology, University of Iowa, Iowa City, IA, USA; Iowa City Veterans Affairs Medical Center, Iowa City, IA, USA*

MONA FONTANA-AYOUB • *OROBOROS INSTRUMENTS, Innsbruck, Austria*

MARLEEN FORKINK • *Department of Biochemistry, Nijmegen Centre for Molecular Life Sciences, Radboud University Medical Centre, Nijmegen, The Netherlands*

KRISTIAN GARDNER • *Mitochondrial Research Group, Institute of Genetic Medicine, Newcastle University, Newcastle-upon-Tyne, UK*

ROSE M. GATHUNGU • *Department of Neurosurgery, Harvard Medical School, Boston, MA, USA; Department of Neurosurgery, Brigham and Women's Hospital, Boston, MA, USA*

KATHRIN GAUPER • *Daniel Swarovski Research Laboratory, Mitochondrial Physiology, Department of Visceral, Transplant and Thoracic Surgery, Medical University of Innsbruck, Innsbruck, Austria*

ERICH GNAIGER • *OROBOROS INSTRUMENTS, Innsbruck, Austria; Daniel Swarovski Research Laboratory, Mitochondrial Physiology, Department of Visceral, Transplant and Thoracic Surgery, Medical University of Innsbruck, Innsbruck, Austria*

VLADIMIR GOGVADZE • *Division of Toxicology, Institute of Environmental Medicine, Karolinska Institutet, Stockholm, Sweden; Faculty of Basic Medicine, MV Lomonosov Moscow State University, Moscow, Russia*

BERTRAND GOUDEAU • *ISM, CNRS, University of Bordeaux, Pessac, France*

GILLES GOUSPILLOU • *Département de Kinanthropologie, Université du Québec à Montréal Montreal, QC, Canada*

WOLFGANG F. GRAIER • *Institute of Molecular Biology and Biochemistry, Center of Molecular Medicine, Medical University of Graz, Graz, Austria*

SANDER GREFTE • *Department of Biochemistry, Nijmegen Centre for Molecular Life Sciences, Radboud University Medical Centre, Nijmegen, The Netherlands*

LUKAS N. GROSCHNER • *Institute of Molecular Biology and Biochemistry, Center of Molecular Medicine, Medical University of Graz, Graz, Austria*

MICHEL HAISSAGUERRE • *INSERM U1045, Centre de Recherche Cardio-Thoracique de Bordeaux & LIRYC, Institut de Rythmologie et Modélisation Cardiaque, Université de Bordeaux, Lormont, France*

FLOOR A. HARMS • *Laboratory of Experimental Anesthesiology, Department of Anesthesiology, Erasmus MC University Medical Center Rotterdam, Rotterdam, The Netherlands*

SONJA HARTWIG • *Institute of Clinical Biochemistry and Pathobiochemistry, German Diabetes Center at the Heinrich-Heine-University Duesseldorf, Leibniz Center for Diabetes Research, Duesseldorf, Germany*

JULIANA HEIDLER • *Daniel Swarovski Research Laboratory, Mitochondrial Physiology, Department of Visceral, Transplant and Thoracic Surgery, Medical University of Innsbruck, Innsbruck, Austria*

JULIA HORILOVA • *Department of Biophotonics, International Laser Center, Bratislava, Slovakia; Department of Biophysics, Faculty of Science, Pavol Jozef Safarik University, Kosice, Slovakia*

JAMES HYNES • *Luxcel Biosciences Ltd., BioInnovation Centre, UCC, Cork, Ireland*

CARMEN INFANTE-DUARTE • *Institute for Medical Immunology, Charité-Universitätsmedizin Berlin, Berlin, Germany; Experimental and Clinical Research Center, a joint cooperation between the Charité-Universitätsmedizin Berlin and the Max-Delbrück Center for Molecular Medicine, Berlin, Germany*

OUTI ITKONEN • *Laboratory Division HUSLAB, Helsinki University Central Hospital, Helsinki, Finland*

PIERRE JAIS • *INSERM U1045, Centre de Recherche Cardio-Thoracique de Bordeaux & LIRYC, Institut de Rythmologie et Modélisation Cardiaque, Université de Bordeaux, Lormont, France*

CLAIRE JEAN-QUARTIER • *Institute of Molecular Biology and Biochemistry, Center of Molecular Medicine, Medical University of Graz, Graz, Austria*

B. Kalyanaraman • *Department of Biophysics, Free Radical Research Center, Medical College of Wisconsin, Milwaukee, WI, USA*

Petra Kerscher • *Department of Chemical Engineering, Auburn University, Auburn, AL, USA*

Carla M. Koehler • *Department of Chemistry and Biochemistry, University of California at Los Angeles, Los Angeles, CA, USA; Molecular Biology Institute, University of California at Los Angeles, Los Angeles, CA, USA; Jonsson Comprehensive Cancer Center, University of California at Los Angeles, Los Angeles, CA, USA; Broad Stem Cell Research Center, University of California at Los Angeles, Los Angeles, CA, USA*

Werner J.H. Koopman • *Department of Biochemistry, Nijmegen Centre for Molecular Life Sciences, Radboud University Medical Centre, Nijmegen, The Netherlands*

Jorg Kotzka • *Institute of Clinical Biochemistry and Pathobiochemistry, German Diabetes Center at the Heinrich-Heine-University Duesseldorf, Leibniz Center for Diabetes Research, Duesseldorf, Germany*

Bruce S. Kristal • *Department of Neurosurgery, Harvard Medical School, Boston, MA, USA; Department of Neurosurgery, Brigham and Women's Hospital, Boston, MA, USA*

Gerhard Krumschnabel • *OROBOROS INSTRUMENTS, Innsbruck, Austria*

Stefan Lehr • *Institute of Clinical Biochemistry and Pathobiochemistry, German Diabetes Center at the Heinrich-Heine-University Duesseldorf, Leibniz Center for Diabetes Research, Duesseldorf, Germany*

Elizabeth A. Lipke • *Department of Chemical Engineering, Auburn University, Auburn, AL, USA*

Alexander R. Lippert • *Department of Chemistry, Center for Drug Discovery, Design, and Delivery, Southern Methodist University, Dallas, TX, USA*

Qian Liu • *Department of Biochemistry, University of Toronto, Toronto, ON, Canada*

Afshan Malik • *Diabetes Research Group, School of Medicine, King's College London, London, UK*

Roland Malli • *Institute of Molecular Biology and Biochemistry, Center of Molecular Medicine, Medical University of Graz, Graz, Austria*

Rong Mao • *ARUP Laboratories, Salt Lake City, UT, USA; Department of Pathology, University of Utah, Salt Lake City, UT, USA*

Pier Luigi Martelli • *Biocomputing Group, CIRI Health Sciences & Technologies (HST), University of Bologna, Bologna, Italy*

Alexander Y. Maslov • *Department of Genetics, Albert Einstein College of Medicine, New York, NY, USA*

Egbert G. Mik • *Laboratory of Experimental Anesthesiology, Department of Anesthesiology, Erasmus MC University Medical Center Rotterdam, Rotterdam, The Netherlands*

Suresh Mishra • *Department of Internal Medicine, University of Manitoba, Winnipeg, MB, Canada; Department of Physiology, University of Manitoba, Winnipeg, MB, Canada*

Pavol Miskovsky • *Centre for Interdisciplinary Biosciences (CIB), Pavol Jozef Safarik University, Kosice, Slovakia; Department of Biophysics, Faculty of Science, Pavol Jozef Safarik University, Kosice, Slovakia*

Zuzana Nadova • *Centre for Interdisciplinary Biosciences (CIB), Pavol Jozef Safarik University, Kosice, Slovakia; Department of Biophysics, Faculty of Science, Pavol Jozef Safarik University, Kosice, Slovakia*

Elizabeth J. New • *School of Chemistry, The University of Sydney, Sydney, NSW, Australia*

Raluca Niesner • *Deutsches Rheuma-Forschungszentrum, Berlin, Germany*

MAHTA NILI • *Department of Pathology and Laboratory Medicine, University of California at Los Angeles, Los Angeles, CA, USA*

STEN ORRENIUS • *Division of Toxicology, Institute of Environmental Medicine, Karolinska Institutet, Stockholm, Sweden*

ELIZABETH J. OSTERLUND • *Department of Biochemistry, University of Toronto, Toronto, ON, Canada*

G. PAULINE PADILLA-MEIER • *Department of Internal Medicine, University of Manitoba, Winnipeg, MB, Canada*

ANNETTE PATTERSON • *Department of Genetics, St. Jude Children Research Hospital, Memphis, TN, USA*

BRENDAN A. I. PAYNE • *Mitochondrial Research Group, Institute of Genetic Medicine, Newcastle University, Newcastle-upon-Tyne, UK*

FABIANA PEROCCHI • *Gene Center, Ludwig-Maximilians-Universität, Munich, Germany; Institute of Human Genetics, Helmholtz Zentrum Munich, Munich, Germany*

VANESSA PFIFFER • *Max Delbrueck Center for Molecular Medicine (MDC), Berlin, Germany*

TED PHAM • *Integrated Nanosystem Research Facility, Electrical Engineering and Computer Science, University of California Irvine, Irvine, CA, USA*

VASILY N. POPOV • *Department of Genetics, Cytology and Bioengineering, Voronezh State University, Voronezh, Russia*

ALESSANDRO PRIGIONE • *Max Delbrueck Center for Molecular Medicine (MDC), Berlin, Germany*

WILBER QUISPE-TINTAYA • *Department of Genetics, Albert Einstein College of Medicine, New York, NY, USA*

MIRIA RICCHETTI • *Team "Stability of nuclear and mitochondrial DNA", Unité de Génétique Moléculaire des Levures, CNRS UMR3525, Institut Pasteur, Paris, France*

PIERRE DOS SANTOS • *INSERM U1045, Centre de Recherche Cardio-Thoracique de Bordeaux & LIRYC, Institut de Rythmologie et Modélisation Cardiaque, Université de Bordeaux, Lormont, France*

CASTRENSE SAVOJARDO • *Biocomputing Group, CIRI Health Sciences & Technologies (HST), University of Bologna, Bologna, Italy*

KATJA SCHEFFLER • *Department of Medical Biochemistry, Oslo University Hospital, University of Oslo, Oslo, Norway; Department of Microbiology, Oslo University Hospital, University of Oslo, Oslo, Norway*

ERIKO SHIMADA • *Department of Chemistry and Biochemistry, University of California at Los Angeles, Los Angeles, CA, USA*

ADAM SIKORA • *Institute of Applied Radiation Chemistry, Lodz University of Technology, Lodz, Poland*

WILLIAM I. SIVITZ • *Department of Internal Medicine/Endocrinology, University of Iowa, Iowa City, IA, USA; Iowa City Veterans Affairs Medical Center, Iowa City, IA, USA*

PAUL M. SMITH • *Institute of Medical Sciences, Ninewells Hospital and Medical School, Dundee University, Dundee, Scotland, UK*

NESO SOJIC • *ISM, CNRS, University of Bordeaux, Pessac, France*

ROGER SPRINGETT • *Department of Radiology, Dartmouth Medical School, Hanover, NH, USA*

IRINA G. STAVROVSKAYA • *Department of Neurosurgery, Harvard Medical School, Boston, MA, USA; Department of Neurosurgery, Brigham and Women's Hospital, Boston, MA, USA*

HAUKE STUDIER • *Becker & Hickl GmbH, Berlin, Germany*

ZUZANA SUMBALOVA • *OROBOROS INSTRUMENTS, Innsbruck, Austria; Pharmacobiochemical Laboratory of 3rd Department of Internal Medicine, Medical Faculty, Comenius University in Bratislava, Bratislava, Slovakia*

EMMANUEL SURANITI • *ISM, CNRS, University of Bordeaux, Pessac, France*

GIANLUCA TASCO • *Biocomputing Group, CIRI Health Sciences & Technologies (HST), University of Bologna, Bologna, Italy*

MICHAEL A. TEITELL • *Department of Pathology and Laboratory Medicine, University of California at Los Angeles, Los Angeles, CA, USA; Molecular Biology Institute, University of California at Los Angeles, Los Angeles, CA, USA; Jonsson Comprehensive Cancer Center, University of California at Los Angeles, Los Angeles, CA, USA; Broad Stem Cell Research Center, University of California at Los Angeles, Los Angeles, CA, USA; California NanoSystems Institute, University of California at Los Angeles, Los Angeles, CA, USA; Department of Bioengineering, University of California at Los Angeles, Los Angeles, CA, USA; Department of Pediatrics, University of California at Los Angeles, Los Angeles, CA, USA*

MARKÉTA TESAŘOVÁ • *Department of Pediatrics and Adolescent Medicine, First Faculty of Medicine, Charles University in Prague, General University Hospital in Prague, Prague, Czech Republic*

FATEME TOUSI • *Department of Neurosurgery, Harvard Medical School, Boston, MA, USA; Department of Neurosurgery, Brigham and Women's Hospital, Boston, MA, USA*

URSULA TURPEINEN • *Laboratory Division HUSLAB, Helsinki University Central Hospital, Helsinki, Finland*

VENKATA SURESH VAJRALA • *ISM, CNRS, University of Bordeaux, Pessac, France*

KATEŘINA VESELÁ • *Department of Pediatrics and Adolescent Medicine, First Faculty of Medicine, Charles University in Prague, General University Hospital in Prague, Prague, Czech Republic*

HELENA L.A. VIEIRA • *Chronic Diseases Research Center (CEDOC), Faculdade de Ciências Médicas, Universidade Nova de Lisboa, Lisbon, Portugal; Instituto de Biologia Experimental e Tecnológica (IBET), Oeiras, Portugal*

JAN VIJG • *Department of Genetics, Albert Einstein College of Medicine, New York, NY, USA*

ANTÓN VILA-SANJURJO • *Dept. Química Fundamental, Facultade de Ciencias, Universidade da Coruña (UDC), A Coruña, Spain*

ALŽBĚTA VONDRÁČKOVÁ • *Department of Pediatrics and Adolescent Medicine, First Faculty of Medicine, Charles University in Prague, General University Hospital in Prague, Prague, Czech Republic*

MARKUS WALDECK-WEIERMAIR • *Institute of Molecular Biology and Biochemistry, Center of Molecular Medicine, Medical University of Graz, Graz, Austria*

DOUGLAS WALLACE • *Center for Mitochondrial and Epigenomic Medicine, Children's Hospital of Philadelphia, University of Pennsylvania, Philadelphia, PA, USA*

WEI WANG • *Department of Medical Biochemistry, Oslo University Hospital, University of Oslo, Oslo, Norway; Department of Microbiology, Oslo University Hospital, University of Oslo, Oslo, Norway*

GENG WANG • *School of Life Sciences, Tsinghua University, Beijing, China*

RYAN R. WHITE • *Department of Genetics, Albert Einstein College of Medicine, New York, NY, USA*

PETER H.G.M. WILLEMS • *Department of Biochemistry, Nijmegen Centre for Molecular Life Sciences, Radboud University Medical Centre, Nijmegen, The Netherlands*

LIPING YU • *NMR Core Facility and Department of Biochemistry, University of Iowa, Iowa City, IA, USA; Iowa City Veterans Affairs Medical Center, Iowa City, IA, USA*

KATAYOUN ZAND • *Integrated Nanosystem Research Facility, Electrical Engineering and Computer Science, University of California Irvine, Irvine, CA, USA*

JIŘÍ ZEMAN • *Department of Pediatrics and Adolescent Medicine, First Faculty of Medicine, Charles University in Prague, General University Hospital in Prague, Prague, Czech Republic*

BORIS ZHIVOTOVSKY • *Division of Toxicology, Institute of Environmental Medicine, Karolinska Institutet, Stockholm, Sweden; Faculty of Basic Medicine, MV Lomonosov Moscow State University, Moscow, Russia*

JACEK ZIELONKA • *Department of Biophysics, Free Radical Research Center, Medical College of Wisconsin, Milwaukee, WI, USA; Institute of Applied Radiation Chemistry, Lodz University of Technology, Lodz, Poland*

Chapter 1

Preparation of "Functional" Mitochondria: A Challenging Business

Stefan Lehr, Sonja Hartwig, and Jorg Kotzka

Abstract

As the powerhouse of the cell, mitochondria play a crucial role in many aspects of life, whereby mitochondrial dysfunctions are associated with pathogenesis of many diseases, like neurodegenerative diseases, obesity, cancer, and metabolic as well as cardiovascular disorders. Mitochondria analysis frequently starts with isolation and enrichment procedures potentially affecting mitochondrial morphology having impact on their function. Due to the complex mitochondrial morphology, the major task is to preserve their structural integrity. Here we critically review a commonly used isolation procedure for mitochondria utilizing differential (gradient) centrifugation and depict major challenges to achieve "functional" mitochondria as basis for comprehensive physiological studies.

Key words Isolation of mitochondria, Differential gradient centrifugation, Mitochondrial integrity

1 Introduction

Since their naming in 1898 by Carl Benda, the importance of mitochondria mediating several fundamental cellular processes is constantly growing. Due to the fact that mitochondria dysfunctions are involved in the pathophysiology of a wider variety of human diseases [1], dissecting mitochondria physiology is a main focus of recent biomedical sciences. Approximately 1 in 200 individuals bears a pathogenic mutation in mitochondrial genes [2], affecting mitochondrial biogenesis and inheritance and therefore containing the risk to develop severe diseases including neurodegenerative diseases, aging, obesity, cancer, and metabolic as well as cardiovascular disorders [3–7]. This underlines the importance to characterize exact composition and function of mitochondria in order to understand their role in cellular metabolism more precisely. In this context a major task of all this associated studies is to preserve the pristine mitochondrial structural integrity for analysis, which is indispensably linked to functionality of the organelles. Due to the complex organelle morphology enclosed by two membranes,

Volkmar Weissig and Marvin Edeas (eds.), *Mitochondrial Medicine: Volume I, Probing Mitochondrial Function*, Methods in Molecular Biology, vol. 1264, DOI 10.1007/978-1-4939-2257-4_1, © Springer Science+Business Media New York 2015

i.e., the outer membrane with a large number of specialized proteins (porins) enabling pass of molecules less than 5,000 Da and the inner membrane exhibiting a folding (cristae) to increase the surface area containing the complex respiratory transport chain, being critical for ATP production, and last but not least the matrix harboring the wide variety of enzymes for metabolic pathways, e.g., citric acid cycle, mitochondrial ribosomes, tRNAs, and mitochondrial DNA, the preparation of mitochondria approximating the in vivo situation is tremendously challenging.

2   Isolated Mitochondria: Addressing Composition and Function

In order to dissect mitochondrial composition as well as function in detail and to allow direct manipulation of mitochondria by exposure to specific substrates and inhibitors, the major part of conducted studies are utilizing isolated mitochondria [8, 9], achieved from diverse cellular and tissue sources. These isolation procedures most frequently are based on the fundamental work of George Pallade's group [10], done more than 60 years ago. They introduced a differential centrifugation work flow enabling separation of almost pure organelles with high yield, which have paved the way for such revolutionary discoveries like the oxidative phosphorylation mechanism [11], discovery of mitochondrial DNA [12], or the description of the mitochondrial ultrastructure [13]. Although today diverse adapted methods are available to face almost all kinds of sample sources and scientific questions [14–21], preparation of functional mitochondria reflecting approximately the in vivo situation is still challenging.

3   First Step: The Inevitable

It is supposed that mitochondria in vivo develop complex tubular branched structures [22, 23], which are significantly different from the relatively homogeneous circular organelles occurring during standard isolation techniques [9]. It has been commonly assumed that isolation of mitochondria inevitably comes along with disruption of the native mitochondrial morphology. Thereby, the mitochondria network is disrupted and sealed again, which probably leads to a partial loss of soluble mitochondrial proteins [24]. Up to now the functional consequences are largely unknown. In this context studies comparing functionality of isolated mitochondria with mitochondria within permealized cells, leaving the organelles in their native surrounding [25], indicate impairments of mitochondrial function, e.g., regarding mitochondrial respiration [24, 26]. Accordingly, it should be carefully considered that during data interpretation the common assumption isolated mitochondria preserve their complete functionality and composition will not be valid in any case.

4   Isolating "Intact" Mitochondria: A Challenging Business

Despite the known limitations, proper isolated mitochondria keep their compartment properties and provide a powerful tool for in-depth analysis, especially when comparing samples achieved under identical conditions [9]. Accordingly, many scientific fields comprising metabolite and protein transport up to dynamic remodeling of mitochondria as well as recent biomedical concerns benefit from isolated, almost pure mitochondria.

The most frequently applied method for isolation is differential centrifugation [8], whereby in a first step, cell or tissue samples are carefully homogenized in an appropriate buffer preventing damaging of the organelles by mechanical forces, chemical reactions, or osmosis. In order to separate components of different sizes and densities, e.g., cellular organelles, the homogenate is subjected to repeated centrifugation consecutively increasing sedimentation forces, enabling a rough fractionation of the cellular environment. In order to achieve pure organelles, i.e., mitochondria, the last purification step is an equilibrium density gradient centrifugation. Samples are centrifuged at high g-forces in a buffer gradient, e.g., sucrose gradient, concentrating the target organelles in a concentration range of comparable density (isopycnic point), resulting in high pure mitochondria.

Although most studies are carried out based on this general isolation strategy, different sample sources and special scientific requirements need specific adaptions of the used protocols [9, 19–21], in order to achieve optimal results regarding purity as well as functionality. In the literature a confusing diversity regarding utilized sedimentation forces, buffer compositions for homogenization, and gradient compositions are available, which should be carefully reviewed before use. In the end the choice of a suitable separation protocol depends on the researcher's requirements regarding organelle purity, yield, activity, and structural integrity. In some cases the most important factor is purity, whereby activity, yield, and preparation time may be less important. In studies exploring cellular compartment metabolism, high activity of the organelles is often the most important requirement. For higher-throughput experiments in comprehensive studies where many samples are to be compared, it is important to shorten the time for sample preparation.

From our point of view, the differential gradient centrifugation approach, closely monitored by appropriate quality control methods, offers the most valuable compromise between applied efforts (man power, expenses) and achievable yield, purity, activity, and functional integrity. In order to achieve valid results, isolation of "functional" mitochondria based on this methodology has some basic requirements and should follow a general work flow (Fig. 1, upper panel) illustrated and described in Chapter 2. This basic protocol provides a useful starting point for the isolation of mitochondria by differential

Differential Gradient Centrifugation Strategy

Sample Homogenization
(e.g. Potter, Douncer, Ultraturrax)

Differential Centrifugation
(e.g. different g-forces)

Gradient Centrifugation
(e.g. linear sucrose gradient)

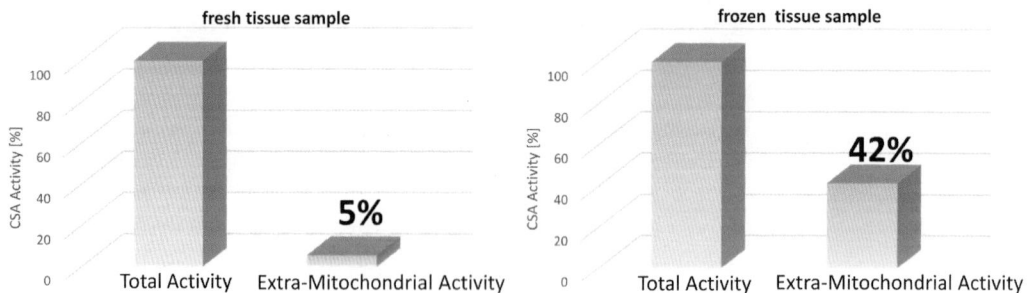

Yield

Purity

Mitochondria Integrity by Citrate Synthase (CS) Activity

fresh tissue sample

frozen tissue sample

5%

42%

Total Activity   Extra-Mitochondrial Activity

Total Activity   Extra-Mitochondrial Activity

CSA Activity [%]

Mitochondria Structural Integrity by electron microscopy

Isolated mitochondria
(fresh tissue sample)

Isolated mitochondria
(frozen tissue sample)

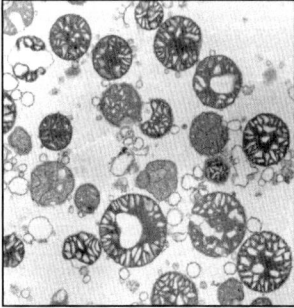

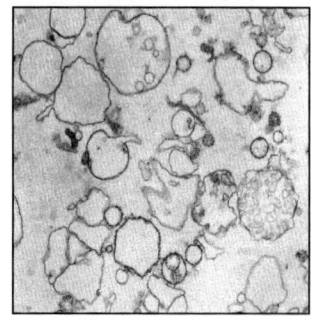

Fig. 1 Impact of sample freezing on mitochondria structural integrity. Centrifugation-based isolation of mitochondria is performed according to a general strategy consisting of three major processing steps shown in the *upper panel*. (1) Careful sample homogenization using a Potter, Douncer, or Ultraturrax device. (2) Consecutive differential centrifugation to separate cellular compartments according to size and density. (3) Increasing organelle purity due to equilibrium density gradient centrifugation, e.g., linear sucrose gradient, which enables concentration of mitochondria in the gradient fraction of comparable density (isopycnic point). To illustrate the importance of choosing appropriate sample material, the *lower panels* show a comparison of isolated mitochondria from fresh and frozen material. Measurement of citrate synthase (CS) activity, an enzyme in situ exclusively located in the mitochondrial matrix, allows monitoring structural integrity of mitochondria during processing. CS release and therefore significant CS activity in the homogenate (extramitochondrial activity) suggest disruption of mitochondria structure. To assign the total CS activity as control, CS activity of complete

gradient centrifugation. It describes preparation in detail and guide through critical steps of the separation method and how to control them regarding yield and organelle functionality. In laboratory routine frequently separation protocols are utilized, which pass the last centrifugation step through a density gradient, which requires an ultracentrifuge and some experience. Fractionation solely by different sedimentation forces, also known as g-forces, is also applied in commercial available isolation kits but anyway results in a decreased enrichment performance [9].

In addition to the discussed differential (gradient) centrifugation strategy, it is important to mention that isolation of mitochondria using integrated zone electrophoresis on a free-flow electrophoretic device [9, 18] represents a relevant alternative for preparing high purity organelles, which are particularly appropriate for comprehensive proteomic studies. The major drawback of this method is that a special instrument, i.e., a free-flow apparatus, is necessary and that processing is time consuming and needs special expertise. Additionally, a protocol was described combining simple differential centrifugation with a final purification of mitochondria utilizing anti-TOM22 magnetic beads. This reproducible protocol has been described to be suitable for various tissues yielding in the isolation of highly pure mitochondria avoiding non-mitochondrial contaminations [27]. Both methods provide pure mitochondria fractions with structural integrity, but come along with much higher costs than applied for differential gradient centrifugation methods.

5   Reproducibility and Quality Control: Guarantee for Successful Analysis

Many recent studies in the field of biomedical research addressing potential mitochondrial dysfunction require analysis of properties in a highly parallel fashion. In order to achieve valid results, therefore, it is crucial to utilize a standardized processing. In this context it would be advantageous to collect the samples successively and store them in a freezer before use. Accordingly, one could think that starting with frozen material for mitochondria isolation would simplify the work flow significantly. But unfortunately, the most striking prerequisite to isolate functional mitochondria is to use fresh (not frozen) material. Investigating the impact of sample

Fig. 1 (continued) lysed mitochondria samples was set as 100 %. Comparing activity levels of CS, in the homogenate of fresh and frozen samples, impressively shows that initial freezing destroys mitochondria structure and induces a release of CS from the mitochondrial matrix. This results in an eightfold increase in CS activity in the homogenate. Accordingly, in case of frozen sample, nearly half of the total CS activity is found outside of the mitochondria. In addition to that, electron microscopy (EM) analysis of gradient fractions shows a more or less complete loss of the typical mitochondrial cristae structure resulting in blank membrane covers

freezing indicates dramatic consequences on mitochondria structure and functionality (Fig. 1, middle and lower panel; unpublished data). Monitoring the activity of citrate synthase (CS) [28], an enzyme normally exclusively located in the mitochondrial matrix, reveals that freezing leads to a strong release of CS from the mitochondrial matrix into the homogenate. In contrast to fresh material, where approximately 5 % of total CS activity is found in the homogenate, in case of frozen material, more than 40 % of CS activity can be assigned to the homogenate. This eightfold increase indicates that the mitochondria structure is significantly disrupted. Corresponding electron microscopy (EM) images confirm these observations and demonstrate loss of structural integrity due to a nearly complete destruction of isolated mitochondria expected morphology (Fig. 1, lower panel). Accordingly, these observations strongly suggest that frozen material in any case is inappropriate for organelle isolation and therefore to study mitochondria composition, function, or physiological behavior. Another challenging point to enable reliable comparison of a huge number of samples is the initial homogenization step. In most laboratories it is performed manually using Potter, Douncer, or Ultraturrax homogenizers, potentially introducing significant variations. The utilized forces for homogenization are a highly subjective parameter, which carefully have to be validated.

Due to the fact that organelle purity and functionality are the striking prerequisites for any study addressing isolated mitochondria, monitoring the whole isolation process is mandatory. Without an appropriate quality control, regarding mitochondria functionality and purity, no reasonable assessment of the organelle sample is possible. In this context it is important to mention that Western blot analysis of organelle-specific proteins (e.g., mitochondria (anti-Tom20), lysosomes (anti-Lamp-1), endoplasmic reticulum (anti-BiP/GRP78), and peroxisomes (anti-catalase)), which is frequently used as a standard method to examine product composition and therefore the isolation success, only allows detection of relative distribution of the dedicated proteins but does not allow assignment of mitochondria functionality. A more meaningful evaluation, allowing to calculate mitochondria purity, can be achieved when additional biochemical assays are used. We recommend measuring activity of marker enzymes specific for mitochondria as well as contaminating cell compartments, i.e., succinate dehydrogenase for mitochondria, glucose-6-phosphatase for endoplasmic reticulum [29], alkaline phosphatase for plasma membrane, acidic phosphatase for lysosomes [30], and catalase for peroxisomes [31]. These control assays are easily applicable with standard laboratory equipment and enable to calculate organelle distribution and to estimate some functional aspects. These assays are also very helpful during the establishing phase of the gradient centrifugation protocols, in order to select the region within the gradient corresponding to most pure and active mitochondria fraction. In order to assess "functional" integrity, measuring JC-1 uptake [32] or oxygen consumption with the principle of a traditional Clark electrode [33] as read out for ATP production should be mandatory. Both methods are appropriate to monitor the status of mitochondrial membrane potential. If possible, additional investigations of the isolated mitochondria by transmission electron microscopy allow assessing structural integrity of the inner and outer membranes as well as the mitochondrial matrix.

Anyone has to mention disclaiming such investigation tools deprives the possibility to make substantiated conclusions about organelle integrity and therefore the functional status of the underlying preparation. Accordingly, investigation of purity and morphology-function relationship should be an inherent part of any organelle fractionation procedures in order to avoid working with inappropriate sample material.

6   Conclusion

Our recent knowledge suggest that available protocols fail to allow isolation of native, functional mitochondria. This has to be considered, when planning and interpreting the experiments. Emerging understanding of structure/function relationship and the effect of morphology changes during isolation may help to improve isolation methods or develop novel strategies for in-depth in situ analysis. Nevertheless, proper isolated mitochondria may approximate the "intact" status and still provide an indispensable tool to address future challenges of mitochondrial participation in the pathophysiology of diverse widespread diseases including neurodegenerative, muscular, cardiovascular, and metabolic disorders and cancer.

References

1. Nunnari J, Suomalainen A (2012) Mitochondria: in sickness and in health. Cell 148:1145–1159

2. Elliott HR, Samuels DC, Eden JA, Relton CL, Chinnery PF (2008) Pathogenic mitochondrial DNA mutations are common in the general population. Am J Hum Genet 83(1030):254–260

3. Schon EA, DiMauro S, Hirano M (2012) Human mitochondrial DNA: roles of inherited and somatic mutations. Nat Rev Genet 13:878–890

4. Patti ME, Corvera S (2010) The role of mitochondria in the pathogenesis of type 2 diabetes. Endocr Rev 31:364–395

5. Wallace DC (2012) Mitochondria and cancer. Nat Rev Cancer 12:685–698

6. Piaceri I, Rinnoci V, Bagnoli S, Failli Y, Sorbi S (2012) Mitochondria and Alzheimer's disease. J Neurol Sci 322:31–34

7. Hill BG (2013) Recent advances in mitochondrial research. Circ Res 113:107–110

8. Frezza C, Cipolat S, Scorrano L (2007) Organelle isolation: functional mitochondria from mouse liver, muscle and cultured fibroblasts. Nat Protoc 2:287–295

9. Hartwig S, Feckler C, Lehr S, Wallbrecht K, Wolgast H, Müller-Wieland D, Kotzka J (2009) A critical comparison between two classical and a kit-based method for mitochondria isolation. J Proteomics 11:3209–3214

10. Hogeboom GH, Schneider WC, Pallade GE (1948) Cytochemical studies of mammalian

tissues. I. Isolation of intact mitochondria from rat liver; some biochemical properties of mitochondria and submicroscopic particulate material. J Biol Chem 172:619–635

11. Mitchell P, Moyle J (1967) Chemiosmotic hypothesis of oxidative phosphorylation. Nature 213:137–139

12. Nass MM, Nass S (1963) Intramitochondrial fibers with DNA characteristics. I. Fixation and electron staining reactions. J Cell Biol 19:593–611

13. Palade GE (1952) The fine structure of mitochondria. Anat Rec 114:427–451

14. Graham JM (2001) Purification of a crude mitochondrial fraction by density-gradient centrifugation. Curr Protoc Cell Biol Chapter 3:Unit 3.4

15. Noguchi T, Fujiwara S (1984) Developmental profiles and properties of hepatic peroxisomal apo- and mitochondrial holoalanine: glyoxylate aminotransferase during chick embryogenesis. J Biol Chem 259:14498–14504

16. Liu Y, He J, Ji S, Wang Q et al (2008) A comparative study of early liver dysfunction in senescence-accelerated mouse using mitochondrial proteomics approaches. Mol Cell Proteomics 7:1737–1747

17. Lehr S, Kotzka J, Avci H, Knebel B et al (2005) Effect of sterol regulatory element binding protein-1a on the mitochondrial protein pattern in human liver cells detected by 2D-DIGE. Biochemistry 44:5117–5128

18. Zischka H, Weber G, Weber PJ, Posch A et al (2003) Improved proteome analysis of *Saccharomyces cerevisiae* mitochondria by free-flow electrophoresis. Proteomics 3:906–916

19. Stahl WL, Smith JC, Napolitano LM, Basford RE (1963) BRAIN MITOCHONDRIA: I. Isolation of bovine brain mitochondria. J Cell Biol 19:293–307

20. Cannon B, Lindberg O (1979) Mitochondria from brown adipose tissue: isolation and properties. Methods Enzymol 55:65–78

21. Mela L, Seitz S (1997) Isolation of mitochondria with emphasis on heart mitochondria from small amounts of tissue. Methods Enzymol 55:39–46

22. Ogata T, Yamasaki Y (1997) Ultra-high-resolution scanning electron microscopy of mitochondria and sarcoplasmic reticulum arrangement in human red, white, and intermediate muscle fibres. Anat Rec 248:214–223

23. Fang H, Chen M, Ding Y, Shang W, Xu J et al (2011) Imaging superoxide flash and metabolism-coupled mitochondrial permeability transition in living animals. Cell Res 21:1295–1304

24. Picard M, Taivassalo T, Ritchie D, Wright KJ, Thomas MT et al (2011) Mitochondrial structure and function are disrupted by standard isolation methods. PLoS One 6:e18317

25. Kuznetsov AV, Veksler V, Gellerich FN, Saks V, Margreiter R, Kunz WS (2008) Analysis of mitochondrial function in situ in permeabilized muscle fibres, tissues and cells. Nat Protoc 3:965–976

26. Saks V, Guzun R, Timohhina N, Tepp K, Varikmaa M et al (2010) Structure-function relationships in feedback regulation of energy fluxes in vivo in health and disease: mitochondrial interactosome. Biochim Biophys Acta 1797:678–697

27. Franko A, Baris OR, Bergschneider E, von Toerne C, Hauck SM et al (2013) Efficient isolation of pure and functional mitochondria from mouse tissues using automated tissue disruption and enrichment with anti-TOM22 magnetic beads. PLoS One 8:e82392

28. Schulman JD, Blass JP (1971) Measurement of citrate synthase activity in human fibroblasts. Clin Chim Acta 33:467–469

29. Pennington RJ (1961) Biochemistry of dystrophic muscle. Mitochondrial succinate-tetrazolium reductase and adenosine triphosphatase. Biochem J 80:649–654

30. Bergmeyer HU (1974) Methoden der enzymatischen analyse. Verlag Chemie, Weinheim

31. Aebi H (1984) Catalase in vitro. Methods Enzymol 105:121–126

32. Reers M, Smiley ST, Mottola-Hartshorn C, Chen A, Lin M, Chen LB (1995) Mitochondrial membrane potential monitored by JC-1 dye. Methods Enzymol 260:406–417

33. Li Z, Graham BH (2012) Measurement of mitochondrial oxygen consumption using a Clark electrode. Methods Mol Biol 837:63–72

Chapter 2

Isolation and Quality Control of Functional Mitochondria

Sonja Hartwig, Jorg Kotzka, and Stefan Lehr

Abstract

Numerous protocols are available being adapted for different cell or tissue types allowing isolation of pure mitochondria trying to preserve their "structural and functional" integrity. In this chapter we intend to provide a more general framework introducing differential isopycnic density gradient centrifugation strategy with a special focus sensitizing for the specific challenges coming along with this method and how to obtain "functional," enriched, "intact" mitochondria. Due to the fact that in any study dealing with these organelles standardized processing is mandatory. Here we describe a strategy addressing quality control of prepared intact mitochondria. The quality control should be an integrated part of all isolation processes. The underlying protocol should be seen as starting point and has to be carefully adjusted to cover different sample types used for the diverse research questions.

Key words Sample pre-fractionation, Mitochondria enrichment, Isopycnic density gradient centrifugation, Marker enzymes

1 Introduction

Mitochondria biology plays a crucial role in many aspects of life, including in pathophysiology of many diseases, like neurodegenerative diseases, obesity, cancer, and metabolic disorders [1]. In order to dissect the specific role of mitochondria various scientific fields benefit from analysis of isolated, almost pure mitochondria. Due to the fact, that there is a strong relationship between mitochondria structure and functionality, there are strong demands to preserve their "structural and functional" integrity during preparation. Today, numerous protocols are available enabling isolation of "pure" mitochondria, including differential gradient centrifugation [2, 3], affinity purification utilizing coupled antibodies [4] or separation by integrated zone-electrophoresis [5]. In this context the differential gradient centrifugation strategy offers the most valuable compromise between applied efforts (man-power, expenses) and achievable yield, purity, activity, and functional integrity. In order to achieve high quality mitochondria appropriate

Volkmar Weissig and Marvin Edeas (eds.), *Mitochondrial Medicine: Volume I, Probing Mitochondrial Function*, Methods in Molecular Biology, vol. 1264, DOI 10.1007/978-1-4939-2257-4_2, © Springer Science+Business Media New York 2015

for further detailed investigations moreover it is mandatory to monitor functionality and enrichment progress during the entire isolation process. Here we describe the preparation of mitochondria in detail and guide through critical steps of the separation method and how to control them regarding yield and organelle functionality.

2 Materials

2.1 Mitochondria Fractionation

1. Potter S homogenizer with glass cylinders (15 ml) including appropriate plunger (for breaking up the tissue in a gentle manner).

2. Ground-in glass douncer (loose fit, for gentle manual homogenization).

3. Centrifuges, corresponding rotors and tubes with capability for $11,000 \times g$-force and $85,000 \times g$-force, e.g. SS34 Rotor for the RC-5b (Sorvall/DuPont) and swing-out rotor SW28 for the OptimaL70 (Beckmann).

4. Homogenization buffer: 225 mM mannitol, 75 mM saccharose, 10 mM Tris/HCl, 0.5 mM EGTA, pH 7.4 and 0.5 mM DTT (*see* **Note 1**).

5. Resuspension buffer: 250 mM saccharose, 10 mM Tris/HCl, 0.5 mM EGTA, pH 7.4, and 0.5 mM DTT (*see* **Note 1**).

2.2 Linear Saccharose Gradient

1. Gradient mixer.

2. Ultracentrifuge tubes (e.g. Beckmann tubes for SW28 rotor).

3. Three different saccharose solutions in resuspension buffer: 24 % (w/w), 54 % (w/w) and 57 % (w/w) (*see* **Note 2**).

2.3 Marker Enzyme Assays

1.5 and 2 ml reaction tubes with corresponding stand, incubator for 37 °C, the usage of a multipette would be a benefit and a photometer with corresponding solvent-resistent cuvette to measure at 405, 410, 490, and 815 nm are needed. For JC-1 Assay a spectrofluorometer with an excitation wavelength of 490 nm and an emission wavelength of 590 nm is needed.

2.3.1 Succinate Dehydrogenase (SDH) Assay

1. INT solution: 2.5 mg/ml p-Iodonitrotetrazolium in 0.05 M Na-dihydrogenphosphate pH 7.5.

2. Na-succinate solution: 0.01 M Na-succinate in 0.05 M Na-dihydrogenphosphate pH 7.5.

3. Stop solution: 4. Ethylacetate:Ethanol:TCA 5:5:1 (v/v/w).

2.3.2 Acidic Phosphatase Assay

1. Nitrophenylphosphate solution: 16 mM *p*-Nitrophenylphosphate in H_2O.

2. Na-acetate solution: 180 mM Na-acetate pH 5.0 (adjustable with acetic acid).

3. Stop solution: 250 mM NaOH.

2.3.3 Basic
Phosphatase Assay

1. Nitrophenylphosphate solution: 16 mM p-Nitrophenylphosphate in H_2O.

2. Na-borate solution: 250 mM Na-borate pH 9.8 (adjustable with NaOH).

3. Stop solution: 250 mM NaOH.

4. 1 M $MgCl_2$.

2.3.4 Catalase Assay

1. Catalase sample buffer: 20 mM Tris/HCl pH 7.0, 1 % BSA, 2 % Triton X-100.

2. Assay buffer: 20 mM Tris/HCl pH 7.0, 1 % BSA, 0.25 % H_2O_2.

3. Titanyl solution: Dissolve 22.5 mg titanium oxy sulfate sulfuric acid hydrate in 100 ml 1 M sulfuric acid (*see* **Note 3**).

2.3.5 Glucose-6-
Phospatase (G-6-Pase)
Assay

1. Saccharose buffer: 250 mM saccharose, 100 mM EDTA pH 7.2.

2. Cacodylate buffer: 100 mM cacodylic acid sodium salt adjusted to pH 6.5.

3. Glucose-6-phospat solution: 100 mM glucose-6-phospat (*see* **Note 4**).

4. Potassium-dihydrogenphosphate solution: 200 μM Potassium dihydrogenphosphate in H_2O.

5. TCA Solution: 8 % trichloroacetic acid in H_2O.

6. Fiske-SubbaRow reagent: Dissolve 0.75 g sodium sulfite in 5 ml H_2O. Also dissolve 6.85 g sodium disulfite and 0.125 g Amino-2-hydroxynaphthalin-4 sulfonic acid in 50 ml H_2O. Mix both solutions and store it in a tightly capped amber bottle (*see* **Note 5**).

7. Ammonium-heptamolybdate solution: 0.48 % ammonium heptamolybdate in H_2O (w/v).

2.3.6 JC-1 Uptake Assay

1. 5× storage buffer: 50 mM HEPES, 1,25 M saccharose, 5 mM ATP, 0.4 mM ADP, 25 mM sodium-succinate, 10 mM K_2HPO_4, 5 mM DTT. Weigh out all ingredients and dissolve it in H_2O 90 % of calculated end volume (e.g. for 100 ml add 90 ml). Adjust pH to 7.5 with concentrated NaOH and fill up with H_2O to 100 % of calculated volume. Pass the buffer through a 0.2 μm filter and store 5–15 ml aliquots at –20 °C.

2. 5× JC-1 assay buffer: 100 mM MOPS, 550 mM KCl, 50 mM ATP, 50 mM $MgCl_2$, 50 mM sodium-succinate, 5 mM

EGTA. Weigh out all ingredients and dissolve it in H_2O 90 % of calculated end volume (e.g. for 50 ml add 45 ml). Adjust pH to 7.5 with concentrated NaOH and fill up with H_2O to 100 % of calculated volume. Pass the buffer through a 0.2 μm filter and store 2–5 ml aliquots at –20 °C.

3. JC-1 stain: Dissolve 25 μg JC-1 (5,5′,6,6′-tetrachloro-1,1′,3,3′tetraethylbenzimidazol carbocyanine iodide) in 25 μl DMSO (*see* **Note 6**) to obtain a solution with 1 μg/μl (resulting concentration is 1.53 mM).

3   Methods

Here we describe a protocol for reproducible mitochondria isolation from mouse liver with isopycnic saccharose gradient centrifugation. The general workflow and some typical electron microscopy images from such a preparation are shown in Fig. 1. Also an example result for enzymatic quality check is presented in Table 1.

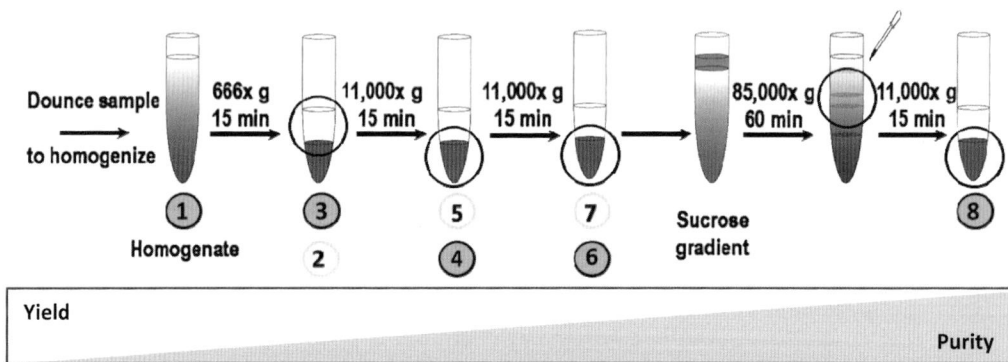

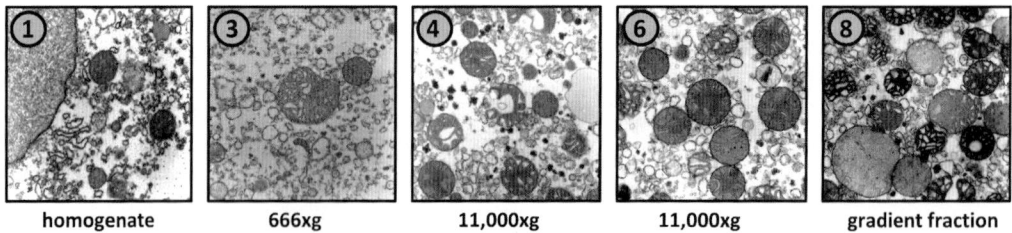

Fig. 1 Mitochondria isolation work flow shows the consecutive steps from homogenization, differential centrifugation up to the density gradient. Purification progress among the different processing steps is visualized by electron microscopy

Table 1

Example of monitoring the subcellular fractionation process

Preperation step		Mitochondria					
		Succinate dehydrogenase				JC-1 uptake	
	Total protein [mg]	Total activity [µmol/min]	yield [%]	Specific activity [µmol/mg/min]	Accumulation factor	Specific activity [Fluor./mg]	Accumulation factor
Homogenate	243±74	12,649±5,076	100	52.3±15.8	1	372±73	1
666 g supernatant	152±56	5,649±5,692	39	33.9±31.3	0.7	308±79	0.8
11,000 g pellet	16±2	2,113±836	17	134.2±52.8	2.6	601±91	1.7
Gradient fraction	5.3±1.6	865±420	7	164.1±51.3	3.2	687±97	1.9

Preperation step	Lysosomes			Plasma membrane			Endoplasmic reticulum			Peroxisomes		
	Acidic phosphatase			Basic phosphatase			Glucose-6-phosphatase			Catalase		
	Total activity [µmol/min]	Yield [%]	Specific activity [µmol/mg/min]	Total activity [µmol/min]	Yield [%]	Specific activity [µmol/mg/min]	Total activity [µmol/min]	Yield [%]	Specific activity [µmol/mg/min]	Total activity [µU/min]	Yield [%]	Specific activity [µU/mg/min]
Homogenate	464±161	100	1.96±0.6	11.95±3.4	100	0.05±0.01	1,405±548	100	6.0±2.0	6,195±2,811	100	25.6±10.1
666 g supernatant	249±59	56	1.73±0.4	6.06±1.7	51	0.04±0.03	627±159	47	4.4±1.0	4,126±1,610	72	27.9±8.8
11,000 g pellet	56±12	13	3.54±0.6	0.51±0.4	4	0.03±0.03	128±60	9	7.8±2.9	517±323	9	31.9±18.1
Gradient fraction	7.8±3.7	1.8	1.51±0.5	0.07±0.01	0.5	0.01±0.003	29±16	2	5.2±1.9	79±48	1.4	14.6±5.2

3.1 Subcellular Fractionation

3.1.1 Preparation of Linear Saccharose Gradient

1. Pipette 4 ml 57 % saccharose solution in a 36 ml ultracentrifuge tube (for SW28 Beckmann rotor) as a *pillow* and place it in an angular tube stand.

2. Build up the gradient mixer, assure that you can stir the solution in the first chamber and put a blunt cannula at the end of the hose.

3. Fill 15 ml low 24 % saccharose solution in the first chamber and 15 ml 54 % saccharose solution in the second.

4. Position the blunt cannula in the angular positioned tube tight above the 57 % *pillow* and fix it with a clip.

5. Open the taps and let the saccharose solutions flow by gravity into the tube (*see* **Note 7**).

6. Pull the blunt cannula carefully out, right the tube and store it at 4 °C until use.

3.1.2 Homogenization

1. Use a fresh liver tissue sample (obtained within 1 h of sacrifice) kept on ice in homogenization buffer.

2. Wash mice liver twice with 2 volumes of homogenization buffer and mince it into pieces with scissors.

3. Transfer liver pieces (in total about 1.5 g) to the 15 ml glass cylinder and add tenfold (w/v) homogenization buffer.

4. Homogenize the liver pieces by ten strokes with the ground-in glass dounce (*see* **Note 8**).

5. Take a 10 % aliquot of the homogenate and store on ice (*see* **Note 9**).

3.1.3 Differential Centrifugation

1. Transfer homogenate to a 15 ml centrifugation tube and centrifuge at $666 \times g$ at 4 °C for 15 min to remove cell debris.

2. After centrifugation transfer the supernatant to a SS34 centrifugation tube fill up to 30 ml with homogenization buffer and centrifuge at $11,000 \times g$ 4 °C for 15 min.

3. Decant the supernatant and put it aside (mainly cytosolic proteins).

4. Wash the pellet with homogenization buffer by pipetting carefully across, to remove the *fluffy layer* (yellowish layer above a darker tawny layer)

5. Resuspend the tawny pellet in 3–5 ml homogenization buffer, take an aliquot, transfer the resuspended pellet to a new SS34 tube and add up to 30 ml with homogenization buffer and centrifuge again at $11,000 \times g$ for 15 min at 4 °C.

6. Decant the supernatant and put it aside.

7. Resuspend the resulting tawny pellet in 2–3 ml resuspension buffer and again take an aliquot.

3.1.4 Isopycnic Density Gradient Centrifugation	1. Layer the resuspended pellet carefully onto a linear sucrose gradient.

1. Layer the resuspended pellet carefully onto a linear sucrose gradient.

2. Carry out centrifugation in a SW28 swing-out bucket rotor at $85,000 \times g$ for 60 min at 4 °C without brakes (*see* **Note 10**).

3. After centrifugation carefully remove the gradient tube from the rotor bucket and the enriched mitochondria show up in the middle of the tube as a light brown-yellowish ring.

4. Collect 2 ml aliquots by pipetting carefully from above in rotary movements with a wide-opening pipette tip (*see* **Note 11**).

5. Determine the mitochondrial activity from the aliquots by measuring succinate dehydrogenase activity (*see* Subheading 3.3.1) and pool aliquots with highest activity (*see* **Note 12**).

6. Dilute pooled aliquots with fivefold volume (v/v) of resuspension buffer and transfer to SS34 tube follow a centrifugation at $11,000 \times g$ for 15 min and 4 °C

7. Resuspend resulting final mitochondria pellet in 1–2 ml resuspension buffer.

8. This mitochondria fraction is ready for further experiments including quality control assays.

3.2 Protein Measurement

Protein measurements from all steps of the mitochondria preparation should be done. Therefore take 10–20 µl of each fraction/aliquot and mix it up with the same amount of 1 M NaOH to denature for protein measurement with standard *Bradford assay*.

3.3 Marker Enzyme Assays

With all these enzyme assays you can measure the protein activity per ml and calculate the specific activity with the values from protein content and therefore protocol your mitochondria enrichment and decrease of other organelles during preparation. For the different enzyme activity assays you need different dilutions of your samples: succinate-dehydrogenase delivers good results with a sample/aliquot dilution of 1:10 and 1:20, acidic phosphatase with 1:8 and 1:16, catalase 1:10, basic phosphatase pure and 1:2 and glucose-6-phosphatase 1:10 (*see* **Note 13**).

3.3.1 Mitochondria: Succinate-Dehydrogenase (SDH) (See Ref. 6)

The succinate-dehydrogenase catalyze the oxidation from succinate to fumarate under release of hydrogen. In vivo FAD would be reduced to $FADH_2$, but under these test assay conditions the p-Iodonitrotetrazolium (INT), an artificial electron acceptor would be reduced to formazan, which turns from colorless to rusty red and can be measured at 490 nm.

1. Produce the INT solution: 100 µl for each sample/dilution is needed.

2. Prepare 2 ml reaction tubes with 20 µl of diluted samples and one only with resuspension buffer as a blank.

3. Add 300 μl Na-succinate solution to each tube and incubate for 10–20 min at 37 °C.

4. Add 100 μl INT solution and incubate 10 min at 37 °C.

5. Stop the enzyme reaction by adding 1 ml stop solution.

6. Centrifuge the tubes for 2 min at 20,000×g to get rid of precipitates.

7. Measure the extinction of supernatants at 490 nm.

8. Calculate the activity per ml and the specific activity:

$$\frac{\Delta E \times V_{E[ml]} / V_{P[ml]}}{\varepsilon_{mol}\left[ml / \mu mol\, cm\right] \times d\left[cm\right] \times t\left[min\right]} = \frac{\left[\mu mol\right]}{\left[ml\, min\right]}$$

ΔE: extinctions difference (measured value – measured blank).

V_E/V_P: enzyme assay volume/added sample volume, i.e. the entire dilution factor of the measured sample in the assay.

ε_{mol}: molar extinctions coefficient [ml/μmol cm] (for INT it is 0.0134).

d: thickness of used cuvette [cm].

t: incubation time Na-succinate solution with sample before adding INT solution.

Calculate the specific activity [μmol/mg min] in your sample by simple dividing through the protein concentration of the measured sample.

3.3.2 Lysosomes: Acid Phosphatase (See Ref. 7)

The acid phosphatase has a working optima at pH 4–5. In this assay the conversion from *p*-nitrophenylphosphate and H_2O to p-nitrophenol is the enzymatic step and with a pH-increase by adding NaOH the "produced" nitrophenol shift to nitrophenolate-ions and the solution gets yellow color and extinction could be measured at 410 nm.

1. Prepare an assay mix with 16 mM p-Nitrophenylphosphate solution and the 180 mM Na-acetate solution (1:1, v/v).

2. Prepare 1.5 ml reaction tubes and add 25 μl of recommended sample dilutions and use resuspension buffer as blank.

3. Add 200 μl assay mix solution to each tube and incubate for 20–30 min at 37 °C.

4. Stop the enzyme reaction by adding 600 μl stop solution to each tube.

5. Centrifuge the tubes for 2 min at 20,000×g to get rid of precipitates.

6. Measure the extinction of supernatants at 410 nm.

7. Calculate the activity per ml and the specific activity:

$$\frac{\Delta E \times V_{E[ml]} / N_{P[ml]}}{\varepsilon_{mol}\left[ml / \mu mol\,cm\right] \times d\left[cm\right] \times t\left[min\right]} = \frac{\left[\mu mol\right]}{\left[ml\,min\right]}$$

ΔE: extinctions difference (measured value – measured blank),

V_E / V_P: enzyme assay volume/added sample volume, i.e. the entire dilution factor of the measured sample in the assay.

ε_{mol}: molar extinctions coefficient [ml/μmol cm] here for converted nitrophenol is 0.521

d: thickness of used cuvette [cm]

t: incubation time

Calculate the specific activity [μmol/mg min] in your sample by simple dividing through the protein concentration of the measured sample.

3.3.3 Plasma Membrane: Basic Phosphatase (See Ref. 7)

The basic phosphatase has a working optima at pH 9–10. In this assay the conversion from p-nitrophenylphosphate and H_2O to p-nitrophenol is the enzymatic step and with a pH-increase by adding NaOH the "produced" nitrophenol shift to nitrophenolate-ions and the solution gets yellow color and extinction could be measured at 410 nm.

1. Prepare an assay mix with nitrophenylphosphate solution and the Na-borate solution (1:1, v/v). Also add 1 M $MgCl_2$ to a final concentration in the assay mix of 2 mM (i.e. 2 μl in 1 ml)

2. Prepare 1.5 ml reaction tubes with 25 μl of pure and diluted samples and use resuspension buffer as blank.

3. Add 200 μl assay mix solution to each tube and incubate for 20–30 min at 37 °C.

4. Stop the enzyme reaction by adding 600 μl stop solution to each tube.

5. Centrifuge the tubes for 2 min at 20,000×g to get rid of precipitates.

6. Measure the extinction of supernatants at 410 nm

7. Calculate the activity per ml and the specific activity:

$$\frac{\Delta E \times V_{E[ml]} / N_{P[ml]}}{\varepsilon_{mol}\left[ml / \mu mol\,cm\right] \times d\left[cm\right] \times t\left[min\right]} = \frac{\left[\mu mol\right]}{\left[ml \times min\right]}$$

ΔE: extinctions difference (measured value – measured blank),

V_E / V_P: enzyme assay volume/added sample volume, i.e. the entire dilution factor of the measured sample in the assay.

ε_{mol}: molar extinctions coefficient [ml/µmol cm] here for converted nitrophenol is 0.521,

d: thickness of used cuvette [cm],

t: incubation time.

Calculate the specific activity [µmol/mg min] in your sample by simple dividing through the protein concentration of the measured sample.

*3.3.4 Peroxisome: Catalase (**See** Ref. 8)*

The Catalase converts hydrogen peroxide to water and hydrogen. Titanoxid sulfat build up a yellow complex with hydrogen peroxide and the extinction of this complex could be measured at 405 nm (*see* **Note 14**).

1. Prepare the titanyl solution fresh (*see* **Note 3**).

2. Test assay buffer by adding 1 ml titanyl solution to 500 µl assay buffer, centrifuge for 5 min at 20,000×g and measure extinction from the supernatant at 405 nm. The value should be between OD 0.5–0.6. If this is not the case you have to prepare the assay buffer fresh (*see* **Note 15**).

3. Mix 10 µl of the sample with 30 µl of the catalase sample buffer and do all further steps on ice (*see* **Note 16**).

4. Add 500 µl from the tested assay buffer and stop the reaction by adding 1 ml titanyl solution exactly after 1 min.

5. Centrifuge samples for 5 min at 20,000×g.

6. Measure the extinction at 405 nm against water and your "positive" blank (*see* **Note 17**).

7. Define E_{405nm} 0.001 = 0.001 U/ml and calculate the difference from extinction measured from sample to the "positive" blank and also calculate the activity and specific activity for catalase as follow:

$$\frac{\Delta E \times V_{E[ml]} / N_{P[ml]}}{\varepsilon_{mol}[ml / U\ cm] \times d[cm] \times t[min]} = \frac{[U]}{[ml\ min]}$$

ΔE: extinctions difference (measured "positive" blank – measured value)

V_E / V_P: enzyme assay volume/added sample volume, i.e. the complete dilution factor of the measured sample in the assay.

ε_{mol}: molar extinctions coefficient [ml/U cm] adaption 0.001 = 1 U (*see* **Note 18**).

d: thickness of used cuvette [cm]

t: incubation time before addition of titanyl solution

Calculate the specific activity [U/mg min] in your sample by simple dividing through the protein concentration of the measured sample.

3.3.5 Endoplasmic Reticulum: Glucose-6-Phophatase (See Ref. 9)

The glucose-6-phosphatase (G-6-Pase) dephosphorylate G-6-P to glucose and free phosphate. Colorimetric measurement of free phosphate content could be done by Fiske-SubbaRow method. In this assay a blue complex is formed, when free phosphate is mixed with ammonium molybdate and 1-amino-2-naphthol-4-sulfonic acid, and could be measured at 815 nm. Important for this enzyme assay is to measure next to a blank the sample without or with substrate. To calculate the nmol/ml concentrations a standard curve with potassium hydrogen phosphate has to be measured.

1. Prepare Fiske-SubbaRow-mix freshly by adding 3.75 ml Fiske-SubbaRow-reagent and 10 ml perchloric acid (60 %) to 86.25 ml ammonium-heptamolybdate solution.

2. For each sample you have to measure two values, one without substrate (endogenous Pi) and one with substrate (G-6-P) for enzymatic dephosphorylation combine in 2 ml tubes:

   For endogenous value

   (a) 100 µl cacodylate buffer
   (b) 100 µl saccharose buffer
   (c) 100 µl H_2O

   For enzymatic value

   (a) 100 µl cacodylate buffer
   (b) 100 µl saccharose buffer
   (c) 100 µl glucose-6-phosphate solution.

3. By adding 100 µl sample to each tube the enzymatic reaction is started.

4. Incubate all samples for 30 min at 37 °C.

5. During incubation prepare a standard curve with the potassium hydrogen phosphate as replicates (0, 20, 50, 100, 150–200 nmol/ml). The volume of each standard sample is 1 ml (*see* **Note 19**).

6. Stop the enzymatic reaction (**step 4**) in the sample tubes by adding 1.5 ml 8 % TCA solution to each tube.

7. Centrifugate the sample for 10 min at $1,000 \times g$ to get rid of precipitates.

8. Transfer 1 ml of the resulting supernatant to a new 2 ml tube and add 1 ml of the fresh prepared Fiske-SubbaRow-mix to each tube and also to the samples of the standard curve.

9. Vortex all samples and incubate for 30 min at RT.

10. Measure the standard curve, basal, and enzymatic samples at 815 nm.

11. Generate a x/y-graph with values from standard curve, y-axis is OD and x-axis the known free phosphate (Pi) amount [nmol].

12. To calculate the produced Pi amount subtract the endogenous from the enzymatic OD, read off the Pi value from the standard curve and calculate the enzymatic activity as follows:

$$\frac{Pi[n\,mol\,/\,ml]\times V_{E}[ml]/V_{P}[ml]}{t[min]} = \frac{[n\,mol]}{[ml\,min]}$$

V_{E}/V_{P}: enzyme assay volume/added sample volume, i.e. the complete dilution factor of the measured sample in the assay.

t: incubation time before addition of titanyl solution.

3.3.6 Mitochondria (Functional Integrity): JC-1 Uptake Assay (See Ref. 10)

Uptake measurement of the fluorescent carbocyanine dye (JC-1) is a surrogate parameter for mitochondrial inner membrane integrity because its only possible if electrochemical proton gradient is formed. Depending upon the transmembrane electric field JC-1 will be uptake into mitochondrial matrix and if concentration raised more than 1 mM a red-orange fluorescence will occur at 590 nm, due to aggregation of dye within the matrix.

1. Prepare 1× buffer from storage and JC-1 assay buffer by 1:5 (v/v) dilution in H_2O.

2. Prepare solution of the samples with storage buffer in a protein concentration of 0.4 mg/ml.

3. From this prepare a dilution row where you apply in total 10, 20, 30, and 40 µg protein in a volume of 100 µl (filled up with storage buffer) in a 2 ml reaction tube.

4. Add 1.9 ml 1× JC-1 assay buffer to each sample of the dilution row.

5. Add 2 µl JC-1-stain to the lid of the reaction tube.

6. Close the tube and vortex directly.

7. Incubate the samples 10 min at RT in the dark.

8. Read the fluorescence of samples in a spectrofluorometer with an excitation wavelength of 490 nm and an emission wavelength of 590 nm.

9. Generate an x/y-graph with values from four samples of the dilution row, x-axis is the amount of protein and y-axis is the fluorescence at 590 nm.

10. Calculate the fluorescence produced in the original sample per mg protein:

$$\frac{\Delta FL \times dil}{V \times C} = \frac{FLU}{mgP}$$

FLU: fluorescence units

mgP: milligram protein

ΔFL: fluorescence (sample) – fluorescence (blank)

dil: dilution factor to prepare 0.4 mg/ml sample dilution

V: Volume of the sample in [ml]

C: protein concentration [mg/ml]

4   Notes

General note: Especially in face of the complexity of the workflow and the enormous biological variability it appears that an extraordinary diligence in each step of the analysis is of great importance. This begins with the first experimental step i.e. sample collection and preparation, which frequently is underestimated. In this context the close collaboration between the different scientific disciplines and the development of Standard Operation Procedures is of particular importance.

1. First do the pH adjustment of buffers with 2 M NaCl and then add required amount of DTT always fresh.

2. To bring high amounts of saccharose in solution quickly, warm up the solutions a little bit (30–40 °C).

3. The titanyl solution needs up to 30 min at RT to get dissolved (clear solution) and is stable and usable at RT for 2 h.

4. Store the solution at 4 °C, it is stable up to 4 weeks.

5. Store the solution in an amber bottle, because it is light sensitive. A benefit is to prepare the solution 2 days before use and pass the solution through a 0.02 μm filter before storing. The solution is perishable, if it gets yellowish, throw it away and make a fresh one.

6. Use anhydrous solvents (i.e. in this case dimethyl sulfoxide (DMSO)) to dissolve cyanine dyes like JC-1.

7. Test the gradient linearity by aliquoting the gradient up to 20 fractions and measure the saccharose concentration with a refractometer.

8. Make sure that the glass douncer reach the bottom of the cylinder and that the solution and the equipment kept on ice. Prevent negative pressure while douncing by using *loose fit* douncer.

9. If you take aliquots from each step of fractionation you can easily perform protein content measurement and marker enzyme assays. The results show the mitochondria enrichment and decrease of other organelles. For example *see* Table 1.

10. The running down without brakes from here used high g-forces increase the standby time, therefore you can go for lunch

11. It is important to use a pipette tip with a wide opening to collect the gradient suspension carefully from the top of the surface.

12. The highest mitochondria content is located in a density of about 42 % saccharose. The refractometer measurement helps, next to SDH activity assay to identify the correct fraction.

13. For marker enzyme assays different solution are needed and best use resuspensions buffer for all samples.

14. In this assay the decrease of catalase substrate is the indicator for enzyme activity. Adding H_2O_2 to titanyl solution results in yellow solution. If Catalase is present the H_2O_2 amount decreases and the solution get colorless. So the decrease of color represents enzymatic activity.

15. H_2O_2 is light sensitive, therefore protect the buffer and assay samples from light.

16. The catalase is one of the quickest known enzymes, therefore the exact compliance of incubation time and temperature of buffers is mandatory.

17. The positive blank is the maximum amount of H_2O_2 that could be measured within the assay.

18. There is no molar extinction coefficient known, so to calculate a relative activity adapt 0.001 OD difference = 1 arbitrary unit.

19. This assay is very sensitive to free phosphate (Pi) so it is mandatory to use always disposable, phosphate free plastic material, because cleaning solution etc. contains and leaves traces of free phosphate and therefore influence or rather damage your assay measurements.

References

1. Nunnari J, Suomalainen A (2012) Mitochondria: in sickness and in health. Cell 148:1145–1159

2. Frezza C, Cipolat S, Scorrano L (2007) Organelle isolation: functional mitochondria from mouse liver, muscle and cultured fibroblasts. Nat Protoc 2:287–295

3. Hartwig S, Feckler C, Lehr S, Wallbrecht K, Wolgast H, Müller-Wieland D, Kotzka J (2009) A critical comparison between two classical and a kit-based method for mitochondria isolation. J Proteomics 2009(11):3209–3214

4. Franko A, Baris OR, Bergschneider E, von Toerne C, Hauck SM et al (2013) Efficient isolation of pure and functional mitochondria from mouse tissues using automated tissue disruption and enrichment with anti-TOM22 magnetic beads. PLoS One 8:e82392

5. Zischka H, Weber G, Weber PJ, Posch A et al (2003) Improved proteome analysis of *Saccharomyces cerevisiae* mitochondria by free-flow electrophoresis. Proteomics 3:906–916

6. Pennington RJ (1961) Biochemistry of dystrophic muscle. Mitochondrial succinate-tetrazolium reductase and adenosine triphosphatase. Biochem J 80:649–654

7. Walter K, Schuett C (1974) In Bergmeyer HU (ed) Methods in enzymatic analysis. Phosphatasen (Book in German), vol 3. Academic, New York, pp 860–864

8. Aebi H (1984) Catalase in vitro. Methods Enzymol 105:121–126

9. Baginski ES, Foa PP, Zak B (1969) Determination of phosphate and phospho-monoesterases in biologic materials. Am J Med Technol 35:475–486

10. Reers M, Smith TW, Chen LB (1991) J-aggregate formation of a carbocyanine as a quantitative fluorescent indicator of membrane potential. Biochemistry 30:4480–4486

Chapter 3

Isolation of Mitochondria-Associated ER Membranes (MAMs) and Glycosphingolipid-Enriched Microdomains (GEMs) from Brain Tissues and Neuronal Cells

Ida Annunziata, Annette Patterson, and Alessandra d'Azzo

Abstract

Subcellular fractionation is a valuable procedure in cell biology to separate and purify various subcellular constituents from one another, i.e., nucleus, cytosol, membranes/organelles, and cytoskeleton. The procedure relies on the use of differential centrifugation of cell and tissue homogenates, but additional purification steps now permit the isolation of inter-organellar membrane contact sites. Here, we outline a protocol tailored for the isolation of mitochondria, mitochondria-associated ER membranes (MAMs) and glycosphingolipid-enriched microdomains (GEMs) from the adult mouse brain, primary neurospheres, and murine embryonic fibroblasts (MEFs).

Key words ER, Mitochondria, MAMs, GEMs, Centrifugation, Brain, MEFs, Neurospheres

1 Introduction

By increasing their degree of complexity, eukaryotes have acquired a complex network of intracellular compartments and membrane-enclosed organelles with distinct biological functions [1]. Increased complexity is achieved mainly by compartmentalization of cellular components within cells and by the ability of intracellular organelles to tether their membranes at specific membrane contact sites. The latter enables the coordinated exchange of information between different organelles, thereby influencing their metabolic activities and functions [1, 2]. To date, the best studied membrane contact sites are those that are formed between the vast membrane network of the endoplasmic reticulum (ER) and the mitochondria at the opening of the IP3R-sensitive Ca^{2+} channel and are known as the mitochondria-associated ER membranes or MAMs [3–6]. The MAMs serve as biological platform for a wide range of diverse activities, which are essential for the survival of the cell, including Ca^{2+} signaling and regulation of

Volkmar Weissig and Marvin Edeas (eds.), *Mitochondrial Medicine: Volume I, Probing Mitochondrial Function*,
Methods in Molecular Biology, vol. 1264, DOI 10.1007/978-1-4939-2257-4_3, © Springer Science+Business Media New York 2015

mitochondrial Ca^{2+} concentration, lipid biosynthesis and transport, and energy metabolism [3, 4, 7–11].

Subcellular fractionation procedures are commonly used to isolate organelles and their membrane contact sites from various cells and tissues [12]. In particular, the development of methods for the isolation and purification of the MAMs has been central to the understanding of the composition and structural organization of these microdomains [3, 4, 6, 13–16]. Several protocols have been described for the purification of these microdomains primarily from liver tissue [16, 17]. However, it is important to keep into consideration the fact that methods applicable for the isolation of these microdomains from one tissue type may not be necessarily suitable for other tissues.

Here, we report a modified protocol for the isolation and purification of mitochondria and MAMs from the adult mouse brain and from primary murine neurospheres as well as murine embryonic fibroblasts (MEFs). In addition, the procedure includes an extra purification step, namely, a Triton X-100 extraction, which allows for the isolation of the glycosphingolipid-enriched microdomain (GEM) fraction from both the mitochondria and the MAMs. These GEM preparations share several protein components with caveolae and lipid rafts, derived from the plasma membrane or other intracellular membranes, and are proposed to function as signaling platforms at specific contact sites [4, 18].

2   Materials

Prepare all solutions using ultrapure water and analytical grade reagents. All the solutions should be prepared the day before the fractionation procedure and stored at 4 °C.

2.1   Equipments

1. 2 ml Dounce all-glass tissue grinders.
2. 15 ml polypropylene conical centrifuge tubes.
3. 30 ml round-bottom glass centrifuge tubes (15 and 30 ml).
4. Ultracentrifuge tubes, ultra-clear, thin wall, 14×89 mm.
5. Parafilm.
6. 26 G needles (26 and 30 G).
7. 3 ml syringes.
8. Disposable borosilicate glass Pasteur pipettes, 9″ (*See* **Note 1**).

2.2   Solutions

1. *Solution A*: 0.32 M sucrose, 1 mM NaHCO$_3$, 1 mM MgCl$_2$, 0.5 mM CaCl$_2$ + protease inhibitors (*see* **Note 2**).
2. *Solution B*: 0.32 M sucrose, 1 mM NaHCO$_3$ + protease inhibitors (*see* **Note 2**).

3. *Hypotonic buffer*: 250 mM sucrose, 20 mM HEPES pH 7.4, 10 mM KCl, 1.5 mM $MgCl_2$, 1 mM EDTA, 1 mM EGTA + protease inhibitors (*see* **Note 2**).

4. *Isolation medium*: 250 mM mannitol, 5 mM HEPES pH 7.4, 0.5 mM EGTA, 0.1 % BSA.

5. *Gradient buffer*: 225 mM mannitol, 25 mM HEPES pH 7.5, 1 mM EGTA, 0.1 % BSA (final concentration).

6. *Discontinuous sucrose gradient*: 3 ml 850 mM sucrose (in 1 mM $NaHCO_3$), 3 ml 1 M sucrose (in 1 mM $NaHCO_3$), 3 ml 1.2 M sucrose (in 1 mM $NaHCO_3$).

7. *30 % Percoll gradient* with *gradient buffer* 8 ml per half brain and place in an ultra-clear Beckman centrifuge tube (*see* **Note 3**).

8. *GEMs extraction buffer*: 25 mM HEPES pH 7.5, 0.15 M NaCl, 1 % Triton X-100 + protease inhibitors (*see* **Note 4**).

9. *GEMs-solubilizing buffer*: 50 mM Tris–HCl pH 8.8, 5 mM EDTA, 1 % SDS.

3 Methods

3.1 Purification and Isolation of Mitochondria and MAMs from Brain Tissue

3.1.1 Removal of Nuclei, Intact Cells, and Cellular Debris

1. Euthanize mouse in a CO_2 chamber.

2. Immediately remove brain, halve it, place each half in 2 ml tubes on ice, and weigh (*see* **Note 5**).

3. Place half brain in a prechilled 2 ml glass Dounce tissue grinder containing 1 ml cold Solution A. Homogenize with 15 total strokes of a large clearance pestle.

4. Transfer to a 15 ml falcon tube on ice, and dilute homogenates up to 10 volumes w/v, e.g., 0.225 g to 2.25 ml.

5. Centrifuge sample at $1,400 \times g$ for 10 min at 4 °C.

6. Carefully remove the supernatant and transfer to a 30 ml round-bottom glass centrifuge tube on ice (*see* **Note 6**).

7. Resuspend the pellet in the same 10 % volumes of *Solution A*. Homogenize in the same grinder, with 3–6 strokes of a small clearance pestle, 1 ml at a time.

8. Transfer to a fresh 15 ml falcon tube and centrifuge at $710 \times g$ for 10 min at 4 °C. This results in a pellet of *nuclei and cell debris*.

9. Carefully remove the supernatant and pool with the supernatant saved in **step 6**.

3.1.2 Isolation of Crude Mitochondria

1. Centrifuge supernatant at $13,800 \times g$ for 10 min at 4 °C.

2. Transfer supernatant to an ultra-clear Beckman centrifuge tube on ice.

3. Resuspend pellet in 10 volumes *Solution A* with 6 strokes of a small clearance pestle. Homogenize in the same grinder, with 3–6 strokes of a small clearance pestle, 1 ml at a time.

4. Centrifuge as **step 1**.

5. Repeat **steps 2–4**.

6. The resulting pellet is an enriched mitochondrial fraction. Pool the supernatants (cytosol and ER), cover with Parafilm, and keep on ice.

7. Homogenize the pellet in 4.8 ml/g of *Solution B* with 6 strokes of a small clearance pestle (*see* **Note 7**).

8. Using a glass Pasteur pipette, prepare a discontinuous sucrose gradient in an ultra-clear Beckman centrifuge tube by adding in sequence to the bottom of the tube the following solution: (a) pellet resuspended in 0.32 M sucrose, (b) 3 ml 850 mM sucrose, (c) 3 ml 1 M sucrose, and (d) 3 ml 1.2 M sucrose. Ensure that no bubbles are formed.

9. Centrifuge at $82,500 \times g$ for 2 h at 4 °C. The resulting gradient will show from top to bottom three defined bands consisting of (a) myelin and other membrane contaminants (0.32–0.85 M interface); (b) ER, Golgi, and plasma membranes (0.85–1 M interface); and (c) synaptosomes (1–1.2 M interface). The pellet, consisting of crude mitochondria, is used for the subsequent purification steps (*see* **Note 8**).

3.1.3 Isolation of MAMs

1. Resuspend the freshly isolated crude mitochondria from one half brain in 2 ml isolation medium.

2. Prepare a *30 % Percoll gradient* with 8 ml *gradient buffer* per half brain, and place it in an ultra-clear Beckman centrifuge tube.

3. Layer the mitochondrial suspension on top of the prepared gradient (*see* **Note 9**).

4. Centrifuge at $95,000 \times g$ 4 °C for 30 min.

5. Remove the heavy fraction (lower band) with a glass Pasteur pipette, and transfer it to a clean round-bottom glass tube on ice. Remove the light fraction (upper band) with another glass Pasteur pipette, and transfer to a separate clean round-bottom glass tube on ice (*see* **Note 10**).

6. Dilute both fractions collected in **step 5** with 10 ml isolation medium and centrifuge at $6,300 \times g$ at 4 °C for 10 min.

7. Discard the supernatant from the heavy fraction obtained in **step 6**, resuspend the pellet with another 10 ml isolation medium, and centrifuge again, as in **step 6**. The resulting pellet will be the *pure mitochondrial fraction*.

8. Transfer the supernatant from the light fraction obtained in **step 6** to an ultra-clear Beckman centrifuge tube on ice. Discard the pellet.

9. Dilute the supernatant with isolation medium and centrifuge at $100,000 \times g$ at 4 °C for 1 h. The resulting pellet will be the *MAMs fraction*. Carefully remove and discard the supernatant.

   At this step, you may simultaneously centrifuge the supernatant saved in **step 6** of Subheading 3.1.2 (isolation of crude mitochondria). The pellet from this centrifugation will be the *ER fraction* and the supernatant the *cytosolic fraction*.

3.2 Purification and Isolation of Mitochondria and MAMs from MEFs and Neurospheres

Start with two subconfluent T75 flasks of MEFs (~20×10^6 cells) and five T75 flasks of neurospheres (~15×10^6 cells) (*see* **Note 11**).

3.2.1 Removal of Nuclei, Intact Cells, Cellular Debris, and Isolation of Crude Mitochondria

1. Harvest both cell cultures, centrifuge, and wash the cell pellets with cold $1\times$ PBS.

2. Keep on ice.

3. Centrifuge cells at $235 \times g$ (Beckman tabletop centrifuge) for 10 min.

4. Remove PBS and resuspend cell pellets in 0.5 ml of *hypotonic buffer* per T75 flask, and transfer in 2 ml Eppendorf tubes.

5. Incubate on ice for 30 min.

6. Disrupt cells using 3 ml syringes by passing them through a 26 G needle for 30 times, followed by a 30 G needle for 20 times (*see* **Note 12**).

7. Centrifuge cells at $750 \times g$ (Eppendorf centrifuge) at 4 °C for 10 min. The pellet contains nuclei and intact cells.

8. Transfer supernatant to clean tube and centrifuge at $10,000 \times g$ (Eppendorf centrifuge) at 4 °C for 20 min. The resulting pellet represents the crude mitochondria or heavy membrane fraction (HM fraction) (*see* **Note 13**).

9. Collect and store supernatant from **step 8** in an ultra-clear Beckman centrifuge tube covered with Parafilm on ice.

3.2.2 Isolation of MAMs

1. Resuspend the freshly isolated crude mitochondria in 2 ml isolation medium.

2. Prepare a *30 % Percoll gradient* with *gradient buffer* (8 ml), and place it in an ultra-clear Beckman centrifuge tube.

3. Layer mitochondrial suspension on top of the prepared gradient (*see* **Note 14**).

4. Centrifuge at $95,000 \times g$ at 4 °C for 30 min.

5. (a) Remove from the gradient the lower band containing purified mitochondria (heavy fraction) using a glass Pasteur pipette, and transfer it to a round-bottom glass tube on ice.

(b) Remove the upper band containing the MAMs (light fraction) with a clean glass Pasteur pipette, and transfer it to a separate round-bottom glass tube on ice (*see* **Note 15**).

6. Dilute both fractions collected in **step 5** with 10 ml of isolation medium and centrifuge at $6,300 \times g$ at 4 °C for 10 min.

7. Discard the supernatant from the heavy fraction obtained in **step 6**, resuspend the pellet with 10 ml isolation medium, and centrifuge again, as in **step 6**. The resulting pellet will be the *pure mitochondrial fraction*.

8. Transfer the supernatant from the light fraction obtained in **step 6** to an ultra-clear Beckman centrifuge tube on ice. Discard the pellet.

9. Dilute the supernatant with isolation medium and centrifuge at $100,000 \times g$ at 4 °C for 1 h. The resulting pellet will be the *MAMs fraction*. Carefully remove and discard the supernatant.

10. Centrifuge the supernatant collected in **step 8** at $100,000 \times g$ (28,500 rpm, Beckman ultracentrifuge) at 4 °C for 1 h. The resulting pellet represents the ER/microsomes fraction. The supernatant represents the cytosolic fraction.

3.3 Extraction of GEMs from Mitochondria and MAMs Isolated from Brain Tissue and Cells

1. Lyse pure mitochondria and/or MAM fractions in 500 μl–1 ml of *extraction buffer* on ice for 20 min.

2. Centrifuge lysates at $15,294 \times g$ at 4 °C for 2 min.

3. Collect the supernatants (Triton X-100 soluble material), and recentrifuge the pellets for 2 min to remove remaining soluble material (*Triton X-100 extracted mitochondria and/or Triton X-100 extracted MAMs*).

4. Solubilize pellets in *solubilizing buffer*. This solubilized material represents the *GEMs fraction*.

3.4 Anticipated Results

The protocols as outlined are highly reproducible. After isolation and purification of mitochondria, ER, MAMs, and GEMs, the individual subcellular fractions are resuspended in a lysis buffer, separated on SDS-polyacrylamide gels, and blotted onto PVDF membranes. Western blots are then probed with a battery of antibodies against specific protein markers that will verify the purity of the individual fractions and their protein composition. Pure mitochondrial preparations should be devoid of both ER and cytosolic markers. The close apposition between ER and mitochondrial membranes at the MAM contact sites explains the presence of both ER and mitochondrial markers (Fig. 1a), such as calreticulin and

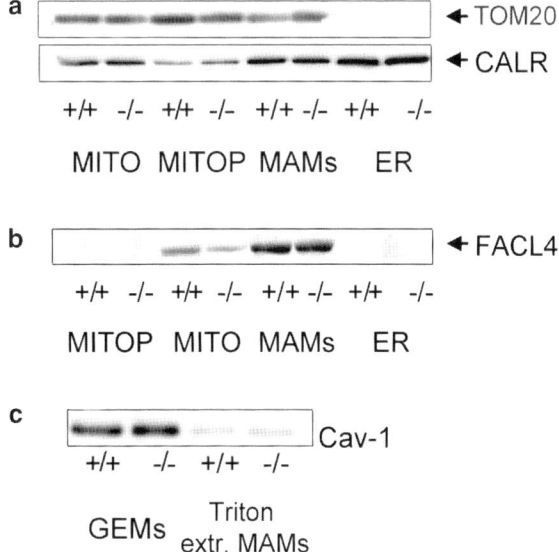

Fig. 1 (**a**) Immunoblots of crude mitochondria (MITO), ER, purified mitochondria (MITO P), and MAMs probed with anti-TOMM20 and anti-calreticulin (CALR), markers of mitochondria and ER, respectively. (**b**) Immunoblots of the same fractions as in (**a**) probed with the anti-FACL4, a marker of the MAMs. (**c**) Immunoblots of GEMs extracted from MAM preparations of β-gal$^{+/+}$, and β-gal$^{-/-}$ brains were probed with caveolin-1 antibody (Cav-1). Adapted from a research that was originally published in *Molecular Cell* [4]

TOMM20, in the purified MAMs. These preparations can also be probed with bona fide markers for the MAMs, i.e., FACL4 (Fig. 1b) and PACS2. As outlined above, the MAMs can be further extracted with Triton X-100 to obtain the GEMs, a procedure that was first developed in our laboratory. These microdomains, which contain components of lipid rafts and/or caveolae, should be caveolin-1 positive (Fig. 1c).

In addition, we have analyzed crude mitochondria by electron microscopy and analyzed the MAMs as the number of ER vesicles juxtaposed to individual mitochondria (Fig. 2).

4   Notes

1. Precool the glass tubes and homogenizers in ice.

2. Add fresh protease inhibitors to the different solutions.

3. Make gradient buffer more concentrated (1.43 times) to achieve correct final concentration after diluting Percoll to 30 %.

4. Add fresh protease inhibitors to the solution.

5. Keep brain halves separate throughout the entire procedure. The following applies to the treatment of a single brain half.

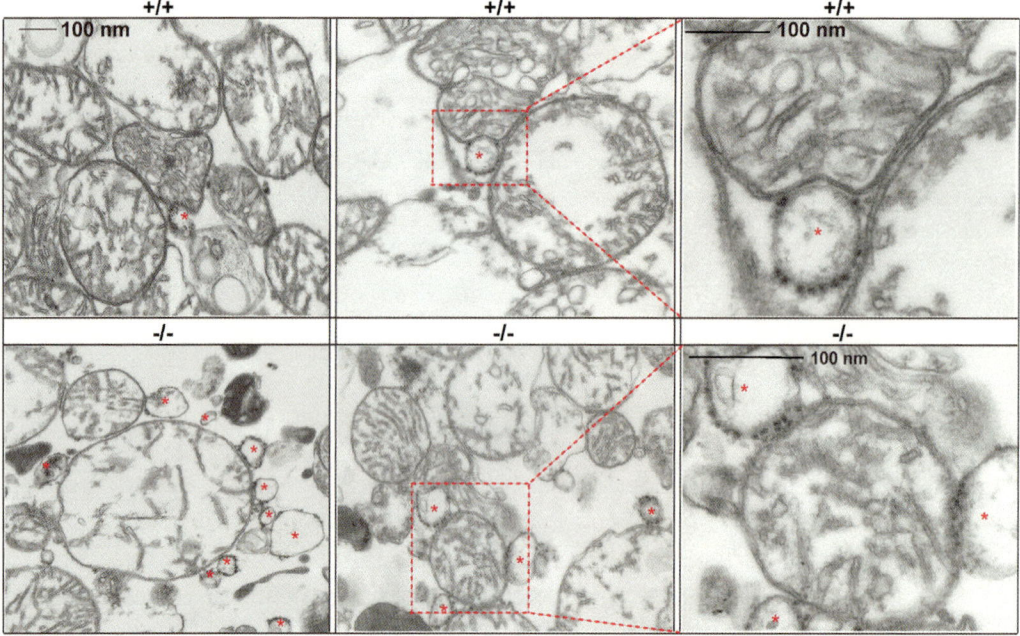

Fig. 2 A representative electron micrographs of crude mitochondria isolated from *β-gal*^{+/+} and *β-gal*^{-/-} brains. *Asterisks* mark ER vesicles juxtaposed to mitochondrial membranes. This research was originally published in *Molecular Cell* [4]

6. Make sure not to disturb the pellet.

7. Gr refers to the original brain weight.

8. Store a small amount of the crude mitochondrial fraction for subsequent tests.

9. Layer mitochondrial suspension on top of the gradient slowly to avoid any bubbles.

10. Remove first the heavy fraction and then the light fraction. A defined, milky band containing purified mitochondria separates at the bottom of the ultracentrifuge tube, while the MAMs fraction appears as a diffuse and broad band above the mitochondrial band.

11. The number of cells is given as starting point. To increase the yield of the obtained fractions, it may be necessary to scale up the initial number of cells to be used for the fractionation experiments.

12. It is crucial to perform this step on ice.

13. Store a small amount of the crude mitochondria for subsequent tests.

14. Layer mitochondrial suspension on top of the gradient slowly to avoid any bubbles.

15. Remove first the heavy fraction and then the light fraction. A defined, milky band containing purified mitochondria separates at the bottom of the ultracentrifuge tube, while the MAMs fraction appears as a diffuse and broad band above the mitochondrial band.

Acknowledgment

We acknowledge the contribution of Renata Sano in setting up the initial protocols. A.d'A. holds the Jewelers for Children (JFC) Endowed Chair in Genetics and Gene Therapy. This work was funded in part by NIH grants GM60905, DK52025, and CA021764, the Assisi Foundation of Memphis, and the American Lebanese Syrian Associated Charities (ALSAC).

References

1. Levine T, Loewen C (2006) Inter-organelle membrane contact sites: through a glass, darkly. Curr Opin Cell Biol 18:371–378

2. Helle SC, Kanfer G et al (2013) Organization and function of membrane contact sites. Biochim Biophys Acta 1833:2526

3. Rizzuto R, Murgia M, Pozzan T (1993) Microdomains with high Ca2+ close to IP3-sensitive channels that are sensed by neighboring mitochondria. Science 262:744–747

4. Sano R, Annunziata I et al (2009) GM1-ganglioside accumulation at the mitochondria-associated ER membranes links ER stress to Ca(2+)-dependent mitochondrial apoptosis. Mol Cell 36:500–511

5. Raturi A, Simmen T (2012) Where the endoplasmic reticulum and the mitochondrion tie the knot: the mitochondria-associated membrane (MAM). Biochim Biophys Acta 1833:213–224

6. Vance JE (1990) Phospholipid synthesis in a membrane fraction associated with mitochondria. J Biol Chem 265:7248–7256

7. Bionda C, Portoukalian J et al (2004) Subcellular compartmentalization of ceramide metabolism: MAM (mitochondria-associated membrane) and/or mitochondria? Biochem J 382:527–533

8. Grimm S (2011) The ER-mitochondria interface: the social network of cell death. Biochim Biophys Acta 1823:327–334

9. Pizzo P, Pozzan T (2007) Mitochondria-endoplasmic reticulum choreography: structure and signaling dynamics. Trends Cell Biol 17:511–517

10. Ardail D, Popa I et al (2003) The mitochondria-associated endoplasmic-reticulum subcompartment (MAM fraction) of rat liver contains highly active sphingolipid-specific glycosyltransferases. Biochem J 371:1013–1019

11. Kornmann B, Currie E et al (2009) An ER-mitochondria tethering complex revealed by a synthetic biology screen. Science 325:477–481

12. Michelsen U, von Hagen J (2009) Isolation of subcellular organelles and structures. Methods Enzymol 463:305–328

13. Annunziata I, Patterson A, d'Azzo A (2013) Mitochondria-associated ER membranes (MAMs) and glycosphingolipid-enriched microdomains (GEMs): isolation from mouse brain. J Vis Exp (73): e50215

14. Myhill N, Lynes EM (2008) The subcellular distribution of calnexin is mediated by PACS-2. Mol Biol Cell 19:2777–2788

15. Simmen T, Aslan JE et al (2005) PACS-2 controls endoplasmic reticulum-mitochondria communication and Bid-mediated apoptosis. EMBO J 24:717–729

16. de Brito OM, Scorrano L (2008) Mitofusin 2 tethers endoplasmic reticulum to mitochondria. Nature 456:605–610

17. Wieckowski MR, Giorgi C et al (2009) Isolation of mitochondria-associated membranes and mitochondria from animal tissues and cells. Nat Protoc 4:1582–1590

18. Pizzo P, Giurisato E et al (2004) Physiological T cell activation starts and propagates in lipid rafts. Immunol Lett 91:3–9

Chapter 4

Fluorescence Analysis of Single Mitochondria with Nanofluidic Channels

Ted Pham, Katayoun Zand, Douglas Wallace, and Peter Burke

Abstract

Single mitochondrial assays are uncovering a new level of biological heterogeneity, holding promises for a better understanding of molecular respiration and mitochondria-related diseases. Here, we present a nanoscale approach to trapping single mitochondria in fluidic channels for fluorescence microscopy. We fabricate the nanofluidic channels in polydimethylsiloxane and bond them onto a glass slide, creating a highly reproducible device that can be connected to external pumps and mounted to a microscope. Having a unique nanoscale cross section, our channels can trap single mitochondria from a purified mitochondrial preparation flown across. Compared with the traditional fluorescence method to monitor single mitochondrial membrane potential with glass slides and open fluidic chambers, our nanofluidic channels reduce background fluorescence, enhance focus, and allow ease in experimental buffer exchanges. Hence, our channels offer researchers a new effective platform to test their hypotheses on single mitochondria.

Key words Mitochondria, Membrane potential, Depolarization, Nanofluidic channels, PDMS, JC-1, TMRM, OXPHOS

1 Introduction

There are many research and clinical motivations driving the development of high-throughput techniques to study the biophysical and biochemical properties of individual (rather than ensemble) mitochondrion. These motivations arise fundamentally from the established relationship between mitochondrial dysfunction and hereditary disorders as well as age-related diseases. Furthermore, while there is a growing body of evidence that mitochondria do not all behave identically, even within the same cell, and that this functional heterogeneity has important clinical manifestations, the advent of high-throughput techniques can aid in quantifying and studying this heterogeneity among different patients, organs, tissues, and even cells, under different chemical conditions, as well as a function of the age of an organism.

Volkmar Weissig and Marvin Edeas (eds.), *Mitochondrial Medicine: Volume I, Probing Mitochondrial Function*, Methods in Molecular Biology, vol. 1264, DOI 10.1007/978-1-4939-2257-4_4, © Springer Science+Business Media New York 2015

A wide range of degenerative disease symptoms have been linked to mitochondrial disorders. It has become increasingly evident that mitochondrial dysfunction contributes to a variety of age-related human disorders, ranging from neurodegenerative and neuromuscular diseases, stroke, and diabetes to ischemia-reperfusion injury and cancer [1, 2]. Multiple diseases have been identified that result from mutations in the mitochondrial DNAs (mtDNAs) (see our website: http://mitomap.org). These genetic diseases have a very different genetics because each human cell contains hundreds of mitochondria and thousands of copies of mtDNAs, which are inherited through the oocyte cytoplasm and thus are maternally transmitted. Furthermore, the mtDNA has a high mutation rate, undoubtedly due to its direct exposure to mitochondrial ROS. When a new mtDNA mutation arises in a cell, this creates a mixed intracellular population of mtDNAs, a state termed heteroplasmy [3]. As the percentage of mutant mtDNAs increases, the mitochondrial energetic capacity declines, ROS production increases, and propensity for apoptosis increases. Consequently, diseases of the mitochondria have a delayed-onset and progressive course [1]. Since the mitochondria are the major endogenous source of ROS and ROS is thought to damage cells and DNA during the aging process, damage to the mitochondria and the mtDNA is likely to be the aging clock [4, 5].

The empirical link between a genetic defect in an electron transport chain (ETC) component and a clinical pathology (specifically Leber's hereditary optical neuropathy, LHON, a neurological disease) was discovered by one of us in 1988 [6]. Since that time, researchers around the world have identified numerous genetic causes of ETC deficiencies, and genetic mitochondrial disease is now thought to occur in 1 in 5,000 births [7]. 1 in 200 individuals is thought [8] to harbor deleterious mutations in ETC coding genes, which do not present clinically, because the heteroplasmy of mitochondria allows both defective and healthy ETC mtDNA to exist in the same individual and even the same cell [3].

In Parkinson's diseases it has been demonstrated that mitophagy, a cellular process that eliminates damaged mitochondria with low membrane potential, is compromised [9, 10]. However, the pathways leading to mitophagy as well as compromising the process are not completely understood and studies are underway.

These three examples (aging, hereditary mitochondrial disorders, and neurodegenerative diseases) provide strong motivation for methods to study the properties of mitochondria at the individual mitochondria level in a platform compatible with high-throughput screening. In this chapter, we present a platform in an attempt to forward the effort of such technology. Specifically, we aim to address the current technological obstacles in studying mitochondrial functional heterogeneity such as background fluorescence, photobleaching, mitochondria movement out of the plane of focus, and difficulties in precise delivery of substrates. Our solution comprises nanofluidic channels whose suitable topography can trap single mitochondria for fluorescence analysis [11]. Using a single-step soft lithography, we create channels with a multiple height profile including a trapezoidal cross section of 500 nm high and 2 μm wide. This particular geometry of our channel can mechanically trap single mitochondria while reducing background fluorescence and eliminating the out-of-focus issue. Connected to an external pump, our channels can deliver controlled amounts of respiratory substrates and inhibitors to the trapped mitochondria. As a result, we present an improved platform to monitor single mitochondrial membrane potential by performing first a quick trapping step then fluorescence microscopy. Interested researchers applying this simple methodology can perform similar assays on single mitochondria using different experimental buffers.

2 Materials

Use ultrapure water with resistivity of 18.2 MΩ*cm to prepare all solutions.

2.1 Nanofluidic Channel Components

1. Chrome photomask with channel patterns (*see* **Note 1**).
2. Four-inch silicon wafer (University Wafer, Boston, MA, USA).
3. Piranha solution: 3 parts sulfuric acid and 1 part hydrogen peroxide (*see* **Note 2**).
4. Hexamethyldisilazane, Sigmacote (Sigma-Aldrich, St. Louis, MO, USA).
5. Positive photoresist Microposit SC1827, Microposit MF-319 developer (MicroChem, Newton, MA, USA).
6. Silicone elastomer and curing agent (Sylgard 184, Dow Corning, Midland, MI, USA).
7. 0.025 OD×0.017 ID, 0.500″ length stainless steel pins (New England Small Tube, Litchfield, NH, USA).

2.2 Fabrication Equipment

1. Karl Suss MA6 aligner.
2. Oxygen plasma cleaner (Harrick Plasma, Ithaca, NY, USA).
3. Spin coater (Laurell Technologies, North Wales, PA, USA).
4. Hexamethyldisilazane (HMDS) priming oven (YES, Livermore, CA, USA).

2.3 Cell Culture Components

1. HeLa cells (ATCC, Manassas, VA, USA).
2. Culture media: Minimum essential medium, 10 % fetal bovine serum. Add 50 mL fetal bovine serum to 500 mL minimum essential medium. Store at 4 °C. 1 % penicillin-streptomycin solution (ATCC, 30-2300) is optional.

3. 0.25 % trypsin-EDTA 1× phenol red. Store at −20 °C.

4. Phosphate-buffered saline of pH 7.2 without calcium and magnesium.

5. Vented tissue culture T-75 flasks.

6. Dounce homogenizer: Pyrex 5 mL Potter-Elvehjem tissue grinder with PTFE pestle (Corning, Tewksbury, MA, USA).

7. Isolation buffer: 210 mM mannitol, 70 mM sucrose, 1 mM EGTA, 5 mM HEPES, and 0.5 % BSA, pH 7.2 (*see* **Note 3**). Store at 4 °C.

8. Basal respiration buffer: 2 mM $MgCl_2$, 10 mM NaCl, 140 mM KCl, 0.5 mM EGTA, 0.5 mM KH_2PO_4, and 20 mM HEPES, pH 7.2 (*see* **Note 4**). Store at 4 °C.

9. Stock solutions of respiratory substrates and inhibitors: 2 mM rotenone, 500 mM succinate, 500 mM malate, 20 mM CCCP, and 10 mM ADP (*see* **Note 5**).

10. Protease inhibitor cocktail (Thermo Scientific, Waltham, MA, USA) (*see* **Note 6**).

11. Pierce™ BCA protein assay kit (Thermo Scientific, Rockford, IL, USA).

2.4 Fluorescent Dyes   Obtain all dyes from Life Technologies, Grand Island, NY, USA.

1. JC-1 assay kit: Store at 4 °C. Make 200 µM JC-1 with DMSO immediately prior to use.

2. TMRM: Make and divide 10 mM TMRM in DMSO to 4 µL aliquots. Store at −20 °C.

3. MitoTracker® Green (MTG): The dye comes in 20 aliquots of 50 µg. Store at −20 °C. Prepare fresh by first thawing then adding 74 µL of DMSO to 1 mM. Dilute in experimental buffer to 100 nM final concentration.

2.5 Microscope Setup   Experiments can be performed with any comparable setup to the following:

1. Inverted microscope Olympus IX-71 with 4×, 20×, and 60× objectives.

2. TRITC and FITC filter sets.

3. X-Cite 120PC Q fluorescence light source (Lumen Dynamics, Mississauga, Ontario, Canada).

4. 12-bit monochromatic QIclick-F-M-12 CCD camera (QImaging, Surrey, British Columbia, Canada).

5. NE-4000 programmable double syringe pump (Pump Systems, Farmingdale, NY, USA).

3   Methods

Perform all procedures at room temperature unless otherwise noted. Use a chemical fume hood to handle hazardous chemicals and a biosafety cabinet for biological samples.

3.1   Silicon Mold Fabrication

1. Clean a silicon wafer with 120 °C piranha solution for 1 h. Rinse the wafer thoroughly with ultrapure water. Dry the wafer with nitrogen gas stream and dehydrate it at 120 °C on a hot plate for 1 h.

2. Prime the wafer with hexamethyldisilazane to improve photoresist adhesion (*see* **Note** 7).

3. Spin coat Microposit SC1827 positive photoresist at 3,500 rpm for 30 s. The photoresist should cover 75 % of the wafer prior to spin coating. Prebake at 90 °C for 30 min in a convection oven.

4. Load the prebaked wafer and the chrome photomask into the Karl Suss MA6 mask aligner. Use soft contact setting and exposure dosage at 160 mJ/cm².

5. Use 300 mL Microposit MF-319 in a glass beaker to develop the exposed wafer. Swirl the beaker gently to promote the developing process. The developing should take around 25 s. Perform occasional checking of the patterns with a microscope and adjust the developing time to ensure accurate development the first time.

6. Postbake is not necessary. The mold is ready to use at this point (Fig. 1a).

3.2   PDMS Device Fabrication

1. Mix silicone elastomer and curing agent thoroughly at a 10:1 weight to volume ratio. Degas PDMS for 40 min in a vacuum desiccator. We use 40 g of silicone elastomer and 4 mL of the curing agent.

2. Place the silicon mold into a Petri dish. Use a disposable plastic pipette to dispense a few drops of Sigmacote onto the mold and let it dry in a fume hood for 40 min. This step aids with the release of PDMS from the mold later.

3. Pour the degased PDMS onto the mold. Blow away any surface bubbles with a plastic pipette. Place the mold in a 60 °C curing oven overnight.

4. After curing, cut and peel individual PDMS devices from the mold (Fig. 1c, d). Punch out inlet and outlet holes in PDMS using 23-gauge stainless steel needles with inner diameter of 0.017″ and outer diameter of 0.025″.

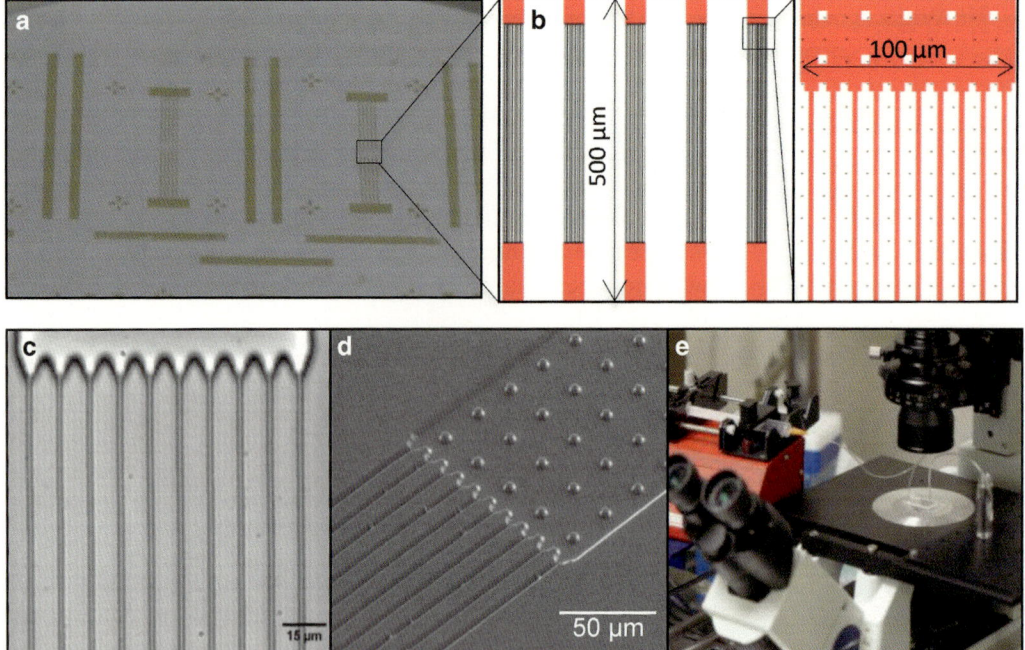

Fig. 1 (**a**) Fabricated mold on a Si wafer. (**b**) An example of the mask design in L-Edit. (**c**) Bright-field image of PDMS channels. (**d**) SEM image of PDMS channels upside down; micro-posts in the larger channel keep the wide part from collapsing. (**e**) A typical experimental setup (Panels **c** and **d** adapted from [7] with permission from American Chemical Society.)

5. Clean glass slides with piranha solution at 120 °C for 1 h on the same day with cutting PDMS devices.

6. To promote adhesion between glass surface and PDMS slaps containing the channels, expose the two surfaces to 70 W oxygen plasma treatment at 100 mTorr for 20 s. Immediately bond the PDMS slap and the glass slide and incubate the device at 70 °C for 20 min (*see* **Note 8**).

3.3 Cell Culture and Mitochondrion Isolation

1. Grow HeLa cells in T-75 flasks. Replace cell culture media every 2 days and passage 1–4 every 4 days. Keep the cells in exponential growth phase.

2. Harvest cells at 95 % confluence in two T-75 flasks on the days of experiment.

3. Wash the cell pellet with phosphate-buffered saline. Perform all subsequent steps on ice at 4 °C.

4. Replace PBS with 1 mL isolation buffer and transfer the solution containing cells to a Dounce homogenizer.

5. Use 30 downward and circular strokes to shear the cells. Do this slowly and on ice (*see* **Note 9**).

6. Add 3 mL isolation buffer to the homogenizer and transfer the homogenate to two 2-mL Eppendorf tubes.

7. Spin at $2,000 \times g$ for 4 min at 4 °C and discard the pellet.

8. Transfer the resulting supernatant to two new Eppendorf tubes.

9. Centrifuge at $12,000 \times g$ for 10 min at 4 °C and discard the supernatant.

10. Resuspend the pellet in each tube with 300 μL isolation buffer and combine into one tube.

11. Centrifuge at $12,000 \times g$ for 10 min at 4 °C. There will be a light-colored sediment containing damaged mitochondria surrounding a brownish-color pellet. Carefully rid the light-colored sediment with a 1–10 μL micropipette tip from the pellet. Discard the supernatant.

12. Resuspend the purified pellet with 1 mL isolation buffer and centrifuge at $12,000 \times g$ for 10 min.

13. Resuspend the pellet with 100–200 μL basal respiration buffer in a small tube (*see* **Note 10**).

14. Dilute to a final protein concentration of around 50 μg/mL.

15. Label mitochondria with fluorescent dyes: Add JC-1 or TMRM to the mitochondrial suspension to a final concentration of 300 nM or 100 nM, respectively. MitoTracker® Green can also be used concurrently with TMRM (*see* **Note 11**).

3.4 Typical Experiment Setup

1. Insert a $0.025 \text{ OD} \times 0.017 \text{ ID}$, 0.500″ length stainless steel pin in the inlet hole of the PDMS chip for chip-to-tube interface.

2. Connect the chip to the syringe pump using a Tygon tube with ID 0.02″ and OD 0.06″.

3. Initially pump respiration buffer to fill the channel. Set the flow rate at 10 μL/h (*see* **Note 12**).

4. Examine the device under a bright-field microscope. Once the channel is filled and the flow is uniform, introduce mitochondrial solution into the channel (Fig. 1e).

5. Monitor the nanochannels with fluorescence microscope, mitochondria will gradually get trapped. Start taking images with the fluorescence microscope (*see* **Note 13**) (Fig. 2).

6. After about 2 min, there will be a few mitochondria immobilized in each channel.

7. To investigate the effect of OXPHOS substrates on the membrane potential, swap and pump in the experimental buffer of choice, and decrease the flow rate to avoid dislodging the trapped mitochondria. Let the solution flow for a few minutes before taking images (*see* **Note 14**).

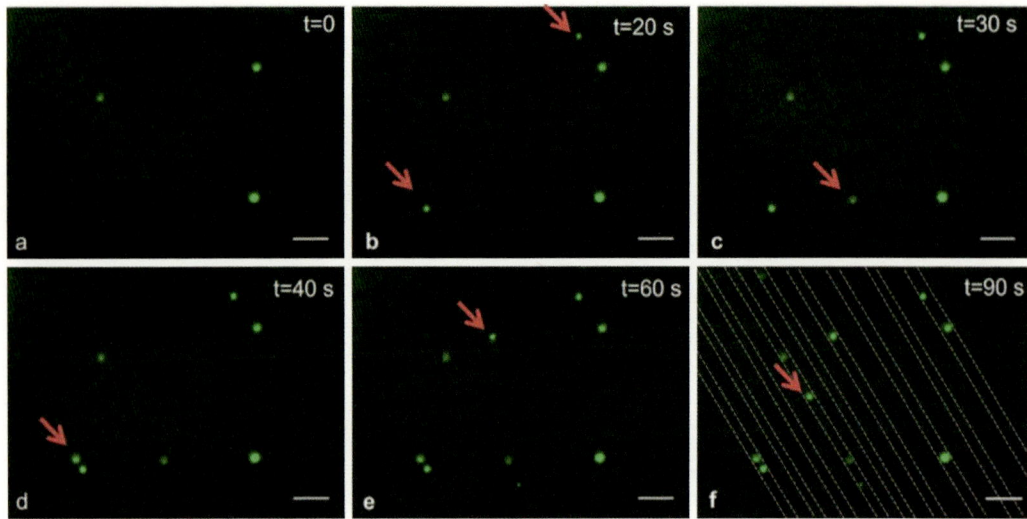

Fig. 2 (**a–f**) Time-lapse images showing sequential addition of mitochondria inside the channels. Each *red arrow* indicates new trapped mitochondria in the respective panel. *Dash lines* in panel (**f**) represent the outlines of the channels. Mitochondria were stained with 100 nM MTG (Reprinted from [7] with permission from American Chemical Society.)

3.5 Image Acquisition and Analysis

1. We use μ-Manager software to control image acquisition and ImageJ to analyze the data.

2. A typical experiment usually lasts 10 min with continuously imaging every 5 s with 2.5 s exposure time.

3. Load time-lapse data into ImageJ. Use a 3×3 median filter to remove noise.

4. Draw and define regions of interest by tracing the perimeter of each mitochondrion. Use the same area of a region of interest and define three background areas in the vicinity of each mitochondrion.

5. Measure the intensity for each mitochondrion and subtract the averaged background noise.

6. For JC-1 assay, it is useful to establish red-to-green intensity ratio to qualitatively compare the mitochondrial membrane potential (*see* **Note 15**) (Fig. 3).

7. For TMRM assay, co-staining the mitochondria with MitoTracker® Green helps with the trapping visualization in green channel. Once trapping is established, switch to red channel and perform time-lapse experiments (*see* **Note 16**) (Fig. 4).

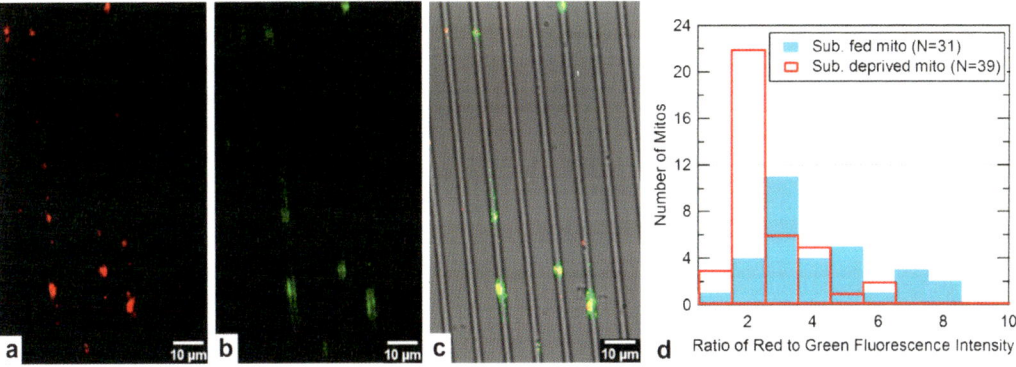

Fig. 3 JC-1 example experiment (**a**). (**b**) Trapped mitochondria which were stained with JC-1; *red* and *green* fluorescence are shown, respectively. (**c**) Overlap of panels (**a**) and (**b**) and the bright-field image of the channels. (**d**) Histogram of substrate fed (+5 mM pyruvate, 5 mM malate) vs. substrate-deprived mitochondria in basal respiration buffer. *N* is the number of observed trapped mitochondria in the channels (Reprinted from [7] with permission from American Chemical Society.)

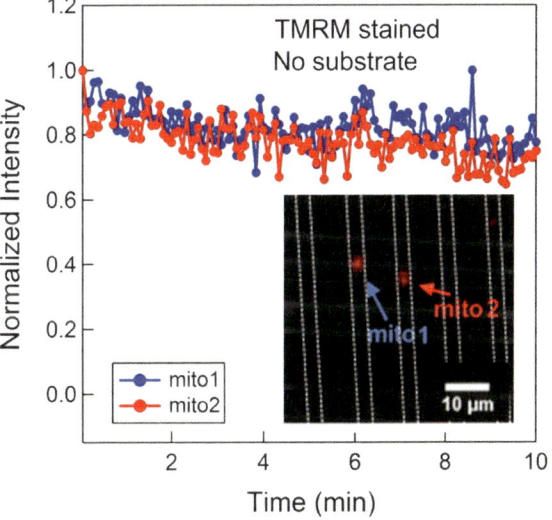

Fig. 4 TMRM example experiment. Two mitochondria are trapped and measured in the channels. Their TMRM fluorescence intensities are tracked and normalized by dividing from the first measurement (Reprinted from [7], with permission from American Chemical Society.)

4   Notes

1. We draw the channel design in L-Edit, a computer-aided design software, and send the exported graphic design system ("*.gds") file to a photomask foundry, e.g., Photo Sciences Inc. A chrome photomask is necessary due to the smallest feature size of 2 μm, which is too small for transparent papermask and

flood exposure to render faithfully. Our design and gds files can be provided upon request (Fig. 1b).

2. Piranha solution is extremely corrosive and must be handled with care and inside a chemical fume hood. Wear adequate and correct personal protective equipment. To clean one silicon wafer, we use 60 mL sulfuric acid and 20 mL hydrogen peroxide. Always add hydrogen peroxide to acid. The hot plate should be set at 120 °C and run for an hour. Bubbles will occur, indicating the cleaning action in effect.

3. To make 250 mL isolation buffer, use 10.25 g mannitol, 6.42 g sucrose, 0.05 g EGTA, and 1.192 g HEPES. Add 200 mL ultrapure water, and adjust pH to 7.2 with 1 M KOH. Fill more ultrapure water to make up 250 mL. Sterile filter the solution. Make 0.5 % BSA by filling 5 g of fat-free BSA (Sigma A6003) with ultrapure water to 100 mL. On days of experiment, add 1 mL 0.5 % BSA to 9 mL isolation buffer to achieve 0.05 % BSA. Store at 4 °C.

4. To make 250 mL basal respiration buffer, use 0.1 g $MgCl_2$, 0.15 g NaCl, 2.609 g KCl, 0.02 g EGTA, 0.017 g KH_2PO_4, and 1.19 g HEPES. Add 200 mL ultrapure water, and adjust pH to 7.2 with 1 M KOH. Fill more water to make up 250 mL. Sterile filter the solution.

5. CCCP is extremely hazardous, make a stock solution with absolute ethanol immediately upon receipt and store at −20 °C. Prepare rotenone, succinate, and ADP stock solutions with ultrapure water and make aliquots of 100 μL for later use. Except rotenone stored at room temperature, other chemicals should be stored at −20 °C. Dilute stock solutions with basal respiration buffer.

6. Because proteases released from the disrupted cells can digest mitochondria, adding the cocktail prolongs the isolated mitochondria's viability and intactness. Add protease inhibitor cocktail to the isolation buffer prior to cell lysing. The stock cocktail comes 100×, use 1 part of stock and 99 parts of the isolation buffer.

7. Hexamethyldisilazane is used before photoresist spinning to make the Si substrate hydrophobic and improve the adhesion between the photoresist and the substrate. HMDS can be spin coated on the substrate or applied onto the substrate by vapor priming method. Best results are achieved by the vapor priming method.

8. Plasma treatment converts PDMS surface from being hydrophobic to hydrophilic. This helps with the adhesion to the glass slide and promotes fluid flow through the nanochannels. Plasma treatment is not, however, permanent. Hence, the devices should be used as soon as they are bonded. To further prolong the hydrophilicity, prime the channels with polyvinyl alcohol and then wash with ultrapure water.

9. Other cell lines will require different numbers of homogenizing strokes. Perform a simple lysing experiment to determine an optimal number for your cell lines. Alternatively, use a motorized homogenizer [12].

10. We store the final crude mitochondria isolation in a small tube to minimize exposure to surface oxygen. Once isolated, the mitochondria have a short useable lifetime of 1–2 h. At this point, mitochondrial protein quantification can be performed. However, because BSA and EGTA interfere with protein measurements, prepare and use an isolation buffer without BSA and resuspend the mitochondrial pellet with basal respiration buffer without EGTA. Use at least 100 μL of sample for the protein assay.

11. Because TMRM and JC-1 are dynamic dyes, whose distribution across the mitochondrial inner membrane is dependent on the mitochondrial membrane potential, experimental buffers should contain the same dye concentration. In contrast, MitoTracker® Green is also mitochondrial specific but does not depend on the mitochondrial membrane potential; hence, other than an additional incubation with the isolated mitochondria for 10 min before use, the dye is not necessary in experimental buffers. The addition of MitoTracker® Green to assays with TMRM is useful to visualize the mitochondria because some of them might not be energized enough for TMRM fluorescence.

12. Use the pump in WITHDRAW mode, that is, the syringe serves as a waste reservoir, pulling solution away from the vials containing experiential buffers and through the channels. We find that this approach is easier than pushing the solutions into the channels and also allows us to quickly change the experimental buffers by swapping out vials instead of syringes.

13. Use a fast exposure time, e.g., 100–500 μs, to monitor the efficiency of trapping. Initially, to get acquainted with the trapping process, use fluorescently labeled microbeads with diameter of 1 μm (Life Technologies, Grand Island, NY, USA).

14. To investigate complex II, for example, flow in 100 nM TMRM, 2 μM rotenone, and 10 mM succinate. To initiate mitochondrial respiration, add 50 μM ADP and to uncouple ATP synthase and the electron transport chain, use 100 nM CCCP. We recommend to prepare experimental buffers with the basal respiration buffer. For other components of oxidative phosphorylation process, respective experimental buffer solutions can be strategically used [13].

15. Since we have only one detector, we image the red and green signals from JC-1 assays asynchronously by manually adjusting the filter cubes. For time-lapse fluorescence imaging, the filter is set to capture the red fluorescence only. If two detectors are

available, simultaneous monitoring of red and green fluorescence can be performed.

16. From our experience, JC-1 works well with comparing mitochondria in treated vs. non-treated assessments. In other applications, JC-1 might not be suitable because of its slow response time and its high propensity to photobleaching [14]. We observe a larger extent of photobleaching in using JC-1 compared with TMRM. TMRM has a faster response time but can easily bind to PDMS if the channels' surface is not hydrophilic enough, increasing the interfering background fluorescence. This observation is one weakness of our channels. Optimization steps such as fine-tuning oxygen plasma treatment and priming the channels with polyvinyl alcohol might reduce the hydrophobicity of PDMS and thus potentially mitigating the problem. If absolute determination of mitochondrial membrane potential is desired, the interested readers can refer to our previous work [15] and others [16].

References

1. Wallace DC (2005) A mitochondrial paradigm of metabolic and degenerative diseases, aging, and cancer: a dawn for evolutionary medicine. Annu Rev Genet 39:359–407

2. Wallace DC, Lott MT, Procaccio V (2007) Mitochondrial genes in degenerative diseases, cancer and aging. In: Rimoin DL, Connor JM, Pyeritz RE, Korf BR (eds) Emery and Rimoin's principles and practice of medical genetics, 5th edn. Churchill Livingstone Elsevier, Philadelphia, PA, pp 194–298

3. Payne BAI, Wilson IJ, Yu-wai-Man P et al (2013) Universal heteroplasmy of human mitochondrial DNA. Hum Mol Genet 22: 384–390

4. Bandy B, Davison AJ (1990) Mitochondrial mutations may increase oxidative stress: implications for carcinogenesis and aging? Free Radic Biol Med 8:523–539

5. Harman D (1972) The biologic block: the mitochondria? J Am Geriatr Soc 20:145–147

6. Wallace DC, Singh G, Lott MT et al (1988) Mitochondrial DNA mutation associated with Leber's hereditary optic neuropathy. Science 242:1427–1430

7. Schaefer AM, Taylor RW, Turnbull DM, Chinnery PF (2004) The epidemiology of mitochondrial disorders–past, present and future. Biochim Biophys Acta 1659:115–120

8. Elliott HR, Samuels DC, Eden JA et al (2008) Pathogenic mitochondrial DNA mutations are common in the general population. Am J Hum Genet 83:254–260

9. Vives-Bauza C, Przedborski S (2011) Mitophagy: the latest problem for Parkinson's disease. Trends Mol Med 17:158–165

10. Gilkerson RW, De Vries RL, Lebot P et al (2012) Mitochondrial autophagy in cells with mtDNA mutations results from synergistic loss of transmembrane potential and mTORC1 inhibition. Hum Mol Genet 21:978–990

11. Zand K, Pham T, Davila A et al (2013) Nanofluidic platform for single mitochondria analysis using fluorescence microscopy. Anal Chem 85:6018–6025

12. Frezza C, Cipolat S, Scorrano L (2007) Organelle isolation: functional mitochondria from mouse liver, muscle and cultured fibroblasts. Nat Protoc 2:287–295

13. Zhang J, Nuebel E, Wisidagama DRR et al (2012) Measuring energy metabolism in cultured cells, including human pluripotent stem cells and differentiated cells. Nat Protoc 7: 1068–1085

14. Brand MD, Nicholls DG (2011) Assessing mitochondrial dysfunction in cells. Biochem J 435:297–312

15. Lim T-S, Davila A, Zand K et al (2012) Wafer-scale mitochondrial membrane potential assays. Lab Chip 12:2719–2725

16. Gerencser AA, Chinopoulos C, Birket MJ et al (2012) Quantitative measurement of mitochondrial membrane potential in cultured cells: calcium-induced de- and hyperpolarization of neuronal mitochondria. J Physiol 590: 2845–2871

Chapter 5

Optical Microwell Arrays for Large-Scale Studies of Single Mitochondria Metabolic Responses

Venkata Suresh Vajrala, Emmanuel Suraniti, Bertrand Goudeau, Neso Sojic, and Stéphane Arbault

Abstract

Most of the methods dedicated to the monitoring of metabolic responses from isolated mitochondria are based on whole-population analyses. They rarely offer an individual resolution though fluorescence microscopy allows it, as demonstrated by numerous studies on single mitochondria activities in cells. Herein, we report on the preparation and use of microwell arrays for the entrapment and fluorescence microscopy of single isolated mitochondria. Highly dense arrays of 3 μm mean diameter wells were obtained by the chemical etching of optical fiber bundles (850 μm whole diameter). They were manipulated by a micro-positioner and placed in a chamber made of a biocompatible elastomer (polydimethylsiloxane or PDMS) and a glass coverslip, on the platform of an inverted microscope. The stable entrapment of individual mitochondria (extracted from *Saccharomyces cerevisiae* yeast strains, inter alia, expressing a green fluorescent protein) within the microwells was obtained by pretreating the optical bundles with an oxygen plasma and dipping the hydrophilic surface of the array in a concentrated solution of mitochondria. Based on the measurement of variations of the intrinsic NADH fluorescence of each mitochondrion in the array, their metabolic status was analyzed at different energetic respiratory stages: under resting state, following the addition of an energetic substrate to stimulate respiration (ethanol herein) and the addition of a respiratory inhibitor (antimycin A). Statistical analyses of mean variations of mitochondrial NADH in the population were subsequently achieved with a single organelle resolution.

Key words Mitochondria, Fluorescence microscopy, Microwells, Optical fibers, Polydimethylsiloxane, Single organelle, Metabolic monitoring

1 Introduction

Methods dedicated to mitochondrial metabolic studies are usually performed with large populations of isolated organelles [1]. In particular, the first developed method in this field, dedicated to oxygen concentration measurements with the so-called Clark electrode, requires large quantities of mitochondria (mg amounts of proteins). Although this method allowed defining standard energetic states observed as variations of the oxygen consumption by a population of typically millions of mitochondria [2], it is now demonstrated

Volkmar Weissig and Marvin Edeas (eds.), *Mitochondrial Medicine: Volume I, Probing Mitochondrial Function*, Methods in Molecular Biology, vol. 1264, DOI 10.1007/978-1-4939-2257-4_5, © Springer Science+Business Media New York 2015

that mitochondria constituting a cell's mitochondrial network are genetically (heteroplasmy) and metabolically heterogeneous [3–5]. Consequently, there is a necessity to monitor metabolic responses from individual mitochondria in order to understand the span of their activities within a cell or an isolated population. Metabolic parameters of major interest include oxygen consumption, ATP production, inner membrane potential usually quoted as $\Delta\Psi$, substrate and cofactors of the respiratory chain, as well as reactive oxygen species (superoxide ion, hydrogen peroxide) formed as side species of oxygen reduction [6, 7].

Metabolic waves involving mitochondria in a single cell have been well described owing to confocal fluorescence microscopy, for different cell types including myocytes wherein the mitochondrial network is organized along the cytoskeleton [8–10]. Sensitive fluorescence measurements were less often used for metabolic studies on single mitochondria isolated from cells. A few reports were specifically focused on assessing mitochondrial membrane potential variations under different conditions of respiratory activation or inhibition [11–13]. These studies were achieved with diluted solutions of mitochondria deposited on the surface of a bare or poly-L-lysine-modified glass coverslip. We have recently reported that the monitoring of single mitochondria could also be successfully achieved within wells created in the biocompatible polydimethylsiloxane (PDMS) polymer [14]. Indeed, mitochondria in the solution were deposited in well structures, which diameter ranged from 40 μm to 2 mm, and observed owing to their NADH autofluorescence by microscopy. This allowed studying responses within small populations of mitochondria, a few tens to a few hundreds simultaneously. However, such well structures should be downsized and organized within an array in order to isolate each single mitochondrion within a cavity and to observe simultaneously a whole population of these individual mitochondria [15].

To achieve that, we took benefit of the well-ordered organization of optical fiber bundles to create highly dense microwell arrays that could match the requirements for individualization, immobilization, and observation by fluorescence of single mitochondria. Such bundles, composed of a few hundreds to a few tens of thousands of single micrometric-diameter optical fibers, have been extensively used for a broad spectrum of bioanalytical applications because of their coherent transmission of light throughout but also because of a relative ease for their surface modification and structuration [16, 17]. In particular, depending on the fiber material's composition, their glass surface can be etched selectively so as to create a micrometric cavity at the end of each optical fiber in the bundle [18, 19]. Etched optical fiber bundles have been then used as arrays for single-cell or single-enzyme traps and parallelized analyses [20–23]. In the present work, we report on the preparation and use of such an array of optical microwells for the entrapment of a large number of single mitochondria and for the kinetic monitoring by fluorescence of their endogenous NADH (reduced nicotinamide adenine dinucleotide) variations under activation and inhibition of their respiratory chain.

2 Materials

2.1 Solutions

1. Growth medium for yeasts: 0.175 % yeast nitrogen base, 0.2 % casein hydrolysate, 0.5 % $(NH_4)_2SO_4$, 0.1 % KH_2PO_4, 2 % lactate (w/v) as carbon source, 20 mg/L L-tryptophan, 40 mg/L adenine hydrochloride, and 20 mg/mL uracil, pH 5.5.

2. Preparation buffer: 0.75 M deionized sorbitol, 0.4 M deionized mannitol, 10 mM Tris-maleate, 10 mM Tris-orthophosphate, and 0.1 % BSA, pH 6.8.

3. Homogenization buffer: 0.5 M deionized mannitol, 2 mM EGTA, 10 mM Tris-maleate, 10 mM Tris-orthophosphate, and 0.2 % BSA, pH 6.8.

4. Resuspension buffer: 0.6 M deionized mannitol, 2 mM EGTA, 10 mM Tris-maleate, and 5 mM Tris-orthophosphate, pH 6.8.

5. Cell weakening buffer: 0.1 M Tris–HCl and 0.5 M β-mercaptoethanol, adjusted at pH 9.3 with NaOH.

6. First washing buffer: 0.5 M KCl and 10 mM Tris–HCl, adjusted at pH 7 with HCl.

7. Digestion buffer: 1.35 M deionized sorbitol, 1 mM EGTA, and 0.1 M dimonophosphate (prepared by mixing disodium phosphate 0.2 M to 40.5 % (v/v), sodium phosphate monobasic, 9.5 % (v/v)), pH 7.4.

8. Second washing buffer: 0.75 M deionized sorbitol, 0.4 M deionized mannitol, 10 mM Tris-maleate, 10 mM Tris-orthophosphate, and 0.1 % BSA, pH 6.8.

9. Homogenization buffer: 0.5 M deionized mannitol, 2 mM EGTA, 10 mM Tris-maleate, 10 mM Tris-orthophosphate, and 0.2 % BSA, pH 6.8.

10. Experimental buffer: 0.6 M deionized mannitol, 2 mM EGTA, 10 mM Tris-maleate, and 5 mM Tris-orthophosphate, pH 6.8.

11. Analytical grade 90 % EtOH diluted at 1 and 11 % (v/v) in the experimental buffer.

12. Antimycin A (AA) stock solution at 5 mg/mL is diluted first 400-fold in two steps in 1 % EtOH–MB, to obtain a 20 μM AA diluted solution (*see* **Note 1**).

13. Glass fiber bundle etching solution: NH_4F, HF, and H_2O (1:3:1 volume ratios).

2.2 Optical Microwell Arrays

1. Silica optical fiber bundles (3 μm mean diameter each; 850 μm mean whole diameter) were purchased from Sumitomo Electric Industries, Ltd. (ref. IGN-08/30).

2. Bundles were cleaved with high precision with a dedicated optical fiber cleaver from Vytran© (ref. LDC 400).

3. Polydimethylsiloxane liquid (PDMS) was purchased from Momentive Performance Materials Inc. (ref. RTV615 kit).

4. Microwell array bundles were treated before use with a low-pressure oxygen plasma generator from Harrick Plasma© (ref. plasma cleaner).

5. The fiber bundle was fixed in a fiber chuck from Siskiyou© (BFC300) for its easier manipulation.

6. The fiber chuck is fixed at a 3D holder/positioner from Newport© (ref. 340-RC) for its micrometric manipulation on the microscope stage.

2.3 Fluorescence Imaging

1. Experiments were performed with an inverted epifluorescence microscope from Leica© (DMI 6000 model) equipped with a camera from Hamamatsu© (ref. ORCA-Flash4.0).

2. Images were collected and analyzed via MetaMorph (Molecular Devices©) or ImageJ (NIH free supply) softwares.

3. Statistical analyses of data distributions are performed using Origin software (version 8.0, OriginLab©).

3 Methods

3.1 Mitochondria Preparation

1. *Saccharomyces cerevisiae* – **strain BY4742 expressing a green fluorescent protein (GFP)**. These yeast cells have been transformed with the plasmid pGAL-CLbGFP containing the presequence of mitochondrial citrate synthase fused to GFP under the control of a GAL1/10 promoter and have been a gift from J. P. di Rago, University of Bordeaux, CNRS, IBGC UMR5095, Bordeaux, France.

2. Cells are grown aerobically at 28 °C in the specific growth medium.

3. Cells are harvested in the exponential growth phase, and their mitochondria are isolated from protoplasts, according to the following procedure.

4. When cell growth exceeds the shift-up, which means in the absence of glucose, yeasts are collected in 1 L pots and then centrifuged at $3,000 \times g$ for 10 min.

5. Yeasts are pelleted and the culture medium is removed. Then 2–3 more washings are done with distilled water during which the cell pellet is resuspended in distilled water at 5 °C and then harvested.

the planned mean coverage to be substantially higher. This issue is less of a problem for short amplicon resequencing where coverage will be constant within one sample, and should be fairly uniform between samples.

6. Careful consideration should be given to the bioinformatic pipeline for sequence analysis. Commercial software is available (e.g., CLC bio Genomics Workbench; NextGENe), but many bioinformatics support units prefer to use "in-house" pipelines. In general terms, the pipeline steps will comprise: raw data QA and modification, genome alignment, variant/indel prediction, variant QC and filtering, and variant annotation. Bidirectional filtering of identified variants is then generally recommended, i.e., the requirement for rare variants detected to have support in both read directions (at similar proportional levels) if they are to be accepted. This is generally considered to eliminate much of the noise from the data.

7. Consideration should be given to the use of adequate "negative" controls for the resequencing experiment. Most commercially available next-generation sequencing kits contain some kind of control which can be run in parallel with the sample to establish that there is not excessive noise. This control is generally phage DNA. We have previously used cloned fragments of mtDNA, which we have argued is a more rigorous control as it attempts to control for any sequence-specific determinants of noise. (Amplicons were generated from cloned DNA fragments (*MT-HV2* clone, nt.16548–771; *MT-CO3* clone, nt.9127–9661; cloned in pGEM-T Easy Vector, Promega)). We have also previously used a nuclear DNA amplicon (*BRCA2*, NC_000013.10, 32907060-32907350) where no variance was expected as a negative control, with similar results.

Acknowledgments

This work was funded by the Medical Research Council (MRC), UK (B. A. I. P.); the National Institute for Health Research (NIHR, UK) Biomedical Research Centre (BRC) awarded to Newcastle University and Newcastle Upon Tyne Hospitals NHS Foundation Trust (B. A. I. P., P. F. C.); the Wellcome Trust (P. F. C.); and the AstraZeneca (K. G.).

References

1. Schon EA, Dimauro S, Hirano M (2012) Human mitochondrial DNA: roles of inherited and somatic mutations. Nat Rev Genet 13:878–890

2. Just RS, Leney MD, Barritt SM et al (2009) The use of mitochondrial DNA single nucleotide polymorphisms to assist in the resolution of three challenging forensic cases. J Forensic Sci 54:887–891

3. Mishmar D, Ruiz-Pesini E, Brandon M et al (2004) Mitochondrial DNA-like sequences in

the nucleus (NUMTs): insights into our African origins and the mechanism of foreign DNA integration. Hum Mutat 23:125–133

4. Ameur A, Stewart JB, Freyer C et al (2011) Ultra-deep sequencing of mouse mitochondrial DNA: mutational patterns and their origins. PLoS Genet 7:e1002028

5. Brodin J, Mild M, Hedskog C et al (2013) PCR-induced transitions are the major source of error in cleaned ultra-deep pyrosequencing data. PLoS One 8:e70388

6. Payne BA, Wilson IJ, Hateley CA et al (2011) Mitochondrial aging is accelerated by anti-retroviral therapy through the clonal expansion of mtDNA mutations. Nat Genet 43:806–810

7. Payne BA, Wilson IJ, Yu-Wai-Man P et al (2013) Universal heteroplasmy of human mitochondrial DNA. Hum Mol Genet 22:384–390

8. Ye J, Coulouris G, Zaretskaya I et al (2012) Primer-BLAST: a tool to design target-specific primers for polymerase chain reaction. BMC Bioinform 13:134

Chapter 7

Single-Cell Analysis of Mitochondrial DNA

Brendan A.I. Payne, Lynsey Cree, and Patrick F. Chinnery

Abstract

Understanding the biology of mitochondrial DNA (mtDNA) at the single-cell level has yielded important insights into inheritance, disease, and normal aging. In nuclear gene disorders of mtDNA maintenance, neurodegeneration, and aging, different somatic mtDNA mutations exist within individual cells and may be missed by techniques applied to whole tissue DNA extract. We therefore provide a method for characterizing mtDNA within single skeletal muscle fibers. During embryogenesis, mtDNA content is subject to a tight bottleneck and this may account for differential segregation of mutant mtDNA in offspring. We also present a method to study this phenomenon by single-cell analysis of embryonic PGCs (primordial germ cells).

Key words DNA, Mitochondrial, Single-cell analysis, Germ cells, Laser capture microdissection

1 Introduction

Mammalian cells contain hundreds to tens of thousands of copies of the mitochondrial genome per cell. This mtDNA is constantly turned over, even in nondividing cells [1]. As a result mtDNA mutations often coexist with wild-type mtDNA within the same cell (a state known as heteroplasmy). In inherited disorders of mtDNA maintenance (e.g., defects of the nuclear gene, *POLG*, encoding the mtDNA polymerase, pol γ), normal aging, and neurodegenerative disease, different somatic (acquired) mtDNA mutations are present within individual cells [2–4]. DNA analysis from a whole tissue extract will therefore fail to reliably detect many somatic mtDNA mutations. Even in the case of an inherited single mtDNA mutation which is present within every cell, the relative amounts of mutant and wild-type mtDNA may vary from cell to cell due to natural drift. This determines whether there is a functional defect of respiratory chain function within a given cell. Therefore, assessment of the heteroplasmy level across the tissue as a whole may not accurately reflect the state within affected cells. Inherited mutations of mtDNA may be subject to differential

Volkmar Weissig and Marvin Edeas (eds.), *Mitochondrial Medicine: Volume I, Probing Mitochondrial Function*,
Methods in Molecular Biology, vol. 1264, DOI 10.1007/978-1-4939-2257-4_7, © Springer Science+Business Media New York 2015

segregation due to a bottleneck effect during oogenesis, with important consequences for the penetrance of that mutation in the offspring [5]. Analysis of mtDNA mutations within single cells is therefore a valuable solution for the study of inherited and acquired mtDNA defects [6].

In order to select single cells for mtDNA analysis, it is common practice to first perform sequential COX (cytochrome *c* oxidase)/ SDH (succinate dehydrogenase) histochemistry on sections of the relevant tissue (e.g., skeletal muscle, the prototype tissue for mitochondrial analyses) [7]. This will identify those cells which have a respiratory chain defect affecting COX function and are thus likely to contain high levels of an mtDNA mutation. COX contains respiratory chain subunits encoded by mtDNA, and cells which contain high levels of an mtDNA defect are therefore frequently deficient in COX function (loss of brown stain). In contrast, SDH is encoded entirely by the nuclear genome, and activity is thus preserved in the face of an mtDNA defect within a cell. It thus provides an effective counterstain (blue). COX/SDH is performed on unfixed (fresh frozen) tissue which therefore allows optimal recovery of nucleic acid, including for quantitative assays. In contrast, in studying the mtDNA bottleneck within single embryonic PGCs (primordial germ cells), we have used stella-GFP transgenic mice, whereby the reporter is expressed exclusively within these cells to facilitate identification and microdissection [5].

Here we describe methods for a range of downstream molecular analyses for the identification and quantification of mtDNA within single cells following microdissection from tissue cryo-sections, or Fluorescence-activated cell sorting-sorted embryonic PGCs.

2 Materials

2.1 Sample Preparation and COX/SDH Histochemistry (Skeletal Muscle)

1. Hazards: 3,3-diaminobenzidine tetrahydrochloride and cytochrome c are not hazardous according to Directive 67/548/EC (source: Sigma-Aldrich MSDS). Sodium azide is very toxic if swallowed. Potential for microbiological hazard from unfixed human tissue on microscope slides.

2. Cryostat.

3. PEN membrane slides (Leica).

4. 5 mM 3,3-diaminobenzidine tetrahydrochloride (*see* **Note 1**).

5. 500 µM cytochrome c.

6. Bovine liver catalase.

7. 1.875 mM nitroblue tetrazolium.

8. 1.3 M sodium succinate.

9. 2.0 mM phenazine methosulfate.

10. 100 mM sodium azide.

11. 37 °C incubator.

12. 0.1 M phosphate buffered saline, pH 7.0.

13. Graded alcohol series for dehydration: 70 % ethanol, 95 % ethanol, 100 % ethanol (2×).

2.2 Single-Cell Laser Microdissection

1. Laser microdissection system (e.g., Leica LMD6500).

2. 200 or 500 µl sterile microtubes, with appropriate sized mounting stage for microscope.

3. Microcentrifuge.

2.3 Sample Preparation (Embryo)

1. C57BL/6J.CBA F1 females and *Stella*-GFP BAC-homozygous C57BL/6J.CBA F1 males maintained in a normal 12 h light/dark cycle.

2. Dulbecco's minimal essential medium (DMEM).

3. 7.5 % fetal calf serum (FCS).

4. 10 mM HEPES.

5. Trypsin.

6. Leica MZ16FA stereomicroscope.

7. Sterile plastic tissue culture dishes.

8. Tungsten needles.

9. Fine scissors.

10. Watchmaker's forceps.

11. 70 % ethanol in a squeeze bottle.

12. 37 °C incubator.

13. Ice.

2.4 Isolation of Single PGCs by FACS

1. 96-well plates.

2. BD FACSAria III sorter (BD Biosciences).

2.5 Lysis

1. Lysis buffer: 50 mM Tris–HCl pH 8.5, 0.5 % Tween-20, 200 µg/ml Proteinase K.

2.6 Real-Time PCR from Single Cells (Skeletal Muscle)

1. iQ SYBR Green supermix (Bio-Rad) (*see* **Note 2**).

2. iCycler real-time PCR machine (Bio-Rad).

3. *MT-ND1* real-time PCR primers: nt. 3458–3481, nt. 3569–3546 (*see* **Note 3**).

4. *MT-ND4* real-time PCR primers: nt. 11144 11165, nt. 11250–11230.

5. *MT-ND1* PCR template primers for quantitative standard: nt. 3017–3036, nt. 4057–4037.

6. *MT-ND4* PCR template primers for quantitative standard: nt. 10534–10553, nt. 11605–11586.

7. QIAquick gel extraction kit (Qiagen).

2.7 Real-Time PCR from Single Cells (PGCs)

1. iQ SYBR Green supermix (Bio-Rad) (*see* **Note 2**).

2. iCycler real-time PCR machine (Bio-Rad).

3. *MT-ND5* real-time PCR primers: nt. 12789–12806, nt. 12876–12857 (*see* **Note 4**).

4. *MT-ND4* real-time PCR primers: nt. 10571–10590, nt. 10643–10624.

5. *MT-ND1* real-time PCR primers: nt. 3006–3025, nt. 3121–3104.

6. *MT-ND5* PCR template primers for quantitative standard: nt. 12704–12723, nt. 13832–13813.

7. *MT-ND4* PCR template primers for quantitative standard: nt. 10333–10352, nt. 11006–10987.

8. *MT-ND1* PCR template primers for quantitative standard: nt. 2785–2804, nt. 3593–3574.

9. QIAquick gel extraction kit (Qiagen).

2.8 Sequencing from Single Cells

1. Nine overlapping primary PCR primer pairs covering the whole mtDNA genome.

2. 36 overlapping secondary PCR primer pairs (4 are nested within each primary PCR product). Secondary PCR primers are M13 tagged to facilitate cycle sequencing. (Primer sequences are available from authors.)

3. AmpliTaq Gold DNA Polymerase kit (Life Technologies).

4. BigDye Terminator v3.1 kit (Applied Biosystems).

3 Methods

3.1 Sample Preparation and COX/ SDH Histochemistry (Skeletal Muscle)

1. 20 μm frozen sections are cut by cryostat onto PEN membrane slides (Leica) (*see* **Notes 5–7**).

2. Allow to air-dry at room temperature for 60 min.

3. Rapidly thaw:

   0.8 ml of 5 mM 3,3-diaminobenzidine tetrahydrochloride.

   0.2 ml of 500 μM cytochrome c.

   Mix and add a few grains of catalase.

   Mix, apply to the section, and incubate for 45 min at 37 °C.

4. Wash in three changes of 0.1 M phosphate buffered saline, pH 7.0.

5. Rapidly thaw:

   0.8 ml of 1.875 mM nitroblue tetrazolium.

   0.1 ml of 1.30 M sodium succinate.

   0.1 ml of 2.0 mM phenazine methosulfate.

   0.010 ml of 100 mM sodium azide.

   Mix, apply to section, and incubate for 40 min at 37 °C.

6. Wash in three changes of 0.1 M phosphate buffered saline, pH 7.0.

7. Dehydrate through graded alcohols: 2 min in 70 % ethanol, 2 min in 95 % ethanol, 2 min in 100 % ethanol, and a further 10 min in a final bath of 100 % ethanol.

8. Slides may be frozen (at −20 or −80 °C) if not required for immediate microdissection.

3.2 Single-Cell Laser Microdissection

1. Single cells are captured by laser microdissection into sterile 200 or 500 µl microtubes using the software provided with the microscope (*see* **Notes 8–10**) (Fig. 1).

2. After collection centrifuge cells in microcentrifuge at 16,000 g for 10 min.

3. Cells may then be stored at −20 or −80 °C until required.

4. Centrifuge again at 16,000 g for 5 min prior to applying lysis buffer.

3.3 Sample Preparation (Embryo)

1. Natural timed matings are set up between C57BL/6J.CBA F1 females in estrus and *Stella*-GFP BAC-homozygous C57BL/6J. CBA F1 males. Noon of the day of the vaginal plug was considered embryonic day E0.5 (*see* **Notes 11–14**).

2. Pregnant female mice are sacrificed humanely at E7.5.

3. Lay the female on its back and wash thoroughly in 70 % ethanol. Cut horns below the oviduct and remove from the mesometrium.

4. Remove the decidua from the uterus.

5. Transfer to a sterile, plastic tissue culture dish containing cold dissection medium (DMEM supplemented with 7.5 % FCS and 10 mM HEPES).

6. Using a stereomicroscope, carefully dissect the embryo free of the deciduum using watchmaker's forceps, and isolate the posterior parts of the E7.5 embryos (containing the GFP-positive PGCs) using tungsten needles.

7. Pool embryos according to age [8].

8. Incubate the dissected region in 0.25 % trypsin at 37 °C for 15 min.

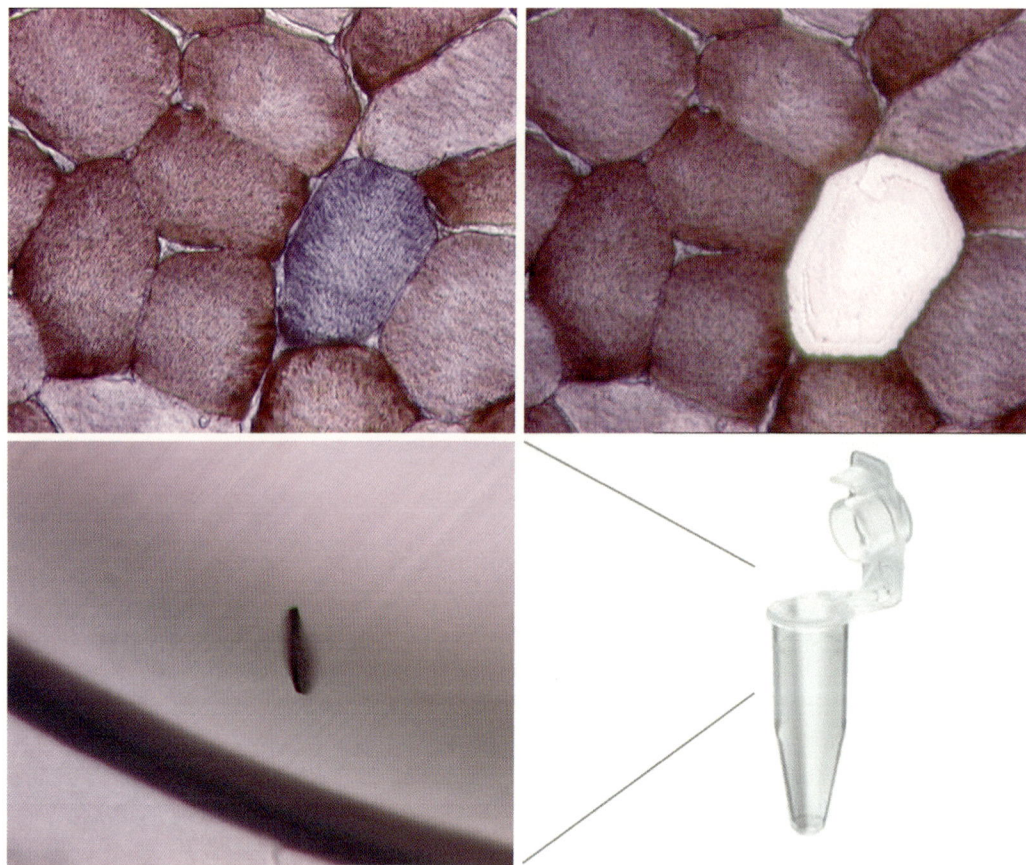

Fig. 1 Laser microdissection of a single COX-deficient skeletal muscle fiber. *Top left*: sequential COX/SDH histochemistry of 20 μm cryo-section of a human lower limb skeletal muscle on PEN membrane slide, showing one COX-deficient fiber (counterstained *blue*). *Top right*: the fiber has been cut by laser microdissection. *Bottom left* and *right*: fiber appears in cap of sterile microtube, ready for lysis and molecular analyses

9. Enzymatic digestion is halted by adding an equal volume of FCS and pipetting to mix.

10. Centrifuge samples at $2,000 \times g$ for 3 min.

11. Resuspend the pellet in DMEM supplemented with 7.5 % FCS and 10 mM HEPES.

12. Keep samples on ice until FACS sorting.

3.4 Isolation of Single PGCs by FACS

1. All FACS are performed using a BD FACSAria III sorter; instrument sensitivity is proved stable by internal QC procedures (*see* **Note 15**).

2. GFP is detected using a 100 mW sapphire laser.

3. Single PGCs are unidirectionally sorted at 20 psi using a 100 μm nozzle with a sort rate of 2,000 events per second into single wells of a 96-well plate.

4. Centrifuge plate at $2,000 \times g$ for 5 min.

5. Plates may then be stored at –20 or –80 °C until required.

6. Centrifuge again at $2,000 \times g$ for 5 min prior to applying lysis buffer.

3.5 Lysis

1. Cells are digested in 20 µl of lysis buffer for 16 h at 55 °C, followed by inactivation of proteinase at 85 °C for 10 min (*see* **Note 16**).

3.6 Real-Time PCR from Single Cells

1. mtDNA content within single cells can be determined using a target template in *MT-ND1*, as this gene is very seldom deleted by large-scale mtDNA deletion mutations. The proportion of mtDNA molecules within a cell that contain large-scale deletions may be determined by comparing with a target template in *MT-ND4* (as this gene is very commonly deleted in large-scale deletion mutations) (*see* **Note 17**).

2. A PCR-generated template is used to produce a quantitative standard series for each of *MT-ND1* and *MT-ND4*, using the primer pairs shown above. Gel-extract the PCR product, and quantify the product by spectrophotometry. Store template at minus 80 °C and store aliquots of working stocks of 10^8 copies/µl at minus 20 °C.

3. Produce a quantitative standard series for each of *MT-ND1* and *MT-ND4* by serial tenfold dilution of the template working stock in PCR-grade water. Run concentrations from 10^7 to 10^2 copies per reaction.

4. Real-time PCR is performed in a 25 µl reaction comprising 1× SYBR Green supermix, 400 nM primers, and 2 µl of cell lysate.

5. Cycling conditions are: 95 °C for 3 min; 40 cycles at 95 °C for 10 s and 62.5 °C for 1 min; melt curve.

6. Perform real-time PCR replicates in quadruplicate for single-cell samples.

3.7 Real-Time PCR from Single Cell (PGCs)

1. mtDNA content within single cells can be determined using a target template in *MT-ND1, ND4,* or *ND5* in mice.

2. A PCR-generated template is used to produce a quantitative standard series for each of *MT-ND1, MT-ND4,* and *MT-ND5*, using the primer pairs shown above. Gel-extract PCR products, and quantify by spectrophotometry. Store template at minus 80 °C and store aliquots of working stocks of 10^8 copies/µl at minus 20 °C.

3. Produce a quantitative standard series for each template by serial tenfold dilution of the template working stocks in PCR-grade water. Run concentrations from 10^7 to 10^2 copies per reaction.

4. Real-time PCR is performed in a 25 µl reaction comprising 1×
SYBR Green Supermix, 500 nM primers, and 2 µl of cell lysate.

5. Cycling conditions are: 95 °C for 3 min; 40 cycles at 95 °C for
10 s and 62.5 °C for 1 min; melt curve.

6. Perform real-time PCR replicates in quadruplicate.

3.8 Sequencing from Single Cells

1. Whole genome sequencing from individual fibers employs a
nested PCR comprising a primary PCR with nine overlapping
primer pairs followed by 36 overlapping secondary PCR primer
pairs (*see* **Notes 18–20**).

2. Primary PCR was performed in a 50 µl volume containing 1×
PCR buffer (10 mM Tris–HCl pH 8.3, 1.5 mM $MgCl_2$,
50 mM KCl, 0.001 % w/v gelatin), 1 mM $MgCl_2$, 0.2 mM
dNTPs, 0.6 µM primers, 1.75 U AmpliTaq Gold (Applied
Biosystems), and 1 µl lysate. PCR conditions were 94 °C for
10 min and 38 cycles at 94 °C for 45 s, 58 °C for 45 s, and
72 °C for 2 min. Final extension was 8 min.

3. Secondary PCR was performed in a 25 µl volume containing
1× PCR buffer (as above), 0.2 mM dNTPs, 0.8 µM primers,
0.65 U AmpliTaq Gold, and 1 µl of primary PCR product.
PCR conditions were as above except for 1 min extension and
30 cycles.

4. Cycle-sequencing was performed using BigDye Terminator
v3.1 kit (Applied Biosystems) and visualized through a 3130×
Genetic Analyzer (Applied Biosystems).

4   Notes

1. COX/SDH reagents should be prepared into ready-to-use aliquots and stored frozen until required.

2. The real-time method presented here is based on DNA binding by SYBR Green. The method could be adapted for use
with a probe-based real-time PCR assay, however with a potential loss of sensitivity for the low-copy-number target.

3. All primer positions for human assays refer to the revised
Cambridge Reference Sequence (rCRS, NC_012920).

4. All primer positions for mouse assays refer to the reference
sequence NC_005089.

5. We give the example of COX/SDH histochemistry on human
skeletal muscle; however, the method can be applied to a wide
variety of mammalian tissues in frozen section. The incubation
times required to obtain a clear distinction between COX and
SDH staining may vary between tissues and species, and optimization may be required for combinations other than that
presented here.

6. In general for skeletal muscle single-cell analyses, transverse sections are required, and care should be taken to ensure the correct orientation of the tissue block. However, the method can also be applied to longitudinal sections if required.

7. We recommend using a precision pipette (e.g., Gilson) to apply the COX and SDH reagents to the sections in order to minimize wastage. Generally 50–100 μl may be required per section, depending on section size. Care should be taken to ensure that the section is fully covered by the stain, and the staining tray should be kept flat to ensure even coverage.

8. Care should be taken to properly calibrate the laser microdissection stage on the microscope (according to the manufacturer's instructions) to ensure optimal alignment of the microtube caps and the slide to optimize successful collection of cells.

9. Some optimization of the laser cutting parameters may be required (e.g., diameter and intensity of beam, speed of cut). In general the lowest diameter and intensity which give an efficient cut should be used to minimize burning of adjacent tissue. Ethanol dehydration performed at the end of the COX/SDH histochemistry protocol causes a slight shrinkage and separation of muscle fibers which facilitates easier laser microdissection.

10. Always check that the cell has been captured into the microtube cap. If the dissected cell is not seen however, that cap should not be reused, in order to avoid the potential for two cells to be present in the same cap.

11. All culture media were prepared fresh on the day of embryo recovery.

12. All culture media components were dissolved in Milli-Q water.

13. pH of the media was adjusted prior to final sterilization by filtration (0.2 μm filter).

14. All animal procedures were performed, under license, in accordance with the UK Home Office Animal Act (1986).

15. FACS equipment should be subject to rigorous QC (quality control).

16. The amount of lysis buffer used can be varied depending on the downstream application required. For example, for real-time PCR, we use 2 μl per reaction with each of two reactions (one per target) performed in quadruplicate.

17. When comparing mtDNA copy number within single skeletal muscle fibers, the content may be expressed per unit volume by measuring the cross-sectional area (automatically recorded by the laser microdissection software) and section thickness (10 or 20 μm).

18. Great care should be taken to avoid trace contamination of samples or reagents with exogenous mtDNA, for example, from the laboratory environment, as this may lead to the detection of spurious "variants." All amplicon generation steps should be prepared in a PCR hood which has been treated with ultraviolet irradiation and DNase.

19. When performing nested PCR for whole mtDNA genome sequencing from single cells, the amount of primary PCR product required to obtain an optimal secondary PCR product may vary between amplicons. The electrophoresis gel image of the primary PCR products should be inspected to give an approximate quantitative of product concentration.

20. We recommend that any mutations detected in single cells should be confirmed by resequencing from the original lysate.

Acknowledgment

This work was funded by the Medical Research Council (MRC), UK (BAIP), and the Wellcome Trust (LC).

References

1. Gross NJ, Getz GS, Rabinowitz M (1969) Apparent turnover of mitochondrial deoxyribonucleic acid and mitochondrial phospholipids in the tissues of the rat. J Biol Chem 244:1552–1562

2. Bua E, Johnson J, Herbst A, Delong B, McKenzie D et al (2006) Mitochondrial DNA-deletion mutations accumulate intracellularly to detrimental levels in aged human skeletal muscle fibers. Am J Hum Genet 79:469–480

3. Bender A, Krishnan KJ, Morris CM, Taylor GA, Reeve AK et al (2006) High levels of mitochondrial DNA deletions in substantia nigra neurons in aging and Parkinson disease. Nat Genet 38:515–517

4. Hudson G, Chinnery PF (2006) Mitochondrial DNA polymerase-gamma and human disease. Hum Mol Genet 15(Spec No 2):R244–R252

5. Cree LM, Samuels DC, de Sousa Lopes SC, Rajasimha HK, Wonnapinij P et al (2008) A reduction of mitochondrial DNA molecules during embryogenesis explains the rapid segregation of genotypes. Nat Genet 40:249–254

6. Payne BA, Wilson IJ, Hateley CA, Horvath R, Santibanez-Koref M et al (2011) Mitochondrial aging is accelerated by anti-retroviral therapy through the clonal expansion of mtDNA mutations. Nat Genet 43:806–810

7. Brierley EJ, Johnson MA, Lightowlers RN, James OF, Turnbull DM (1998) Role of mitochondrial DNA mutations in human aging: implications for the central nervous system and muscle. Ann Neurol 43:217–223

8. Downs KM, Davies T (1993) Staging of gastrulating mouse embryos by morphological landmarks in the dissecting microscope. Development 118:1255–1266

Chapter 8

A High-Throughput Next-Generation Sequencing Assay for the Mitochondrial Genome

Shale Dames, Karen Eilbeck, and Rong Mao

Abstract

Next-generation sequencing (NGS) is an effective method for mitochondrial genome (mtDNA) sequencing and heteroplasmy detection. The following protocol describes an mtDNA enrichment method up to library preparation and sequencing on Illumina NGS platforms. A short command line alignment script is available for download via FTP.

Key words Mitochondrial disorders, mtDNA, Next-generation sequencing, NGS, Massively parallel sequencing, Long range PCR, DNA enrichment, Bioinformatics, Heteroplasmy, Homoplasmy

1 Introduction

Mitochondrial disorders are attributed to mutations in the human nuclear and mitochondrial genomes [1, 2]. Mitochondria contain their own genome and genetic code, and replicate independently of the host cell. Unlike the nuclear genome, which usually contains two sets of autosomes and sex chromosomes in a cell, a mitochondrion may have between 2 and 15 copies of mtDNA, and a cell may contain as many as 5,000 mitochondria [3, 4]. Due to the high copy number of mtDNA, the usual allelic ratios and nomenclature used for the nuclear genome are different for mtDNA. These varying allele frequencies are known as heteroplasmy and homoplasmy. Heteroplasmy, as defined herein, is any allelic ratio less than 95 %, and homoplasmy is any allelic ratio greater than 95 %. The ability to reproducibly detect varying levels of heteroplasmy is important for adequate diagnosis, treatment, and prognosis of mitochondrial disorders [5–7].

Many bioinformatics software packages rely on probabilistic zygosity calls for diploid organisms, where a mutation is classified

Electronic supplementary material: The online version of this chapter (doi:10.1007/978-1-4939-2257-4_8) contains supplementary material, which is available to authorized users.

Volkmar Weissig and Marvin Edeas (eds.), *Mitochondrial Medicine: Volume I, Probing Mitochondrial Function*, Methods in Molecular Biology, vol. 1264, DOI 10.1007/978-1-4939-2257-4_8, © Springer Science+Business Media New York 2015

as homozygous or heterozygous. The following protocol uses freeware software and read coverage data to calculate heteroplasmy levels for a given variant. It should be noted that the lowest level of heteroplasmy that can be reproducibly detected depends on many factors, including the background error rate of a given sequencing platform, the quality and depth of the sequence data, and the composition of the DNA sequence itself. For clinical assays, we currently verify all reportable mutations by a secondary method—either Sanger sequencing or variant-specific PCR. To date the lowest level of heteroplasmy reproducibly detected using NGS and verified by a secondary method is 5.9 %. The theoretical lowest level of heteroplasmy detection has dropped from 1.0 to 2.0 % using older HiSeq chemistries to between ~0.2 and 0.5 % using current Illumina NGS HiSeq chemistries and background error rate subtraction (data not shown) [4, 7, 8]. Background error rates increase with lower average coverage. In order to obtain lower background error rates, average coverage of >1,000-fold is required.

The human mtDNA is a circular, 16,596 base pair (bp) haploid genome composed of 37 genes. This protocol employs long range PCR for enrichment of the mtDNA and NGS to detect heteroplasmy levels greater than 10 % (*see* **Note 1**). The entire mtDNA is amplified by long range PCR (LRPCR) using five overlapping primer pairs (Fig. 1). The amplicons are subsequently processed to create libraries for sequencing using Illumina HiSeq or MiSeq platforms. After sequencing, the fastq reads are aligned to the mtDNA Cambridge reference sequence NC_012920 and variants reported using Burrows Wheeler Alignment (BWA), samtools, bcftools, and vcftools [9–12]. The pipeline described is a basic version designed to report positional information and heteroplasmy levels. The pipeline can be easily modified to include annotations and other user-defined information deemed relevant. The output from the informatics analysis can be imported into the Integrative Genomics Viewer (IGV) for visualization as a .bam file (a binary version of a .sam file, which contains sequence and quality information, among other data) [13] and associated .vcf (variant call format). A second tab-delimited text file describing the variants can be viewed in a word processing program or spreadsheet for review.

2 Materials

2.1 Required Equipment

The protocol has been validated using the following equipment:

1. Long range PCR amplification: Applied Biosystems® GeneAmp® PCR System 9700 (Life Technologies, Grand Island, NY) (*see* **Note 2**).

2. DNA electrophoresis: Owl™ EasyCast™ B1A Mini Gel Electrophoresis Systems (Thermo Scientific, Waltham, MA).

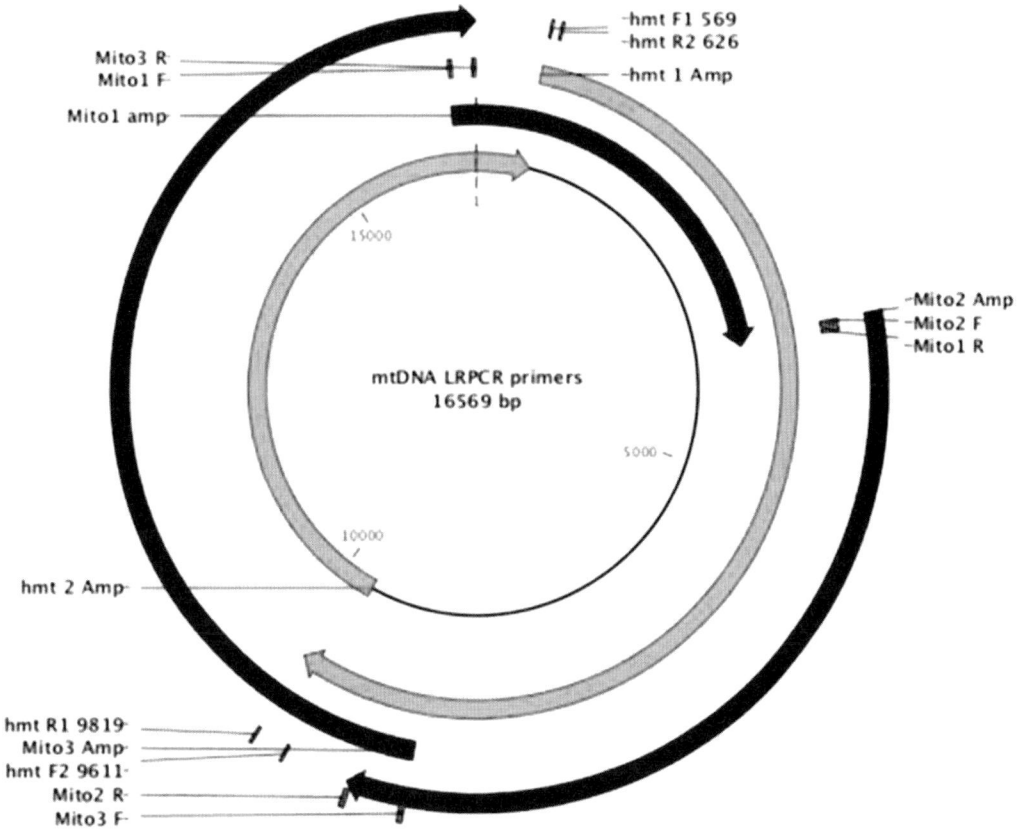

Fig. 1 Diagram of primer binding sites and amplicon sizes relative to mtDNA reference sequence NC_012920

3. Gel imaging: BIO-RAD Gel Doc 1000 imaging system (BIO-RAD, Hercules, CA).

4. DNA quantification: NanoDrop ND 8000 Spectrophotometer (NanoDrop, Wilmington, DE).

5. Amplicon sonication: Covaris S2 focused ultrasonicator (Covaris, Woburn, MA).

6. Post-amplification high-resolution electrophoresis: Agilent 2100 Bioanalyzer (Agilent Technologies, Santa Clara, CA).

7. Illumina HiSeq and MiSeq high-throughput sequencers (Illumina Inc. San Diego, CA).

8. Sequencing alignment and variant reporting: Computer with Mac OS X or Linux OS.

2.2 Oligonucleotides   All primers require a 5′ amino-C6 block modification to reduce sequence overrepresentation of amplicon ends (Integrated DNA Technologies, Coralville, IA). Order primers desalted without specialized purification. Primer sequences are shown in Table 1.

Table 1
Primers for mtDNA long range PCR amplification

Primer	Sequence 5′ → 3′	Amp length
Mito1 F	*ACATAGCACATTACAGTCAAATCCCTTCTCGTCCC	3,968 bp
Mito1 R	*TGAGATTGTTTGGGCTACTGCTCGCAGTGC	
Mito2 F	*TACTCAATCCTCTGATCAGGGTGAGCATCAAACTC	5,513 bp
Mito2 R	*GCTTGGATTAAGGCGACAGCGATTTCTAGGATAGT	
Mito3 F	*TCATTTTTATTGCCACAACTAACCTCCTCGGACTC	7,814 bp
Mito3 R	*CGTGATGTCTTATTTAAGGGGAACGTGTGGGCTAT	
hmt F1	*AACCAAACCCCAAAGACACC	9,289 bp
hmt R1	*GCCAATAATGACGTGAAGTCC	
hmt F2	*TCCCACTCCTAAACACATCC	7,626 bp
hmt R2	*TTTATGGGGTGATGTGAGCC	

Five sets of primers are used to amplify the mtDNA are listed under the "Primer" column. The sequence of each primer and the amplicon length for each primer pair are also displayed. *Asterisk* denotes the requirement of a 5′ amino-C6 modification

2.3 Consumables and Reagents

The polymerase, reaction buffer, and deoxynucleotide triphosphates (dNTPS) reagents are supplied from Clontech Laboratories, Inc., Mountain View, CA (catalogue number RR002M).

1. DNA polymerase: *TaKaRa LA Taq* 5 units/μL.
2. PCR reaction buffer: 10× LA PCR Buffer II (Mg^{2+} plus).
3. Deoxynucleotide triphosphates: 2.5 mM each dNTP.
4. DNA marker: 1 Kb Plus ladder (Life Technologies, Grand Island, NY).
5. 0.2 mL PCR tubes and caps.
6. Molecular grade water.
7. 1 % agarose Tris/Borate/EDTA (TBE) gel.
8. TBE buffer.
9. microTUBE AFA Fiber Pre-Slit Snap-Cap 6×16 mm (catalogue number 520045, Covaris).

2.4 Software

Alignments and variant calling of the mtDNA has been validated using the following freeware applications.

1. Practical Extraction and Report Language (perl, www.perl.org). Versions of perl that have been tested include v5.10.0, v5.10.1, and v5.12.3.
2. Burrows-Wheeler Aligner (BWA, Version 0.1.1-r104) http://sourceforge.net/projects/bio-bwa/files/.
3. Samtools (Version 0.1.18 (r982:295)) http://sourceforge.net/projects/samtools/files/.

4. vcftools (Version 0.1.10) http://vcftools.sourceforge.net/.

5. Integrative Genomics Viewer (IGV, version 2.2 or later). www.broadinstitute.org.

3 Methods

3.1 DNA Isolation

Human mtDNA is co-isolated with nuclear DNA and requires no specialized enrichment protocol. Commercially available DNA extraction methods, such as the QIAamp DNA Blood Mini Kit (QIAGEN, Valencia, CA) or PureGene DNA extraction (QIAGEN), yield DNA suitable for mtDNA amplification (*see* **Note 3**).

3.2 Long Range PCR

1. Determine the concentration of genomic DNA using a spectrophotometer. The optimum range of DNA input is between 10 and 100 ng per PCR reaction (*see* **Note 4**).

2. Prepare primer working stocks at 10 μM each primer pair in molecular grade water (*see* **Note 5**).

3. For each sample five PCR amplifications are required. Add reagents in the order shown in Table 2.

4. PCR cycling conditions: 95 °C (2:00) + [95 °C (0:15) + 68 °C (10:00)] × 30 cycles + 68 °C (20:00) + 4 °C (∞).

5. Purify PCR amplicons individually using a QIAquick PCR Purification kit (QIAGEN) or similar primer removal protocol. Resuspend purified PCR amplicons in 50 μL of molecular grade water (*see* **Note 6**).

Table 2
Long Range PCR Master Mix setup

Reagent	Final concentration	1× Reaction (μL)
Molecular grade water	NA	(30.5)
10× LA PCR Buffer II (Mg²⁺ plus)	1×	5
dNTPs (2.5 mM each)	400 nM each	8
Working primers (10 μM each)	1 μM each	5
DNA (10–100 ng)	Input dependent	(1)
TaKaRa LA Taq (5 U/μL)	0.05 U/μL	0.5
	Total→	50

Reagents are shown in the *left column*, the final concentration per reaction in the *center column*, and the volume per reaction in the *right column*. Volumes in *parenthesis* () are variable depending upon concentration and volume of DNA input. Water should be adjusted accordingly up to a final total volume of 50 μL

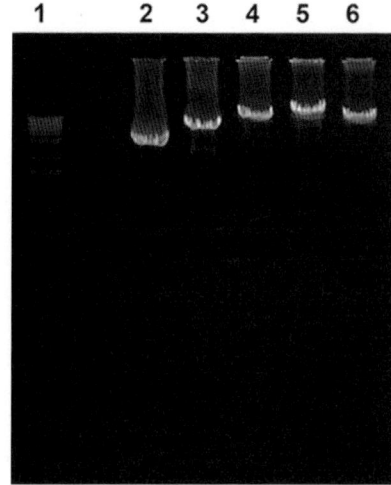

Fig. 2 Long Range PCR Amplicons. A representative 1 % TBE agarose gel picture of the five long range mtDNA amplicons. *Lane 1* 1 Kb Plus ladder, *lane 2* Mito1 amplicon (3,969 bp), *lane 3* Mito2 amplicon (5,513 bp), *lane 4* Mito3 amplicon (7,814 bp), *lane 5* hmt1 amplicon (9,289 bp), and *lane 6* hmt2 (7,262 bp)

3.3 Post-amplification Gel Electrophoresis

1. Add 5 μL of 1 Kb + DNA ladder prepared following manufacturer's directions into a single well on a 1 % TBE agarose gel.

2. Mix 3 μL of each purified amplicon with loading buffer and pipette into unique wells in the 1 % TBE agarose gel.

3. Run gel at 100 V to discriminate amplicon length. The length of time will depend on gel size and well orientation.

4. A representative long rang PCR amplicon gel picture is shown in Fig. 2.

3.4 Equimolar Amplicon Pooling

1. For each amplicon perform three spectrophotometer quantifications using the Nanodrop 8000. Each reading should be ±2 % of each other in ng/μL for a given amplicon to ensure accuracy.

2. Record the ng/μL for each amplicon and average.

3. Calculations for equimolar pooling are based on the Nanodrop measurements. For the sonication step required for Illumina NGS sequencing, 10 μg in a total volume of 105 μL of amplicon DNA input is required.

4. Ten μg of the total mass of amplicons calculates out to 443 fmoles. Therefore, 443 fmoles of each amplicon is needed for equimolar pooling. Table 3 indicates the amount of amplicon DNA in ng for pooling.

5. Mix the appropriate amount of each amplicon in a 1.5 mL and volume up to 105 μL in molecular grade water.

Table 3
Amplicon equimolar pooling

Amplicon	Size (bp)	Required (ng)
Mito1	3,968	1,160
Mito2	5,513	1,612
Mito3	7,814	2,285
Hmt1	9,289	2,716
Hmt2	7,626	2,230
	Total→	~10,000

The DNA quantities shown in *column three* are required for Illumina NGS sequencing of the mtDNA. Divide the required quantity of DNA by the Nanodrop readings to obtain the desired ratio for subsequent amplicon sonication

Table 4
Covaris sonication parameters

Parameter	Setting
Duty cycle	10 %
Intensity	5
Cycle/burst	200
Time	60 s
Cycles	2
Temperature	9 °C

Parameters required to sonicate mtDNA amplicons to a range of 200–1,000 bp

3.5 Amplicon Sonication

Set up the Covaris S2 at least 30 min prior to starting the sonication to allow for time to degas the instrument and the chiller to equilibrate to 9 °C. This protocol has been optimized to produce dsDNA lengths between 200 and 1,000 bp.

1. Load the 10 μg of equimolarly pooled amplicons DNA volume up to 105 μL to a Covaris microTUBE.

2. Place the microTUBE into the sonication horn following manufacturer's instructions.

3. Sonicate the pooled amplicons in the Covaris S2 using the settings found in Table 4.

3.6 Sonication Quality Control

Examine the sonicated DNA by 1 % agarose TBE gel or an Agilent Bioanalyzer to determine the quality of the shearing (*see* **Note 7**).

1. 1 % TBE agarose gel: Add 3 μL of 1 Kb + DNA ladder prepared following manufacturer's directions into a single well on a 1 % TBE agarose gel

2. Add 5 μL of equimolarly pooled, sonicated DNA mixed with loading buffer and pipette into unique wells in the 1 % TBE agarose gel.

   Run gel at 100 V to discriminate amplicon length. The length of time will depend on gel size and well orientation.

3. Optional → Agilent BioAnalyzer or LabChip GX: follow manufacturer's instructions loading 1 μL of equimolarly pooled, sonicated DNA.

4. A representative pseudogel picture of the equimolarly pooled, sonicated DNA is shown in Fig. 3.

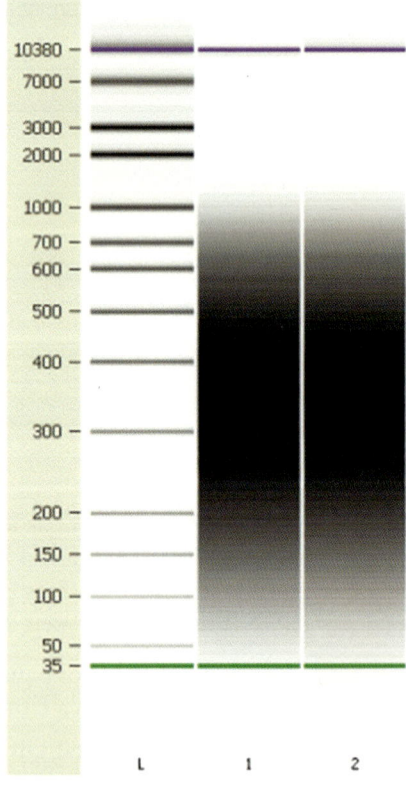

Fig. 3 Pseudogel picture of sonicated amplicons. A representative pseudogel picture of the equimolarly pooled and sonicated DNA amplicons. *Lane L*: ladder with sizes in bp, *lanes 2 and 3*: two different equimolarly pooled samples post-sonication. The majority of the sheared DNA should be between 200 and 1,000 bp without any un-sonicated amplicon detected

3.7 Next-Generation Sequencing

Illumina TruSeq library preparation, real-time PCR library quantification, loading/running the Illumina sequencing platforms, and conversion of raw data to fastq files are beyond the scope of this protocol. However some caveats unique to this experiment should be noted to the scientists who pool samples and run the sequencer.

1. The mitochondrial genome is very small (16.5 Kb) and requires very little library to obtain adequate sequencing depth.

2. Reports have shown that 400-fold coverage is adequate to detect heteroplasmy levels lower than 10 %; however the greater the sequence depth, the lower the background error rate if very low levels of heteroplasmy wish to be detected.

3. We have shown empirically that 1,000-fold average coverage provides reproducible results when detecting >10 % heteroplasmy levels (assuming that all amplicons are equimolarly pooled).

4. When pooling mtDNA samples, add enough library to obtain an estimated sequence output of 82.5 Mb per mtDNA. This number represents 5,000-fold coverage, which generally allows for real-time PCR or pipetting variability, and PCR duplicate removal.

3.8 Bioinformatic Analysis

A directory, mtDNA_Informatics, may be downloaded (extras. springer.com) that contains the required program and reference files, a perl script for alignment and variant calls, as well as two examples at http://topaz.genetics.utah.edu/mtDNA_Informatics/. Inside the directory are instructions on how to run the perl script through a UNIX/LINUX terminal. The bioinformatic analysis pipeline is a basic version designed for low-level heteroplasmy detection and variant reporting. The output will provide a .bam file for visual analysis, a raw .vcf file, and a basic .txt file (*see* **Note 8**). The output of the basic text file includes positional information for variants, heteroplasmy levels, coverage, and strand bias. The .txt file can be opened in a spreadsheet program, such as Excel, and variants can be manually checked on a website such as MitoMap (www.mitomap. org) or the Mitochondrial Disease Sequence Data Resource Consortium (MSeqDR, mseqdr.org). These basic files can be further annotated using different programs to provide information such as gene symbols, dbSNP rs numbers, m. (mitochondrial chromosomal position) and p. (gene-specific protein nomenclature based on position) numbers etc... Note that when annotating coding genes, a different amino acid table is used for mtDNA—Vertebrate Mitochondrial Code (transl_table=2, *see* **Note 9**).

4   Notes

1. Very low heteroplasmy levels can be detected using this protocol as described in the introduction. Clinically all variants require a secondary detection method such as variant-specific PCR, digital PCR, or cloning to confirm.

2. Many vendors provide an ABI 9700 emulation mode that can be used during amplification. If final PCR products appear suboptimal on a different thermocycler, the first modification to the PCR cycling conditions should be to limit the temperature ramp rate to a maximum of 5.0 °C/s.

3. Any DNA extraction technique that does not inhibit PCR amplification should be appropriate. Techniques that might linearize mtDNA should be avoided.

4. The protocol was optimized for 10–100 ng of genomic DNA input. However, due to the high copy number of mtDNA in a given genomic DNA sample, lower DNA input will yield acceptable results. Adequate PCR results have been obtained from 200 sorted myeloid cells.

5. When possible, *always* resuspend any stock oligonucleotide at a 100 µM concentration—it will make downstream calculations for experiments easier. For example, in order to make the Mito 1 primer working stock, add 20 µL of Mito1 F add 20 µL of Mito1 R to 160 µL of molecular grade water. The 10 µM working stocks can be kept at –20 °C up to 7 days.

6. Alternatively, if exact equimolar ratios are not required, amplicons can be pooled into a single PCR cleanup reaction. A rough estimate of pre-PCR pooling can be determined by spectrophotometer readings and mixing based on weight (*see* Table 3).

7. If there is un-sheared, high-molecular-weight DNA still observed re-sonicate the sample for an additional cycle. The desired size range of the band will depend on end user's library preparation requirements. When preparing libraries we perform a size selection between 300 and 600 base pairs.

8. General file type definitions. Some of these file types have specific information included based on sequencing platforms or programs used to create the file types.

Extension	Function
fastq	A text file containing nucleotide sequences and quality scores, as well as cluster, positional, lane, index, and read pair information. Use for .sai generation during bwa alignment
sai	Creates suffix array coordinates for the reads from the fastq file and contains positional and quality information

Extension	Function
bam	A compressed, binary version of a sam file generated during samtools view
sam	A sequence alignment/map tab-delimited text file. These files contain multiple predefined pragmas including quarry template name, sequence, quality, cigar length, as well as others [13]. Data from the different tags associated with a given sequence read are used during vcf generation
bai	Indexed bam file used for fast random access
vcf	Variant call format (1000 Genomes [14]). This file is human-readable and contains predefined and user-defined information. The required headers include #CHROM (chromosome), POS (chromosomal position), ID (unique identifier such as dbSNP rs number of other user-defined ID), REF (reference allele), ALT (alternate allele), QUAL (numeric quality score of variant), FILTER (does variant pass quality metrics), and INFO (user selectable pragmas that are used to provide information specified by the user from the sam file or other input file type, depth of coverage is included in our INFO output as DP4)
bcf	A compressed, binary version of the vcf file generated from samtools mpileup and bcf tools view
.gz	Compressed or "zipped" file

9. If further annotation of the .vcf files is required, the addition of read groups will be required. Read groups can be added during the bwa sampe step or at different steps using programs not discussed. Generic read groups are included in the downloadable perl script.

Acknowledgements

The authors would like to thank Barry Moore, Rebecca Margraf, and Jacob Durtschi for help in testing the command line script and setting up the server link for the mtDNA Informatics directory.

References

1. Dames S et al (2013) The development of next-generation sequencing assays for the mitochondrial genome and 108 nuclear genes associated with mitochondrial disorders. J Mol Diagn 15(4):526–534

2. Wong LJ (2012) Mitochondrial syndromes with leukoencephalopathies. Semin Neurol 32(1):55–61

3. Castle JC et al (2010) DNA copy number, including telomeres and mitochondria, assayed using next-generation sequencing. BMC Genomics 11:244

4. Li M et al (2012) Fidelity of capture-enrichment for mtDNA genome sequencing:

   influence of NUMTs. Nucleic Acids Res 40(18):e137

5. Payne BA et al (2013) Universal heteroplasmy of human mitochondrial DNA. Hum Mol Genet 22(2):384–390

6. Ballana E et al (2008) Detection of unrecognized low-level mtDNA heteroplasmy may explain the variable phenotypic expressivity of apparently homoplasmic mtDNA mutations. Hum Mutat 29(2):248–257

7. Voelkerding KV, Dames SA, Durtschi JD (2009) Next-generation sequencing: from basic research to diagnostics. Clin Chem 55(4):641–658

8. Margraf RL et al (2011) Variant identification in multi-sample pools by illumina genome analyzer sequencing. J Biomol Tech 22(2): 74–84

9. Danecek P et al (2011) The variant call format and VCFtools. Bioinformatics 27(15): 2156–2158

10. Li H, Durbin R (2009) Fast and accurate short read alignment with Burrows-Wheeler transform. Bioinformatics 25(14):1754–1760

11. Li H et al (2009) The Sequence Alignment/Map format and SAMtools. Bioinformatics 25(16):2078–2079

12. Robinson JT et al (2011) Integrative genomics viewer. Nat Biotechnol 29(1):24–26

13. http://samtools.github.io/hts-specs/SAMv1.pdf

14. http://www.1000genomes.org/wiki/Analysis/Variant%20Call%20Format/vcf-variant-call-format-version-41

Chapter 9

Rapid Mitochondrial DNA Isolation Method for Direct Sequencing

Wilber Quispe-Tintaya, Ryan R. White, Vasily N. Popov, Jan Vijg, and Alexander Y. Maslov

Abstract

Standard methods for mitochondrial DNA (mtDNA) extraction do not provide the level of enrichment for mtDNA sufficient for direct sequencing and must be followed by long-range-PCR amplification, which can bias the sequence results. Here, we describe a reliable method for the preparation of mtDNA-enriched samples from eukaryotic cells ready for direct sequencing. This protocol utilizes a conventional miniprep kit, in conjunction with a paramagnetic bead-based purification step.

Key words Next-generation sequencing, Mitochondrial DNA, Paramagnetic beads, Solid-phase reversible immobilization, Sequencing libraries

1 Introduction

In mammalian cells, mitochondria are often present in thousands of copies, depending on the cell type. Mitochondrial genomes lack histone protection and reside in close proximity to reactive oxygen species; coupled with the limited fidelity of mtDNA replication and repair machineries, ensure a much higher mutation rate in the mitochondrial genome than in the nuclear genome [1, 2], leading to heterogeneity within the mtDNA population [3, 4]. However, any deleterious effects of random mutations in mtDNA are compensated by the presence of multiple mitochondria in each cell. This decreased selection pressure allows mutated mtDNA to accumulate over time, making mtDNA a powerful indicator of detrimental effects of endogenous and environmental damaging agents, as well as overall somatic deterioration. It is also known that inherited mutations in the mtDNA can cause human diseases or mitochondrial disorders such as maternally inherited diabetes and deafness [5, 6], mitochondrial myopathy [7], and even accelerated aging [8]. Currently, next-generation sequencing

Volkmar Weissig and Marvin Edeas (eds.), *Mitochondrial Medicine: Volume I, Probing Mitochondrial Function*, Methods in Molecular Biology, vol. 1264, DOI 10.1007/978-1-4939-2257-4_9, © Springer Science+Business Media New York 2015

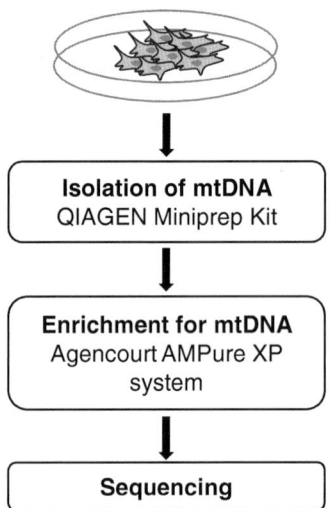

Fig. 1 Workflow for mtDNA isolation

(NGS) approaches are widely used for analysis of mtDNA [9, 10]. However, despite the presence of multiple mitochondrial genomes in each cell, mtDNA only comprises a small portion of total cellular DNA, thus making it necessary to enrich samples for mtDNA before sequencing. Current methods for enrichment either require special equipment (ultracentrifugation in CsCl density gradient), application of relatively expensive kits, or PCR amplification of mtDNA from total cellular DNA. This last and most commonly used method is relatively cheap and efficient, but may lead to artifacts as high number of PCR amplification cycles are often needed for sufficient enrichment. This can lead to mis-interpretation of results and, ultimately, incorrect conclusions.

To overcome these limitations, we have designed a fast, cost-effective, and reliable method for preparation of samples highly enriched for mtDNA (Fig. 1a). This method includes two steps— (1) isolation of total cellular DNA enriched for supercoiled mtDNA using a conventional bacterial miniprep kit followed by (2) an additional purification using solid-phase reversible immobilization on paramagnetic beads.

2   Materials

2.1   Total Cellular DNA Enriched for Supercoiled Mitochondrial DNA Extraction Using the Qiagen QIAprep Spin Miniprep Kit

1. Phosphate-buffered saline 1×, pH 7.4. Store at 4 °C.
2. Cell lifter.
3. Hemocytometer
4. Benchtop centrifuge.
5. QIAprep Spin Miniprep Kit (Qiagen, Germantown, MD, USA) (*see* **Notes 1–5**).

6. Tabletop microcentrifuge.

7. Sterile, nuclease-free microcentrifuge tubes and pipet tips.

2.2 Mitochondrial DNA Purification Using the Agencourt AMPure XP System

1. Agencourt AMPure XP purification system (Beckman Coulter, Brea, CA, USA) (*see* **Note 6**).

2. The DynaMag™-2 magnet (Life technologies, Grand Island, NY, USA).

3. DNA LoBind tube, 1.5 ml (Eppendorf, Hauppauge, NY, USA) (*see* **Note 7**).

4. Nuclease-free water.

5. 70 % ethanol (v/v).

6. Vortex mixer.

7. TE buffer 1×: 10 mM Tris–HCl, 1 mM EDTA, pH 8.0.

3 Methods

Culture cells until they become confluent. As with any biological samples, care should be taken in handling the material. All extraction and purification steps should be carried out at room temperature (15–25 °C), unless otherwise specified.

3.1 Total Cellular DNA Enriched for Supercoiled Mitochondrial DNA Extraction

1. Collect the cultured cells into 10 ml of PBS using cell lifters (*see* **Note 8**).

2. Centrifuge at $500 \times g$ for 5 min at room temperature to pellet the cells.

3. Carefully remove supernatant and resuspend pellet in 5–10 ml of PBS. Keep the cells on ice (*see* **Note 9**).

4. Use a cell counter chamber to calculate the cell concentration and use a maximum of 17 million cells for each extraction. Transfer the proper amount of cells to a microcentrifuge tube (*see* **Note 2**).

5. Centrifuge at $16,000 \times g$ for 1 min at room temperature in a tabletop microcentrifuge.

6. Remove supernatant and resuspend the pellet in 300 μl buffer P1 containing RNase A. Mix well by pipetting up and down until no cell clumps remain (*see* **Note 10**).

7. Add 300 μl buffer P2 and mix gently by inverting the tube 4–6 times (*see* **Note 11**).

8. Add 420 μl buffer N3 and mix immediately by inverting the tube 4–6 times (*see* **Note 12**).

9. Centrifuge at 13,000 rpm for 10 min.

10. Transfer the supernatant to the QIAprep spin column placed in a 2 ml collection tube (*see* **Note 13**).

11. Centrifuge at 13,000 rpm for 30–60 s and discard flow-through.

12. Add 500 μl wash buffer PB to the column and centrifuge at 13,000 rpm for 30–60 s (*see* **Note 14**). Discard the flow-through.

13. Add 750 μl wash buffer PE to the column and centrifuge at 13,000 rpm for 30–60 s (*see* **Note 15**). Discard the flow-through.

14. Centrifuge at top speed for 1 min to remove residual wash buffer.

15. Place the QIAprep spin column into a new LoBind 1.5 ml microcentrifuge tube.

16. To elute the DNA, add 100 μl buffer EB directly to the membrane and allow it to sit for 1 min, and then centrifuge at 13,000 rpm for 1 min (*see* **Note 16**).

17. Place the tube containing eluted DNA on ice.

18. Check DNA concentration using Qubit or NanoDrop (*see* **Note 2**).

19. Proceed to DNA purification steps.

3.2 Mitochondrial DNA Purification

1. For each sample, prepare a LoBind 1.5 ml microcentrifuge tube, transfer the eluted DNA, and add nuclease-free water to a volume of 50 μl when needed (*see* **Note 17**).

2. Mix the Agencourt AMPure XP bottle well before use (*see* **Note 18**).

3. Add the Agencourt AMPure XP paramagnetic beads in a 0.4× proportion (v/v) to the sample (*see* **Note 19**).

4. Mix thoroughly by quickly vortexing 5–10 times.

5. Incubate at room temperature for 5 min to allow time for the DNA to bind to beads.

6. Place the sample in a magnetic stand for 5 min to separate beads from the solution (optional: leave tubes open during this step to ensure the pellet is undisturbed for the next step).

7. While still in the magnetic stand, remove the supernatant without disturbing the beads.

8. Add 500 μl of freshly prepared 70 % ethanol, close the tube lids, and wash the beads by rotating the tube 180° back and forth (*see* **Note 20**).

9. Remove the supernatant and repeat the previous step for a total of two washes.

10. Remove the supernatant, quick spin for 5 s, and then place the tube back into the magnetic stand. Remove all excess ethanol.

Allow the open tube to dry for 5 min at room temperature (*see* **Note 21**).

11. Add 25 μl of 0.1× TE to elute the mtDNA, remove the sample from the magnetic stand, and mix by vortexing 5–10 times.

12. Apply a short spin of 10–15 s and place the sample back in the magnetic stand for 5 min.

13. Without disturbing the beads, transfer the cleared solution containing the mtDNA to a new LoBind 1.5 ml microcentrifuge tube.

14. Check the mtDNA for concentration and quality (*see* **Note 22**).

15. Store the enriched mtDNA at –20 °C (*see* **Note 23**).

4   Notes

1. The QIAprep column contains a silica membrane that adsorbs DNA in the presence of high salt. We assumed, as have others [11, 12], that since mtDNA properties are similar to those of bacterial DNA (i.e., it is supercoiled and its size is in the range of conventional plasmids), a common miniprep kit can be applied for extraction and enrichment of mtDNA from eukaryotic cells.

2. Based on our results using mouse embryonic fibroblasts (MEFs) and human dermal fibroblasts (HDFs), we recommend using no more than 17×10^6 cells per QIAprep column. Overloading the column will significantly increase the presence of total cellular DNA with your mtDNA. You are expected to collect ~100 ng of total cellular DNA enriched for mtDNA.

3. Add RNase A to buffer P1 for a final concentration of 100 μg/ml. Mix and store at 2–8 °C.

4. Add four volumes of ethanol (96–100 %) to concentrated buffer PE, as indicated in the bottle.

5. Buffers P2, N3, and PB contain irritants. Take appropriate laboratory safety measures and wear gloves when handling.

6. This system is based on solid-phase reversible immobilization (SPRI) beads. The beads are in a mixture of polyethylene glycol (PEG) and salt. PEG causes the negatively charged DNA to bind with the carboxyl groups on the bead surface. As the immobilization is dependent on the concentration of PEG and salt in the reaction, the volumetric ratio of beads to DNA is critical in size-selection protocols.

7. We are interested to recover as much of mtDNA as possible. The DNA LoBind Tube provides significant less sample loss of genetic material caused by interaction with the plastic surface of the tube.

8. Using cell lifters to detach and collect cells can avoid the probability of cell damage as a result of long periods of trypsin application.

9. Always keep the cells on ice until the DNA extraction process starts. You want to slow the cellular process that rises on cell death.

10. Buffer P1 is a resuspension buffer. It includes RNase A and should be stored at 2–8 °C.

11. Buffer P2 is an alkaline lysis buffer and should be checked for salt precipitation before use. Redissolve any precipitate by warming to 37 °C.

12. Buffer N3 performs the neutralization step in this alkaline lysis procedure. This buffer contains chaotropic agents (guanidine hydrochloride, potassium acetate) that set up binding conditions for the QIAprep Miniprep column's silica-gel membrane.

13. The QIAprep columns use a silica-gel membrane for selective adsorption of DNA in high-salt buffer and elution in low-salt buffer.

14. Buffer PB removes efficiently endonucleases and enables an efficient binding to the silica-gel membrane. This buffer contains guanidine hydrochloride and isopropanol.

15. Salts are efficiently removed by a brief wash step with buffer PE.

16. Before use, we heat the buffer EB to 65 °C to increase elution efficiency.

17. As mentioned before (*see* **Note 6**), the concentration of PEG and salt in the solution is critical in size-selection protocols so it can help to increase the volume of DNA you are working with by adding H_2O to make the pipetting easier.

18. This will help to resuspend any magnetic particles that may have settled. The reagent should appear homogenous and consistent in color.

19. The lower the ratio of beads/PEG/salt/DNA, the larger DNA fragments will be at elution (*see* **Note 6**).

20. Be sure to remove all of the ethanol from the bottom of the tube as it is a known PCR inhibitor or can affect other downstream applications.

21. Take care not to over-dry the beads, as this will significantly decrease elution efficiency.

22. The enriched mtDNA is ready for direct sequence analysis. Also, an optional limited PCR amplification can be applied if necessary; *see* ref. 13.

23. For long-term freezer storage, transfer the purified samples to new LoBind microcentrifuge tubes.

Acknowledgment

This research was supported by NIH grant P01 AG017242 (JV) and by the Ministry of Education and Science of Russian Federation grant 14.B37.21.1966 (VNP and AYM).

References

1. Brown WM, George M Jr, Wilson AC (1979) Rapid evolution of animal mitochondrial DNA. Proc Natl Acad Sci U S A 76(4):1967–1971

2. Baklouti-Gargouri S, Ghorbel M, Ben MA, Mkaouar-Rebai E, Cherif M, Chakroun N, Sellami A, Fakhfakh F, Ammar-Keskes L (2013) Mitochondrial DNA mutations and polymorphisms in asthenospermic infertile men. Mol Biol Rep 40(8):4705–4712

3. Greaves LC, Reeve AK, Taylor RW, Turnbull DM (2012) Mitochondrial DNA and disease. J Pathol 226(2):274–286

4. Larsson NG, Clayton DA (1995) Molecular genetic aspects of human mitochondrial disorders. Annu Rev Genet 29:151–178

5. Mezghani N, Mnif M, Mkaouar-Rebai E, Kallel N, Charfi N, Abid M, Fakhfakh F (2013) A maternally inherited diabetes and deafness patient with the 12S rRNA m.1555A>G and the ND1 m.3308 T>C mutations associated with multiple mitochondrial deletions. Biochem Biophys Res Commun 431(4):670–674

6. Van Den Ouweland JM, Lemkes HH, Trembath RC, Ross R, Velho G, Cohen D, Froguel P, Maassen JA (1994) Maternally inherited diabetes and deafness is a distinct subtype of diabetes and associates with a single point mutation in the mitochondrial tRNA(Leu(UUR)) gene. Diabetes 43(6):746–751

7. Bindoff LA, Engelsen BA (2012) Mitochondrial diseases and epilepsy. Epilepsia 53 Suppl 4:92–97

8. Khrapko K, Vijg J (2009) Mitochondrial DNA mutations and aging: devils in the details? Trends Genet 25(2):91–98

9. Ameur A, Stewart JB, Freyer C, Hagstrom E, Ingman M, Larsson NG, Gyllensten U (2011) Ultra-deep sequencing of mouse mitochondrial DNA: mutational patterns and their origins. PLoS Genet 7(3):e1002028

10. Mondal R, Ghosh SK (2013) Accumulation of mutations over the complete mitochondrial genome in tobacco-related oral cancer from northeast India. Mitochondrial DNA 24(4):432–439

11. Defontaine A, Lecocq FM, Hallet JN (1991) A rapid miniprep method for the preparation of yeast mitochondrial DNA. Nucleic Acids Res 19(1):185

12. Peloquin JJ, Bird DM, Platzer EG (1993) Rapid miniprep isolation of mitochondrial DNA from metacestodes, and free-living and parasitic nematodes. J Parasitol 79(6):964–967

13. Quispe-Tintaya W, White RR, Popov VN, Vijg J, Maslov AY (2013) Fast mitochondrial DNA isolation from mammalian cells for next-generation sequencing. Biotechniques 55(3):133–136

Chapter 10

Analysis of Mitochondrial DNA and RNA Integrity by a Real-Time qPCR-Based Method

Wei Wang*, Ying Esbensen*, Katja Scheffler, and Lars Eide

Abstract

This chapter describes the use of real-time qPCR to analyze the integrity of mitochondrial nucleic acids quantitatively. The method has low material requirement, is low cost, and can detect modifications with high resolution. The method is specifically designed for mitochondrial RNA and DNA, but can be easily transferred to other high-copy number cases. This procedure describes analyses of brain nucleic acids, but other tissues or cells can be analyzed similarly.

Key words mtDNA mutagenesis, RNA quality, Nucleic acid integrity

1 Introduction

Stability of DNA and RNA is essential for maintaining cellular function, and accumulation of somatic mutations is the underlying cause of cancer as well as other age-associated disorders. Maintenance of DNA is important to preserve genome stability and secure proliferation. Although RNA represents the secondary level, new data show that the stability and quality of RNA has impacts on many cellular conditions, including metabolism, neurological functions, and longevity. Accurate measurement of alterations on nucleic acids is therefore of interest for many disciplines.

Mitochondria are involved in many cellular functions; mitochondria proliferate independently of the host cell and continue to synthesize mitochondrial DNA (mtDNA) and RNA (mtRNA) in postmitotic cells, like neurons. There is an apparent heterogeneity to the stability of mtDNA and mtRNA. While the stability of mtDNA is expected to be higher than mtRNA, in vivo studies have demonstrated the stability of the small ribosomal mitochondrial RNA (12S) to be in the order of months [1]. Mitochondrial dysfunction is

---

*Authors contributed equally.

Volkmar Weissig and Marvin Edeas (eds.), *Mitochondrial Medicine: Volume I, Probing Mitochondrial Function*,
Methods in Molecular Biology, vol. 1264, DOI 10.1007/978-1-4939-2257-4_10, © Springer Science+Business Media New York 2015

associated with human diseases, as most diseases are caused by bioenergetic compromise and mtDNA damage and mutations have been correlated with disease and aging. The mtDNA exists as multi-copy plasmid in the cell, and mtDNA mutagenesis may result in mitochondrial disease, dependent on the heteroplasmic state (fraction of mutated mtDNA molecules in the cell). Because of the multicopy situation of mtDNA as well as mtRNA, assessment of the integrity requires capturing of sufficient amount of molecules determined by the site-specific error frequency. Traditional cloning and sequencing strategies are less useful to determine site-specific mutation frequency and are now being replaced by deep-sequencing approaches. However, the resolution of these techniques is insufficient to match the mutation frequency of mtDNA, which has been found to be lower than 1×10^{-6} per nt, as determined by the random mutation capture method (RMC) [2].

The cited RMC method is technically challenging, and we previously described a modification of this method, where we used real-time qPCR to detect the number of templates prior to enzymatic digestion by TaqI to select mutated restriction enzymes sites. By including S1 nuclease in sufficient amounts to remove single-stranded DNA, we were able to obtain comparable mtDNA mutation frequencies as the RMC method [3]. Here, we include a similar strategy to assess errors in mtRNA, by analyzing the downstream cDNA produced by high-fidelity reverse transcriptases from total RNA. We provide quality control that our laboratory has established to validate the analyses. The concept of the assay is illustrated in Fig. 1.

2   Materials

1. NanoDrop spectrophotometer (Epoch, BioTek Instrument).
2. 7900HT Fast Real-Time PCR System.
3. Eppendorf Mastercycler ProS PCR.
4. RNAlater solution.
5. FastPrep®-24 Instrument.
6. RNA isolation kit.
7. DNA isolation kit.
8. rDNase I (2 units/μl).
9. 70 % ethanol.
10. 100 % ethanol.
11. Nuclease-free water.
12. Tungsten Carbide Beads, 3 mm.
13. Tissue dissection tools.

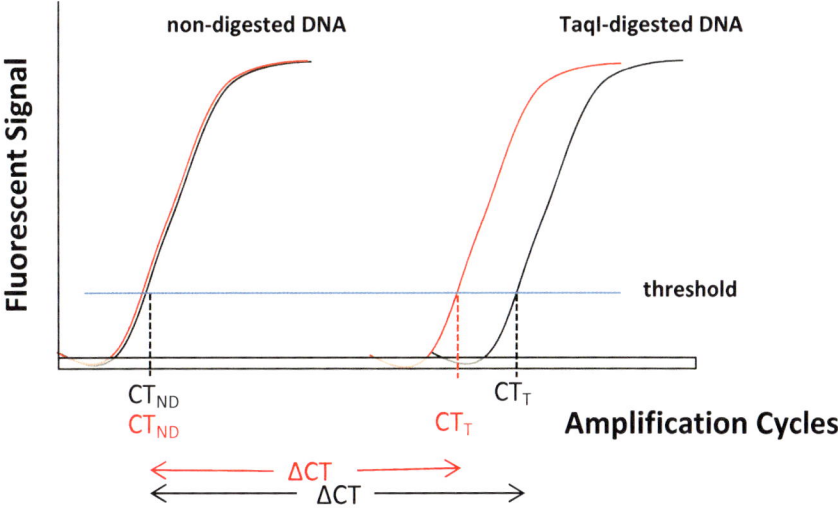

Fig. 1 Real-time qPCR quantification of TaqI site-mutated templates. The difference in CT values (ΔCT) obtained by amplification of non-digested (CT_{ND}) and TaqI-digested (CT_T) DNA represents the logarithmic ratio between mutated and non-mutated TaqI recognition sequences. DNA is either total DNA amplified by mtDNA-specific primers or cDNA prepared from reverse transcriptase synthesis of random primed RNA. The *black* and *red lines* indicate DNA with different mutation frequencies

14. S1 nuclease containing 10× S1 nuclease buffer and S1 nuclease dilution buffer (20,000 units, conc. 400–1,500 units/μl).

15. TaqI restriction enzyme (20,000 units/ml).

16. cDNA synthesis kit.

17. RNase H (1,000 units, 10 units/μl).

18. HotStart PCR Kit.

19. Power SYBR Green PCR mixture.

20. 0.2 ml thin-walled PCR 8-tube strip.

21. Primers used for analyzes of seven sites in the mtDNA is provided in Table 1.

22. TaqI reaction mixture: 5 μl 10× NEBuffer 3, 0.5 μl 100× BSA, 5 μl 20 units/μl TaqI, 34.5 μl H_2O.

3   Methods

3.1   mtDNA Mutation Frequency

3.1.1   Extraction of Total DNA

Total DNA is isolated from homogenized material (here, brain) according to the manufacturer's protocols. The DNA is quantified by NanoDrop analyses. For preparation of samples, for storage, and for stability, *see* **Notes 1** and **2**.

3.1.2   Removal of Interfering Single-Stranded DNA

1. Adjust template DNA concentration to 40 ng/μl.

2. S1 nuclease is diluted to 1 unit/μl prior to use.

Table 1
Oligonucleotides for quantification of mutation frequencies

Loci	Amplicon (nt)	Sequence (5′-3′)
12S	206	actcaaaggacttggcggta agcccatttcttcccatttc
Nd1	133	ttacttctgccagcctgacc cggctgcgtattctacgtta
Nd3	82	gcattctgactcccccaaat gacgtgcagagcttgtaggg
Nd5	111	tcagacccaaacatcaatcg cccttctcagccaatgaaaa
Nd6	154	aacaaccaaccaaaaaggctta gctgggtgatctttgtttgc
CoxI	117	ctgagcgggaatagtgggta aaagcatgggcagttacgat
Cytb	120	cagccttttcatcagtaacaca ctcgtccgacatgaaggaat

3. S1 nuclease treatment: 1 μl 10× S1 buffer + 2 μl S1 nuclease (1 unit/μl) + 2 μl H_2O + 5 μl total DNA (40 ng/μl), and incubate at 37 °C for 15 min.

After reaction, keep sample on ice (*see* **Note 3**).

3.1.3 TaqI Digestion and Real-Time qPCR Analysis

1. 5 μl reaction mix from Subheading 3.1.2 (100 ng of double-stranded DNA) is added to 45 μl of TaqI reaction mixture for TaqI digestion and incubated for 60 min at 65 °C (*see* **Note 4**).

2. Subsequent to TaqI digestion, make a 2 ng/μl dilution of the digested sample and a corresponding dilution in TaqI reaction mixture without TaqI.

3. 6 ng (3 μl) of either TaqI-digested or non-digested DNA is subjected to real-time qPCR analyses, using selected primers (Table 1).

4. To ensure complete digestion of non-mutated TaqI restriction sites, TaqI is added to qPCR reaction mixture (1 U/well) and an additional step of 65 °C for 15 min is included prior to the standard qPCR program (*see* **Note 5**).

5. The qPCR is run in a 7900HT Fast Real-Time PCR System (Applied Biosystems) using the Power SYBR® Green PCR Master Mix. 50 μl of TaqI reaction mixture is sufficient to screen ~5.4×10^7 copies of mtDNA molecules in a 96-well qPCR plate.

6. To confirm complete TaqI digestion of nonmutant molecules, 1 μl of TaqI is added to 10 μl of digested qPCR product in a new 0.2 ml tube and incubated at 65 °C for 10 min. The sensitivity to TaqI (=erroneously measured mutations) is analyzed by 2.5 % agarose gel electrophoresis (*see* **Note 6**).

3.1.4 Mutation Frequency Calculation, Verification of Results

1. The resulting ΔCT (CT value for digested DNA minus CT for non-digested DNA; *see* Fig. 1) is calculated (*see* **Note 7**).

2. The logarithmic ratio between mutated TaqI sites and template, non-digested DNA is transformed to linearity and divided by four in order to obtain mutation frequency per nt: Mutation frequency = $1/(2 \exp(\Delta CT))/4$.

3.2 mtRNA Error Frequency

3.2.1 Extraction of DNA-Free Total RNA

Total RNA is isolated from homogenized brain samples using the RNeasy Kit according to the manufacturer's protocols with modifications.

1. Prepare 600 μl RLT lysis buffer containing 6 μl β-mercaptoethanol in a 1.5 ml Eppendorf tube.

2. Take about 20–30 mg of homogenized brain tissue and add the RLT buffer.

3. Apply to FastPrep®-24 Instrument with a setting of 5 M/s for 10 s.

4. Briefly spin down the tube to erase bubbles.

5. Follow the protocol instruction to isolate RNA.

6. The eluted total RNA is treated with rDNase I (*see* **Note 8**): 5 μl 10× buffer + 1 μl rDNase I + 44 μl RNA, incubate at 37 °C for 30 min.

7. Add 5 μl DNase inhibitor, incubate at room temperature for 2 min, and spin at 10,000 rpm for 2 min. Transfer RNA to new Eppendorf tube.

8. Quantification of isolated total RNA by NanoDrop.

9. If applicable, check RNA quality by Bioanalyzer.

3.2.2 cDNA Synthesis and Amplification of Mitochondrial cDNA

The total RNA was reversely transcribed into first-strand cDNA using cDNA reverse transcription kit.

1. 500–1,000 ng total RNA is reverse-transcribed according to the manufacturer's protocols.

2. After the reverse transcription reaction, synthesized cDNA is treated with RNase H (10 units) at 37 °C for 20 min (*see* **Note 9**).

3. Adjust the cDNA concentration to 20 ng/μl with nuclease-free water.

4. Subsequently, in order to create double-stranded cDNA for TaqI restriction enzyme digestion, 100 ng of cDNA was amplified

with a HotStart PCR system in a total volume of 50 µl reaction mixture: 5 µl 10× HotStart PCR buffer + 1 µl 10 mM dNTPs + 1 µl 10 µM forward primer + 1 µl 10 µM reverse primer + 0.5 µl 5 units/µl Tag DNA polymerase + 36.5 µl H₂O + 5 µl 20 ng/µl cDNA.

5. The PCR reaction is run with: 95 °C for 10 min, then 95 °C for 15 s, and 60 °C for 1 min for 8–15 cycles (*see* **Note 10**).

6. Following PCR reaction, the products are diluted as tenfold dilutions.

7. From an appropriate dilution of the PCR products, analyze 3 µl in additional real-time qPCR with a standard curve (*see* **Note 11**), and calculate accurate amounts in each sample, to make a 20 ng/µl dilution of PCR product (*see* **Note 12**).

8. Proceed as in Subheading 3.1.3 to perform real-time qPCR analyses of mutated cDNA.

3.2.3 Evaluation of Reverse Transcription Error Rate

Reverse transcriptases are highly error-prone replicases because of absent proofreading activity [4]. In the present method, we have used High Capacity cDNA Reverse Transcription Kit for cDNA synthesis. MultiScribe™ Reverse Transcriptase is a recombinant Moloney murine leukemia virus (rMoMuLV) reverse transcriptase, and the error rate is 3×10^{-5} per nt, as provided by the manufacturer. By comparing with another commercially available cDNA synthesis kit (AccuScript High-Fidelity 1st-strand cDNA Synthesis Kit, in which a Moloney murine leukemia virus reverse transcriptase derivative is used in the combination of a proofreading 3′-5′ exonuclease [5]), we have obtained similar mtRNA error rates (Fig. 2).

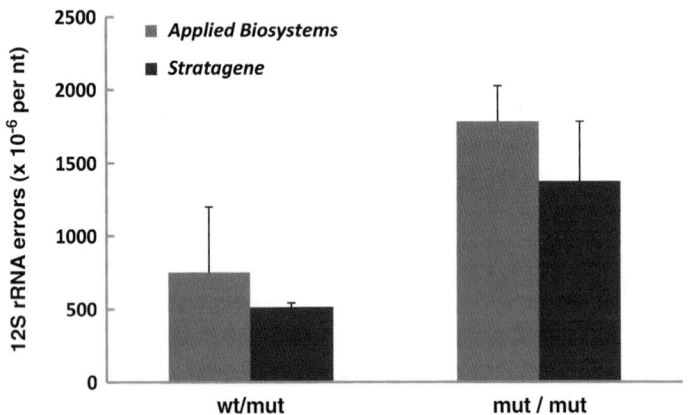

Fig. 2 Comparison of results obtained by two different high-fidelity reverse transcriptases. Total RNA was isolated from heterozygous (wt/mut) and homozygous (mut/mut) mutator mouse brains (kind gift from Nils-Göran Larsson, Cologne). cDNA was synthesized by reverse transcriptases from two providers and analyzed for errors in the TaqI site in 12S rRNA as described. Figure shows mean value with SD

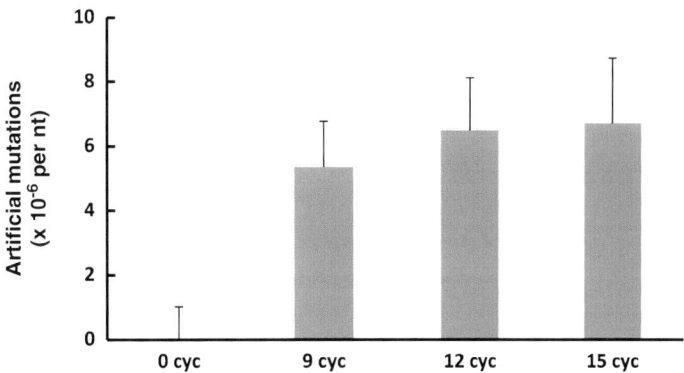

Fig. 3 Estimation of errors introduced by DNA polymerase during PCR amplification. The 12S rDNA region was amplified in a PCR reaction encompassing different cycle numbers. Subsequent to amplification, 100 ng DNA (PCR product) was analyzed for mutation frequency as described in Subheading 3.1.3, subtracted by the mutation frequency of the same DNA that had not been amplified. Data are means with SD from three independent experiments

3.2.4 Evaluation of Error Rates of DNA Polymerase

The fidelity of PCR DNA polymerase additionally influences the results. In order to evaluate the error rates introduced by DNA polymerase, the 12S rDNA region was amplified in PCR reaction with increasing cycles. The amplified products were subsequently analyzed for mutation frequency. In Fig. 3, the estimated values are plotted against the number of cycles. Zero cycles represent the values obtained by standard mutation quantification procedure described above (from Subheading 3.1.3) and are subtracted from the other values in order to present artificial mutations. In summary, the amplification step is associated with less than seven artificial mutations per million nt.

3.2.5 Calibration of the Method and Verification of the Results

To confirm the sensitivity and linearity of the method, a serial dilution of mixed intact and mutant DNA was used. Intact DNA is represented by 12S rDNA PCR product amplified from wt mouse brain as in Fig. 3, and mutant DNA is obtained by amplification of TaqI-resistant 12S rDNA fragment. The TaqI recognition sequence is present in the intact DNA but not in the mutated DNA (Fig. 4a). The intact and mutated PCR products were mixed in different ratios, and the subsequent mutation frequencies in the mixed samples determined and presented as ΔCT values (nonlinearized relation). Thus, the slope of the curve should ideally fit the natural logarithm of 10, or 2.30. The figure demonstrates that the resolution of the assay is close to a mutation frequency of 1×10^{-6} per nt for PCR-amplified products (Fig. 4b).

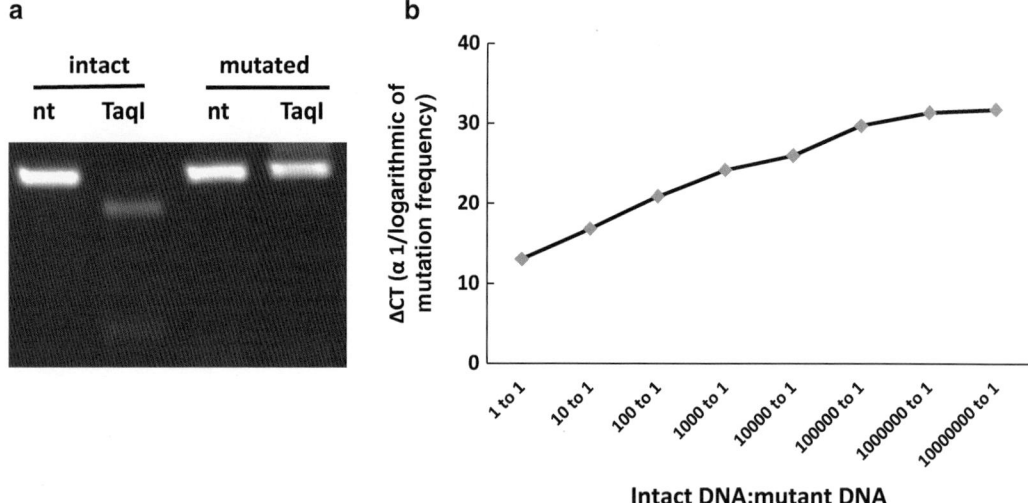

Fig. 4 Validation of method. (**a**) 12S rDNA PCR products amplified from intact DNA (from 1-month-old mouse brain) and mutant DNA (TaqI-resistant DNA from homozygous mutator mouse) were treated with 20 units of TaqI at 65 °C for 10 min and analyzed for TaqI digestion by gel electrophoresis. (**b**) Mixture of intact and mutant DNA was assessed by the mutagenesis method, and the resulting logarithmic difference (ΔCT) between the ratio of TaqI-sensitive and insensitive PCR products was presented as a function of the mutant fraction

4   Notes

1. Tissue stored in RNAlater can be used for DNA and RNA isolation. While the RNAlater can preserve RNA for up to 1 month at 4 °C, it is recommended to start nucleic acid analyses within 2 weeks to minimize the risk of RNA degradation. It is also possible to remove the RNAlater and store the samples in –80 °C for long-term storage. Alternatively, this long-term storage can be done after homogenization.

2. After homogenization, tissue is spread to the wall of the tube and dries out quickly. It is important to collect the sample to the bottom of the tube by spinning down immediately.

3. Isolated total DNA contains single-stranded mtDNA that arise during replication and transcription, which would escape digestion by the double-stranded requiring TaqI enzyme and cause artificial mutation signals during real-time qPCR. S1 nuclease treatment is added to remove this potential error source prior to restriction enzyme digestion.

4. While data sheet indicates 100 % activity of TaqI restriction enzyme in NEBuffer 3, we have found that TaqI enzyme is working equivalently well in Power SYBR Green PCR Mixture. Therefore, TaqI digestion can be performed in the

Power SYBR Green PCR Mixture by corresponding adjustment of reaction mixture volume. In this case, the lowest DNA concentration will be ~5 ng/µl for total 100 ng DNA.

5. Complete TaqI digestion is essential for accurate determination of mutant molecules. To ensure complete digestion, we have introduced double treatments with TaqI enzyme. After 100 ng of DNA is digested in 50 µl of TaqI reaction mixture, aliquot of 6 ng of DNA is digested again for 15 min within the qPCR mixture. This additional TaqI treatment in qPCR buffer significantly reduced the false-positive signal caused by potential inefficient cleavage of intact DNA molecules. In the current protocol, we routinely perform this two-round TaqI digestion, which results in a complete digestion of mouse brain total DNA and cDNA. However, determination of TaqI efficiency is required for different batches of enzyme purchased, different loci of mtDNA due to secondary structure, and different tissue samples such as the liver and heart.

6. The determination of absolute error frequency is dependent of complete TaqI digestion, which can be verified by either sequencing analysis or agarose gel electrophoresis of qPCR products as suggested by previous reports [6, 7].

7. CT values of non-digested DNA serve as loading control of DNA amounts, and CT values of digested DNA represent mutant molecules resistant to TaqI cutting.

8. DNA contamination in the RNA isolation would mask the readout of low error frequency where CT values of TaqI-digested samples reach 32–34. In order to eliminate this potential problem, we verified that double DNase treatment during RNA isolation efficiently removed the remaining DNA. Following two steps of on-column DNase treatment during RNA isolation and an additional rDNase I treatment, qPCR with RNA indicated that residual DNA has been completely removed with CT values reaching 36–39 or undetermined.

9. DNA polymerase has reverse transcriptase activity [8] and the remaining RNA existing in RNA:cDNA hybrids following cDNA synthesis would act as a template for Taq DNA polymerase, causing false-positive signals during PCR amplification. Treatment of cDNA with RNase H significantly removes unconverted RNA.

10. As in the HotStart PCR reaction, cDNA is amplified with gene-specific primers; individual genes need to be measured separately. The cycle number of the HotStart PCR needs to be adjusted based on transcriptional levels of each gene. For example, the ribosomal RNA (12S rRNA) has a high transcription level which requires less than eight cycles to obtain enough PCR products.

11. In order to obtain accurate amounts of amplified products for subsequent TaqI digestion, a qPCR with standard curve is required. Based on the exponential increase of the PCR reaction, DNA amounts are estimated and a serial 1:10 dilution of PCR products is made. An appropriate amount (normally at dilution of 1:1,000–10,000 depending on the transcriptional level of each gene) is selected and applied to qPCR quantification.

12. Based on standard curve, prepare 20 ng/µl of HotStart PCR products for TaqI digestion. If it is not possible for every samples, choose a lower dilution or adjust cDNA concentration to 10 ng/µl. It is essential to make accurate amounts of cDNA from the same dilution for each sample. The mutation frequency is calculated by the ΔCT of non-digested and TaqI-digested molecules. If the mutation frequency falls below 10 per million, the ΔCT would reach 15–18. To ensure that the CT value of TaqI-digested sample lies within the trustable range for qPCR, it is necessary to adjust the CT value of non-digested sample to 15–16. In our experience, 6 ng of non-digested DNA as template normally gives a CT value around 16. This CT value can also be used as loading control of DNA amounts.

Acknowledgment

The work has been supported by the Norwegian Research Council.

References

1. Metodiev MD, Lesko N, Park CB et al (2009) Methylation of 12S rRNA is necessary for in vivo stability of the small subunit of the mammalian mitochondrial ribosome. Cell Metab 9:386–397

2. Vermulst M, Bielas JH, Kujoth GC et al (2007) Mitochondrial point mutations do not limit the natural lifespan of mice. Nat Genet 39: 540–543

3. Halsne R, Esbensen Y, Wang W et al (2012) Lack of the DNA glycosylases MYH and OGG1 in the cancer prone double mutant mouse does not increase mitochondrial DNA mutagenesis. DNA Repair (Amst) 11:278–285

4. Roberts JD, Bebenek K, Kunkel TA (1998) The accuracy of reverse transcriptase from HIV-1. Science 242:1171–1173

5. Arezi B, Hogrefe HH (2007) Escherichia coli DNA polymerase III epsilon subunit increases Moloney murine leukemia virus reverse transcriptase fidelity and accuracy of RT-PCR procedures. Anal Biochem 360:84–91

6. Wright JH, Modjeski KL, Bielas JH et al (2011) A random mutation capture assay to detect genomic point mutations in mouse tissue. Nucleic Acids Res 39:e73

7. Vermulst M, Wanagat J, Kujoth GC et al (2008) DNA deletions and clonal mutations drive premature aging in mitochondrial mutator mice. Nat Genet 40:392–394

8. Martel F, Grundemann D, Schomig E (2002) A simple method for elimination of false positive results in RT-PCR. J Biochem Mol Biol 35:248–250

Chapter 11

Mitochondria-Targeted RNA Import

Geng Wang, Eriko Shimada, Mahta Nili, Carla M. Koehler, and Michael A. Teitell

Abstract

The import of a modest number of nucleus-encoded RNAs into mitochondria has been reported in species ranging from yeast to human. With the advent of high-throughput RNA sequencing, additional nucleus-encoded mitochondrial RNAs are being identified. Confirming the mitochondrial localization of candidate RNAs of interest (e.g., small noncoding RNAs, miRNAs, tRNAs, and possibly lncRNAs and viral RNAs) and understanding their function within the mitochondrion is assisted by in vitro and in vivo import assay systems. Here we describe these two systems for studying mitochondrial RNA import, processing, and functions.

Key words Mitochondria, RNA import, Polynucleotide phosphorylase (PNPASE)

1 Introduction

Mitochondria import a range of nucleus-encoded RNAs, with different species having different substrate specificities [1]. Some of these RNAs are processed within mitochondria and have functions different from their cytosolic or nuclear counterparts [2]. Studying nucleus-encoded mitochondrial RNAs emphasizes a few general approaches. First, RNA localization inside mitochondria can be confirmed using organelle fractionation after the substrate RNA is imported in vitro. Second, the processing or cleavage of imported RNAs can be studied using northern blotting or in vitro import of radiolabeled RNA substrates of known length. Specific RNA processing sites and the addition of nucleotides can be identified using RNA ligation, followed by semiquantitative RT-PCR and sequencing, as described previously [3]. Finally, the functions of imported RNAs can be studied using approaches that are specific to the RNA species of interest. In addition to mechanistic and functional studies, mitochondrial RNA import pathway(s)

Volkmar Weissig and Marvin Edeas (eds.), *Mitochondrial Medicine: Volume I, Probing Mitochondrial Function*,
Methods in Molecular Biology, vol. 1264, DOI 10.1007/978-1-4939-2257-4_11, © Springer Science+Business Media New York 2015

and signal sequence(s) can be co-opted to import RNAs of interest into mitochondria to rescue defects caused by mtDNA mutations or to change the mitochondrial genome expression profile [4, 5]. Here, we describe two mitochondrial RNA import systems for these studies. One system is an in vitro import assay that uses in vitro-transcribed RNA substrates and either isolated yeast or mammalian mitochondria. A second, more challenging, and currently less efficient and less well-understood system is an in vivo import assay that utilizes an RNA import signal sequence in exogenously expressed DNA to target nucleus-encoded RNAs for import into mitochondria. The RNA substrates for in vitro import can be either radiolabeled or unlabeled and detected using autoradiography or RT-PCR, respectively. For yeast mitochondria, exogenous expression of the mammalian RNA import protein, polynucleotide phosphorylase (PNPASE), generates a system with enhanced import efficiency for substrate RNAs [6]. The efficiency of the in vivo import system varies markedly depending on the RNA to be imported; unfortunately, the precise rules for efficient RNA import in vivo have not been fully elucidated. In vivo, pre-mitochondrial processing of target RNA in the nucleus and/or cytosol and trafficking of RNA to the mitochondrion needs to be carefully considered, as these factors seem to greatly affect RNA import efficiency [5].

2   Materials

All solutions should be prepared using RNase-free water and the reagents used should be analytical grade. Unless indicated otherwise, all reagents should be prepared and stored at room temperature.

2.1   In Vitro Mitochondrial RNA Import

2.1.1   In Vitro Transcription

1. MEGAscript® SP6 Kit from Ambion (catalog number: AM1330) is used for in vitro transcription. The kit contains an enzyme mix, 10× reaction buffer, and solutions of ATP, CTP, GTP, and UTP (see **Note 1**).

2. α-P^{32}-labeled CTP, 6,000 Ci/mmol, 10 mCi/mL (see **Note 2**).

2.1.2   RNA Isolation

1. RNase-free water. Store at 4 °C.

2. Trizol reagent (Invitrogen). Store at 4 °C.

3. Chloroform.

4. 75 % Ethanol (RNAse-free). Store at 4 °C.

5. Isopropyl alcohol.

2.1.3   In Vitro RNA Import

1. 2× Import buffer for yeast mitochondria: 1.2 M sorbitol, 100 mM KCl, 100 mM HEPES, 20 mM $MgCl_2$, pH 7.1. Store at −20 °C.

2. 2× Import buffer for mammalian mitochondria: 0.45 M mannitol, 0.15 M sucrose, 20 mM HEPES, 50 mM KCl, 10 mM $MgCl_2$, pH 7.4. Store at –20 °C (*see* **Note 3**).

3. 100 mM ATP. Store at –20 °C.

4. 100 mM DTT (prepare fresh).

5. 500 mM NADH. Store at –20 °C.

6. 500 mM sodium succinate (prepare fresh).

7. 10 mg/mL RNase A. Store at –20 °C.

8. 10 mg/mL proteinase K. Store at –20 °C.

9. SDS buffer: 1 % SDS, 100 mM NaCl, 10 mM Tris–HCl, pH 7.4.

2.1.4 Urea-Polyacrylamide Gel Electrophoresis (Urea-PAGE)

1. 5× Tris-borate-EDTA (TBE) buffer.

2. 40 % Acrylamide/bisacrylamide (29:1) solution. Store at 4 °C.

3. Ammonium persulfate: 10 % solution in water (prepare fresh).

4. N,N,N,N-tetramethyl-ethylenediamine (TEMED).

5. Formamide loading buffer: 95 % deionized formamide, 5 mM EDTA, 0.025 % (w/v) SDS, 0.025 % (w/v) bromophenol blue. Store at –20 °C (*see* **Note 4**).

2.2 In Vivo Mitochondrial RNA Import

1. 2× HEPES-buffered saline (HBS): 50 mM HEPES, 280 mM NaCl, 1.5 mM Na_2HP0_4, pH 7.1.

2. 1 M $CaCl_2$.

2.2.1 Transient Transfection

2.2.2 Mitochondria Isolation

1. Mitoprep buffer: 0.225 M mannitol, 0.075 M sucrose, 20 mM HEPES, pH 7.4. Store at 4 °C.

2. 0.5 M EDTA. Store at 4 °C.

3. 0.2 M PMSF. Store at –20 °C.

2.2.3 RNase A Treatment, RNA Isolation, and DNase Treatment

1. 10 mg/mL digitonin (prepare fresh).

2. 10 mg/mL RNase A. Store at –20 °C.

3. SDS buffer: 1 % SDS, 100 mM NaCl, 10 mM Tris–Cl, pH 7.4.

4. RNase-free water. Store at 4 °C.

5. Trizol reagent (Invitrogen). Store at 4 °C.

6. Chloroform.

7. 75 % RNase-free ethanol. Store at 4 °C.

8. Isopropyl alcohol.

9. DNase I, RNase-free (Thermo Scientific).

2.2.4 RT-PCR

1. One-Step RT-PCR Kit (e.g., Promega).

3 Methods

Carry out all procedures at room temperature unless otherwise specified.

3.1 In Vitro Mitochondrial RNA Import

3.1.1 In Vitro Transcription

1. Prepare a DNA template that contains the SP6 polymerase promoter site for in vitro transcription.

2. Perform in vitro transcription using the MEGAscript® Kit from Ambion (*see* **Note 5**). Mix all of the components at room temperature in a 0.5 mL Eppendorf tube. The following is an example of a 20 μL reaction (*see* **Note 6**):

ATP	1.5 μL
CTP	1.0 μL
GTP	1.5 μL
UTP	1.5 μL
10× Reaction buffer	2.0 μL
α-P^{32}-CTP	7.5 μL
Linear template DNA[a]	0.1–1.0 μg
Enzyme mix	2.0 μL
Nuclease-free water	to 20 μL

[a]Use 0.1–0.2 μg PCR-product template or ~1 μg linearized plasmid

3. Pipette the mixture up and down or flick the tube gently. Then centrifuge the tube briefly to collect the reaction mixture at the bottom of the tube. Incubate at 37 °C for 2–4 h. If the transcripts are less than 500 ribonucleotides (nt), a longer incubation time (up to 16 h) may be advantageous.

3.1.2 RNA Isolation

1. After incubation, add 400 μL Trizol reagent and 170 μL chloroform to the mixture and vortex for 1 min.

2. Centrifuge the sample at $12,400 \times g$ for 5 min at room temperature.

3. Transfer the upper aqueous phase into a fresh Eppendorf tube carefully without disturbing the interphase. Add 300 μL isopropanol to the aqueous solution and mix by pipetting or inverting the tube.

4. Centrifuge the sample at $20,000 \times g$ for 10 min at 4 °C (*see* **Note 7**).

5. Carefully remove the supernatant. Add 600 μL ice-cold 75 % ethanol to the tube and invert the tube ten times.

6. Centrifuge at $20,000 \times g$ for 2 min at 4 °C. Remove the supernatant (*see* **Note 8**).

Therefore, although there are some protocols for MtDNA quantification [19, 20], these do not meet the two main criteria of specificity and reproducibility as they fail to take account either the co-amplification of nuclear regions and/or the dilution effect.

In the current article, we focus on studies using peripheral blood as it is often easier to obtain and store small amounts of whole blood samples for clinical studies. It is not always practical to fractionate blood samples, both because of issues involved with collection of such samples and because of the volumes that may be needed. Peripheral blood is comprised of a mixture of cells, and variations in the number or the relative amounts of different sub-populations of blood cells might interfere with quantitative measures such as Mt/N. In addition, circulating cell-free MtDNA may also be present in serum or plasma, and in this case, the source of this MtDNA may not be clear. Nevertheless, many studies using peripheral blood have shown that disease-associated changes can be detected (reviewed in [4]). The method we describe here requires storage of a very small volume (10–100 µl) of blood for Mt/N assessment and as such could be easily used for existing or new clinical samples.

2   Materials

All solutions were prepared using ultrapure water and stored at room temperature unless stated otherwise.

2.1   Blood Collection and Storage

1. BD Vacutainer® Plus Blood Collection 3 ml tubes containing K_2EDTA.
2. 1.5 ml Eppendorf tubes.

2.2   DNA Extraction

1. DNeasy Blood & Tissue Extraction Kit (Qiagen, UK).
2. Phosphate-buffered saline (PBS), pH 7.2, 50 mM potassium phosphate, 150 mM sodium chloride.
3. Nuclease-free water.
4. Ethanol absolute, 200 proof molecular biology grade.
5. 1.5 ml Eppendorf tubes.

2.3   Dilution Standards Preparation

1. DreamTaq PCR Master Mix (Thermo Scientific).
2. 0.5× Tris-base boric acid-EDTA (TBE) buffer.
3. Agarose—electrophoresis grade.
4. Ethidium bromide.
5. 100 Base-Pair DNA Ladder (Thermo Scientific).
6. Blue/Orange 6 × loading dye (Thermo Scientific).
7. QIAquick Gel Extraction Kit (Qiagen).

8. Ethanol absolute, 200 proof molecular biology grade.

9. Isopropanol, 99.5 + %, extra pure, Acros Organics.

10. Transfer RNA (tRNA).

11. 1.5 ml Eppendorf tubes.

2.4 qPCR

1. Quantifast SYBR Master Mix (2×) (Qiagen, UK).

2. 1.5 ml tubes.

3. DNAse, RNase-free water (Thermo Scientific).

4. 96-Well PCR plates suitable for specific qPCR machine according to the manufacturer's instructions; here we used Roche 480 Light Cycler plates, white with sealing (Thermal Seal RT2 RR from Alpha Labs).

2.5 Equipment Used

1. Microcentrifuge (Biofuge, Heraeus).

2. Real-time PCR machine (Roche LC 480, Roche).

3. Thermal Cycler (LifePro Thermal Cycler, Bioer).

4. Bath sonicator (Kerry, Pulsatron 55).

5. Gel imaging system (Syngene Bioimaging system, Scientific Laboratories).

6. Spectrophotometer (NanoDrop ND-1000, Lab Tech International).

3 Methods

3.1 Collection and Storage of Blood Samples

1. Collect peripheral blood in 3 ml capacity BD Vacutainer® Plus Blood Collection Tubes and invert the tubes gently eight to ten times for thorough mixing of blood with the anticoagulant.

2. As soon as possible and preferably within 2–4 h, transfer the blood sample in 100 μl aliquots into sterile Eppendorf tubes and store at –80 °C (*see* **Note 1**).

3.2 Extraction of DNA from Blood Samples

1. To prepare genomic DNA from the whole blood sample, use a column-based method (Qiagen DNeasy Blood & Tissue Kit), and the instructions below are modified for this kit (*see* **Note 2**).

2. Thaw the blood samples on ice and mix by vortexing. Add 20 μl of proteinase K (>600 mAU/ml) and add 100 μl of PBS.

3. To lyse the cells, add 200 μl of lysis buffer (buffer AL, without added ethanol). Mix thoroughly by vortexing, and incubate at 56 °C for 10 min.

4. Add 200 μl ethanol (96–100 %), and mix thoroughly by vortexing.

5. Pipette the mixtures into the DNeasy Mini spin column placed in a 2 ml collection tube and follow the manufacturer's method for washing and drying the column.

6. To elute the DNA, add 50 μl of elution buffer (buffer AE) directly onto the DNeasy membrane. Incubate at room temperature for 1 min, and then centrifuge for 1 min at $5000 \times g$ and collect the eluent in a fresh tube. For maximum DNA yield, repeat elution once again in 50 μl buffer AE to make 100 μl of total volume.

7. You now have genomic DNA in a total volume of 100 μl. Use 1–2 μl to determine the concentration and quality of the DNA using the NanoDrop or equivalent instrument (*see* **Note 3**).

3.3 Pretreatment/ Fragmentation of DNA (See Note 4)

1. Transfer 30 μl of DNA into a clean 1.5 ml Eppendorf tube and place in a Bath sonicator (Kerry, Pulsatron 55) which uses $38 \text{ kHz} \pm 10 \%$ for 5–10 min.

2. If you do not have access to a sonicating bath, we recommend shearing the DNA by passing through a needle of 21 gauge, 0.8 mm diameter. For this take 30 μl of genomic DNA and add 70 μl of TE buffer, and then shear 100 μl of this diluted DNA (*see* **Note 5**).

3. After sonication, determine the concentration of DNA again and adjust it to between 1 and 10 ng/μl. It is a good idea to have all the different samples under study adjusted to the same concentration.

4. You can now check the samples by amplifying hMito and hB2M genes using PCR and specific primers given in Table 1. Electrophorese the products on a gel. You should obtain clear bands and any problems at this stage would suggest that you should not proceed to the qPCR until they are resolved. Store the samples at 4 °C and avoid freezing for the duration of the study. Prior to use, gently vortex samples and spin briefly.

3.4 Preparation of Standards for Absolute Quantification

1. Here we describe how to make dilution standards for absolute quantification. Alternatively, to use the relative quantification method, this section can be skipped to go straight to the real-time PCR method described in the next section.

Table 1
Oligonucleotide sequences for Mt/N determination using real-time qPCR (adapted from [15])

Primer name	Primer sequence 5′ → 3′	Amplicon bp	Accession no
hMitoF3	CACTTTCCACACAGACATCA	127	NC_012920
hMitoR3	TGGTTAGGCTGGTGTTAGGG		
hB2MF1	TGTTCCTGCTGGGTAGCTCT	187	NG_012920
hB2MR1	CCTCCATGATGCTGCTTACA		

2. Amplify the hMito and hB2M PCR products with the appropriate primers (*see* Table 1) using human DNA isolated from a control sample. Prepare the reaction mixture by adding 12.5 μl of DreamTaq PCR Master Mix, 0.5 μl of forward and reverse primer (10 μM), and 10.5 μl RNase-free H$_2$O. Add 1–2 μl of DNA template. Gently vortex the samples and spin down. Using a Thermal Cycler (such as LifePro Thermal Cycler, Bioer), perform PCR with initial denaturation at 94 °C for 15 min (1 cycle), followed by denaturation at 94 °C for 30 s, annealing at 60 °C for 30 s, and extension 72 °C for 1 min and 30 s (30 cycles). Final extension is at 72 °C for 7 min.

3. Check the PCR products by electrophoresing 10 μl of the reaction on a 2 % agarose gel. If they are of the expected size with no other artifacts present, then electrophorese 20 μl of the PCR products out, visualize briefly under UV light to identify, and excise the bands.

4. Purify the DNA from the excised band using QIAquick Gel Extraction Kit (QIAGEN) according to the manufacturer's instructions, and elute in 20–30 μl of elution buffer.

5. Determine the concentration of each PCR product using the NanoDrop or a UV spectrophotometer and confirm using a gel-based method if needed (*see* **Note 6**).

6. Calculate the copy number per 1 μl of the purified DNA as follows (*see* **Notes 7** and **8**):

   (a) For hMito: Copy number $= C \times 7.18 \times 10^9/\mu l$.

   (b) For hB2M: Copy number $= C \times 4.87 \times 10^9/\mu l$
       (where $C =$ concentration of DNA in ng/μl).

7. Dilute the PCR product with calculated copy numbers to prepare a stock solution of each of hMito and hB2M PCR products containing 1×10^9 copies per μl of the DNA (*see* **Note 9**).

8. You will now prepare dilution standards for each of the two PCR products in tenfold dilution for 2–8 log. Prepare 7 tubes labeled 1×10^2 to 1×10^8 each containing 90 μl of water containing 10 μg/ml transfer RNA (tRNA). Transfer a 10 μl aliquot of the stock solution (1×10^9 copy numbers) into the tube labeled 10^8. Discard tip. Mix the tube and then transfer a 10 μl aliquot from this tube into the 10^7 tube. Repeat until you reach 10^2 tube.

9. The dilution standards labeled 10^{2-8} should be stored at –20 °C (*see* **Note 10**).

3.5 Real-Time Quantitative PCR

1. In this section, the template DNA prepared in Subheading 3.2 and the dilution standards prepared in Subheading 3.4 (if using absolute quantification) are used for qPCR to determine the amount of MtDNA and nuclear DNA in the sample. Here we

use QuantiFast SYBR Green (Qiagen) but this can be substituted for an alternative real-time qPCR assay.

2. For each DNA sample under investigation, carry out the reaction in triplicate.

3. Thaw QuantiFast SYBR Green PCR Master Mix (2× concentrated), template DNA (10 ng/μl), standards (dilutions: 10^2–10^8), primers (10 μM), and RNase-free water on ice, and vortex and briefly spin down individual solution before preparing reaction mix.

4. Set up the reaction mix for each PCR assay containing 5 μl of QuantiFast SYBR Master Mix, 0.5 μl of forward and reverse primer (500 nM final concentration each), and 2 μl template DNA; make up the reaction volume to 10 μl with RNase-free water (*see* **Note 11**).

5. Carry out the qPCR reaction in a real-time qPCR machine using the following conditions: preincubation at 95 °C for 5 min (1 cycle); denaturation at 95 °C for 10 s; annealing and extension at 60 °C for 30 s (repeat denaturation and extension steps for 40 cycles); melting at 95 °C for 5 s, 65 °C for 60 s, and 95 °C continues (melt curve analysis –1 cycle); and, the last step, cooling at 40 °C for 30 s (*see* **Note 12**).

6. In order to obtain copy numbers for tested samples, set up the threshold above the noise and on the linear/exponential part of amplification curves. When running tenfold dilutions standards and after setting up the threshold (*see* **Note 13**), values for each standard will be plotted against their concentration, giving you also values for reaction efficiency (recommended co efficiency of reaction [R^2] is 1.0) (*see* Figs. 3a, c and 4a, c).

7. Based on the threshold, the software extrapolates values for each sample from standard curve and reports them in a table as copy numbers and Ct values (Ct stands for cycle threshold = number of cycles required to cross the threshold line) for each gene.

8. Check the specificity of the primers using melt curve analysis; you should have one clear peak for each PCR product (*see* Figs. 3b and 4b). You can also electrophorese the qPCR products on the gel to confirm the specificity and size (*see* Fig. 1a, b).

9. The resulting data can be analyzed using the software associated with the real-time qPCR instrument.

10. Export data, copy numbers (if doing absolute quantification), and Ct values to Excel spread sheets for analysis: For copy numbers, this should comprise of the 3 replicate values for hMito (sample hMito1 + hMito2 + hMito3) and hB2M (hB2M1 + hB2M2 + hB2M3) (*see* Table 2). Similarly for Ct,

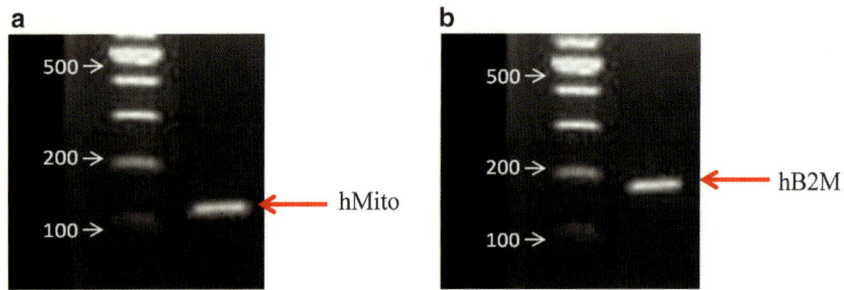

Fig. 1 Agarose gel electrophoresis of purified hMito and hB2M PCR products. The PCR products obtained from primers hMito F3and hMito R3 (**a**) and primers hB2M F1 and hB2M R1 (**b**) were gel purified using spin column and electrophoresed on a 2 % agarose gel to check that the purification had been successful. The 127 bph hMito and 189 bp hB2M PCR products are indicated by the *red arrows*. Ladder bands are marked with *white arrows* and labeled with appropriate molecular weight sizes

Table 2
Absolute copy number for mitochondrial (hMito1–hMito3) and hB2M (hB2M1–hB2M3) genes detected in ten different blood samples

Samples	hMito 1	hMito 2	hMito 3	Average hMito	hB2M1	hB2M2	hB2M3	Aver hB2M	Mt/N
1	62100	85800	83200	77033	7460	7490	7500	7483	10.3
2	73600	64000	63500	67033	6740	6270	5650	6220	10.8
3	101000	86900	102000	96633	5530	6180	6410	6040	16.0
4	326000	326000	326000	326000	2530	2070	2200	2267	143.8
5	148000	153000	179000	160000	30600	34200	38100	34300	4.7
6	67300	62500	66700	65500	3820	3360	3440	3540	18.5
7	423000	411000	408000	414000	8640	8230	8230	8367	49.5
8	314000	313000	325000	317333	10400	10300	8840	9847	32.2
9	836000	664000	702000	734000	19100	22900	22100	21367	34.4
10	249000	253000	252000	251333	18700	18600	15500	17600	14.3

Columns 2–4 represent replicates of mitochondrial gene copy number; columns 6–8 represent replicates of hB2M copy number. Columns 5 and 9 represent average copy number of mitochondrial and hB2M genes, respectively. Mitochondrial DNA content (MT/N ratio) is shown in column 10

3.6 How to Analyze Your Results and Check Specificity/Accuracy of the Experiment

enter the triplicate values for hMito and hB2M in separate columns (*see* Table 3).

1. For analysis using the absolute quantification method, calculate Mt/N by dividing the sum of hMito copy numbers (hMito1 + hMito2 + hMito3) by the sum of the hB2M copy numbers (hB2M1 + hB2M2 + hB2M3) (*see* **Note 14**).

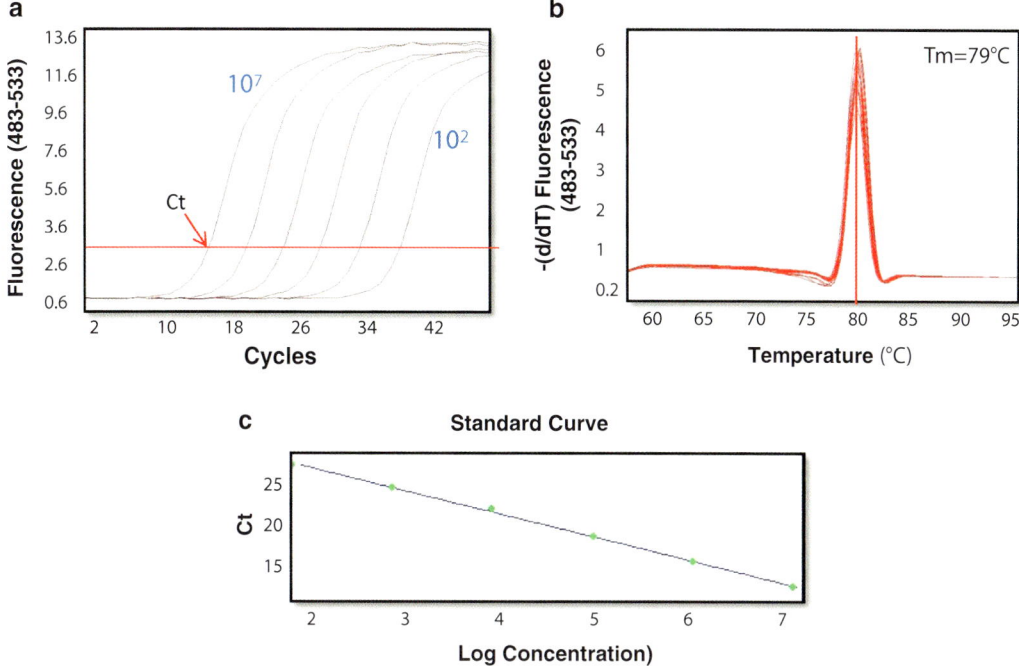

Fig. 4 Amplification of hB2M in tenfold dilutions and standard curve generation. Dilution standards of hB2M were prepared and 2 μl of each standard was used in qPCR reaction and amplified. Fluorescence data was acquired once per cycle, and amplification curve (**a**) was generated showing dilution series from 10^2 to 10^7 copies of hB2M. Melting point analysis (**b**) represents specificity of multiplied product as one single melt peak is visible at 79 °C. (**c**) Standard curve showing the crossing point (Ct) for each of the tenfold dilutions plotted against a log concentration. Threshold line is shown in *red*

Preparation of qPCR and Analysis of Results

11. Samples should be quantified in duplicate or triplicate in the presence of 5–6 appropriate dilution standards. To avoid pipetting errors, prepare master mix containing common reagents for all qPCRs and add these before the template. A blank (no template control) should be included in every qPCR run in order to detect occurrence of contamination.

12. For melting curve analysis, use the analysis step built into the software of the real-time PCR instrument.

13. We recommend using automatic setup of threshold and calculation of baseline; however, default settings might not always be accurate and it is a good idea to monitor this in every run.

14. As mammalian cells contain a single genome with two copies of nuclear genome, the real MtDNA content per cell will be $Mt/N \times 2$.

Acknowledgments

We would like to thank Kiran Hoolsy-Chandani for critical reading of the manuscript. We are grateful to mitoDNA Service Lab (mitodna@kcl.ac.uk) for their help in the development and high-throughput testing of this protocol. SA is supported by a King's overseas PhD scholarship and AC is supported by a KCL PhD studentship.

References

1. Wojtczak L, Zabłocki K (2008) Basic mitochondria physiology in cell viability and death. In: Dykens JA, Will Y (eds) Drug-induced mitochondria dysfunction. Wiley, Hoboken, NJ, pp 3–36

2. Michel S, Wanet A, De Pauw A, Rommelaere G, Arnould T, Renard P (2012) Crosstalk between mitochondrial (dys)function and mitochondrial abundance. J Cell Physiol 227: 2297–2310

3. Wallace DC (1999) Mitochondrial diseases in man and mouse. Science 283:1482–1488

4. Malik AN, Czajka A (2013) Is mitochondrial DNA content a potential biomarker of mitochondrial dysfunction? Mitochondrion 13: 481–492

5. Uranova N, Orlovskaya D, Vikhreva O, Zimina I, Kolomeets N, Vostrikov V, Rachmanova V (2001) Electron microscopy of oligodendroglia in severe mental illness. Brain Res Bull 55:597–610

6. Piko L, Matsumoto L (1976) Number of mitochondria and some properties of mitochondrial DNA in the mouse egg. Dev Biol 49:1–10

7. Duran HE, Simsek-Duran F, Oehninger SC, Jones HW Jr, Castora FJ (2011) The association of reproductive senescence with mitochondrial quantity, function, and DNA integrity in human oocytes at different stages of maturation. Fertil Steril 96:384–388

8. Selak MA, Lyver E, Micklow E, Deutsch EC, Onder O, Selamoglu N, Yager C, Knight S, Carroll M, Daldal F, Dancis A, Lynch DR, Sarry JE (2011) Blood cells from Friedreich ataxia patients harbor frataxin deficiency without a loss of mitochondrial function. Mitochondrion 11:342–350

9. Rodriguez-Enriquez S, Kai Y, Maldonado E, Currin RT, Lemasters JJ (2009) Roles of mitophagy and the mitochondrial permeability transition in remodeling of cultured rat hepatocytes. Autophagy 5:1099–1106

10. Bogenhagen DF (2011) Mitochondrial DNA nucleoid structure. Biochim Biophys Acta 1819:914–920

11. Falkenberg M, Larsson NG, Gustafsson CM (2007) DNA replication and transcription in mammalian mitochondria. Annu Rev Biochem 76:679–699

12. Hock MB, Kralli A (2009) Transcriptional control of mitochondrial biogenesis and function. Annu Rev Physiol 71:177–203

13. Williams RS (1986) Mitochondrial gene expression in mammalian striated muscle. Evidence that variation in gene dosage is the major regulatory event. J Biol Chem 261: 12390–12394

14. Malik AN, Shahni R, Iqbal MM (2009) Increased peripheral blood mitochondrial DNA in type 2 diabetic patients with nephropathy. Diabetes Res Clin Pract 86:e22–e24

15. Malik AN, Shahni R, Rodriguez-de-Ledesma A, Laftah A, Cunningham P (2011) Mitochondrial DNA as a non-invasive biomarker: accurate quantification using real time quantitative PCR without co-amplification of pseudogenes and dilution bias. Biochem Biophys Res Commun 412:1–7

16. Chiu RW, Chan LY, Lam NY, Tsui NB, Ng EK, Rainer TH, Lo YM (2003) Quantitative analysis of circulating mitochondrial DNA in plasma. Clin Chem 49:719–726

17. Hammond EL, Sayer D, Nolan D, Walker UA, Ronde A, Montaner JS, Cote HC, Gahan ME, Cherry CL, Wesselingh SL, Reiss P, Mallal S (2003) Assessment of precision and concordance of quantitative mitochondrial DNA assays: a collaborative international quality assurance study. J Clin Virol 27:97–110

18. Kam WW, Lake V, Banos C, Davies J, Banati R (2013) Apparent polyploidization after gamma irradiation: pitfalls in the use of quantitative polymerase chain reaction (qPCR) for the estimation of mitochondrial and nuclear DNA

gene copy numbers. Int J Mol Sci 14: 11544–11559

19. Venegas V, Halberg MC (2012) Measurement of mitochondrial DNA copy number. Methods Mol Biol 837:327–335

20. Dimmock D, Tang LY, Schmitt ES, Wong LJ (2010) Quantitative evaluation of the mito-chondrial DNA depletion syndrome. Clin Chem 56:1119–1127

21. Guo W, Jiang L, Bhasin S, Khan SM, Swerdlow RH (2009) DNA extraction procedures meaningfully influence qPCR-based mtDNA copy number determination. Mitochondrion 9:261–265

Chapter 13

mTRIP: An Imaging Tool to Investigate Mitochondrial DNA Dynamics in Physiology and Disease at the Single-Cell Resolution

Laurent Chatre and Miria Ricchetti

Abstract

Mitochondrial physiology and metabolism are closely linked to replication and transcription of the genome of the organelle, the mitochondrial DNA (mtDNA). However, the characterization of mtDNA processing is poorly defined at the single-cell level. Here, we describe mTRIP (*m*itochondrial *t*ranscription and *r*eplication *i*maging *p*rotocol), an imaging approach based on modified fluorescence in situ hybridization (FISH), which simultaneously reveals mitochondrial structures engaged in mtDNA initiation of replication and global mitochondrial RNA (mtRNA) content at the single-cell level in human cells. In addition, mTRIP can be coupled to immunofluorescence for in situ protein tracking, or to MitoTracker, thereby allowing simultaneous labelling of mtDNA, mtRNA, and proteins or mitochondria, respectively. Altogether, qualitative and quantitative alterations of the dynamics of mtDNA processing are detected by mTRIP in human cells undergoing physiological changes, as well as stress and dysfunction, with a potential for diagnostic of mitochondrial diseases.

Key words Mitochondrial DNA, FISH, Imaging, mTRIP, Mitochondrial disease, Transcription, DNA replication, Single-cell

1 Introduction

Mitochondria are highly dynamic organelles that can fuse and divide to produce a variety of morphologies ranging from individual entities to interconnected tubular networks, which are functional to cell growth and cell physiology [1–3]. In eukaryotes, mitochondria play a central role in the energy metabolism that is regulated by the nuclear genome and, importantly, by the organelle genome. Individual mitochondria carry multiple copies of double-stranded circular mtDNA, packed into nucleoid structures, and autonomously replicated and transcribed [4–6]. The human mtDNA is a 16.5 kbp molecule coding for two ribosomal RNAs, 22 transfer RNAs, and 13 protein-coding genes that are transcribed into polycistronic precursor RNAs and then processed to mature mRNAs, rRNAs, and tRNAs [7] (Fig. 1). The dynamics of mtDNA

Volkmar Weissig and Marvin Edeas (eds.), *Mitochondrial Medicine: Volume I, Probing Mitochondrial Function*,
Methods in Molecular Biology, vol. 1264, DOI 10.1007/978-1-4939-2257-4_13, © Springer Science+Business Media New York 2015

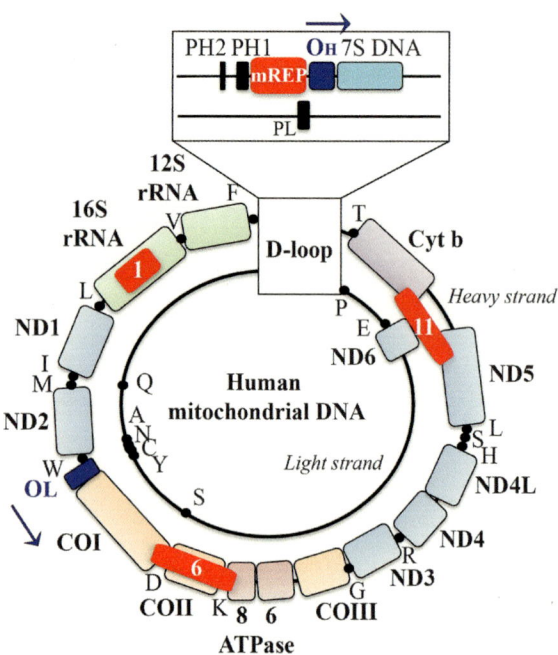

Fig. 1 Schematic representation of the human mitochondrial DNA and probes used in mTRIP. The heavy (H) and light (L) strands of the mitochondrial genome are indicated in the *external* and *inner circle*, respectively, with the position and name of single genes within. tRNA genes are indicated with a *black dot* and the *corresponding letter*. All genes are located on the H-strand, with the exception of *ND6* and several tRNAs, located on the L-strand. Magnification of the D-loop region (which contains H-strand origin of replication (O_H) and promoters of both H (PH1 and PH2) and L (PL) strands) is shown on *top*. *Blue arrows* indicate the direction of DNA replication from O_H and O_L. The position of probe mREP and probes 1, 6, and 11 is indicated with a *red box*. mTRANS is a mix of probes 1, 6, and 11, which targets rRNA as well as several coding genes and tRNA on both strands

replication are still debated, and distinct models are proposed [8–10]. Nevertheless, in these models initiation of replication takes place at the O_H origin, and the nascent heavy (H) strand may subsequently pause about 700 nucleotides downstream generating the 7S DNA, which produces a characteristic triple displacement loop or D-loop. mtDNA replication and transcription have been extensively characterized by biochemical approaches, which assess values within large cell populations (reviewed by [5, 6]). In addition, super-resolution imaging recently revealed, at the single-cell level, that mtDNA replication occurs only in a subset of nucleoids [11, 12]. These findings underscore the urgency of combined protein and DNA analysis of mitochondrial substructures. However, actual FISH procedures may damage protein epitopes and are therefore not suitable also for immunofluorescence analysis, thereby limiting the simultaneous track of mitochondrial nucleic acids and proteins.

mTRIP, a novel imaging protocol based on modified FISH, identifies mitochondrial entities engaged in the mtDNA initiation of replication and can simultaneously reveal the mitochondrial RNA (mtRNA) content in single human cells. Moreover mTRIP can be coupled to immunofluorescence, making possible the combined detection of mtDNA initiation of replication, mtRNA content, and proteins in single human cells [13] (Fig. 2). mtDNA initiation

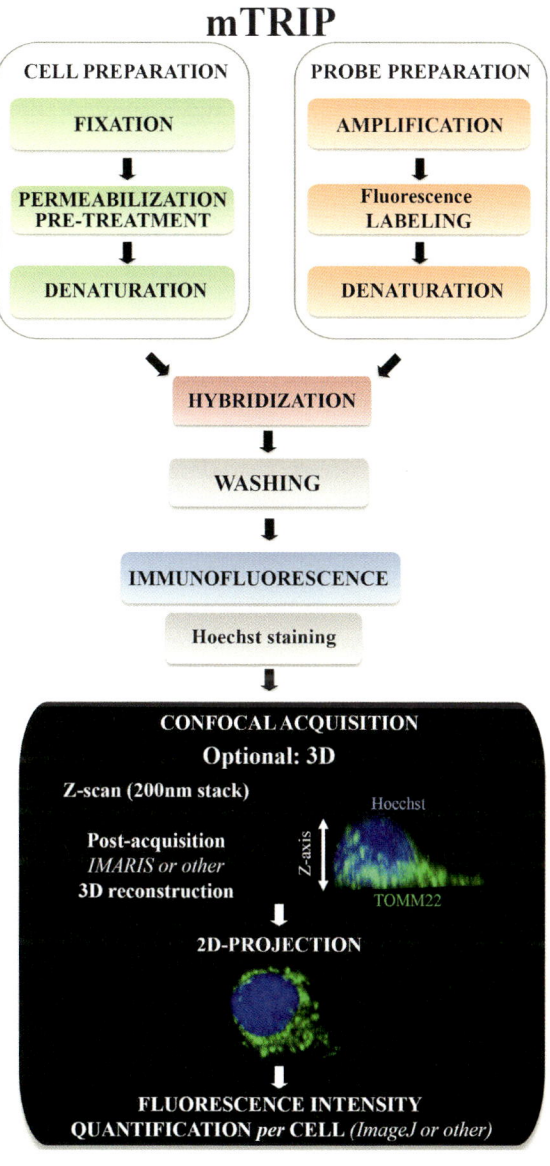

Fig. 2 Scheme of mTRIP labelling. Summary and chronology of steps of mTRIP labelling described in detail in this document. After labelling, 3D confocal acquisition and the subsequent treatment of acquisitions are shown. 3D acquisition is necessary for fluorescence quantification of the entire cell volume, but also 2D acquisition of a single plan (and thereby fluorescence quantification of this single plan) can be performed (not shown here)

of replication is identified by the mREP probe, which targets a specific DNA region located upstream of the replication origin O_H and within in the regulatory D-loop. Probe mREP recognizes only DNA in an open structure; therefore, the double-stranded mitochondrial genome that is not engaged in replication initiation is not detected by mREP. mTRANS, a mix of three DNA probes located on different regions of the mitochondrial genome, detects specific mtRNAs which are considered representative of global mtRNA levels (Fig. 1). We generally associate mTRIP analysis of mitochondrial nucleic acids with immunofluorescence of a mitochondrial outer membrane protein (TOM22 or TOMM22) to identify the mitochondrial network. These labellings allow the detection and quantification of at least three classes of mitochondrial subpopulations: (1) replication initiation active and transcript positive (Ia-Tp), (2) replication initiation silent and transcript positive (Is-Tp), and (3) replication initiation silent and transcript negative (Is-Tn) [13], revealing that individual mitochondria are strongly heterogeneous within human cells (Fig. 3a–d). This heterogeneity is functional to physiological changes, as during the cell cycle [14], and can be also affected during stress and disease [13]. mTRIP can be also associated with MitoTracker labelling, which detects mitochondria in living cells, based on its binding to thiol groups present in the organelle (Fig. 3e–h). Other probes than mREP and mTRANS can be produced that target specific regions of interest in the mitochondrial genome (*see* ref. 13). Described and novel mTRIP probes, combined with a variety of immunofluorescence and mitochondrial markers, should help decipher mtDNA dynamics and their impact under physiological changes, including stress, and in disease (*see* **Notes 1** and **2**).

2 Materials

Prepare all solutions using ultrapure water when needed. Follow all waste disposal regulations when disposing waste materials.

2.1 Probe Preparation and Labelling

1. Lysis buffer to extract total genomic DNA: 0.2 % SDS, 5 mM EDTA, 0.2 mg/ml Proteinase K in 1× phosphate-buffered saline (PBS) buffer.

2. Sodium acetate 3 M, pH 5.2, cold ultrapure isopropanol and ultrapure water.

3. Taq DNA polymerase (we recommend LA Taq DNA polymerase TaKaRa).

4. Primers for PCR amplification of the mREP probe from total genomic DNA:

    Coordinates of all primers are according to NC_012920 GenBank.

    mREP (98 nucleotides, nt): coordinates 446–544.

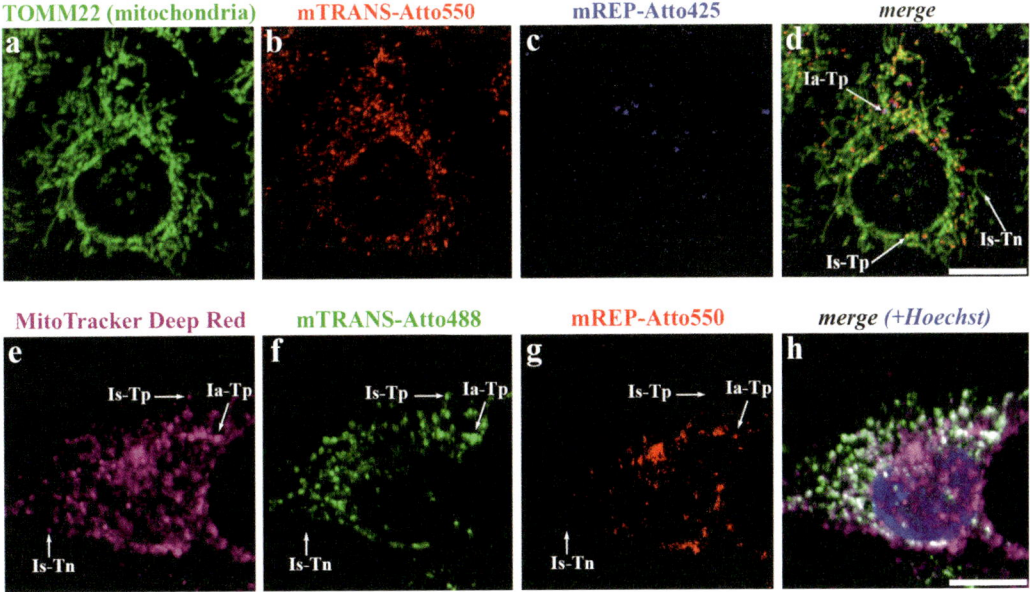

| TOMM22 (mitochondria) | mTRANS-Atto550 | mREP-Atto425 | *merge* |
| MitoTracker Deep Red | mTRANS-Atto488 | mREP-Atto550 | *merge (+Hoechst)* |

Fig. 3 mTRIP colabelling with either immunofluorescence of mitochondrial proteins or MitoTracker. Panels (**a–d**): colabelling of TOMM22 immunofluorescence (*green*, panel **a**), mTRANS (*red*, panel **b**), and mREP (*blue*, panel **c**) probes shows mitochondrial initiation of replication, transcription, and mitochondrial network, respectively. In merge, mREP essentially colocalizes with mTRANS (*purple*; *arrow*, Ia-Tp, replication initiation active and transcript positive), and independent mTRANS labelling (*orange*; *arrow*, Is-Tp, replication initiation silent and transcript active), as well as replication initiation silent and transcript negative mitochondrial structures (*green*; *arrow*, Is-Tn) are also observed. Panels (**e–h**): colabelling of MitoTracker Deep Red (*purple*, panel **e**), mTRANS (*green*, panel **f**), mREP (*red*, panel **g**), and merge (panel **h**, which was also counterstained with Hoechst, blue, for nucleus detection). MitoTracker labelling results in a more diffused signal than immunofluorescence and mTRIP; therefore, colocalization with other markers appears less defined in panel (**h**) than in panel (**d**). Ia-Tp, Is-Tp, and Is-Tn mitochondrial structures are detected also with this colabelling (presence or absence of labelling is shown in panels (**e–g**) rather than in merge, because of the limited resolution of the combined fluorescence signal by visual detection). Scale bars = 10 μm

Forward 5′-ACATTATTTTCCCCTCCC-3′.
Reverse 5′-GGGGTATGGGGTTAGCAG-3′.

5. Primers for PCR amplification of the mTRANS probe from total genomic DNA:

mTRANS is an equimolar mix of three DNA probes: probe 1, probe 6, and probe 11 (coordinates: 1905-2866, 7400-8518, and 13416-14836, respectively).

Primers for PCR amplification of DNA probe 1 (961 nt):
Forward 5′-ACCAGACGAGCTACCTAAGAACAG-3′.
Reverse 5′-CTGGTGAAGTCTTAGCATGT-3′.
Primers for PCR amplification of DNA probe 6 (1,118 nt):
Forward 5′-CTACCACACATTCGAAGAACC-3′.
Reverse 5′-CGTTCATTTTGGTTCTCAGGG-3′.
Primers for PCR amplification of DNA probe 11 (1,420 nt):
Forward 5′-CATACCTCTCACTTCAACCTC-3′.
Reverse 5′-TGAGCCGAAGTTTCATCATGC-3′.

6. PCR product clean-up system kit (*see* **Note 3**).

7. Labelling of purified mREP and mTRANS PCR products. It is strongly recommended to label the DNA probes by Nick translation using Atto425 or Atto488 or Atto550 or Atto647 NT labelling kit (Jena Bioscience) (*see* **Note 4**).

   Labelled and purified mREP and mTRANS probes must be aliquoted and stored at –20 °C in the dark.

2.2 Cell Treatment

1. Tissue culture plates 6-well (12-well or 24-well plates are reasonable alternatives for culture of rare primary cells or slow-growing cells).

2. Microscope glass cover slips 18 mm diameter.

3. 2 % paraformaldehyde (PFA) in 1× PBS (mix 5 ml of 16 % PFA with 35 ml of 1× PBS).

4. 2 % PFA can be stored at 4 °C for a couple of weeks.

5. Permeabilization buffer: 0.5 % Triton X-100 in 1× PBS (mix 50 μl of 100 % Triton X-100 with 9,550 μl 1× PBS).

6. 20× saline sodium citrate (SSC) buffer: 150 mM NaCl, 15 mM Na3citrate×2H$_2$O, pH 7.0. For SSC preparation, dissolve 175.3 g of NaCl and 88.2 g of sodium citrate in 800 ml of ultrapure water. Adjust the pH to 7.0 with 14 N solution of HCl. Adjust the volume to 1 l with additional ultrapure water. Sterilize by autoclaving. Store at room temperature for up to 6 months.

7. Pretreatment buffer for permeabilized cells: 50 % formamide/2× SSC in 1× PBS.

   Store the solution at room temperature for up to 6 months in the dark.

8. Denaturation buffer for permeabilized cells: 70 % formamide/2× SSC in 1× PBS.

   Store the solution at room temperature for up to 6 months in the dark.

2.3 mTRIP Hybridization

1. Parafilm.

2. Hybridization buffer: 10 % dextran sulfate/50 % formamide/2× SSC in 1× PBS.

   Store the solution at room temperature for up to 6 months in the dark (*see* **Note 5**).

3. 100 ng/μl salmon sperm DNA: dilute stock salmon sperm DNA with 1× PBS.

   Store the solution at –20 °C.

4. Washing buffer A: 2× SSC in 1× PBS.

5. Washing buffer B: 1× SSC in 1× PBS.

6. Washing buffer C: 0.1× SSC in 1× PBS.

2.4 mTRIP Coupled to Immunofluorescence

1. Parafilm.

2. Blocking buffer: 5 % bovine serum albumin (BSA) in 1× PBS. The solution can be stored at 4 °C for several weeks (*see* **Note 6**).

3. Primary antibody (*see* **Note 7**): we recommend the use of unconjugated rabbit polyclonal anti-TOMM22 to label mitochondria.

4. Fluorescent-conjugated secondary antibody (*see* **Note 8**): we recommend the use of Cy5 (or similar)-conjugated goat secondary anti-rabbit antibody.

5. Hoechst 33342.

2.5 mTRIP Coupled to MitoTracker®

MitoTracker® probes, Invitrogen. Select a MitoTracker probe with appropriate spectral characteristics for successive labelling with mTRIP. Here we used MitoTracker® Deep Red (655 nm).

2.6 Imaging Equipment

1. Confocal microscope.

2. Optional for 3D imaging: 3D reconstruction imaging software.

3. Quantification—imaging software such as ImageJ.

3 Methods

mTRIP principles are summarized in Fig. 2.

3.1 Extraction of Total Genomic DNA

It is strongly recommended to use primary cells such as IMR90 at early passages (and not cancer-derived or immortalized cell lines) (*see* **Note 9**).

1. Add 200 μl of fresh lysis buffer to the cell pellet at 50 °C for 3 h.

2. Incubate at 50 °C for 3 h.

3. Add 20 μl of sodium acetate 3 M, pH 5.2.

4. Add 200 μl of cold ultrapure isopropanol.

5. Incubate at −20 °C for 15 min.

6. Centrifuge at $16,100 \times g$ at 4 °C for 30 min.

7. Carefully discard the supernatant.

8. Add 500 μl of 70 % ethanol.

9. Centrifuge at $16,100 \times g$ at 4 °C for 5 min.

10. Carefully discard the supernatant.

11. Dry the DNA pellet for 5 min at room temperature. The pellet should be transparent after drying.

12. Resuspend the DNA pellet in 100 μl of ultrapure water.

13. Incubate at 4 °C overnight.

14. Quantify the total genomic DNA. Dilute in ultrapure water, if needed.

15. Storage: −20 °C.

3.2 Preparation and Labelling of the DNA Probe

1. Amplify the mREP probe by PCR using at least 100 ng/μl of total genomic DNA and LA Taq DNA polymerase (recommendation: annealing temperature, 56 °C; extension, 68 °C).

2. Amplify the mTRANS component probe 1 by PCR using at least 100 ng/μl of total genomic DNA and LA Taq DNA polymerase (recommendation: annealing temperature, 56 °C; extension, 68 °C).

3. Amplify the mTRANS component probe 6 by PCR using at least 100 ng/μl of total genomic DNA and LA Taq DNA polymerase (recommendation: annealing temperature, 56 °C; extension, 68 °C).

4. Amplify the mTRANS component probe 11 by PCR using at least 100 ng/μl of total genomic DNA and LA Taq DNA polymerase (recommendation: annealing temperature, 56 °C; extension, 68 °C).

5. Purify the PCR products using 2 % agarose electrophoresis gel and conventional PCR product clean-up kit.

6. Estimate the amount of the purified PCR product. It is essential to get 1 μg of PCR product at a concentration of 75 ng/μl or higher. If this amount of DNA product is not reached, one can concentrate the DNA by sodium acetate and isopropanol precipitation and then resuspend in the appropriate volume of ultrapure water to reach the required concentration. Note that precipitation of highly diluted DNA is not recommended as it may result in salt concentration higher than necessary, in spite of intense washing.

   Be careful as the following steps are dedicated to DNA probe fluorescent labelling by Nick translation (NT) and are crucial for mTRIP. Consider that no other technique of DNA labelling than Nick translation has been tested in this context.

7. Choose the fluorescence label for mREP and mTRANS DNA probes (*see* **Note 10**). If mTRIP is coupled to immunofluorescence and Hoechst 33342, resulting in four color channels, the recommended choice is mREP labelled with Atto550 NT kit, mTRANS labelled with Atto488 NT kit, immunofluorescence with a Cy5-conjugated secondary antibody, and Hoechst.

8. Preparation of the mREP DNA probe:

   (a) Resuspend 1 μg of pure PCR product in a maximum volume of 14 μl in a 0.2 ml thin-wall 8-tube strip with caps for regular thermal cycler (PCR machine).

      If the volume is smaller, adjust to 14 μl with PCR-grade water from the kit.

9. Preparation of the mTRANS DNA mix probes:

    (a) Mix 333 ng of mTRANS probe 1, 333 ng of mTRANS probe 6, and 333 ng of mTRANS probe 11 (equimolar amounts of the three probes) for a final amount of 1 μg of DNA mix, in maximum volume of 14 μl in a 0.2 ml thin-wall 8-tube strip with caps for regular thermal cycler (PCR machine).

    If the volume is smaller, adjust to 14 μl with PCR-grade water from the kit.

    The following steps are identical for mREP and mTRANS fluorescence labelling:

10. Gently mix 14 μl of template DNA with 2 μl of 10× NT labelling buffer on ice.

11. Add 2 μl of Atto488 or Atto550 NT labelling mix.

12. Add 2 μl of 10× enzyme mix.

13. Ensure sample homogeneity by gentle mixing.

14. Incubate the mix at 15 °C for 90 min, for example, in a regular thermal cycler, in the dark.

15. Add 5 μl of NT Kit Stop buffer to stop the reaction.

16. Transfer the reaction mixture in 1.5 ml Eppendorf tube.

17. Add 2 μl of 3 M pH 5.2 sodium acetate.

18. Add 14 μl of ultrapure cold isopropanol.

19. Gently mix the sample.

20. Incubate at −20 °C for 15 min in the dark.

21. Centrifuge at $16,100 \times g$ at 4 °C for 30 min.

22. Discard the supernatant.

23. Add 500 μl of 70 % ethanol.

24. Centrifuge at $16,100 \times g$ at 4 °C for 5 min.

25. Discard the supernatant.

26. Add 500 μl of 70 % ethanol.

27. Centrifuge at $16,100 \times g$ at 4 °C for 5 min.

28. Discard the supernatant.

29. Resuspend the fluorescent-labelled DNA probe in 50 μl of ultrapure water to get a final concentration of 20 ng/μl.

30. Store the labelled DNA probe at −20 °C in the dark.

3.3 Pre-hybridization of the DNA Probe

To reduce bench timing, Subheading 3.3 can be done in parallel with Subheading 3.4.

1. Choice 1: one probe, preparation of the reaction mixture (in a 1.5 ml tube):

    If only one probe (e.g., mREP or mTRANS) is used, the reaction mix is done in a final volume of 25 μl per cover slip.

Gently mix 2 μl of 20 ng/μl of fluorescent DNA probe (40 ng DNA probe is used per cover slip at a final concentration of 1.6 ng/μl) with 4 μl of 100 ng/μl salmon sperm DNA (400 ng of salmon sperm DNA is used for 40 ng of DNA probe and per cover slip) and 19 μl of hybridization buffer.

2. Choice 2: two probes, preparation of the reaction mixture (in a 1.5 ml tube):

If two probes (mREP *and* mTRANS) are used, the reaction mix is done in a final volume of 25 μl per cover slip. Gently mix 2 μl of 20 ng/μl fluorescent mREP probe (40 ng of DNA probe is used per cover slip; final concentration 1.6 ng/μl) and 2 μl of 20 ng/μl fluorescent mTRANS probe (40 ng DNA probe is used per cover slip; final concentration 1.6 ng/μl) with 8 μl of 100 ng/μl of salmon sperm DNA (800 ng of salmon sperm DNA is used for 2× 40 ng of DNA probes and per cover slip) and 13 μl of hybridization buffer.

3. Denaturation: incubate the reaction mixture at 80 °C for 10 min in the dark.

4. Precooling step: transfer the denatured reaction mixture-containing tube at 37 °C for at least 30 min in the dark. This time can be extended to a maximum of 1 h at 37 °C in the dark. This step helps lowering unspecific binding of salmon sperm DNA versus specific binding of DNA probes. Keep denatured reaction mixture under these conditions until the end of Subheading 3.4.

3.4 Preparation of Cells on Cover Slip

1. Culture cells under appropriate culture conditions in (6-well or 12-well) culture plates containing a clean cover slip at the bottom of each well. Ideally cells should be grown at about 50–60 % confluence to avoid plans with packed cells during microscope analysis.

If mTRIP is coupled to MitoTracker labelling, incubate live cells with MitoTracker® (e.g., Mitotracker® Deep Red) for 1 h in appropriate culture conditions (*see* **Note 11**).

2. Discard the culture medium.

3. Wash cells once with 1× PBS.

4. Add 2 % PFA at room temperature.

5. Incubate for 20 min at room temperature.

6. Discard PFA.

7. Wash 2× with 1× PBS. At this step, PFA-fixed cells can be stored in 1× PBS for several weeks at 4 °C.

8. Discard PBS from the cover slip.

9. Permeabilize cells with 0.5 % Triton X-100 in 1× PBS for 5 min at 4 °C.

10. Discard Triton X-100.

11. Wash 4× with 1× PBS.

   Optional: if required, at this step permeabilized cells can be incubated with specific nucleases for the time recommended by the manufacturer (generally 1 h at 37 °C). After incubation, wash 4× with 1× PBS (*see* **Note 12**).

12. Discard PBS.

13. Pretreat the permeabilized cells by adding pretreatment buffer (50 % formamide/2× SSC in 1× PBS).

14. Incubate at room temperature for 30 min.

15. Discard the pretreatment buffer.

16. Add 70 % formamide/2× SSC in 1× PBS.

17. Put the 6-well culture plate (with, at the bottom of each well, a cover slip with pretreated permeabilized cells) on the top of a metal block heater at 75 °C for 5 min to denature the sample.

18. Immediately transfer the 6-well culture plate on ice (4 °C) and keep on ice at least 1 min but not longer than 10 min (*see* **Note 13**).

3.5 mTRIP Hybridization

1. Put a 25 μl drop of denatured and precooled reaction mixture which contains DNA probe(s) on an appropriate surface of parafilm (at least 3.6 × 3.6 cm for 18 mm diameter cover slip). For multiple labellings, use a single surface of parafilm with enough room for multiple and well-separated drops.

2. Put the cover slip upside down, with the face containing cells (samples previously undergone denaturation) directly on the drop.

3. Incubate at 37 °C for 15 h either at the top of a heating metal block in the dark (cover with plastic top to set dark) or on a humid chamber at 37 °C in the dark.

3.6 mTRIP Washing

1. After 15 h incubation, remove the cover slip from parafilm and put it back into a clean cell culture plate (6-well plate).

   The following steps must be performed in the dark:

2. Wash 2× with 2× SSC in 1× PBS for 2 min at room temperature (RT) with gentle shaking.

3. Wash 2× with 1× SSC in 1× PBS for 2 min at RT with gentle shaking.

4. Wash 2× with 0.1× SSC in 1× PBS for 2 min at RT with gentle shaking.

5. Wash 2× with 1× PBS for 2 min at RT with gentle shaking.

   At this point, either label the sample by immunofluorescence, and in this case go to Subheading 3.7 (mTRIP coupled to immunofluorescence), or do not label the sample by immunofluorescence (this is also the case for MitoTracker labelling), and in this case:

6. Incubate the cover slip for 1 h in the dark with 10 μg/ml Hoechst 33342 in 1× PBS.

7. Wash 5× with 1× PBS.

8. Mount the cover slip on a clean and dry glass slide using the favorite mounting buffer or a drop of 50 % glycerol in 1× PBS.

9. Seal the slide with varnish and keep the mounted slide in a dark and clean box either at room temperature if confocal analysis is done the same day or at 4 °C until confocal analysis is done (it is recommended to analyze the slide not later than 2 weeks from labelling).

3.7 mTRIP Coupled to Immunofluorescence

Follow the immunofluorescence (IF) protocol suggested below (*see* **Note 14**):

1. Incubate the cover slip with 5 % BSA in 1× PBS at RT for 1 h in the dark.

2. Wash 2× in 1× PBS for 2 min at RT with gentle shaking.

3. Dilute the primary antibody (anti-TOMM22 (1:200), recommended for assessing the mitochondrial network) in 1 % BSA and 1× PBS.

4. Put a 50 μl drop of diluted primary antibody mix on an appropriate surface of parafilm.

5. Put the cover slip upside down, with the face containing cells directly on the drop.

6. Incubate at RT for 1 h in the dark. The incubation time depends on the primary antibody.

7. Wash 3× in 1× PBS for 2 min at RT with gentle shaking.

8. Dilute secondary antibody (1:1,000) in 1 % BSA and 1× PBS, and add 10 μg/ml of Hoechst 33342.

9. Put a 50 μl drop of diluted secondary antibody mix on an appropriate surface of parafilm.

10. Put the cover slip upside down, with the face containing cells directly on the drop.

11. Incubate at RT for 1 h in the dark.

12. Wash 5× in 1× PBS for 2 min at RT with gentle shaking.

13. Mount the cover slip on a clean and dry slide using the favorite mounting buffer or a drop of 50 % glycerol in 1× PBS.

14. Seal the slide with varnish and keep the mounted slide in a dark and clean box either at room temperature if confocal analysis is done the same day or at 4 °C until confocal analysis is done (it is recommended to analyze the slide not later than 2 weeks from labelling).

3.8 Imaging Acquisition and Fluorescence Intensity Quantification

1. Acquisition for subsequent quantification is done with confocal microscope (*see* **Note 15**).

2. One can do either 2-dimensional (2D) or 3-dimensional (3D) acquisition. In the last case, 3D reconstruction of the image should follow 3D acquisition.

3. With the image format (.tiff) of the acquired image, one can quantify the fluorescence intensity per cell using the ImageJ software.

   A 3D-reconstructed human cell stained by mTRIP is shown in Fig. 3 (panels b, c, f, g).

4 Notes

1. mTRIP tool is covered by patent applications: EP2500436 and WO2012123588 "Method, probe and kit for DNA in situ hybridization and use thereof."

2. mTRIP can be used with paraformaldehyde-fixed human cells and paraffin-embedded human tissue sections (method not described here).

3. *Wizard® SV Gel, PCR Clean-Up System (Promega)*, and *NucleoSpin® Extract II (Macherey-Nagel)* worked well in our hands.

4. Labelling of DNA probes with other techniques than Nick translation has not been tested. We do not exclude that other DNA probe label techniques are compatible with mTRIP.

5. It is strongly recommended to prepare the hybridization buffer several days (at least 2 days) before the experiment as dextran sulfate takes a long time to dissolve.

6. Alternative blocking buffers can also be used.

7. Unconjugated rabbit polyclonal anti-TOMM22 is used to illuminate the mitochondrial network. However, you can use other primary antibodies for detecting the mitochondrial network or other cellular structures or proteins of interest.

8. The choice of the secondary antibody-conjugated fluorescence depends on the available spectra after exclusion of the fluorophore(s) used for DNA probes (mTRIP). All combinations compatible with available fluorophores and the equipment (fluorescence filters) of your microscope can be used.

9. Cancer and immortalized cell lines may have mutated and/or highly heteroplasmic mitochondrial genomes.

10. The fluorescence color of the DNA probes can be different from the ones used here, and for this appropriate fluorophores must be selected.

11. In this case, use mREP Atto550 and mTRANS Atto488 for mTRIP labelling and Hoechst.

12. Treatment with DNaseI (recommended 100 U/ml) or RNase A (recommended 100 μg/ml) allows identification of the DNA and RNA components of the mTRIP signal, respectively. Treatment with RNase H (recommended 100 U/ml) allows identification of RNA/DNA structures, usually at replication origins and in transcription bubbles. For combined use of nucleases and the respective readouts on mitochondrial nucleic acids by mTRIP, *see* ref. [13].

13. Do not keep denatured samples at room temperature, as this condition will interfere with the hybridization step. Keep the samples always on ice until hybridization.

14. The favorite IF procedure, if different from the one described here, can alternatively be performed.

15. Reliable quantification requires fluorescence measurement of the cell volume. Cell surface (2D acquisition) can also be acquired instead of cell volume, but the corresponding quantification will not measure the total cell fluorescence.

Acknowledgment

This work was supported by the Association Nationale contre le Cancer (ARC 4022 and SFI20111204038), PTR-Institut Pasteur (PTR217), DARRI-Institut Pasteur (project P790319), Association Française contre les Myopathies (AFM, project 2012-0905 n° 16290), and Agence Nationale pour la Recherche (ANR 11BSV202502). L. C. was the recipient of a Bourse Roux Institut Pasteur and was supported by Mr. François R. Lacoste.

References

1. Chan DC (2006) Mitochondrial fusion and fission in mammals. Annu Rev Cell Dev Biol 22:79–99

2. Lee S, Kim S, Sun X, Lee JH, Cho H (2007) Cell cycle-dependent mitochondrial biogenesis and dynamics in mammalian cells. Biochem Biophys Res Commun 357(1):111–117

3. Mitra K, Wunder C, Roysam B, Lin G, Lippincott-Schwartz J (2009) A hyperfused mitochondrial state achieved at G1-S regulates cyclin E buildup and entry into S phase. Proc Natl Acad Sci U S A 106(29):11960–11965

4. Bonawitz ND, Clayton DA, Shadel GS (2006) Initiation and beyond: multiple functions of the human mitochondrial transcription machinery. Mol Cell 24(6):813–825

5. Falkenberg M, Larsson NG, Gustafsson CM (2007) DNA replication and transcription in mammalian mitochondria. Annu Rev Biochem 76:679–699

6. Scarpulla RC (2008) Transcriptional paradigms in mammalian mitochondrial biogenesis and function. Physiol Rev 88(2):611–638

7. Ojala D, Montoya J, Attardi G (1981) tRNA punctuation model of RNA processing in human mitochondria. Nature 290(5806): 470–474

8. Chang DD, Clayton DA (1985) Priming of human mitochondrial DNA replication occurs at the light-strand promoter. Proc Natl Acad Sci U S A 82(2):351–355

9. Clayton DA (1991) Replication and transcription of vertebrate mitochondrial DNA. Annu Rev Cell Biol 7:453–478

10. Holt IJ, Lorimer HE, Jacobs HT (2000) Coupled leading- and lagging-strand synthesis of mammalian mitochondrial DNA. Cell 100(5):515–524

11. Kukat C et al (2011) Super-resolution microscopy reveals that mammalian mitochondrial nucleoids have a uniform size and frequently contain a single copy of mtDNA. Proc Natl Acad Sci U S A 108(33):13534–13539

12. Brown TA et al (2011) Superresolution fluorescence imaging of mitochondrial nucleoids reveals their spatial range, limits, and membrane interaction. Mol Cell Biol 31(24):4994–5010

13. Chatre L, Ricchetti M (2013) Large heterogeneity of mitochondrial DNA transcription and initiation of replication exposed by single-cell imaging. J Cell Sci 126(Pt 4):914–926

14. Chatre L, Ricchetti M (2013) Prevalent coordination of mitochondrial DNA transcription and initiation of replication with the cell cycle. Nucleic Acids Res 41(5):3068–3078

Chapter 14

Simultaneous Quantification of Mitochondrial ATP and ROS Production

Liping Yu, Brian D. Fink, and William I. Sivitz

Abstract

Several methods are available to measure ATP production by isolated mitochondria or permeabilized cells but with a number of limitations, depending upon the particular assay employed. These limitations may include poor sensitivity or specificity, complexity of the method, poor throughput, changes in mitochondrial inner membrane potential as ATP is consumed, and/or inability to simultaneously assess other mitochondrial functional parameters. Here we describe a novel nuclear magnetic resonance (NMR)-based assay that can be carried out with high efficiency in a manner that alleviates the above problems.

Key words ATP, Mitochondria, Superoxide, Reactive oxygen species, H_2O_2, NMR, Bioenergetics

1   Introduction

Here we describe a novel, highly sensitive, and specific nuclear magnetic resonance (NMR)-based ATP assay that can be carried out with reasonably high throughput using small amounts of mitochondrial isolates or permeabilized cells (*see* **Note 1**). There are several advantages of this method, including the ability to simultaneously carry out fluorescent measurements. For example, it is possible to quantify the production of reactive oxygen species (ROS) simultaneously with ATP production measurement, as we discuss below. Another major advantage of the assay is that it avoids the problem of changing mitochondrial membrane potential ($\Delta\Psi$) while ADP is converted to ATP, as occurs in conventional assays. In contrast to conventional assays, $\Delta\Psi$ in our assay is clamped at fixed levels determined by the amount of ADP added. We have used this assay to study muscle mitochondrial function in rats made diabetic with the islet cell toxin, streptozotocin [1], and to study liver and heart mitochondrial function in mice fed high dietary fat [2].

Volkmar Weissig and Marvin Edeas (eds.), *Mitochondrial Medicine: Volume I, Probing Mitochondrial Function*,
Methods in Molecular Biology, vol. 1264, DOI 10.1007/978-1-4939-2257-4_14, © Springer Science+Business Media New York 2015

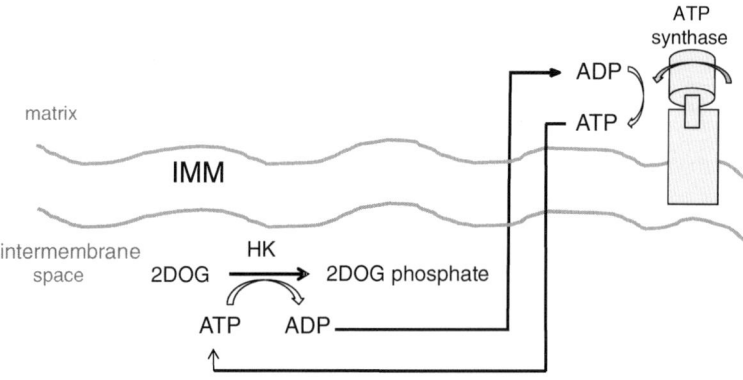

Fig. 1 The 2-deoxyglucose (2DOG) energy clamp. Saturating amounts of 2-deoxyglucose (2DOG) and hexokinase (HK) recycle ATP back to ADP by rapidly and irreversibly converting 2DOG into 2-deoxyglucose phosphate (2DOGP). The resulting ADP availability is clamped at levels determined by the amount of ADP added. *IMM* inner mitochondrial membrane

1.1 2DOG ATP Energy Clamp

To assess mitochondrial functional parameters at fixed $\Delta\Psi$, we use excess 2-deoxyglucose (2DOG) (*see* **Note 2**) and hexokinase (HK) to generate an "ATP energy clamp" (Fig. 1). The conversion of 2DOG to 2DOG phosphate (2DOGP) occurs rapidly and irreversibly, thereby effectively clamping ADP concentrations and $\Delta\Psi$ dependent on the amount of exogenous ADP added (*see* **Note 3**). This enables titration of membrane potential at different fixed values with mitochondria in respiratory states ranging from state 4 (no ADP, maximal potential) to state 3 (high levels of ADP, reduced potential) (Fig. 2).

1.2 Use of the 2DOG ATP Energy Clamp to Quantify ATP Production in Isolated Mitochondria and Simultaneous Assessment of H_2O_2 Production

Mitochondria are added to individual wells of 96-well plates in a total volume of typically 60 μL and incubated at 37 °C in respiratory buffer in the presence of HK, 2DOG, and the desired concentration of ADP. [6-^{13}C]2DOG is used in two-dimensional (2D) ^{13}C/^{1}H-HSQC NMR-based assays, while unlabeled 2DOG is used in one-dimensional (1D) ^{1}H NMR-based assays (see below). After incubation for the desired time, the contents of the microplate wells are removed to tubes on ice containing oligomycin to inhibit ATP synthase. The tubes are then centrifuged and the supernatants are held at –20 °C until NMR analysis. To prepare the NMR sample for measurement of ATP production by mitochondria, the assay supernatants are added to 5 mm (OD) standard NMR tubes along with appropriate NMR buffer (see below).

ATP production rates are calculated based on the percent conversion of 2DOG to 2DOGP as determined by NMR, the initial 2DOG concentration, incubation volume, and incubation time.

In order to simultaneously assess H_2O_2 production, mitochondrial incubations can be carried out in the presence of 10-Acetyl-3,7-dihydroxyphenoxazine (DHPA). Moreover, it is possible to

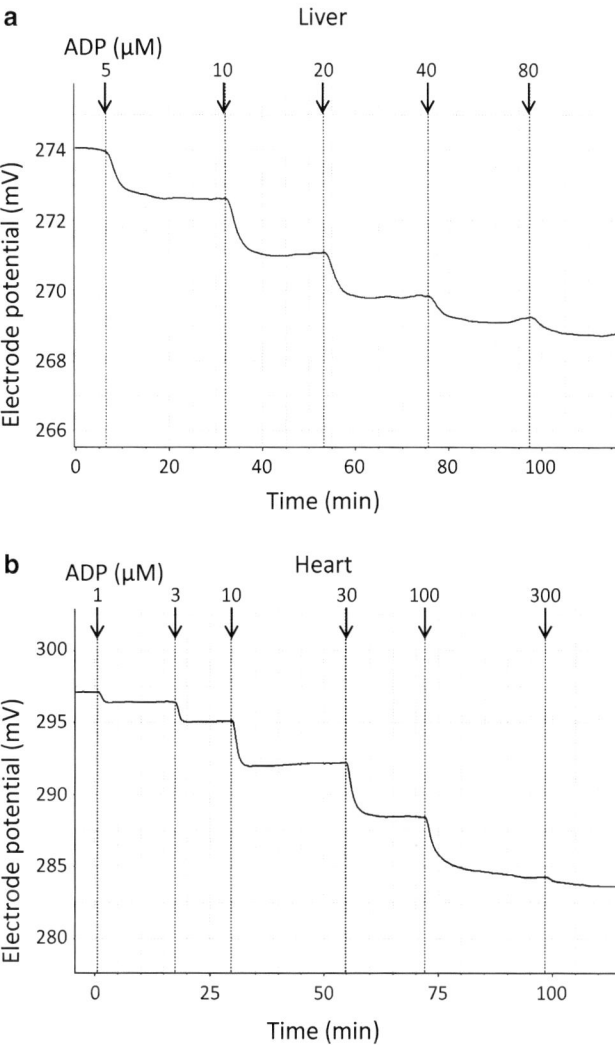

Fig. 2 Computer tracings of inner membrane potential vs. time obtained by incubating normal mouse liver mitochondria, 0.5 mg/mL (panel **a**), or heart mitochondria, 0.25 mg/mL (panel **b**), fueled by the combined substrates, 5 mM succinate +5 mM glutamate +1 mM malate. ADP was added in incremental amounts to generate the final total recycling nucleotide phosphate concentrations (indicated by *arrows*). After each addition, a plateau potential was reached, consistent with recycling at a steady ADP concentration and generation of a stepwise transition from state 4 to state 3 respiration. Note that the potential shown on the *y*-axis represents electrode potential (not mitochondrial potential). The actual $\Delta\Psi$ follows a similar pattern after calculation using the Nernst equation based on the distribution of tetraphenylphosphonium (TPP) external and internal to mitochondria

use other fluorescent probes to assess mitochondrial parameters such as membrane potential and calcium uptake as long as the probes do not interfere with the NMR detection.

Thus we describe a novel means to assess ATP production. The assay clearly has several advantages over existing methods (*see* **Note 4**).

2   Materials

2.1   Isolation of Mitochondria

1. Isolation medium: 0.25 M sucrose, 5 mM HEPES (pH 7.2), 0.1 mM EDTA, 0.1 % BSA (fatty acid-free).

2. Purification medium: 30 % v/v Percoll®. Dilute three parts of Percoll® with seven parts of isolation medium. 2.4 mL of Percoll® +5.6 mL of isolation medium = 8 mL, sufficient for two centrifuge tubes. Keep on ice.

3. Beckman XL-80 ultracentrifuge or similar instrument, pre-cooled to 4 °C, SW60 swinging bucket rotor with caps and greased O-ring seals, polyallomer centrifuge tubes ~4.2 mL max capacity per tube.

2.2   Assay Incubation

1. Respiration medium: 0.3 % fatty acid-free BSA, 120 mM KCl, 1 mM EGTA, 5 mM KH_2PO_4, 2 mM $MgCl_2$, 10 mM HEPES, pH 7.2.

2. Microplate with 96 wells (e.g., a Costar #3792 black round-bottom plate).

3. 5 U/mL hexokinase (HK).

4. 5 mM 2-deoxyglucose (2DOG) or [6-^{13}C]2DOG.

5. 10-Acetyl-3,7-dihydroxyphenoxazine (DHPA or Amplex Red).

6. Horseradish peroxidase (HRP).

2.3   Sample Preparation for NMR Spectroscopy

1. Sample dilution buffer: 120 mM KCl, 5 mM KH_2PO_4, 2 mM $MgCl_2$, pH 7.2.

2. Deuterium oxide (D_2O).

3. Standard 7 in. (length) × 5 mm (outer diameter) NMR tubes.

2.4   NMR Spectroscopy

1. NMR spectrometer equipped with a dual or triple resonance probe and capable of acquiring ^{1}H and ^{1}H/^{13}C HSQC NMR spectra.

2. NMR spectrometer equipped with an auto sample changer, thus capable of continuous data acquisition of multiple samples, for example, 60 samples.

3   Methods

3.1 Isolation of Mitochondria from the Tissue

1. Carry out all procedures on ice or at 4 °C.

2. Harvest the tissue and rinse in isolation medium.

3. Homogenize up to 1 g of tissue in 10–15 mL of isolation medium using a Potter-Elvehjem-type tissue grinder in an ice bucket. Fibrous tissues should be minced with scissors to aid tissue disruption. The Teflon pestle is mounted on a drill set to approximately 300 rpm. Four to six passes are typically required. Optionally, a subsequent pass of the homogenate through a ground-glass-style homogenizer can increase the mitochondrial yield from fibrous tissues.

4. Centrifuge homogenate at $500 \times g$ for 10 min (low-speed spin) (*see* **Note 5**).

5. Transfer the supernatant to Sorvall-type tube (Oakridge screw cap). Discard the pellet and centrifuge at $10,000 \times g$ for 10 min (high-speed spin). Discard supernatant.

6. Wash the mitochondrial pellet with isolation medium without BSA.

7. Resuspend the final pellet at ~50 % v/v in isolation medium without BSA.

3.2 Further Purification of Isolated Mitochondria (See Note 6)

1. Add 3.8 mL of 30 % Percoll® solution to each SW60 polyallomer tube on ice.

2. Resuspend mitochondrial crude prep pellets in 0.1 mL of isolation medium containing BSA.

3. Lay the mitochondria on top of the Percoll® solution and insert the tubes into the buckets.

4. Use a balance to precisely equalize the mass of the buckets + tubes + lids, adding isolation medium containing BSA to adjust the mass. Ensure that the contents of the tubes are within 3 mm of the top of the tubes.

5. Securely attach the bucket caps. A liquid sample within a sealed bucket will be protected from the vacuum applied to the centrifuge chamber.

6. Hang the buckets on the precooled SW61 rotor at 4 °C and spin for 30 min at 30,000 rpm ($\sim 90,000 \times g$).

7. The pure mitochondria band appears near the bottom of the tube, just above a clear and dense mass of Percoll®. Remove all contaminating fractions above the mitochondria band with a pipet.

8. Transfer the mitochondria band to a 1.5 mL centrifuge tube.

9. Add 1 mL of isolation medium without BSA. Spin in a microfuge at $8{,}000 \times g$ for 5 min at 4 °C. Remove the supernatant. If subjecting the mitochondria to endogenous Ca^{2+} depletion, then stop here and proceed with depletion protocol. Otherwise, resuspend the pellet in 1 mL of isolation medium without BSA and spin in a microfuge at $8{,}000 \times g$ for 5 min at 4 °C.

10. Resuspend the final washed pellet in BSA-free isolation medium and keep on ice.

3.3 Assay Incubation

1. Warm a 96-well microplate to 37 °C for 10 min. If carrying out simultaneous fluorescent detection (e.g., for reactive oxygen species), warm the plate in the plate reader and set the plate reader gain as desired.

2. Preload all reagents (before adding mitochondria) to wells with 1.2× respiration medium (upon subsequent addition of mitochondria, the medium will be 1×) in a total volume of 50 μL. For rat hind limb muscle mitochondria [1], we used the following final or 1× assay concentrations: 5 mM succinate +5 mM glutamate +1 mM malate (or other mitochondrial fuel selection and concentrations as desired), 5 U/mL hexokinase (HK), 5 mM 2-deoxyglucose (2DOG) (using [6-^{13}C]2DOG for 2D NMR detection or unlabeled 2DOG for 1D NMR detection; see below), and ADP at desired concentration (up to 100 μM). For simultaneous assay of reactive oxygen species, add 20 μM 10-acetyl-3,7-dihydroxyphenoxazine (DHPA or Amplex Red, Invitrogen) plus 5 U/mL of horseradish peroxidase (HRP).

3. Wells containing no added substrate should be present on the plate to serve as background control. Reserve some wells for inclusion of an H_2O_2 standard curve, typically containing a range of H_2O_2 concentrations up to a maximum of 5–10 μM. These standard curve wells do not have added mitochondria.

4. Two to three wells should be included in the plate to serve as positive controls and/or standards for NMR quantification of the amount of 2DOGP. These wells have the same final total volume of 60 μL as the others but only contain 5 mM 2DOG or [6-^{13}C]2DOG (depending on whether 1D or 2D NMR methods are used for measurement, respectively; see below), 5 U/mL hexokinase, and 10 mM ATP in 1× respiration medium. The use of twofold molar excess of ATP with respect to 2DOG is to ensure full conversion of 2DOG to 2DOGP.

5. To start the assay, add 10 μL of 6× concentrated mitochondria suspended in 1× respiration medium. The well has a total volume of 60 μL now. The plate reader program is started and

fluorescence determined at a frequency of once per minute or greater. Final mitochondrial concentrations are typically 0.1–0.5 mg/mL.

6. After incubation typically for 5–30 min, harvest the wells by transferring the well contents to 500 μL tubes containing 1 μL of 120 μM oligomycin. Then immediately centrifuge the tubes at $10,000 \times g$ for 4 min at 4 °C.

7. Transfer the supernatants to new labeled tubes and hold them at −20 °C until NMR sample assembly.

3.4 Fluorescent Assessment of H_2O_2 Production

1. As indicated above, the assay wells contain DHPA (20 μM) and HRP (5 U/mL).

2. Monitor DHPA fluorescence over the duration of the assay at 544 nm excitation and 590 nm emission.

3. Determine the average fluorescence over the duration of the assay for all points in the standard curve. Subtract the average background fluorescence from all values for the standard curve. The background value is defined as fluorescence in the absence of H_2O_2 (the zero H_2O_2 point in the standard curve).

4. For all wells containing mitochondria, determine the slope of fluorescence as a function of time (*see* **Note 7**). It is not necessary to subtract a background fluorescence value in order to determine the slope for these wells since the background is constant.

5. Use a curve fitting software to assess background-subtracted standard curve fluorescence as a function of the molar amount of H_2O_2 present in the well. Then use the fitted curve equation to convert slope values for the wells containing mitochondria to molar values (e.g., pmol per min).

3.5 Processing the Well Contents for NMR-Based ATP Assay

1. Add 0.39 mL of sample dilution buffer, 50 μL of deuterium oxide (D_2O), and 40 μL of assay well supernatant to a standard 7 in. (length) × 5 mm (outer diameter) NMR tube.

2. Deliver the prepared NMR samples (kept at 4 °C) to the NMR facility.

3.6 NMR Spectroscopy for Quantifying ATP Production

1. In our studies, we use a Bruker Avance II 500 MHz NMR spectrometer equipped with a 5 mm TXI triple resonance non-cryoprobe operating at 37 °C. The spectrometer is also equipped with an automatic sample changer capable of holding a maximum of 60 samples. The amount of ATP produced by the mitochondria is quantified by measuring the amount of 2DOGP produced from 2DOG in the presence of hexokinase as described above.

2. For precise measurement of the amount of 2DOGP formed, duplicate or triplet control samples are prepared during the assay (*see* Subheading 3.3). The control samples have the same

total volume of 60 μL as the other samples but only contain 5 mM 2DOG or [6-^{13}C]2DOG (depending on whether 1D or 2D NMR methods are used for measurement, respectively. *See* **Note 8**), 5 U/mL hexokinase, and 10 mM ATP in 1× respiration medium. These control samples are then subjected to the same protocol for NMR sample preparation (*see* Subheading 3.5). Due to the presence of excess ATP, these control samples have the 2DOG being fully converted into 2DOGP. Therefore, these control samples serve as standards for quantification of 2DOGP formation within the mitochondrial samples and are also used to check for assay reproducibility since duplicate or triplet control samples are used.

3. Load the control samples and mitochondrial samples onto the sample changer. We can run 60 samples continuously without interruption at this spectrometer.

4. Use a control sample or mitochondrial sample to lock, tune, and shim. Save the optimized shimming parameters which serve as the starting shimming setting for the subsequent automatic robot run. Also, use this sample to calibrate ^{1}H and ^{13}C channel pulse widths (^{13}C pulse calibration is needed only when [6-^{13}C]2DOG is used).

5. For Bruker TopSpin software, start the automation program ICONNMR. Set up the automatic robot run by choosing 1D ^{1}H NMR experiment if unlabeled 2DOG is used or choosing 2D ^{13}C/^{1}H HSQC NMR experiment if [6-^{13}C]2DOG is used in the samples. Enter appropriate pulse program parameters such as pulse power level, pulse width, relaxation delay, number of scans, water presaturation parameters, number of t1 increments, etc. Also, set to tune ^{1}H channel and shim on every sample. It takes about 25 min to collect either a 1D ^{1}H spectrum or a 2D ^{13}C/^{1}H HSQC spectrum for each sample. Start the robot run.

6. Transfer the collected NMR data to another Linux computer where the acquired data are processed by using NMRPipe package [3] and analyzed using NMRView [4].

7. Using the assigned ^{1}H and ^{13}C NMR resonances of 2DOG and 2DOGP [1], measure the peak intensities in NMRView of 2DOGP in the processed control and mitochondrial samples.

8. Quantify the amount of 2DOGP present in the mitochondrial samples by comparing the peak intensity of 2DOGP of the mitochondrial sample with that of the control/standard samples. Since a fixed amount of 2DOG (5 mM) is used in the assay incubation (*see* Subheading 3.3), the amount of 2DOGP formed corresponds to the amount of reduction of 2DOG and

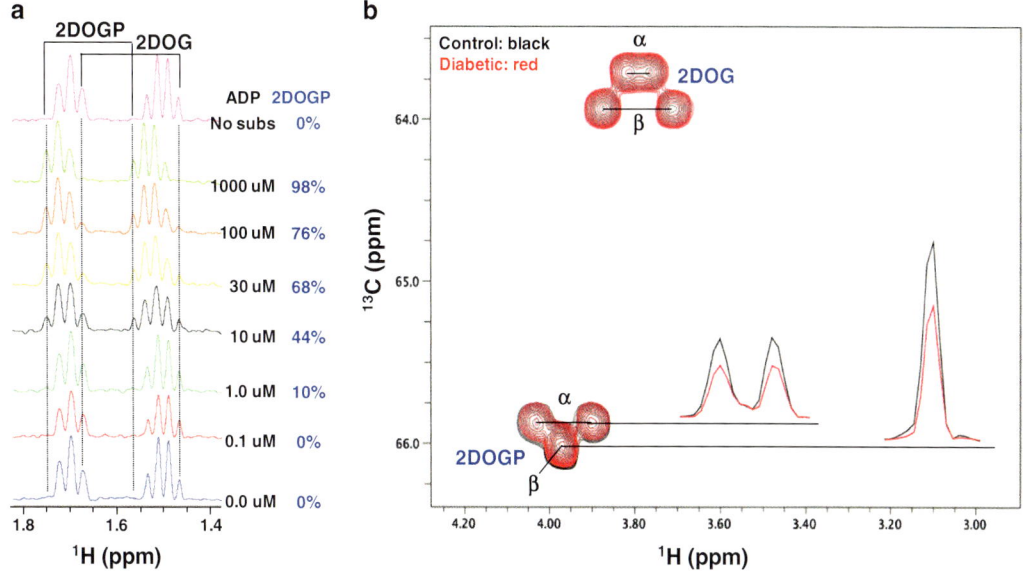

Fig. 3 Quantification of ATP generated by mitochondria via measurement of 2DOGP converted from 2DOG by 1D ¹H NMR method (**a**) and 2D ¹³C/¹H HSQC NMR method (**b**). In panel **a**, representative data are derived from the incubation of isolated gastrocnemius muscle mitochondria from a control rat. Mitochondria (0.1 mg/mL) are incubated for 20 min in respiration medium containing 5 mM 2DOG, 5 U/mL hexokinase, and variable concentrations of ADP and fueled with 5 mM succinate, 5 mM glutamate, and 1 mM malate. It clearly shows that the percent 2DOGP formed by the mitochondrial incubation increases with increasing amount of ADP used in the incubation. No subs means that substrates of succinate, glutamate, and malate are not added. In panel **b**, overlay of representative 2D ¹H/¹³C HSQC spectra is shown for the H6/C6 region of 2DOG and 2DOGP of the samples that contain mitochondria from a control (*black*) and diabetic (*red*) rat. Mitochondria (0.1 mg/mL) are incubated for 20 min in respiration medium containing 5 mM [6-¹³C]2DOG, 5 U/mL hexokinase, and 1 mM ADP and fueled with 5 mM succinate, 5 mM glutamate, and 1 mM malate. The cross peaks are labeled and 1D slices through the cross peaks of 2DOGP are included and clearly show that the control mitochondria produce more 2DOGP or ATP than the diabetic mitochondria

thus can be expressed as percentage of conversion from 2DOG to 2DOGP. The amount of 2DOGP determined for the mitochondrial samples corresponds to the amount of ATP produced by the mitochondria. Representative data derived from the analysis by 1D and 2D NMR methods are shown in Fig. 3.

3.7 Calculation of ATP Production Rates

1. Determine the percent conversion of 2DOG to 2DOGP. From the known initial 2DOG concentration and percent conversion; determine the molar amount of 2DOGP formed which equals to the molar amount of ATP generated. Calculate ATP production rates in the microplate assay wells based on the volume in the microplate, the amount used in NMR, and incubation time.

4 Notes

1. The cytoplasm of permeabilized cells is replaced by the respiratory media enabling assessment of mitochondrial function independent of cytoplasmic events. The methods described herein are for isolated mitochondria. Permeabilized cells can also be used by adapting the methodology beginning in Subheading 3.3.

2. Cytoplasmic hexokinase is competitively inhibited by 2-deoxyglucose, but this is not an issue in isolated mitochondria or permeabilized cells wherein the cytoplasm is replaced by the incubation medium.

3. The 2DOG clamp has been used in the past to assess mitochondrial-bound HK activity at constant ADP [5], but not to assess mitochondrial physiology or ATP production as described herein. Mitochondrial HK is not an issue in our assay since the amount of added HK is in excess.

4. There are several advantages to our ATP assay. First, $\Delta\Psi$ is clamped, allowing assessment of ATP as a function of its direct driving force (i.e., $\Delta\Psi$). Second, the assay is sensitive enough to measure ATP production using small amounts of mitochondria. We found [1] that the 1D ^{1}H NMR method is 34-fold more sensitive and the 2D ^{1}H/^{13}C HSQC NMR method is 41-fold more sensitive when compared to 1D ^{31}P NMR for ATP detection. The higher sensitivity of the 2D NMR method is due to the fact that the chemical shifts of the two H6 protons of the β-anomeric form of 2DOGP are degenerate, resulting in detection of one single C6/H6 HSQC cross peak with high intensity. Third, both the 1D and 2D NMR spectra are highly specific. Fourth, throughput is quite good since we add mitochondria to multiple wells of a 96-well plate, incubate, spin off the mitochondria, and then save the samples for NMR analysis. Fifth, a powerful aspect is that we can assess mitochondrial ROS simultaneously with ATP quantification by NMR since the ROS probe does not interfere with mitochondrial ATP production or with NMR detection of 2DOGP for ATP quantification [1]. Therefore, the ATP assay described herein has clear advantages over conventional methods. Fluorescent and bioluminescent measurements are sensitive, but lack specificity and may be confounded by background interference or variations in light emission [6]. Quantifying ATP by phosphorous NMR is not sensitive enough and requires long acquisition times unless large amounts of mitochondria are used. High-pressure liquid chromatography is precise, but cumbersome. The ATP/O ratio is often considered representative of ATP production. However, ATP itself is not measured and the ratio can be altered by any condition that affects uncoupling, ATP synthase, or respiration.

5. To increase the mitochondrial yield with fibrous tissues, save the first low-speed pellet for resuspension and regrinding using the ground-glass-type homogenizer. Spin the homogenate at low speed. Combine the supernatants from both low-speed spins prior to the high-speed spin.

6. Crude mitochondria pellets obtained by standard methods can be further purified using a self-generating Percoll® gradient. We use a published method [7] described for liver mitochondria (which we adapted for heart and skeletal muscle mitochondria) by using a centrifugal force of $95,000 \times g$ to establish the gradient. Others recommend only $30,000 \times g$. In our initial attempts by using $30,000 \times g$, we did not get acceptable results as evidenced by loose, fluffy, and diffuse bands or mitochondria that stayed only at the top of the tube. When we increased the centrifugal force to $95,000 \times g$, we obtained excellent separation of mitochondria from contaminants. In this way the mitochondrial band (near the bottom of tube) is clearly separated from less dense contaminants and broken mitochondria (upper and middle bands, respectively).

7. The observed increase in Amplex Red fluorescence over time is usually quite linear. The slope for each well can be converted to molar H_2O_2 per unit time per mg of mitochondrial protein with the aid of the standard curve.

8. The 2D NMR method is preferred since it is more sensitive and has almost no background interference.

Acknowledgments

This work was supported by Veterans Affairs Medical Research Funds, by the National Institute of Health [5R01HL073166], and by the Iowa Affiliate Fraternal Order of the Eagles.

References

1. Yu L, Fink BD, Herlein JA, Sivitz WI (2013) Mitochondrial function in diabetes: novel methodology and new insight. Diabetes 62: 1833–1842

2. Yu L, Fink BD, Herlein JA, Oltman CL, Lamping KG, Sivitz WI (2013) Dietary fat, fatty acid saturation and mitochondrial bioenergetics. J Bioenerg Biomembr 46:33–44

3. Delaglio F, Grzesiek S, Vuister GW, Zhu G, Pfeifer J, Bax A (1995) NMRPipe: a multidimensional spectral processing system based on UNIX pipes. J Biomol NMR 6:277–293

4. Johnson BA, Blevins RA (1994) NMR view: a computer program for the visualization and analysis of NMR data. J Biomol NMR 4: 603–614

5. Da Silva WS, Gomez-Puyou A, De Gomez-Puyou MT, Moreno-Sanchez R, De Felice FG, De Meis L, Oliveira MF, Galina A (2004) Mitochondrial bound hexokinase activity as a preventive antioxidant defense: steady-state ADP formation as a regulatory mechanism of membrane potential and reactive oxygen species generation in mitochondria. J Biol Chem 279:39846–39855

6. Manfredi G, Spinazzola A, Checcarelli N, Naini A (2001) Assay of mitochondrial ATP synthesis in animal cells. Methods Cell Biol 65:133–145

7. Hovius R, Lambrechts H, Nicolay K, de Kruijff B (1990) Improved methods to isolate and subfractionate rat liver mitochondria. Lipid composition of the inner and outer membrane. Biochim Biophys Acta 1021:217–226

Live-Cell Assessment of Mitochondrial Reactive Oxygen Species Using Dihydroethidine

Marleen Forkink, Peter H.G.M. Willems, Werner J.H. Koopman, and Sander Grefte

Abstract

Reactive oxygen species (ROS) play an important role in both physiology and pathology. Mitochondria are an important source of the primary ROS superoxide. However, accurate detection of mitochondrial superoxide especially in living cells remains a difficult task. Here, we describe a method and the pitfalls to detect superoxide in both mitochondria and the entire cell using dihydroethidium (HEt) and live-cell microscopy.

Key words MitoSOX, Membrane potential, Imaging

Abbreviations

FCCP	Carbonyl cyanide-p-trifluoromethoxyphenylhydrazone
HEt	Dihydroethidium
HT	HEPES-Tris
mito-HEt	Mito-dihydroethidium
ROS	Reactive oxygen species
TPP	Triphenylphosphonium
$\Delta\psi$	Mitochondrial membrane potential

1 Introduction

Reactive oxygen species (ROS) are generally known to damage cellular macromolecules but are also increasingly recognized as signaling molecules [1–3]. Mitochondria are an important source of mitochondrial ROS in both physiology and pathology, even though they might not be the main cellular producers [4]. The primary ROS superoxide ($O_2^{.-}$) is produced by membrane-bound parts of the respiratory chain and a number of soluble mitochondrial proteins. Although $O_2^{.-}$ cannot traverse membranes and has

Volkmar Weissig and Marvin Edeas (eds.), *Mitochondrial Medicine: Volume I, Probing Mitochondrial Function*,
Methods in Molecular Biology, vol. 1264, DOI 10.1007/978-1-4939-2257-4_15, © Springer Science+Business Media New York 2015

only limited reactivity with biological targets, it might function as a redox signal in mitochondria [5–7]. Since mitochondrial signaling and cellular function are intricately linked, it is important to measure mitochondrial $O_2^{\cdot-}$ in the proper context—i.e., the living cell. However, it has proven difficult to develop fluorescent probes or sensors that are specific for $O_2^{\cdot-}$ and sensitive enough to compete with superoxide dismutase (SOD), which converts $O_2^{\cdot-}$ to hydrogen peroxide at an extremely high rate. For example, a recent development in $O_2^{\cdot-}$ detection involved the use of circularly permuted YFP (cpYFP) to detect superoxide flashes [8–10]. However, critical reevaluation of the data suggests that, under certain conditions, cpYFP might not only respond to changes in $O_2^{\cdot-}$ levels but also to pH changes [11–13]. Quantifying the oxidation of dihydroethidium (HEt) is another widely applied strategy to detect $O_2^{\cdot-}$ in living cells. HEt is membrane permeable and reacts with $O_2^{\cdot-}$ to form the specific fluorescent product 2-hydroxyethidium (2-OH-Et$^+$) (Fig. 1a). In addition, when HEt acts as a hydride acceptor, its oxidation leads to the formation of the nonspecific fluorescent product ethidium (Et$^+$) [14, 15].

Whereas the fluorescence excitation peaks of 2-OH-Et$^+$ (Fig. 1b, upper blue curve) and Et$^+$ (lower blue curve) overlap at around 500 nm, 2-OH-Et$^+$ has one additional peak at 396 nm which has been used for more specific detection of 2-OH-Et$^+$ [16]. Still, this assay should be considered semiquantitative due to the unknown reactivity of $O_2^{\cdot-}$ with SOD and the possible oxidation of

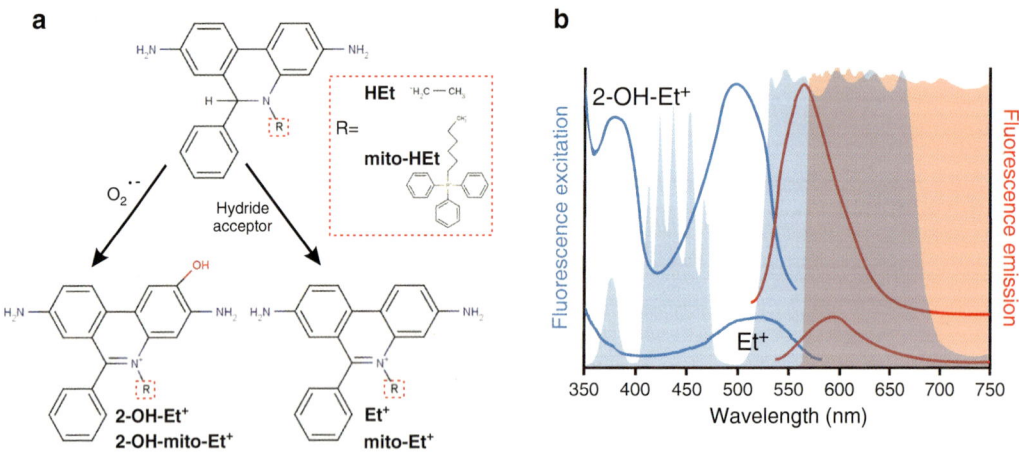

Fig. 1 Chemical structures and excitation spectra of HEt and mito-HEt. (**a**) HEt and mito-HEt specifically react with $O_2^{\cdot-}$ to form the reaction product 2-OH-(mito-)Et$^+$ and (mito-)Et$^+$ as a nonspecific by-product. The inset shows the HEt- and mito-HEt-specific groups. Figure was adapted from Ref. 21. (**b**) Fluorescence excitation (*blue*) and emission (*red*) spectra of 2-OH-Et$^+$ (*top traces*) and Et$^+$ (*lower traces*) superimposed with the excitation and emission filters used in our microscope setup. Figure was adapted from Ref. 22

HEt by cytochrome c and because HEt can catalyze $O_2^{\cdot-}$ dismutation [16, 17]. Although HEt oxidation can occur throughout the cell, both 2-OH-Et$^+$ and Et$^+$ are positively charged and therefore accumulate in the nucleus where they intercalate with nucleic acids and in the mitochondrial matrix due to the inside-negative membrane potential ($\Delta\psi$) of this organelle. Utilizing $\Delta\psi$, the more mitochondria-specific $O_2^{\cdot-}$ probe MitoSOX red (i.e., mito-HEt) was designed, which consists of HEt extended with a cationic triphenylphosphonium (TPP) side group [16]. Here, we describe the critical loading procedure for different intact cells and methods to measure HEt-sensitive ROS using live-cell microscopy.

2 Materials

1. Cells cultured on a glass coverslip (ø24 mm, Thermo Scientific, Etten-Leur, the Netherlands) placed in a 35-mm CELLSTAR tissue culture dish (Sigma-Aldrich) or disposable incubation chamber (WillCo Wells BV) (*see* **Note 1**).

2. A microscope system with the following setup or similar; a monochromator (Polychrome IV, TILL Photonics) allowing excitation with 405 and/or 490 nm light, a 525DRLP dichroic mirror (Omega Optical Inc.) and 565ALP emission filter (Omega Optical Inc.), and an image capturing device (f.i. Cool SNAP HQ monochrome CCD camera (Roper Scientific)). The microscope should be equipped with an environmental control system to sustain cell viability.

3. 5× HEPES buffer: 662.2 mM NaCl, 50 mM HEPES, 21.06 mM KCl, 6.1 mM $MgCl_2$ (*see* **Note 2**).

4. 1× HEPES-Tris buffer: 10 mM HEPES, 132 mM NaCl, 4.2 mM KCl, 1.2 mM $MgCl_2$, 1.0 mM $CaCl_2$, 5.5 mM D-glucose. Adjust the pH to 7.4 using Tris-base (*see* **Note 3**).

5. HEt stock solution (31.7 mM in DMSO): Dissolve 1 mg of HEt powder (Invitrogen) in 100 µl DMSO. Prepare 10 µl aliquots in brown Eppendorf tubes, overlay with N_2 gas, and store at –20 °C (*see* **Note 4**).

6. HEt working solution (5 mM in DMSO): Thaw an aliquot of HEt stock solution and add 53.4 µl of DMSO to yield the 5 mM HEt working solution.

7. Mito-HEt working solution (500 µM in DMSO): Dissolve 50 µg of MitoSOX powder (Invitrogen) in 132 µl DMSO. Prepare 4 µl aliquots in brown Eppendorf tubes, overlay with N_2 gas, and store at –20 °C.

3 Methods

3.1 Imaging of Dihydroethidium Oxidation

1. Seed the cells at such densities that the cells are 70–80 % confluent at the time of imaging. This allows subtraction of the background signal (*see* **Note 5**).

2. Transfer the cells to an incubator close to the microscope system at least 1 h prior to imaging (*see* **Note 1**).

3. Aliquot 2 μl of HEt working solution in Eppendorf tubes.

4. Take 1 ml of medium from the culture dish, add it to the aliquoted HEt, and vortex for 5–10 s to prepare a final concentration of 10 μM (*see* **Note 6**).

5. Replace the rest of the medium from the dish by the medium/HEt solution.

6. Incubate the cells in a 37 °C, 5 % CO_2 incubator for exactly 10 min (*see* **Note 7**).

7. Wash the coverslip three times with 1 ml phosphate buffered saline (PBS).

8. Replace the PBS by pre-warmed HT buffer and mount the coverslip in a Leiden chamber on the microscope [18]. When using oil-based objectives, be careful to remove all excess buffer from the bottom of the coverslip with a tissue to avoid optical artifacts.

9. Start loading of a next coverslip as soon as loading of the current one is complete.

10. Preferably, take images at both 405 and 490 nm excitation. Set the exposure time to 100 ms.

11. Record at least ten different images, each containing ~15 cells (*see* **Note 8** and **9**).

3.2 Imaging of Mito-dihydroethidium Oxidation

1. Seed the cells at such densities that the cells are 70–80 % confluent at the time of imaging. This allows subtraction of the background signal.

2. Transfer the cells to an incubator close to the microscope system at least 1 h prior to imaging.

3. Thaw the 4 μl mito-HEt contents of a brown Eppendorf tube.

4. Take 1 ml of medium from the culture dish, add it to mito-HEt, and vortex for 5–10 s to prepare a final concentration of 2 μM (*see* **Note 10**).

5. Replace the rest of the medium from the dish by the medium/mito-HEt solution.

6. Incubate the cells in a 37 °C, 5 % CO_2 incubator for exactly 10 min (*see* **Note 7**).

7. Wash the coverslip three times with 1 ml (PBS).

8. Replace the PBS by pre-warmed HT buffer and mount the coverslip in a Leiden chamber on the microscope [18]. When using oil-based objectives, be careful to remove all excess buffer from the bottom of the coverslip with a tissue to avoid focusing problems.

9. Start loading of a next coverslip as soon as loading of the current one is complete.

10. Preferably, take images at both 405 and 490 nm excitation. Set the exposure time to 500 ms.

11. Record an image sequence of one field of view, acquiring one image every 5 or 10 s for a total of at least 2 min (*see* **Note 11**).

3.3 Image Analysis

1. Load the raw images into an image analysis program such as MetaMorph® (Molecular Devices Corporation, Palo Alto, CA, USA), Image-Pro Plus (Media Cybernetics), or the open source program ImageJ (http://rsb.info.nih.gov/ij/).

2. Draw circular regions of interest (ROIs) in (a) a mitochondria-dense area, (b) surrounding the nucleus, and (c) just outside every cell to correct for background intensity (*see* **Note 9**).

3. Export the average ROI gray value to a spreadsheet program such as Excel (Microsoft) and calculate the background-subtracted values of the mitochondria and the nucleus of each cell.

4 Notes

1. Due to the positive charge of mito-HEt and the reaction products of HEt oxidation, the fluorescence intensity measured in the mitochondria is $\Delta\psi$ dependent (Fig. 2). Since the mitochondrial membrane potential is very sensitive to environmental changes (e.g., temperature, pH), we advise to culture the cells in separate dishes and to allow cells to recover in an incubator close to the microscope system at least 1 h prior to imaging. In addition, we have noticed that HEt can react with residues on a number of glass-bottom culture dishes, leading to fluorescent "spots" in the background. Therefore, always check background levels for disturbances of that kind.

2. We store a large volume of 5× HEPES buffer at 4 °C, which can be stored for at least 6 months.

3. We advise to make the 1×HT buffer supplemented with the required substrates at the day of imaging. We routinely use 5.5 mM D-glucose for human skin fibroblasts and myoblasts/fibers, whereas HEK293 cells are imaged in HT buffer containing 25 mM D-glucose. In addition, one can consider supplementing pyruvate and/or glutamine. If necessary, the HT buffer can be stored for ~1 week at 4 °C.

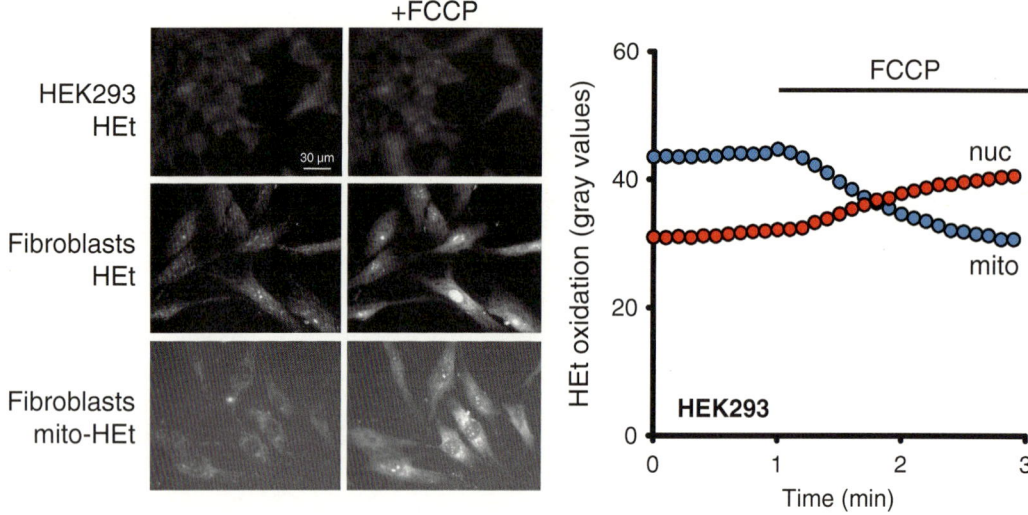

Fig. 2 Localization of HEt oxidation products is dependent on the membrane potential. Addition of the mito-chondrial protonophore carbonyl cyanide-p-trifluoromethoxyphenylhydrazone (FCCP, 0.5 μM) acutely induces the translocation of fluorescent HEt and mito-HEt oxidation products from the mitochondria to the nucleus in HEK293 cells and human skin fibroblasts

4. Always keep HEt solutions protected from light and air, since this might cause extracellular HEt oxidation.

5. Proper background correction close to the cells is of particular importance when the fluorescent signal is close to the background intensity. Such a background correction is not possible at a confluency higher than 80 %. Therefore, adjust the seeding condition according to the number of days the cells are cultured.

6. We prefer to load the cells with HEt in the collected cell culture medium to detect ROS levels in the exact cell culture conditions. However, we have noticed lower fluorescence intensities in medium-loaded cells as compared to cells loaded with HEt in the HT buffer. This is possibly caused by the binding of the HEt to proteins present in the medium (i.e., serum) but might also be due to increased fluorescence levels in HT medium (Fig. 3a).

7. The HEt incubation time should be determined experimentally for each cell line separately. To be able to semiquantitatively measure HEt oxidation, the increase in fluorescence intensity from oxidation products should be linear. To that end, we mount unloaded cells onto the microscope system in HT buffer, start imaging every 10 s, and add the required HEt in HT buffer in 1:1 ratio. We then calculate the maximum time of incubation in which the increase is still linear. An incubation time of 10 min is most often used for HEK293 and human skin fibroblasts (Fig. 3b). However, in some cell types such as primary mouse

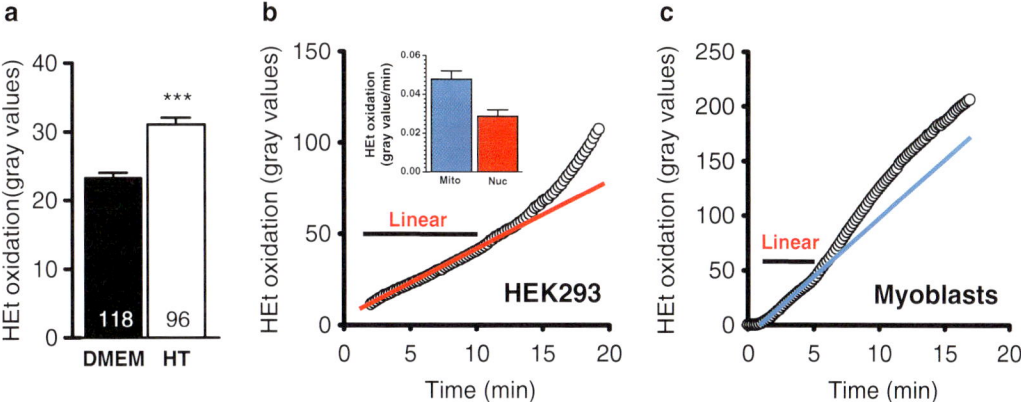

Fig. 3 HEt loading procedure in intact cells. (**a**) HEK293 cells were incubated with HEt for exactly 10 min in the collected cell culture medium (DMEM) or washed and incubated for exactly 10 min in HT buffer. HEt oxidation was measured at 490 nm. The data presented represent the mean ± standard error of the mean (SEM) of two different experiments. Statistical significance was assessed using a Mann–Whitney test. Numerals within bars indicate the number of analyzed cells. (**b**) Representative trace of nuclear HEt oxidation fluorescence intensity at 490 nm excitation light in HEK293 cells. Up to approximately 10 min, the increase remains linear. Note that the mitochondrial increase occurs faster in these cells than in the nucleus. (**c**) Representative trace of mitochondrial HEt oxidation fluorescence intensity at 490 nm in myoblasts. The increase is linear during 4.5 min

myoblasts/myotubes, it appears to be safer to use shorter incubation times because the signal increase deviates from linearity within 10 min (Fig. 3c).

8. When trying to image HEt oxidation at 405 nm, we advise to use the 490 nm light to locate cells and find a good focus, because in our experimental setup, fluorescence intensity is higher at 490 nm than at 405 nm excitation (Fig. 4a). Moreover, we routinely only measure fluorescence from 490 nm excitation since, in some cell types, the 405 nm signal is too low for reliable quantification. However, relative changes in emission fluorescence observed following 405 and 490 nm excitation are similar, justifying the use of 490 nm as a measure of HEt oxidation [19].

9. It is sufficient to have only a portion of the cells in the field of view, since only a certain area of the cell and not the entire cell is measured.

10. We advise to use a mito-HEt concentration that is as low as possible, but still revealing detectable levels of fluorescence. In theory, the positive charge accumulating in the mitochondrial matrix might interfere with mitochondrial bioenergetics and therefore report incorrect HEt oxidation values. In addition, Robinson et al. reported nuclear fluorescence to occur at concentrations as low as 2 μM and therefore advised to use between 0.1 and 2.5 μM mito-HEt [16].

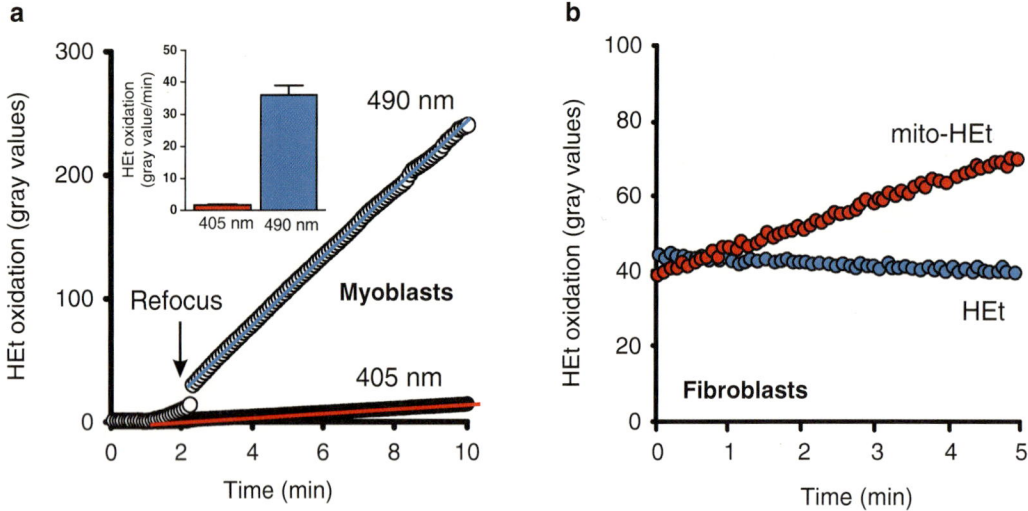

Fig. 4 Fluorescence intensity from 405 to 490 nm excitation and measurement of HEt versus mito-HEt oxidation. (**a**) Simultaneous measurement of HEt oxidation using 405 and 490 nm excitation light in myoblasts. (**b**) Human skin fibroblasts were incubated with 10 μM HEt or 1 μM mito-HEt, washed, and imaged every 5 s using 490 nm excitation light. Staining with HEt yields end point signals, whereas mito-HEt signals continue to increase in time

11. Removing excess HEt after incubation effectively removes all non-oxidized HEt, and therefore the signal remains stable for at least 10 min of continuous imaging [20]. The TPP moiety of mito-HEt induces accumulation of non-oxidized mito-HEt in the mitochondrial matrix that is insensitive to washing. Consequently, the fluorescence signal continues to increase after washing, and therefore we advise to measure the slope of the increase after loading and washing (Fig. 4b).

Acknowledgments

This research was supported by a grant from the Netherlands Organization for Scientific Research (NWO, No: 911-02-008), the Energy4All Foundation, the NWO Centers for Systems Biology Research initiative (CSBR09/013V), and a grant from the Institute for Genetic and Metabolic Disease (IGMD) of the Radboud University Medical Center (RUMC) to W.J.H.K. We are grateful to Dr.A. S. De Jong (Dept. of Biochemistry, RUMC) for performing the HEt and mito-HEt experiments on human skin fibroblasts.

References

1. Finkel T (2012) Signal transduction by mitochondrial oxidants. J Biol Chem 287: 4434–4440

2. Murphy MP, Holmgren A, Larsson NG (2011) Unraveling the biological roles of reactive oxygen species. Cell Metab 13:361–366

3. Distelmaier F, Valsecchi F, Forkink M et al (2012) Trolox-sensitive reactive oxygen species regulate mitochondrial morphology, oxidative phosphorylation and cytosolic calcium handling in healthy cells. Antioxid Redox Signal 17:1657–1669

4. Brown GC, Borutaite V (2012) There is no evidence that mitochondria are the main source of reactive oxygen species in mammalian cells. Mitochondrion 12:1–4

5. Zhou L, Aon M, Almas T et al (2010) A reaction–diffusion model of ROS-induced ROS release in a mitochondrial network. PLoS Comput Biol 6:e1000657

6. Tormos KV, Anso E, Hamanaka RB et al (2011) Mitochondrial complex III ROS regulate adipocyte differentiation. Cell Metab 14: 537–544

7. Murphy M (2009) How mitochondria produce reactive oxygen species. Biochem J 417:1–13

8. Wang W, Fang H, Groom L et al (2008) Superoxide flashes in single mitochondria. Cell 134:279–290

9. Pouvreau S (2010) Superoxide flashes in mouse skeletal muscle are produced by discrete arrays of active mitochondria operating coherently. PLoS One 5:e13035

10. Fang H, Chen M, Ding Y et al (2011) Imaging superoxide flash and metabolism-coupled mitochondrial permeability transition in living animals. Cell Res 21:1295–1304

11. Muller FL (2009) A critical evaluation of cpYFP as a probe for superoxide. Free Radic Biol Med 47:1779–1780

12. Schwarzländer M, Murphy MP, Duchen MR et al (2012) Mitochondrial "flashes": a radical concept repHined. Trends Cell Biol 22: 503–508

13. Wei-Lapierre L, Gong G, Gerstner BJ et al (2013) Respective contribution of mitochondrial superoxide and pH to Mt-cpYFP flash activity. J Biol Chem 288:10567–10577

14. Zhao H, Kalivendi S, Zhang H et al (2003) Superoxide reacts with hydroethidine but forms a fluorescent product that is distinctly different from ethidium: potential implications in intracellular fluorescence detection of superoxide. Free Radic Biol Med 34: 1359–1368

15. Zhao H, Joseph J, Fales HM et al (2005) Detection and characterization of the product of hydroethidine and intracellular superoxide by HPLC and limitations of fluorescence. Proc Natl Acad Sci U S A 102:5727–5732

16. Robinson KM, Janes MS, Pehar M et al (2006) Selective fluorescent imaging of superoxide in vivo using ethidium-based probes. Proc Natl Acad Sci U S A 103:15038–15043

17. Benov L, Sztejnberg L, Fridovich I (1998) Critical evaluation of the use of hydroethidine as a measure of superoxide anion radical. Free Radic Biol Med 25:826–831

18. Ince C, Beekman RE, Verschragen G (1990) A micro-perfusion chamber for single-cell fluorescence measurements. J Immunol Methods 128:227–234

19. Forkink M, Smeitink JAM, Brock R et al (2010) Detection and manipulation of mitochondrial reactive oxygen species in mammalian cells. Biochim Biophys Acta 1797: 1034–1044

20. Koopman W, Verkaart S, Visch H et al (2005) Inhibition of complex I of the electron transport chain causes O_2-mediated mitochondrial outgrowth. Am J Physiol Cell Physiol 288: C1440–C1450

21. Zielonka J, Vasquez-Vivar J, Kalyanaraman B (2008) Detection of 2-hydroxyethidium in cellular systems: a unique marker product of superoxide and hydroethidine. Nat Protoc 3: 8–21

22. Zielonka J, Kalyanaraman B (2010) Hydroethidine- and MitoSOX-derived red fluorescence is not a reliable indicator of intracellular superoxide formation: another inconvenient truth. Free Radic Biol Med 48:983–1001

Chapter 16

Detection and Differentiation Between Peroxynitrite and Hydroperoxides Using Mitochondria-Targeted Arylboronic Acid

Jacek Zielonka, Adam Sikora, Jan Adamus, and B. Kalyanaraman

Abstract

The development of boronic probes enabled reliable detection and quantitative analysis of hydrogen peroxide and peroxynitrite. The major product, in which boronate moiety of the probe is replaced by the hydroxyl group, is however common for both oxidants. Here, we describe how *ortho*-isomer of mitochondria-targeted phenylboronic acid can be used to detect and differentiate peroxynitrite-dependent and peroxynitrite-independent probe oxidation. This method highlights the detection and quantification of both the major, phenolic product and the minor, peroxynitrite-specific nitrated product of probe oxidation.

Key words Hydrogen peroxide, Peroxynitrite, Mitochondria-targeted probes, Boronic probes, *o*-MitoPhB(OH)$_2$, HPLC-MS

1 Introduction

Boronate-based probes were developed over the last decade for the detection of hydrogen peroxide (H_2O_2) and peroxynitrite ($ONOO^-$) in biological systems [1, 2]. Mitochondria-targeted boronate probes were developed to monitor hydrogen peroxide in mitochondria [3–6]. Boronates react directly and stoichiometrically with both H_2O_2 and $ONOO^-$ [7, 8]. This is in contrast to more classical fluorogenic probes (e.g., dichlorodihydrofluorescein, DCFH, and dihydrorhodamine, DHR), which need the catalysts (e.g., iron, heme proteins) or react with the products of $ONOO^-$ decomposition [9, 10]. Additionally, in contrast to boronates, both DCFH and DHR form intermediate radical species that reduce oxygen to superoxide radical anion [10, 11]. However, most boronates lack specificity needed to distinguish between hydrogen peroxide and peroxynitrite in a complex biological system, as discussed below [2, 7, 12]. In a simple system,

Volkmar Weissig and Marvin Edeas (eds.), *Mitochondrial Medicine: Volume I, Probing Mitochondrial Function*,
Methods in Molecular Biology, vol. 1264, DOI 10.1007/978-1-4939-2257-4_16, © Springer Science+Business Media New York 2015

when bolus amounts of the reactants are quickly mixed, this limitation can be overcome due to significant differences of the reaction rate constants. The rate constant of the reaction of boronates with peroxynitrite is six orders of magnitude higher than that of the analogous reaction with hydrogen peroxide [2, 7]. For example, under the conditions when the reaction with $ONOO^-$ is completed within 100 ms, the reaction with H_2O_2 was not completed even after 12 h [7]. The situation is, however, different when the oxidants react with the probe in biological systems, with the oxidants continuously produced during the incubation with the probe. Therefore, we proposed the use of specific inhibitors of oxidants formation and/or specific scavengers of the oxidants to differentiate between different species responsible for oxidation of boronate probes in cells [2, 13]. Recently, we reported a formation of both major, phenolic product (*o*-MitoPhOH) and minor, nitrated product (*o*-MitoPhNO$_2$) during the reaction of peroxynitrite with *ortho*-isomer of mitochondria-targeted phenylboronic acid (*o*-MitoPhB(OH)$_2$, Fig. 1) [14]. This particular nitrated product is specific for $ONOO^-$ reaction and is not produced by any other oxidant. Here, we show how this unique chemistry can be utilized to selectively detect peroxynitrite in cellular systems [15]. The method is based on

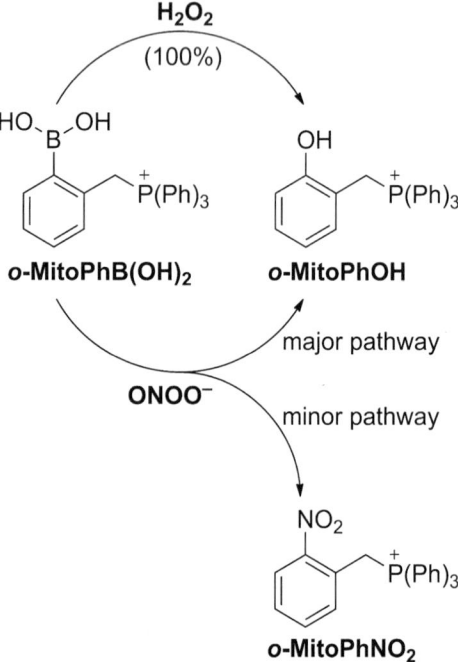

Fig. 1 Scheme of oxidation of *o*-MitoPhB(OH)$_2$. *o*-MitoPhB(OH)$_2$ is oxidized to *o*-MitoPhOH by both $ONOO^-$ and H_2O_2. However, *o*-MitoPhNO$_2$ is formed only in reaction between *o*-MitoPhB(OH)$_2$ and $ONOO^-$ as the minor product

probing the oxidants ($ONOO^-$ and H_2O_2) in cells by incubation of cells with o-MitoPhB(OH)$_2$ followed by extraction and HPLC-MS/MS analysis of the products formed. We describe the protocol for probe preparation, processing of biological samples, and HPLC-MS-based analysis of the products formed.

2 Materials

2.1 Components for the Synthesis of o-MitoPhB(OH)$_2$

1. 2 g of triphenylphosphine.
2. 200 ml of anhydrous diethyl ether.
3. 1 g of 2-(bromomethyl)phenylboronic acid.
4. 100 ml of ethanol.
5. Argon gas.

2.2 Cell Incubation Components

1. Cell growth medium (for RAW 264.7 cells): DMEM containing 10 % FBS, 100 U/ml penicillin, and 0.1 mg/ml streptomycin.
2. Assay medium: HBSS supplemented with 25 mM HEPES (pH 7.4) and dtpa (0.1 mM) (*see* **Note 1**).
3. Solution of 50 mM o-MitoPhB(OH)$_2$ (for the description of synthesis, *see* Subheading 3.1) in DMSO.

2.3 Cell Extraction Components

1. Cell lysis buffer: 10 ml DPBS containing 0.1 % Triton X-100 spiked with 4 µM ethyltriphenylphosphonium cation (TPP$^+$-C$_2$), place on ice (can be stored at 4 °C) (*see* **Note 2**).
2. 100 ml of ice-cold acetonitrile.
3. 10 ml of ice-cold acetonitrile containing 4 µM TPP$^+$-C$_2$, place on ice (can be stored at 4 °C).
4. Protein assay reagent (Bradford reagent).
5. BSA in lysis buffer: 20 mg/ml BSA. Prepare a series of BSA solutions by serial dilutions with the final concentrations of 0.5, 1.0, 1.5, 2, 3, 4, 5, 7, and 10 mg/ml. Keep the solutions on ice.

2.4 HPLC Analysis Components

1. HPLC mobile phase: 0.1 % formic acid in water (mobile phase A) and 0.1 % formic acid in acetonitrile (mobile phase B) (*see* **Note 3**).
2. Solution of o-MitoPhNO$_2$ (50 mM) in DMSO (*see* **Note 4**).
3. Solution of o-MitoPhB(OH)$_2$ (50 mM) in DMSO.
4. Solution of o-MitoPhOH (50 mM) in DMSO (*see* **Note 4**).
5. Water-to-acetonitrile (3:1) mixture spiked with 1 µM TPP$^+$-C$_2$. This will be a solvent for the preparation of standards for HPLC calibration.

6. Mixture of the standards of *o*-MitoPhB(OH)$_2$, *o*-MitoPhOH, and *o*-MitoPhNO$_2$ (0.1 mM each) in water-to-acetonitrile (3:1) mixture containing 0.1 % formic acid spiked with 1 μM TPP$^+$-C$_2$. Prepare serial dilutions down to 1 nM concentration. Use the solvent for standards prepared in the previous step.

3 Methods

3.1 Preparation of o-MitoPhB(OH)$_2$

1. The preparation of the *o*-MitoPhB(OH)$_2$ probe should be carried out inside a hood with well-working ventilation!

2. Prepare a solution of triphenylphosphine (1.048 g, 4.0 mmol) in anhydrous diethyl ether (30 ml).

3. Add 2-(bromomethyl)phenylboronic acid (0.856 g, 4.0 mmol) with constant stirring under argon atmosphere.

4. Stir the reaction mixture for 46 h (using magnetic stirrer) at room temperature under argon atmosphere.

5. A crystalline precipitate of bromide salt of *o*-MitoPhB(OH)$_2$ should be obtained.

6. Filter the suspension. Keep the solid and discard the filtrate.

7. Wash the solid with anhydrous diethyl ether (30 ml). Filter the suspension again and keep the solid.

8. Dissolve the solid in ethanol and precipitate the compound back by addition of diethyl ether (1:1). Filter the powder and keep the solid. Repeat this step once more.

9. Leave the solid to dry. A white powder should be obtained with the melting point of 225–226 °C. Test the identity and purity of the synthesized probe (*see* **Note 5**).

3.2 Cell Incubation with the Probe

1. Prepare the cells according to the experimental conditions to be tested for peroxynitrite formation (*see* **Note 1**).

2. Add *o*-MitoPhB(OH)$_2$ to obtain the final concentration of 50 μM (*see* **Notes 6** and **7**).

3. Incubate the cells for 1 h.

4. Collect an aliquot of the medium (0.1 ml) in 1.5 ml microcentrifuge tube and freeze in liquid nitrogen.

5. Remove the rest of medium and wash the cells twice with ice-cold DPBS.

6. Add 1 ml of ice-cold DPBS and harvest the cells, place the cell suspension in 1.5 ml microcentrifuge tube, and spin down the cells by quick centrifugation. Remove the supernatant and freeze the cell pellet in liquid nitrogen.

7. Frozen cell pellets and media can be stored at –80 °C for at least 1 week before analysis.

3.3 Extraction of the Products

3.3.1 Cell Pellets

1. Preload one set of 1.5 ml microcentrifuge tubes with 0.1 ml of acetonitrile and place on ice.

2. Preload a second set of 1.5 ml microcentrifuge tubes with 0.1 ml of 0.1 % formic acid in water and place on ice.

3. Prepare a clear-bottom 96-well plate for protein assay and place on ice.

4. Place the tubes with frozen cell pellets on ice.

5. Add 0.2 ml of the lysis buffer spiked with TPP$^+$-C$_2$ (4 µM) and lyse the cells by ten syringe strokes using a 0.5 ml insulin syringe with the needle 28 G $\times$ 0.5 in. (0.36 mm $\times$ 13 mm).

6. Transfer 0.1 ml of the cell lysates into the tubes containing ice-cold acetonitrile, vortex for 10 s, and place on ice. Transfer 3 $\times$ 2 µl of the cell lysate aliquots into 3 wells on 96-well plate for the protein assay.

7. Incubate the mixtures of cell lysates with acetonitrile for 30 min on ice.

8. During incubation, measure the protein concentration in the cell lysates using Bradford assay and plate reader with absorption detection (*see* **Note 8**).

9. Vortex the tubes again for 5 s and centrifuge for 30 min at 20,000 $\times g$ at 4 °C.

10. Place the tubes back on ice and transfer 0.1 ml aliquots of the supernatants into the second set of tubes, containing 0.1 % formic acid in water.

11. Vortex the tubes for 5 s and centrifuge for 15 min at 20,000 $\times g$ at 4 °C.

12. Transfer 0.15 ml of the supernatants into HPLC vials preloaded with conical inserts, seal the vials, and place on ice. Once all solutions have been transferred, place the vials in HPLC autosampler precooled to 4 °C.

3.3.2 Media

1. Preload one set of 1.5 ml microcentrifuge tubes with 0.1 ml of 0.1 % formic acid in water and place on ice.

2. Place the tubes with frozen media on ice.

3. Add 0.1 ml of the ice-cold acetonitrile spiked with TPP$^+$-C$_2$ (4 µM) to each tube, vortex for 10 s, and place on ice.

4. Incubate the mixtures of media with acetonitrile for 30 min on ice.

5. Vortex the tubes again for 5 s and centrifuge for 30 min at 20,000 $\times g$ at 4 °C.

6. Place the tubes back on ice and transfer 0.1 ml aliquots of the supernatants into the tubes containing 0.1 % formic acid in water.

7. Vortex the tubes for 5 s and centrifuge for 15 min at $20,000 \times g$ at 4 °C.

8. Transfer 0.15 ml of the supernatants into HPLC vials pre-loaded with conical inserts, seal the vials, and place on ice. Once all solutions have been transferred, place the vials in HPLC autosampler precooled to 4 °C.

3.4 HPLC-MS/MS Analysis of the Extracts

1. Install the column Kinetex Phenyl-Hexyl 50 mm × 2.1 mm, 1.7 μm (Phenomenex) in the HPLC-MS/MS system. The column should be equipped with a UHPLC column filter or guard column to extend the column lifetime.

2. Equilibrate the column with the mobile phase (75 % of mobile phase A and 25 % mobile phase B).

3. Set up the HPLC-MS/MS method and detection parameters according to Tables 1 and 2, respectively (*see* **Note 9**).

4. Test the system by three injections of standards (10 μM) for the reproducibility of retention times and peak intensities for all analytes and internal standard (TPP⁺-C$_2$), as shown in Fig. 2.

Table 1
HPLC method parameters

Flow rate	0.5 ml/min		
Gradient	0 min	75 % A	25 % B
	2 min	68.3 % A	31.7 % B
	2.5 min	0 % A	100 % B
	3 min	0 % A	100 % B
	3.5 min	75 % A	25 % B
Diverter valve	0 min	Waste	
	0.7 min	Detector	
	2.5 min	Waste	

Table 2
MS/MS detection parameters

Analyte	Dominant MRM transition	Reference MRM transition	Reference MRM to dominant MRM intensities ratio	Retention time (min)
TPP⁺-C$_2$	291.10 > 182.95	291.10 > 107.95	0.55	1.12
o-MitoPhB(OH)$_2$	397.00 > 135.00	397.00 > 379.05	0.80	1.55
o-MitoPhOH	369.00 > 107.10	369.00 > 183.05	0.40	1.85
o-MitoPhNO$_2$	397.90 > 262.05	397.90 > 351.10	0.25	1.98

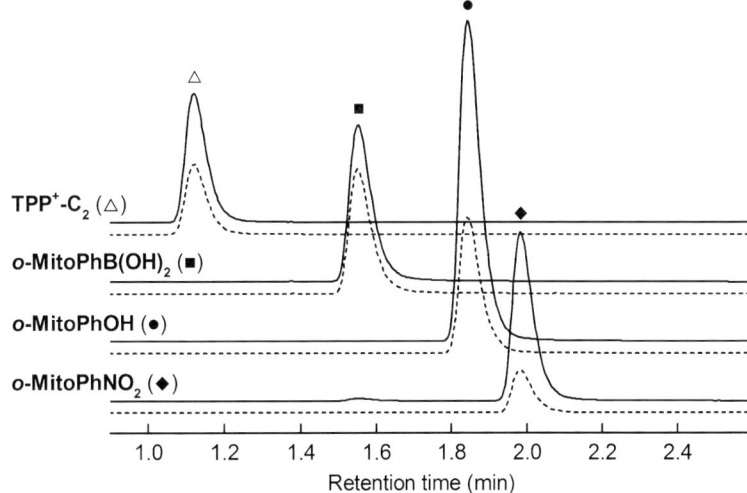

TPP$^+$-C$_2$ ($\triangle$)

o-MitoPhB(OH)$_2$ ($\blacksquare$)

o-MitoPhOH ($\bullet$)

o-MitoPhNO$_2$ ($\blacklozenge$)

Retention time (min)

Fig. 2 HPLC-MS/MS detection of o-MitoPhB(OH)$_2$, o-MitoPhOH, and o-MitoPhNO$_2$. The chromatograms have been obtained with the described method by the injection of 10 μl of the mixture of o-MitoPhB(OH)$_2$, o-MitoPhOH, and o-MitoPhNO$_2$ (1 μM each). For each compound, two MRM transitions are shown: dominant (*solid lines*) and reference (*dashed lines*). The traces have not been scaled and thus represent the actual intensities of each MRM transition in the equimolar mixture

5. Run the analysis of the batch of samples.

6. Include the system and column wash with water-to-methanol (1:1) mixture at the end of batch.

7. Quantify each analyte based on the specific multiple reaction monitoring (MRM) transitions and calibration curves constructed in the concentration range relevant to the samples analyzed (*see* **Note 10**).

8. When appropriate, normalize the concentrations of analytes to the protein levels in cell lysates, as determined by Bradford method.

9. Increase in peak intensities of both o-MitoPhOH and o-MitoPhNO$_2$ (Fig. 3) indicates the formation of ONOO$^-$, while formation of o-MitoPhOH, but not o-MitoPhNO$_2$ (Fig. 4), indicates the presence of other oxidants, most commonly H$_2$O$_2$ (*see* **Note 11**).

4 Notes

1. The medium used may be selected according to experimental design and cell culture needs. However, it is preferred that when monitoring extracellular ONOO$^-$ and/or H$_2$O$_2$, the components of the medium capable of scavenging those

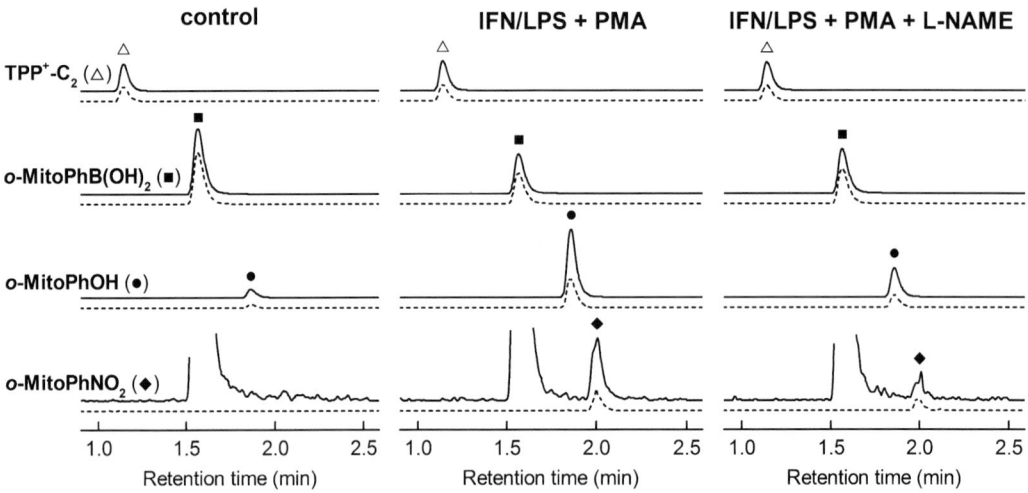

Fig. 3 HPLC-MS/MS analysis of the extracts of RAW 264.7 macrophages activated to produce peroxynitrite. To produce peroxynitrite, RAW 264.7 cells were pretreated overnight with interferon-γ (IFN, 50 U/ml) and lipopolysaccharide (LPS, 1 μg/ml) followed by addition of phorbol 12-myristate 13-acetate (PMA, 1 μM) in the presence or absence of L-NAME (1 mM) [13]. During stimulation with PMA, o-MitoPhB(OH)$_2$ (50 μM) was present and incubated for 1 h. Cell pellets were collected and processed as described in the protocol. The intensities of o-MitoPhOH signals were multiplied by a factor of 5 and intensities of o-MitoPhNO$_2$ signals were multiplied by 1,000 to fit the same scale as of o-MitoPhB(OH)$_2$. *Solid lines* represent the dominant transitions used for quantification, and the *dashed lines* represent the reference transitions used for confirmation of peak identity

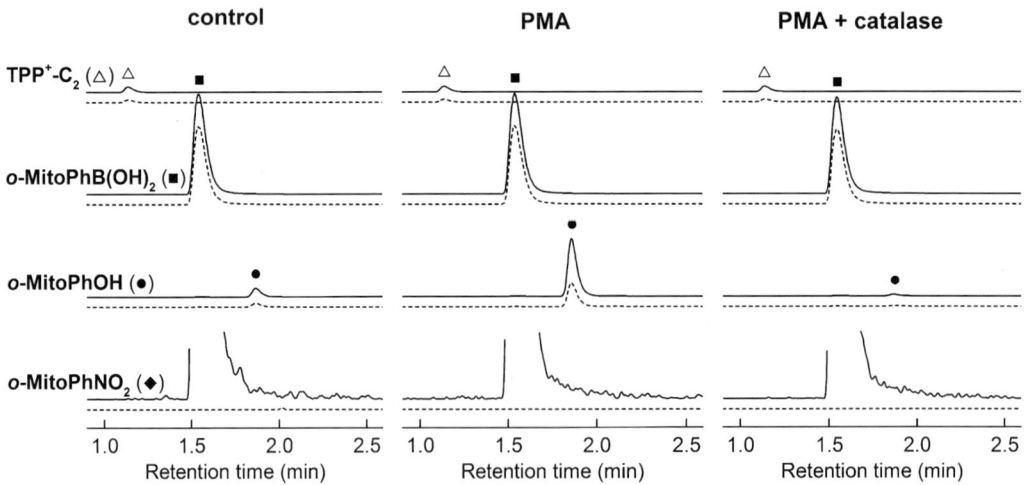

Fig. 4 HPLC-MS/MS analysis of the media of neutrophil-like cells activated to produce hydrogen peroxide. To produce hydrogen peroxide, HL60 cells differentiated for 4 days with all-*trans* retinoic acid were stimulated with PMA (1 μM) in the presence and absence of catalase (1 kU/ml) and co-treated with o-MitoPhB(OH)$_2$. Aliquots of media were collected after 30 min and processed and analyzed as described in the protocol, but with twofold lower concentration of the internal standard (TPP$^+$-C$_2$). The intensities of the peaks for different analytes have been scaled in the same manner as described for Fig. 3

oxidants should be avoided, if possible. For example, pyruvate in the medium may efficiently compete with boronate probes for H_2O_2.

2. Ethyltriphenylphosphonium cation (TPP^+-C_2) is used as an internal standard to compensate for any sample loss during the preparation of samples for analysis and for any changes in detector sensitivity over the course of batch analysis. If significant changes in the peak area of TPP^+-C_2 are observed, normalize the results to TPP^+-C_2 peak area.

3. LC-MS (preferably UHPLC-MS) grade solvents and formic acid should be used. After preparation, mobile phase should be passed through 0.2 μm filter. Prepare only the amount of mobile phase, which is necessary for the experiment. Do not store the mobile phase for longer than 2–3 days to avoid any growth of the biological matter.

4. The standards of the products, o-MitoPhOH and o-MitoPhNO$_2$, are commercially available but can be also synthesized. The phenolic product can be prepared by reaction o-MitoPhB(OH)$_2$ with hydrogen peroxide, followed by addition of catalase to remove excess H_2O_2. The nitrated product can be synthesized in an analogous protocol as described for o-MitoPhB(OH)$_2$, but starting with 2-nitrobenzyl bromide instead of 2-(bromomethyl)phenylboronic acid.

5. Purity of synthesized o-MitoPhB(OH)$_2$ should be tested by HPLC and the compound repurified, if needed. The identity should be confirmed by nuclear magnetic resonance (NMR) analysis. ^{1}H NMR (Bruker 250 MHz, DMSO-d$_6$) δ_{ppm}: 8.18 (s, ^{2}H, *B-OH*), 7.91–7.98 (m, ^{7}H), 7.72–7.79 (m, ^{6}H), 7.34–7.38 (m, ^{2}H), 7.07–7.12 (m, ^{1}H), 5.43 (d, ^{2}H, J = 15.5 Hz, *CH$_2$*). Mass spectrometry analysis: 396.1561 ([M − 1]$^+$, 22 %), 397.1529 ([M]$^+$, 100 %), 398.1568 ([M + 1]$^+$, 25 %), 399.1569 ([M + 2]$^+$, 4 %) [14].

6. The concentration of o-MitoPhB(OH)$_2$ used for probing of ONOO$^-$ should be chosen so as it does not interfere with mitochondrial function. We did not observe significant effects of 50 μM o-MitoPhB(OH)$_2$ on the rate of oxygen consumption by RAW 264.7 cells.

7. When exposing cells to o-MitoPhB(OH)$_2$, it is preferred to add the medium containing the probe, rather than directly adding a solution of concentrated o-MitoPhB(OH)$_2$ in DMSO, to avoid local exposure of cells to high concentrations of DMSO.

8. If plate reader with absorption detection is not available, protein measurements can be carried out using regular spectrophotometer. The volume of cell lysate needed for the assay may be higher, depending on the volume of the spectrophotometer cell.

9. To protect the detector, only portion of eluate is flowed into mass detector. This is achieved by using a diverter valve, which directs the flow into waste before 0.7 min and after 2.5 min after injection. Between 0.7 and 2.5 min, the flow is directed into the detector and the signals recorded.

10. For each compound, two MRM transitions (pairs of parent ion/daughter ion) are recorded. The primary (dominant) transition is used for quantification, while the secondary (reference) transition is used for confirmation of the identity of the analyte. If the ratio of reference to dominant MRM transitions (Table 2) is outside the range allowed (typically ± 30 %), the peak is rejected and not used for quantification. Due to similar m/z values of o-MitoPhB(OH)$_2$ and o-MitoPhNO$_2$, there is a small peak of the boronate appearing in the channel of the nitro derivative (Fig. 2, a peak at 1.55 min), which may show up as the dominant peak, when o-MitoPhB(OH)$_2$ is in high excess comparing to o-MitoPhNO$_2$ (Fig. 3). This peak is not however observed in the reference channel, exemplifying the usefulness of the reference transitions for peak identification.

11. The identity of the oxidizing species can be tested by the application of specific inhibitors and/or scavengers [13, 15]. Confirmation of peroxynitrite involvement may be obtained by testing the inhibitory effects of inhibitors of nitric oxide synthase, for example, L-NAME, as shown in Fig. 3. Similarly, the inhibitory activity of catalase on the yield of o-MitoPhOH (Fig. 4) indicates the involvement of H$_2$O$_2$.

Acknowledgments

This work was supported by NIH grant R01 HL063119 (B.K.). Support from a grant coordinated by JCET, No. POIG. 01.01.02-00-069/09 (supported by the European Union from the resources of the European Regional Development Fund under the Innovative Economy Programme), is acknowledged. A.S. was supported by a grant IP2011 049271 from the Ministry of Science and Higher Education within the "Iuventus Plus" program.

References

1. Lippert AR, Van de Bittner GC, Chang CJ (2011) Boronate oxidation as a bioorthogonal reaction approach for studying the chemistry of hydrogen peroxide in living systems. Acc Chem Res 44:793–804

2. Zielonka J, Sikora A, Hardy M, Joseph J, Dranka BP, Kalyanaraman B (2012) Boronate probes as diagnostic tools for real time monitoring of peroxynitrite and hydroperoxides. Chem Res Toxicol 25:1793–1799

3. Cocheme HM, Quin C, McQuaker SJ, Cabreiro F, Logan A, Prime TA, Abakumova I, Patel JV, Fearnley IM, James AM, Porteous CM, Smith RA, Saeed S, Carre JE, Singer M, Gems D, Hartley RC, Partridge L, Murphy MP (2011) Measurement of H$_2$O$_2$ within

living Drosophila during aging using a ratiometric mass spectrometry probe targeted to the mitochondrial matrix. Cell Metab 13: 340–350

4. Cocheme HM, Logan A, Prime TA, Abakumova I, Quin C, McQuaker SJ, Patel JV, Fearnley IM, James AM, Porteous CM, Smith RA, Hartley RC, Partridge L, Murphy MP (2012) Using the mitochondria-targeted ratiometric mass spectrometry probe MitoB to measure H_2O_2 in living Drosophila. Nat Protoc 7:946–958

5. Dickinson BC, Chang CJ (2008) A targetable fluorescent probe for imaging hydrogen peroxide in the mitochondria of living cells. J Am Chem Soc 130:9638–9639

6. Dickinson BC, Lin VS, Chang CJ (2013) Preparation and use of MitoPY1 for imaging hydrogen peroxide in mitochondria of live cells. Nat Protoc 8:1249–1259

7. Sikora A, Zielonka J, Lopez M, Joseph J, Kalyanaraman B (2009) Direct oxidation of boronates by peroxynitrite: mechanism and implications in fluorescence imaging of peroxynitrite. Free Radic Biol Med 47: 1401–1407

8. Sikora A, Zielonka J, Lopez M, Dybala-Defratyka A, Joseph J, Marcinek A, Kalyanaraman B (2011) Reaction between peroxynitrite and boronates: EPR spin-trapping, HPLC analyses, and quantum mechanical study of the free radical pathway. Chem Res Toxicol 24:687–697

9. Kalyanaraman B (2011) Oxidative chemistry of fluorescent dyes: implications in the detection of reactive oxygen and nitrogen species. Biochem Soc Trans 39:1221–1225

10. Wardman P (2007) Fluorescent and luminescent probes for measurement of oxidative and nitrosative species in cells and tissues: progress, pitfalls, and prospects. Free Radic Biol Med 43:995–1022

11. Zielonka J, Kalyanaraman B (2012) Methods of investigation of selected radical oxygen/nitrogen species in cell-free and cellular systems. In: Pantopoulos K, Schipper HM (eds) Principles of free radical biomedicine. Nova Science Publishers, New York, NY

12. Zielonka J, Sikora A, Joseph J, Kalyanaraman B (2010) Peroxynitrite is the major species formed from different flux ratios of co-generated nitric oxide and superoxide: direct reaction with boronate-based fluorescent probe. J Biol Chem 285:14210–14216

13. Zielonka J, Zielonka M, Sikora A, Adamus J, Joseph J, Hardy M, Ouari O, Dranka BP, Kalyanaraman B (2012) Global profiling of reactive oxygen and nitrogen species in biological systems: high-throughput real-time analyses. J Biol Chem 287:2984–2995

14. Sikora A, Zielonka J, Adamus J, Debski D, Dybala-Defratyka A, Michalowski B, Joseph J, Hartley RC, Murphy MP, Kalyanaraman B (2013) Reaction between peroxynitrite and triphenylphosphonium-substituted arylboronic acid isomers: identification of diagnostic marker products and biological implications. Chem Res Toxicol 26:856–867

15. Zielonka J, Joseph J, Sikora A, Kalyanaraman B (2013) Real-time monitoring of reactive oxygen and nitrogen species in a multiwell plate using the diagnostic marker products of specific probes. Methods Enzymol 526: 145–157

Chapter 17

Time-Resolved Spectrometry of Mitochondrial NAD(P)H Fluorescence and Its Applications for Evaluating the Oxidative State in Living Cells

Julia Horilova, Hauke Studier, Zuzana Nadova, Pavol Miskovsky, Dusan Chorvat Jr., and Alzbeta Marcek Chorvatova

Abstract

Time-resolved fluorescence spectrometry is a highly valuable technological tool to detect and characterize mitochondrial metabolic oxidative changes by means of endogenous fluorescence (Chorvat and Chorvatova, Laser Phys Lett 6: 175–193, 2009). Here, we describe the detection and measurement of endogenous mitochondrial NAD(P)H (nicotinamide adenine dinucleotide (phosphate)) fluorescence directly in living cultured cells using fluorescence lifetime spectrometry imaging after excitation with 405 nm picosecond (ps) laser. Time-correlated single photon counting (TCSPC) method is employed.

Key words Mitochondrial oxidative state, Endogenous NAD(P)H fluorescence, FLIM, Time-resolved spectrometry, Energy metabolism

1 Introduction

Fluorescence lifetime imaging and spectroscopy has made an important impact in the field of monitoring the mitochondrial metabolic processes leading to energy production in living cells and tissues [1–4]. Naturally occurring endogenous fluorophores and their fluorescence (also called autofluorescence) offer the possibility to study noninvasive mitochondrial metabolic processes. Despite their lower fluorescence yield when compared to fluorescence probes or dyes, endogenous fluorophores are highly advantageous for noninvasive tracking of changes in mitochondrial metabolic state by natural means. To evaluate mitochondrial metabolism, NAD(P)H and flavoproteins were proposed to be the marker of metabolic oxidative state thanks to their principal role in energy production [5–7]. Addition of the time resolution into fluorescence spectroscopy enabled a deeper insight into metabolic processes thanks to the ability to distinguish fluorophores

Volkmar Weissig and Marvin Edeas (eds.), *Mitochondrial Medicine: Volume I, Probing Mitochondrial Function*,
Methods in Molecular Biology, vol. 1264, DOI 10.1007/978-1-4939-2257-4_17, © Springer Science+Business Media New York 2015

involved in energy production and their forms [8]. Fluorescence lifetime is an internal property of the fluorophore. In general, it depends on the ability of a fluorophore to transfer energy to its environment—and thus on the physicochemical characteristics of such environment, while being independent on the fluorophore concentration [9, 10].

Endogenous NAD(P)H fluorescence is an intrinsic property of living cells, stimulated after excitation of the sample by UV (ultraviolet) light [1, 7, 11], and results predominantly from mitochondria [12–14]. It is therefore a good practice to verify the mitochondrial localization of the recorded confocal images of cellular autofluorescence. NAD(P)H is best excited in the range from 250 to 400 nm [11–13, 15, 16]. Its fluorescence occurs with maximum at 440–460 nm. At these wavelengths, fluorescence of flavins (starting to emit at around 480–490 nm) can affect the recording of NAD(P)H fluorescence. Polarization filter can serve to distinguish polarized molecular states [17]. Time-resolved approach identified two fluorescence decay components: 0.4–0.6 ns for "free" form of NAD(P)H and 1.8–2.0 ns for NAD(P)H most likely "bound" to enzymes of the mitochondrial ETC (electron transport chain), in other words "in use NAD(P)H" [11–13, 15, 18]. After donating its electrons, NAD(P)H is oxidized into NAD^+ (oxidized nicotinamide adenine dinucleotide), a non-fluorescing molecule. For the best results, it is crucial to darken the experimental setup as much as possible: enclose it into a light-impermeable box/cover, or prevent the outside light to enter the laboratory.

Despite the fact that NADH and NADPH are not spectroscopically distinguishable, they are involved in different metabolic pathways and bind to different enzymes. NADH is an electron donor for the mitochondrial electron transport chain resulting in the creation of the proton-motive force, the proton gradient driving the ATP (adenosine triphosphate) synthesis. NADH serves as an oxidizing/reducing agent by keeping the $NAD^+/NADH$ ratio high. NADPH takes place in anabolic pathways as strong reducing agent with $NADP^+/NADPH$ ratio kept low and contributes to ROS (reactive oxygen species) elimination. They can therefore be distinguished by metabolic modulation [11, 19]. Modulation of the processes of energy production enables to study their roles in the cell metabolic state and can also provide a "calibration" of the metabolic oxidative state. OXPHOS (oxidative phosphorylation) can be either inhibited to reduce the oxidation, leaving the cells in a fully reduced state, or uncoupled from energy production, thus inducing a fully oxidized state [11, 20].

Both NAD(P)H and flavins can serve as metabolic oxidative state modulators. Except for the marginal collagen fluorescence, there are no endogenous fluorophores emitting in the regions of NAD(P)H. On the other hand, there are fluorophores absorbing and emitting in the spectral regions of flavins (lipofuscins, melanin, riboflavin, porphyrins, etc.), and there are also several types and forms of flavins emitting in a wide spectral band from 480 to approximately 560 nm [8, 10, 21]. NAD(P)H is therefore more unambiguous for interpretation. However, due to UV-induced damage and lack of technological means for an in vivo NAD(P)H diagnostics, flavin fluorescence is often used for in vivo recordings [10]. The ability to detect and to distinguish between different forms of NAD(P)H by time-resolved spectroscopy is useful for closer insight into the mitochondrial metabolism of various tissues [8] and thus during either physiological (aging, differentiation) or pathophysiological conditions linked with changes in mitochondrial metabolic oxidative state such as inflammations, transplanted tissue rejections, hypoxia, and/or cancer. Here, we describe NAD(P)H fluorescence lifetime microscopy imaging (FLIM) in a model cell line: U87-MG (human glioblastoma cell line).

2 Materials

Prepare all solutions using ultrapure water (produced by purifying the deionized water to attain a resistivity of 18 MΩ cm at 25 °C):

1. U87-MG, human glioma cell line (Cell Lines Services, Germany).

2. Dulbecco's Modified Eagle's Medium (DMEM) high glucose (4,500 mg/L).

3. GlutaMAX™ heat-inactivated fetal bovine serum.

4. Penicillin/streptomycin and trypsin, 0.05 % (1×) with EDTA 4Na.

5. Phosphate buffered saline (PBS) solution: dissolve one PBS tablet (in 1 L of deionized H_2O). Dissolving one tablet yields 140 mM NaCl, 10 mM phosphate buffer, and 3 mM KCl, pH 7.4 at 25 °C. Sterilize PBS solution by autoclaving (*see* **Note 1**).

6. Carbonyl cyanide *m*-chlorophenylhydrazone (CCCP): 20 mM stock solution in ethanol and store at −20 °C.

7. Carbonyl cyanide 4-(trifluoromethoxy)phenylhydrazone (FCCP): 2 mM stock solution in DMSO and store at −20 °C.

3 Methods

Carry out all procedures at room temperature (*see* **Notes 2** and **3**). Time-resolved confocal FLIM system DCS-120 with two parallel detection channels, equipped with hybrid detectors, was used [2] (*see* **Note 4**). The scanner was mounted on an inverted Axio Observer microscope (Zeiss, Germany), transforming it into a time-resolved confocal laser scanning microscope. TCSPC electronics Simple-Tau with SPC-152 module (Becker&Hickl, Germany) was employed. The microscope was furnished with a Zeiss Neofluar 63×/NA1.25 oil immersion objective. For excitation of NAD(P)H, 405 nm picosecond (ps) diode laser with single mode fiber coupler was used (*see* **Note 5**). The setup was kept in the darkness.

3.1 Cell Preparation for Microscopy

1. Harvest cells, routinely maintained in DMEM containing 10 % FBS and 1 % antibiotics, by low-speed centrifugation ($100 \times g$, 5 min, 4 °C).

2. Seed the cells at concentration approximately 10^5/mL onto glass cover slip bottoms.

3. Incubate the cells at 37 °C in a humidified atmosphere of 5 % CO_2 until the cells are 60–80 % confluent.

4. For microscopy recording, replace culture medium by the sterile PBS (*see* **Note 1**).

3.2 Fluorescence Lifetime Imaging of NAD(P)H Fluorescence in Living Cells

1. Put cover slips with cells onto a microscopy stage.

2. Prepare recording setup: for NAD(P)H measurements, based on its excitation/emission characteristics (as detailed in the Introduction), use excitation at 405 nm (or below, *see* **Note 5**) with an appropriate emission filter. To collect maximum fluorescence, we used LP 435 nm. See **Note 6** on how to gather spectrally resolved data.

3. Create recording protocol: collect high-resolution time-resolved images of cells for 120 s with ADC (analog-to-digital converter) resolution 256 to record image size of 256×256 pixels (*see* Fig. 1 for illustration). To improve data resolution, *see* **Note 7**. For recording of fast processes, *see* **Note 8**.

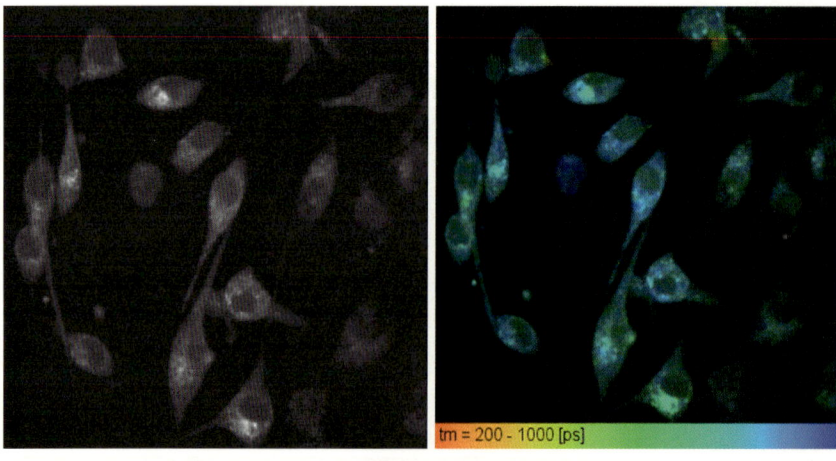

Fig. 1 Intensity (*left*) and FLIM (*right*) image of NAD(P)H fluorescence recorded in U87-MG cells in control conditions after excitation with 405 nm picosecond laser diode and emission with LP 435 nm. Fluorescence decay image scale is between 0.2 and 1 ns

4. When the recording is finished, save data in a corresponding file.

5. Verify the mitochondrial localization of the endogenous fluorescence and absence of the nuclear one from the recorded confocal images.

6. Use cells for up to 20–25 min or until the deterioration of their viability.

3.3 Viability Testing

1. Fluorescence probe Rhodamine 123 can be employed to test the cell viability, as well as to evaluate the presence of mitochondria in living cells (*see* **Note 9**). Apply Rhodamine 123 in PBS at final concentration of 1 μM to the cells. Leave for 15 min, and then replace with fresh PBS.

2. Put cover slips with cells onto a microscopy stage.

3. Prepare recording setup: with excitation/emission of 500/530 nm for Rhodamine 123 [22]; we have used excitation with 473 nm picosecond laser and LP 485 nm.

4. Create recording protocol: collect high-resolution time-resolved images of cells for 120 s with ADC resolution 256 (*see* Fig. 2 for illustration; the same recording setup and protocol can be employed for the measurement of flavin endogenous fluorescence). Verify the mitochondrial localization of the Rhodamine 123 fluorescence from the recorded confocal images.

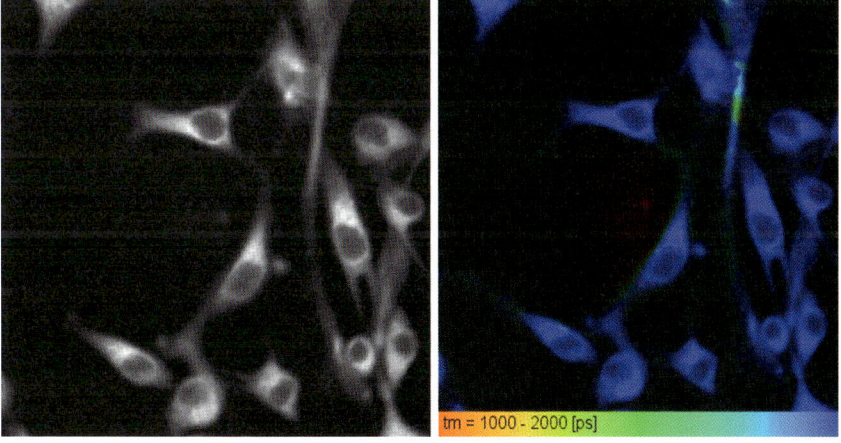

Fig. 2 Intensity (*left*) and FLIM (*right*) image of Rhodamine 123-stained U87-MG cells after excitation with 473 nm picosecond laser diode and emission with LP 485 nm. Fluorescence decay image scale is between 1 and 2 ns

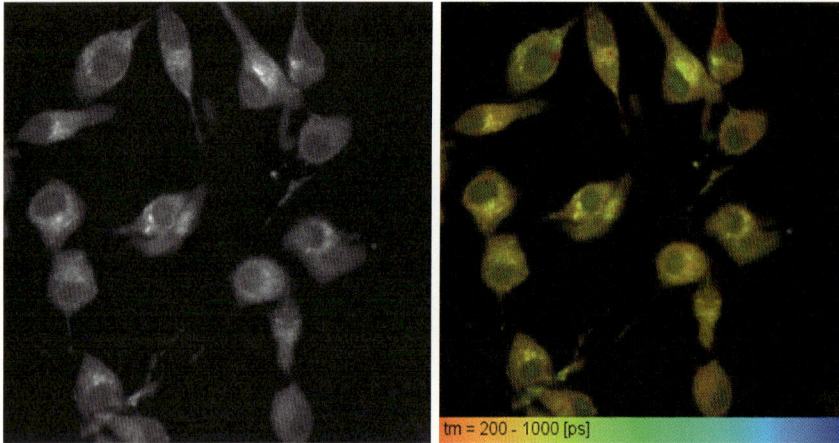

Fig. 3 Intensity (*left*) and FLIM (*right*) image of NAD(P)H fluorescence recorded in U87-MG cells in a reduced state, in the presence of CCCP (1 μM) after excitation with 405 nm picosecond laser diode and emission with LP 435 nm. Fluorescence decay image scale is between 0.2 and 1 ns

3.4 Reducing Mitochondrial Oxidation by OXPHOS Inhibition

1. Prepare CCCP at 1–5 μM in PBS.

2. Apply CCCP to unstained cells by gently replacing PBS medium with PBS + CCCP.

3. Leave CCCP to act for 5 min at room temperature (*see* **Note 2**): cells are now in a fully reduced state.

4. Verify that fluorescence intensity has increased and fluorescence lifetime shifted toward faster lifetimes (*see* example at Fig. 3). *See* **Notes 10–12** for testing of other metabolic modulators.

5. Record corresponding images using exactly the same protocol that was employed in control conditions to insure comparability.

6. Use cells for up to 20–25 min or until the deterioration of their viability.

3.5 Increasing Mitochondrial Oxidation by Uncoupling of Respiratory Chain

1. Prepare FCCP at 0.5–1 μM in PBS.

2. Apply FCCP to cells by gently replacing PBS medium with PBS + FCCP.

3. Leave FCCP to act for 5 min at room temperature (*see* **Note 2**): cells are now in a fully oxidized state following uncoupling of ATP synthesis.

4. Verify that fluorescence intensity has decreased and fluorescence lifetime shifted toward longer lifetimes (*see* example at Fig. 4). *See* **Note 10** for the use of other metabolic uncouplers.

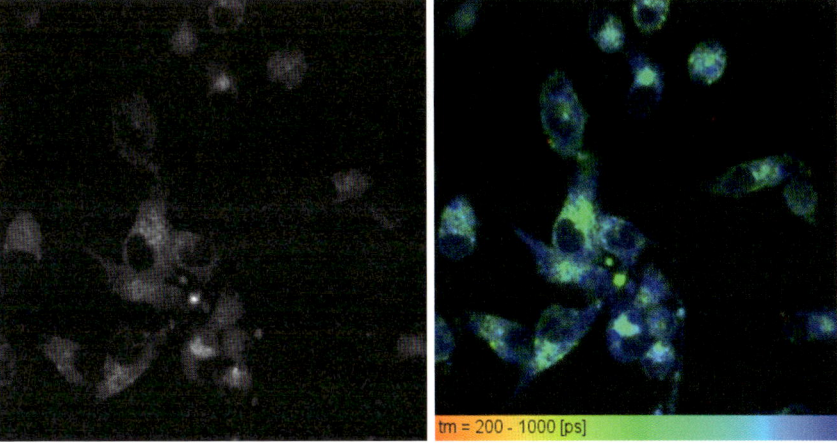

tm = 200 - 1000 [ps]

Fig. 4 Intensity (*left*) and FLIM (*right*) image of NAD(P)H fluorescence recorded in U87-MG cells in an oxidized state in the presence of FCCP (1 μM) after excitation with 405 nm picosecond laser diode and emission with LP 435 nm. Fluorescence decay image scale is between 0.2 and 1 ns

5. Record corresponding images using exactly the same protocol that was employed in control conditions to insure comparability.

6. Use cells for up to 20–25 min or until the deterioration of their viability.

3.6 Analysis

Use SPCImage software to perform first estimation of intensity and fluorescence decays in the recorded images:

1. Send data to SPCImage.

2. Calculate decay matrix (for a selected channel).

3. Use the predefined instrument response function (IRF) (or *see* **Note 13** to use the recorded IRF).

4. To improve image, use export color-coded images.

5. If the result is not good (e.g., bright specks are present in the image), go to "Intensity" tab and turn the "Autoscale" off.

6. Keep an eye on the goodness of the fit represented by chi-squared value (χ^2): usually, it should not be higher than 1.2 (*see* **Note 14** on how to improve χ^2).

7. For advanced analysis, refer to linear unmixing of individual components [11] or phasor approach [23].

4   Notes

1. PBS can be replaced by a self-designed external solution, depending on the type of the cultured and/or isolated cells used [11]. Remember to check that the applied solution does not contain molecules fluorescing in the detected spectral

range, most commonly the phenol red, and that the medium is able to maintain the cells alive for the necessary measurement time, without affecting their metabolism and/or viability. For example, DMEM without phenol red, FluoroBrite™ DMEM medium can be successfully used.

2. Procedures can be carried out at 35–36 °C using an onstage incubator for tracking temperature-dependent changes. In such case, do not forget to use preheated solutions for cell preparation and verify that the level of photobleaching did not increase.

3. Oxygenation of the solution by different O_2/N_2 mixture can be insured by an onstage incubator/controller.

4. Alternative to TCSPC, frequency-modulated (amplitude domain) approach can be employed to evaluate fluorescence lifetimes [3, 4].

5. Excitation by 405 nm ps laser can be replaced by shorter wavelength, namely, 375 nm ps laser [11]. Use an appropriate LP (long-pass) or BP (band-pass) filter. If lower wavelengths are used, it is necessary to verify UV-induced damage to the cell [24].

6. To gather spectrally resolved data, specific narrow BP filters can be employed accordingly. For the most effective detection of NAD(P)H fluorescence, we recommend to use either BP filter to record wavelengths in the range 440–480 nm or LP filter to detect wavelengths longer than 440 nm to cut off the excitation pulse and its reflections, as well as the Raman peak of water. For example, to record spectrally resolved NAD(P)H fluorescence, BP 440 ± 10 nm and BP 470 ± 10 nm can be used to evaluate "bound" and "free" NAD(P)H, respectively. Another possibility is to use a monochromator in front of the detector with the advantage of choosing the exact wavelength detected, but with the disadvantage of recording lower signal intensity.

7. Always verify the state of the recorded cells before and after recording. Discard all recordings during which the cells presented change in their shape and/or state. Consequently, although shorter recordings are more suitable, the length of the recording has to be appropriate to reach a sufficient number of photons for multiexponential decay data analyses (for more details, refer to [2] or the Becker & Hickl TCSPC handbook http://www.becker-hickl.de/literature.htm).

8. The recording time can be adjusted depending on the rapidity of the studied processes. For fast processes, synchronized recording can be considered [25].

9. Other fluorescence probes for viability testing can also be used, depending on the cell type and/or setup characteristics.

10. Effect of metabolic modulators can vary depending on the cell type and/or pathway that it employs. To test the oxidative phosphorylation, we have employed CCCP and FCCP to inhibit and/or uncouple the ATP production by the respiratory chain, respectively. At the same time, CCCP can be replaced by rotenone (the inhibitor of the complex I) and/or cyanides (NaCN or KCN, the inhibitors of the complex IV), while FCCP can be replaced by 2,4-dinitrophenol [11].

11. In cells preferentially using glycolysis instead of oxidative phosphorylation [26, 27], modulators mentioned in **Note 9** would induce only limited effect and other pathways, such as modulation of NADH production [11, 23], and/or regulation of oxidative stress [19, 23], etc., need to be tested.

12. Sensitivity of cells to oxidative stress can be tested by applying compounds such as hydrogen peroxide [19, 23].

13. For precise evaluation of the fluorescence lifetimes, measure the IRF [28] and apply it for the fluorescence lifetime data analysis. The Becker-Hickl SPCImage analysis software does not require the measured IRF; fast lifetime verification can be done with the predefined IRF. However, IRF measurement should always be performed on the new sample and verified regularly to achieve a precise lifetime evaluation. IRF measurement is also needed when another analysis software is employed that requires specification of the number of time windows and/or the number of windows to be analyzed.

14. Goodness of the fit can be enhanced (the χ^2 lowered) by changing the offset and scatter parameters in the SPCImage software, but also by improving the recorded image, for example, by insuring that less background light is entering the setup—darkening of the setup can significantly improve the goodness of the fit.

Acknowledgments

Authors acknowledge support from Integrated Initiative of European Laser Research Infrastructures LASERLAB-EUROPE III (grant agreement no 284464, EC's Seventh Framework Programme), EC's Seventh Framework Programme CELIM 316310 project, the Research Grant Agency of the Ministry of Education, Science, Research and Sport of the Slovak Republic VEGA No 1/0296/11, and the Slovak Research and Development Agency under the contract APVV-0242-11.

References

1. Chorvat D Jr, Chorvatova A (2009) Multi-wavelength fluorescence lifetime spectroscopy: a new approach to the study of endogenous fluorescence in living cells and tissues. Laser Phys Lett 6:175–193

2. Becker W (2005) Advanced time-correlated single photon counting techniques. Springer, New York, NY

3. Lakowicz JR (1999) Principles of fluorescence spectroscopy introducing the phase-modulation methods. Springer, New York, NY

4. Lakowicz JR (2006) Principles of fluorescence spectroscopy. Springer, New York, NY

5. Chance B, Cohen P, Jobsis F, Schoener B (1962) Intracellular oxidation-reduction states in vivo. Science 137:499–508

6. Chance B, Ernster L, Garland PB, Lee CP, Light PA, Ohnishi T, Ragan CI, Wong D (1967) Flavoproteins of the mitochondrial respiratory chain. Proc Natl Acad Sci U S A 57:1498–1505

7. Chance B, Nioka S, Warren W, Yurtsever G (2005) Mitochondrial NADH as the bell-wether of tissue O2 delivery. Adv Exp Med Biol 566:231–242

8. Chorvatova A, Chorvat D Jr (2014) Review of tissue fluorophores and their spectroscopic characteristics. In: Marcu L, French P, Elson D (eds) Fluorescence lifetime spectroscopy and imaging for tissue biomedical diagnostics. CRC Press Publ, Boca Raton, FL, pp 47–84

9. Tadrous PJ, Siegel J, French PM, Shousha S, Lalani e-N, Stamp GW (2003) Fluorescence lifetime imaging of unstained tissues: early results in human breast cancer. J Pathol 199:309–317

10. Berezin MY, Achilefu S (2010) Fluorescence lifetime measurements and biological imaging. Chem Rev 110:2641–2684

11. Chorvatova A, Mateasik A, Chorvat D Jr (2013) Spectral decomposition of NAD(P)H fluorescence components recorded by multi-wavelength fluorescence lifetime spectroscopy in living cardiac cells. Laser Phys Lett 10: 125703

12. Blinova K, Carroll S, Bose S, Smirnov AV, Harvey JJ, Knutson JR, Balaban RS (2005) Distribution of mitochondrial NADH fluorescence lifetimes: steady-state kinetics of matrix NADH interactions. Biochemistry 44: 2585–2594

13. Sud D, Zhong W, Beer DG, Mycek MA (2006) Time-resolved optical imaging provides a molecular snapshot of altered metabolic function in living human cancer cell models. Opt Express 14:4412–4426

14. Mayevsky A, Rogatsky GG (2007) Mitochondrial function in vivo evaluated by NADH fluorescence: from animal models to human studies. Am J Physiol Cell Physiol 292:C615–C640

15. Konig K, Riemann I (2003) High-resolution multiphoton tomography of human skin with subcellular spatial resolution and picosecond time resolution. J Biomed Opt 8:432–439

16. Jamme F, Kascakova S, Villette S, Allouche F, Pallu S, Rouam V, Refregiers M (2013) Deep UV autofluorescence microscopy for cell biology and tissue histology. Biol Cell 105:277–288

17. Vishwasrao HD, Heikal AA, Kasischke KA, Webb WW (2005) Conformational dependence of intracellular NADH on metabolic state revealed by associated fluorescence anisotropy. J Biol Chem 280:25119–25126

18. Lakowicz JR, Szmacinski H, Nowaczyk K, Johnson ML (1992) Fluorescence lifetime imaging of free and protein-bound NADH. Proc Natl Acad Sci U S A 89:1271–1275

19. Chorvatova A, Aneba S, Mateasik A, Chorvat D, Comte B (2013) Time-resolved fluorescence spectroscopy investigation of the effect of 4-hydroxynonenal on endogenous NAD(P)H in living cardiac myocytes. J Biomed Opt 18:67009

20. Romashko DN, Marban E, O'Rourke B (1998) Subcellular metabolic transients and mitochondrial redox waves in heart cells. Proc Natl Acad Sci U S A 95:1618–1623

21. Feenstra KA (2002) Long term dynamics of proteins and peptides. Ponsen & Looijen, Wageningen, pp 119–143

22. Schneckenburger H, Stock K, Lyttek M, Strauss WS, Sailer R (2004) Fluorescence lifetime imaging (FLIM) of rhodamine 123 in living cells. Photochem Photobiol Sci 3:127–131

23. Stringari C, Nourse JL, Flanagan LA, Gratton E (2012) Phasor fluorescence lifetime microscopy of free and protein-bound NADH reveals neural stem cell differentiation potential. PLoS One 7:e48014

24. Chorvatova A, Mateasik A, Chorvat D Jr (2011) Laser-induced photobleaching of NAD(P)H fluorescence components in cardiac cells resolved by linear unmixing of TCSPC signals. Proc of SPIE 7903:790326-1–790326-9

25. Chorvat D Jr, Abdulla S, Elzwiei F, Mateasik A, Chorvatova A (2008) Screening of cardiomyocyte fluorescence during cell contraction

by multi-dimensional TCSPC. Proc SPIE 6860:686029-1–686029-12

26. Warburg O (1956) On the origin of cancer cells. Science 123:309–314

27. Gatenby RA, Gillies RJ (2004) Why do cancers have high aerobic glycolysis? Nat Rev Cancer 4:891–899

28. Chorvat D Jr, Chorvatova A (2006) Spectrally resolved time-correlated single photon counting: a novel approach for characterization of endogenous fluorescence in isolated cardiac myocytes. Eur Biophys J Biophys Lett 36:73–83

Chapter 18

Novel Methods for Measuring the Mitochondrial Membrane Potential

Roger Springett

Abstract

The mitochondrial membrane potential is a critical parameter for understanding mitochondrial function, but it is challenging to quantitate with current methodologies which are based on the accumulation of cation indicators. Recently we have introduced a new methodology based on the redox poise of the *b*-hemes of the *bc*₁ complex. Here we describe the thermodynamic framework and algorithms necessary to calculate the membrane potential from the measured oxidation states of the *b*-hemes.

Key words Mitochondria, Membrane potential, bc_1 complex, Redox potential

1 Introduction

The membrane potential ($\Delta\Psi$) across the inner mitochondrial membrane is a critical parameter for understanding mitochondrial energetics, but the quantitation, particularly in cells, is difficult and current methods rely on measuring the accumulation of a membrane-permeable cation. The free energy change on import of the cation, ΔG, expressed in millivolts (1 kJ/mol = 10.36 mV), is given by

$$\Delta G = -\frac{RT}{F}\ln\left(\frac{C_{\mathrm{c}}}{C_{\mathrm{m}}}\right) - \Delta\psi, \qquad (1)$$

where R is the gas constant, T is the absolute temperature, F is the Faraday constant, and C_{c} and C_{m} are the concentrations (strictly activity) of the cation in the cytosol (or extra-mitochondrial space) and matrix, respectively. This allows the membrane potential to be calculated once the system has reached equilibrium and $\Delta G = 0$.

In isolated mitochondria, $\Delta\Psi$ has been quantified using radioactive rubidium (Rb⁺) [1] or tetraphenylphosphonium (TPP⁺) [2]. In the former, the inner membrane is made permeable to Rb⁺ with

Volkmar Weissig and Marvin Edeas (eds.), *Mitochondrial Medicine: Volume I, Probing Mitochondrial Function*,
Methods in Molecular Biology, vol. 1264, DOI 10.1007/978-1-4939-2257-4_18, © Springer Science+Business Media New York 2015

valinomycin, and the mitochondria and suspension medium are separated using a filter so that the quantity of Rb$^+$ in each can be measured from the radioactivity. The method has the disadvantage that measurements cannot be made in K$^+$ buffers due to the presence of valinomycin and only supplies a point measurement and not a continuous trace. TPP$^+$ is a membrane-permeable cation, and a continuous trace of the concentration in the medium can be measured with a TPP+ electrode. The concentration in the matrix can then be estimated from the loss in the medium allowing $\Delta\Psi$ to be calculated. Both methods require the relative volume of the matrix: the former to convert radioactivity to concentration and the latter to calculate the matrix concentration from the loss in the medium. The matrix volume, which can change depending on the respiratory state of the mitochondria [3], can be measured from the space occupied by water but not by sucrose, but not simultaneously with the measurement of Rb$^+$ distribution.

Cells are more problematic due to the presence of the plasma membrane which also maintains a potential ($\Delta\Phi$, positive outside) which leads to the accumulation of the label in the cytosol according to

$$\Delta G = -\frac{RT}{F}\ln\left(\frac{C_e}{C_c}\right) - \Delta\Phi, \quad (2)$$

where C_e is the concentration of the label in the extracellular space. Quantitation relies on the accumulation of membrane-permeable cationic fluorophores, typically tetramethylrhodamine methyl ester (TMRM$^+$) measured with fluorescence flow cytometry [4] or fluorescence microscopy [5, 6]. Further complications are nonspecific binding of the fluorophore (only the free pools contribute to the concentrations in Eqs. 1 and 2), possible enhanced fluorescent quantum efficiency associated with binding, fluorescence quenching due to aggregation at higher concentrations, and spectral shifts in the excitation and emission spectra on binding. Quantitation has been attempted with flow cytometry [4] where the effects of the plasma membrane were assumed to be negligible at low concentrations of fluorophore, an assertion not supported by Eq. 2. Optical resolution of the fluorescence of matrix from the cytosol has been attempted using fluorescence microscopy by enhancing the spatial resolution by deconvoluting the image with the spatial response of the imaging system, but this technique still cannot resolve the cristae and matrix signals [5]. A recent attempt to calibrate the TMRM$^+$ signal [6] found that it accumulated with an effective charge of only 0.71 rather than 1.0, a physical impossibility explained by the presence of nonspecific binding.

Recently we have introduced a new method of quantitating the membrane potential in cells and mitochondria using the redox

2. Fiber Delivery System (Opotek, Carlsbad, CA, USA) consisting of 50 mm plano-convex lens, x–y fiber mount, and a fiber with a length of 2 m and a core diameter of 1,000 μm (FOA-Inline, Avantes b.v., Eerbeek, the Netherlands).

3. Two 1,000 μm optical fibers both with a length of 1 m (P1000-2-VIS-NIR, Ocean Optics, Dunedin, FL, USA).

4. A stainless steel rod with a length of 5 cm and a diameter of 10 mm.

5. Gated microchannel plate photomultiplier tube (MCP-PMT R5916U series, Hamamatsu Photonics, Hamamatsu, Japan), *see* **Note 1**.

6. A gated socket assembly (E3059-501, Hamamatsu Photonics, Hamamatsu, Japan).

7. A thermoelectric cooler (C10373, Hamamatsu Photonics, Hamamatsu, Japan).

8. A regulated high-voltage DC power supply (C4848-02, Hamamatsu Photonics, Hamamatsu, Japan).

9. Oriel Fiber Bundle Focusing Assembly (Model 77799, Newport, Irvine, CA, USA).

10. Plano-convex lens with focal length of 90 mm (BK-7, OptoSigma, Santa Ana, CA, USA).

11. Electronic shutter (04 UTS 203, Melles Griot, Albuquerque, NM, USA).

12. OEM Shutter Controller Board (59 OSC 205, Melles Griot, Albuquerque, NM, USA).

13. Longpass filter, 590 nm (OG590, Newport, Irvine, CA, USA).

14. Broadband band-pass filter, 600–750 nm (Omega Optical, Brattleboro, VT, USA).

15. Amplifier (Amplifier C6438-01, Hamamatsu Photonics, Hamamatsu, Japan) with an input impedance of $50\,\Omega$, 500 times voltage amplification, and a bandwidth around 50 MHz.

16. PC-based data acquisition system containing a 10 MS/s simultaneous sampling data acquisition board (NI-PCI-6115, National Instruments, Austin, TX, USA) (Fig. 1(3)).

17. BNC interface (BNC-2090A, National Instruments, Austin, TX, USA).

18. Control software, for example, written in LabVIEW (version 10.0 or higher, National Instruments, Austin, TX, USA).

2.2 Animals

Adult Wistar rats with a bodyweight of 275–325 g (Charles River, Wilmington, MA, USA). All animal care and handling should be in accordance with the National Institutes of Health (NIH) Guide for the Care and Use of Laboratory Animals or alternatively national law. In our case the experiments were approved by the Animal Research Committee of the Erasmus MC University Medical Center Rotterdam.

**2.3   Chemicals,
Fluids, and Reagents**

1. Hydrophilic cremor lanette (Lanette creme I FNA, Bipharma, Weesp, the Netherlands).
2. 5-aminolevulinic acid, *see* **Note 2**.
3. Ketamine 90 mg/kg.
4. Medetomidine 0.5 mg/kg.
5. Atropine 0.05 mg/kg.
6. Hair removal cream.
7. Heparin solution 10 IE/ml.
8. Voluven.
9. Ringer solution.
10. Alcohol preps.

2.4   Surgical Tools

1. Tube: enteral feeding tube L. 40 cm, 06 Fr, ref 310.06, Vygon, Ecouen, France, cut to a length of 1.5 cm with a marking at 1 cm.
2. Catheters: 15 cm fine bore polythene tubing PE50, 0.58 mm ID, 0.96 OD (Portex Smiths Medical, Kent, UK) connected to a 22 G×½″ needle (Luer stub, Instech laboratories Inc., Plymouth Meeting, PA, USA).
3. Syringes (1 ml).
4. Suture# 4-0 (Silkam® black, B. Braun, Aesculap, Tuttlingen, Germany).
5. Heating pad with rails for connection fiber standard and BP transducer.
6. Spring scissors (Vannas-Tübingen FST curved 8.5 cm, no. 15004-08).
7. Iris scissors (Strabismus FST, 9 cm, no. 14075-09).
8. Forceps (Graefe forceps FST, 1×2 teeth tip, width 0.8 mm, no. 11053-10).
9. Forceps fine (Dumont #7, Dumostar standard tip, no. 11297-00).
10. Hemostats (Halsted-Mosquito FST, curved, no. 13009-12).
11. Clamp (Schwartz Micro Serrefines FST straight, no. 18052-01).
12. Hot Beads Sterilizer (FST 250 sterilizer, no. 18000-45).
13. Sterile towel 45×75 cm (BARRIER® Mölnlycke Health Care, Göteborg, Sweden).
14. Operation microscope.

**2.5   Ventilation
and Monitoring**

1. Dräger Babylog with anesthetic breathing system pediatric.
2. Siemens Monitor SC 9000 XL.
3. CAPNOSTAT® CO_2.

4. CO_2 cuvette neonatal.

5. Pritt poster buddies Henkel NL BV (for fixation of the CO_2 cuvette).

6. Oxygen 100 % and air.

7. Data acquisition system (PowerLab).

8. Disposable BP transducer connected to a stopcock.

9. Simulator and tester Delta-Cal.

10. T-type Implantable Thermocouple probe in use with ML312 T-type Pod.

11. T copper–constantan thermocouple in use with ML290 Thermometer for T-type thermocouple.

12. Infusion pump.

13. 20 ml syringes.

14. Stopcocks.

15. Extension line L. 100 cm.

3 Methods

3.1 Delayed Luminescence Setup

Couple the excitation source of the pulsed tunable laser into the Fiber Delivery System, and connect it through an *x–y* connector with the excitation probe. Connect the excitation fiber with the reflection fiber by mounting the terminal tips of both fibers into a small aluminum rod with a separation of 1 mm between the fibers (Fig. 2).

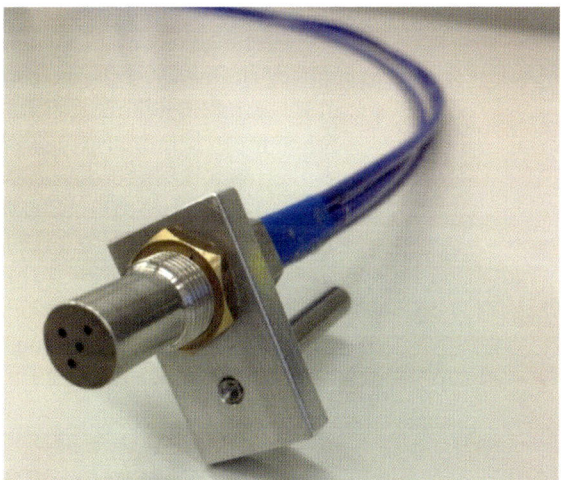

Fig. 2 The measuring probe. Excitation and emission fibers are mounted in a small aluminum rod, allowing application of local pressure on the measurement volume. The shown probe has two additional fibers for spectroscopic measurements

The MCP-PMT should preferably have enhanced (infra)red sensitivity. To this end, our device has been custom adapted with an enhanced red-sensitive photocathode having a quantum efficiency of 24 % at 650 nm. The MCP-PMT should be mounted on a gated socket assembly and cooled to −30 °C by the thermoelectric cooler. The MCP-PMT can be operated at a voltage range between 2,000 and 2,700 V. Fit the emission branch of the reflection probe into a fiber bundle focusing assembly, and couple it to the MCP-PMT by an optic assembly, which consists of a filter holder, a plano-convex lens (90 mm focusing distance), and an electronic shutter. The shutter protects the MCP-PMT, which is configured for the "normally on" mode. Filter out the PpIX emission light by a combination of a 590 nm longpass filter and a broadband (675–25 nm) band-pass filter fitted in the filter holder. Convert the output current of the photomultiplier to a voltage by an amplifier with an input impedance of $50\,\Omega$, 500 times voltage amplification, and a bandwidth around 50 MHz. The data acquisition is performed by a PC-based data acquisition system containing a 10 MS/s simultaneous sampling data acquisition board. The amplifier is coupled to the DAQ board by a BNC. The data acquisition should preferably be performed at the highest possible rate of 10 mega samples per second, allowing for digital post-processing (e.g., low-pass filtering) if necessary. Control of the setup and analysis of the delayed luminescence signals can be performed with software written in LabVIEW, using standard virtual instruments (VIs).

MitoPO$_2$ can be calculated from the delayed fluorescence lifetime by using the Stern–Volmer relationship:

$$\text{mitoPO}_2 = \left(1/\tau - 1/\tau_0\right)/k_q, \tag{1}$$

in which k_q is the quenching constant (830 mmHg/s) and τ_0 is the lifetime in the absence of oxygen (0.8 ms) [12]. So the task is now to recover the delayed fluorescence lifetime from the photometric signal by using an adequate fitting procedure.

Since mitochondrial oxygen tension, and thus delayed fluorescence lifetimes, is heterogenous in respiring tissues, we do not advice to use a simple mono-exponential fitting procedure to the signal. This will lead to a bias toward longer lifetimes and thus results in underestimation of the average mitoPO$_2$. A convenient way to recover a good estimation of the average mitoPO$_2$ within the measuring volume is fitting a rectangular distribution to the delayed luminescence data [13]:

$$Y(t) = Y_0 \exp\left[-\left(1/\tau_0 + k_q \langle \text{mitoPO}_2 \rangle\right)t\right] \sinh\left(k_q \delta t\right)/k_q \delta t, \tag{2}$$

where $Y(t)$ is the delayed fluorescence signal, t is the time from the start of the fit, Y_0 is the initial signal intensity at $t=0$, <mitoPO$_2$> is the average mitochondrial oxygen tension, and δ is the half-width of the PO$_2$ distribution.

The oxygen disappearance curve, i.e., the kinetics of mitoPO$_2$ after cessation of local blood flow, can be analyzed by an adapted Michaelis–Menten approach. In contrast to in vitro respirometry in oxygen-closed reaction vessels, you have to take oxygen back diffusion into the measuring volume in account when performing analysis in vivo. We have previously shown that in skin the following formula can be used if autoconsumption of oxygen by the measuring technique is neglectable [11]:

$$dP_n / dn = -\left(V_0 \cdot P_n\right)/\left(P_{50} + P_n\right) + Z\left(P_0 - P_n\right). \tag{3}$$

In this equation, P_n is the measured PO$_2$ after excitation flash number n, P_0 is the mean PO$_2$ before stop-flow, Z is the inflow coefficient of oxygen, P_{50} is the PO$_2$ at which cellular oxygen consumption is reduced to $1/2V_0$, and dP_n/dn is the rate of oxygen disappearance.

3.2 Animal Preparation (Fig. 3)

1. Anesthetize the rat with an intraperitoneal (IP) injection of a mixture of ketamine, medetomidine, and atropine.

2. After the rat reaches a surgical level of anesthesia, shave the skin of the abdomen and of the tracheal region.

3. Place the animals in a supine position and perform a tracheotomy. Insert a tube into the trachea until the tip is approximately 1 cm above the carina, and fixate the tube with a suture around the trachea.

4. Connect the tube to the ventilation system, which consists of a pediatric breathing system, breathing hoses, a CO$_2$ cuvette, and a CO$_2$ monitor.

5. Fixate the CO$_2$ cuvette with Pritt poster buddies.

6. Start on the following standard ventilator settings: inspiration time (Ti) 0.34 s, expiration time (Te) 0.68 s, ventilation frequency (fset) 59 bpm, inspiration-to-expiration ratio (I:E ratio) 1:2, inspired oxygen fraction (FIO$_2$) 40 %, inspiratory pressure (Pins) 16 mbar, and positive end-expiratory pressure (PEEP) 3.3 mbar. Adjust the ventilator settings based on end-tidal PCO$_2$ and oxygenation needs, and keep the arterial PCO$_2$ between 35 and 45 mmHg.

7. Insert a polyethylene catheter into the right jugular vein. Connect the catheter via an extension with an infusion pump for intravenous administration of fluids and maintenance anesthetics with ketamine.

8. Insert a similar catheter in the left femoral artery, and connect it via the BP transducer with PowerLab to monitor arterial blood pressure and heart rate.

9. Perform every hour an analysis of the arterial blood gas, and if necessary adjust your ventilation settings.

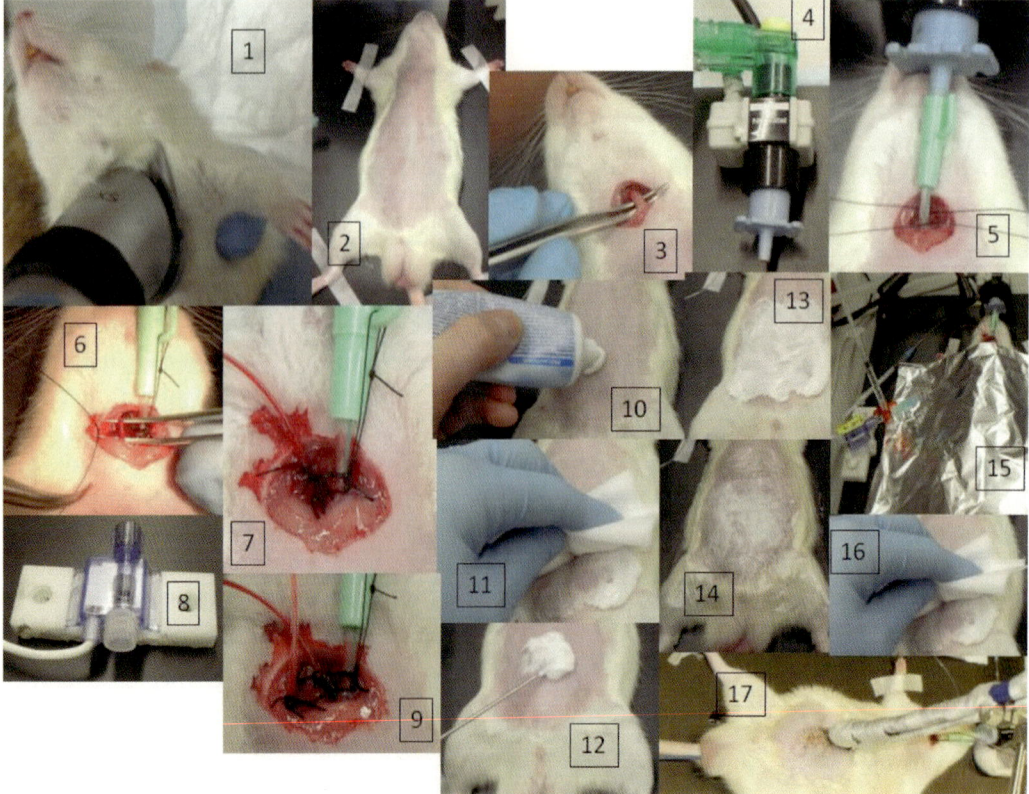

Fig. 3 Animal preparation. (1) Shave the skin of abdomen and tracheal region. (2) Supine position. (3) Tracheotomy. (4) CO_2 cuvette. (5) Connect tube to ventilation system. (6) Carotid artery. (7) Catheter into carotid artery. (8) BP transducer. (9) Catheter into jugular vein. (10) Apply hair removal crème. (11) Remove hair removal crème. (12 and 13) Apply 5-aminolevulinic acid crème. (14) Adhesive film. (15) Cover area with aluminum foil. (16) Remove 5-aminolevulinic acid crème. (17) Start experiments

10. Measure the body temperature with a rectally inserted thermometer, and keep the temperature at 38 ± 0.5 °C by adjustments of the heating pad; *see* **Note 3**.

3.3 Experimental Procedures

1. Prepare the ALA cream by mixing hydrophilic lanette crème with 2.5 % (mass percentage) 5-aminolevulinic acid; *see* **Note 4**.

2. Prepare the abdominal skin of the rats by shaving and removing remaining hairs with a hair removal cream. Remove the hair removal cream after 5 min, and clean the abdominal skin with alcohol preps.

3. Apply the ALA cream topically on the prepared abdominal skin of the rats, and leave it on for 3 h.

4. To avoid evaporation, cover the exposed skin with an adhesive film, and prevent premature light exposure of PpIX by covering the area with aluminum foil.

5. After 3 h, there is a sufficient amount of PpIX converted in the mitochondria to start the mitochondrial oxygenation measurements.

6. Remove the ALA cream before the start of the measurements.

7. Perform the mitochondrial respiration measurements by repeated measurements of mitoPO$_2$. Use the following settings for the PpIX-TSLT software: laser output of the excitation branch, 200 μJ/pulse; pulsfrequentie, 1 Hz; and measurement time, 60 s; *see* **Note 5**.

8. Start for the first 10 s with the measurement of the initial mitoPO$_2$; do this by keeping the probe just above the ALA-treated skin surface.

9. After the first 10 s, apply local pressure with the measurement probe for a period of approximately 30 s. This local pressure blocks the oxygen supply by temporally occluding the microvessels. After the flow arrest in the microvessels, you will observe that the mitoPO$_2$ gradually drops; this is due to the ongoing oxygen consumption inside the mitochondria.

10. Release the pressure on the abdominal skin after 30 s (you have to obtain a new steady mitoPO$_2$ reading at a much lower level), and hold the measurement probe in a stationary position above the measured area. Figure 4 shows a typical example of time course of mitoPO$_2$, and Fig. 5 shows the result of fitting Eq. 3 to another example; *see* **Note 6**.

4   Notes

1. Gating of the photomultiplier tube is essential to prevent saturation of the detector and electronics because of intense prompt fluorescence.

2. Store the stock ALA at 6 °C, and beware for this very hydrophilic compound to attract water.

3. We find that it is essential to keep the body temperature of the rats around the 38 °C.

4. It's best to prepare the ALA cream fresh each time.

5. Reduce the excitation energy and the number of measurements to a usable minimum to prevent oxygen autoconsumption by the measuring method and excessive photoexcitation of the measuring volume.

6. We find that repeated oxygen consumption measurements are possible. However, we advise to change measurement sites on the ALA-treated skin surface for repeated measurements.

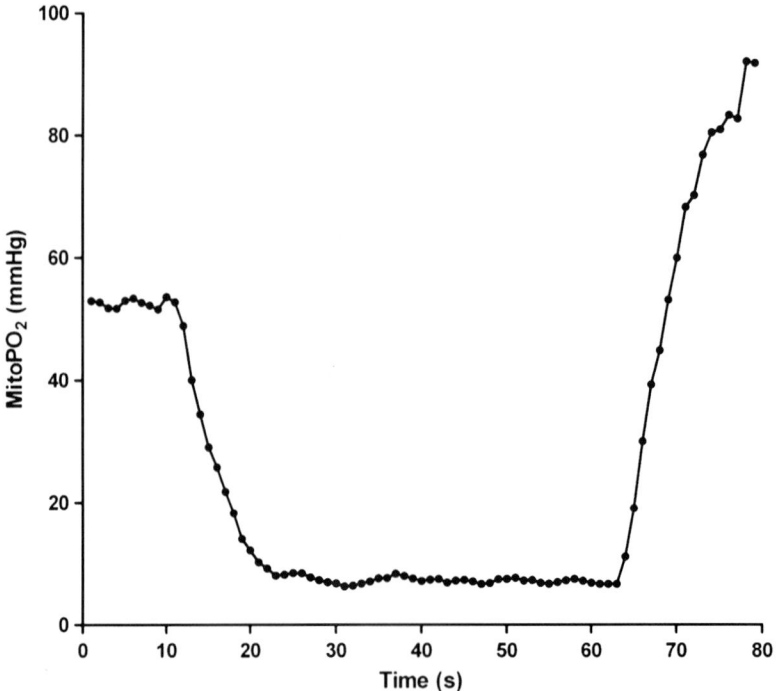

Fig. 4 Dynamic mitoPO$_2$ measurement. Typical time course of mitoPO$_2$ before, during (starting at 10 s), and after release (at 63 s) of local pressure with the measuring probe

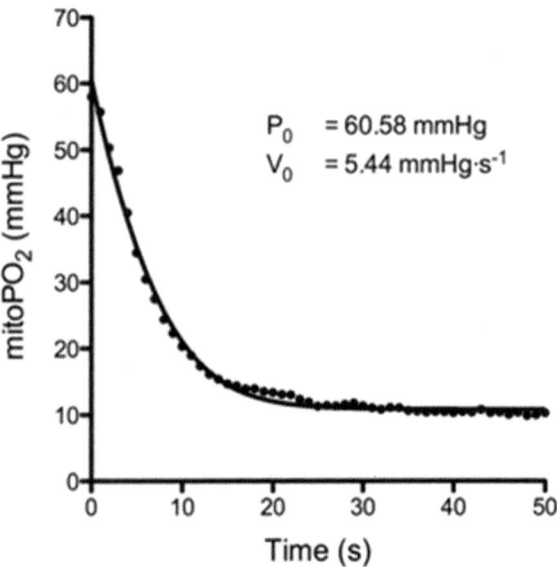

Fig. 5 In vivo respirometry. Typical example of applying Eq. 3 to the mitochondrial oxygen disappearance signal. Recorded data (*globules line*) and the analysis of the data (*straight line*)

References

1. Ishii N (2007) Role of oxidative stress from mitochondria on aging and cancer. Cornea 26(9 Suppl 1):S3–S9

2. Ballinger SW (2005) Mitochondrial dysfunction in cardiovascular disease. Free Radic Biol Med 38(10):1278–1295

3. Fink MP (2002) Cytopathic hypoxia. Is oxygen use impaired in sepsis as a result of an acquired intrinsic derangement in cellular respiration? Crit Care Clin 18(1):165–175

4. Galley HF (2011) Oxidative stress and mitochondrial dysfunction in sepsis. Br J Anaesth 107(1):57–64

5. Garrabou G et al (2012) The effects of sepsis on mitochondria. J Infect Dis 205(3):392–400

6. Edeas M, Weissig V (2014) Strategies, innovations & clinical applications. In: Edeas M, Weissig V (eds). Targeting mitochondria, Berlin

7. Mik EG et al (2006) Mitochondrial PO2 measured by delayed fluorescence of endogenous protoporphyrin IX. Nat Methods 3(11): 939–945

8. Mik EG et al (2009) Mitochondrial oxygen tension within the heart. J Mol Cell Cardiol 46(6):943–951

9. Mik EG et al (2008) In vivo mitochondrial oxygen tension measured by a delayed fluorescence lifetime technique. Biophys J 95(8): 3977–3990

10. Mik EG (2013) Measuring mitochondrial oxygen tension: from basic principles to application in humans. Anesth Analg 117(4):834–846

11. Harms FA et al (2013) Cutaneous respirometry by dynamic measurement of mitochondrial oxygen tension for monitoring mitochondrial function in vivo. Mitochondrion 13(5):507–514

12. Harms FA et al (2012) Validation of the protoporphyrin IX-triplet state lifetime technique for mitochondrial oxygen measurements in the skin. Opt Lett 37(13):2625–2627

13. Golub AS, Tevald MA, Pittman RN (2011) Phosphorescence quenching microrespirometry of skeletal muscle in situ. Am J Physiol Heart Circ Physiol 300:H135–H143

Chapter 21

Imaging Mitochondrial Hydrogen Peroxide in Living Cells

Alexander R. Lippert, Bryan C. Dickinson, and Elizabeth J. New

Abstract

Hydrogen peroxide (H_2O_2) produced from mitochondria is intimately involved in human health and disease but is challenging to selectively monitor inside living systems. The fluorescent probe MitoPY1 provides a practical tool for imaging mitochondrial H_2O_2 and has been demonstrated to function in a variety of diverse cell types. In this chapter, we describe the synthetic preparation of the small-molecule probe MitoPY1, methods for validating this probe in vitro and in live cells, and an example procedure for measuring mitochondrial H_2O_2 in a cell culture model of Parkinson's disease.

Key words Hydrogen peroxide, Fluorescent probes, Boronate, Fluorescence microscopy, Mitochondria, Triphenylphosphonium cations

1   Introduction

Mitochondria are vital organelles that not only house the electron transport chain and oxidative phosphorylation machinery but also play key regulatory roles for many processes including apoptosis [1], metabolism, and proliferation [2]. Given their fundamental importance in human biology, there has been extensive interest in studying the mitochondria for their role in disease genesis and as potential therapeutic targets [3]. Mitochondria are an important site for the production of reactive oxygen species (ROS), a family of distinct oxygen derivatives with unique physical properties and biological activities. These ROS are generated as presumably undesired side products of the mitochondrial electron transport chain and often mediate deleterious effects associated with disease and aging [4–6]. However, some ROS are now known to function as signaling agents to modulate cellular processes. For example, hydrogen peroxide (H_2O_2) has emerged as an important ROS due, in part, to its ability to mediate signaling cascades involved in redox homeostasis [7]. Determining H_2O_2 levels in mitochondria in living cells in real time, however, remains a significant challenge.

Volkmar Weissig and Marvin Edeas (eds.), *Mitochondrial Medicine: Volume I, Probing Mitochondrial Function*,
Methods in Molecular Biology, vol. 1264, DOI 10.1007/978-1-4939-2257-4_21, © Springer Science+Business Media New York 2015

Small-molecule fluorescent probes have enabled the imaging of a wide range of analytes [8]. Boronate-based probes, in particular, offer a method for the selective imaging of H_2O_2 by using the chemoselective and biocompatible oxidation of boronates into phenols [9]. For example, the first-generation probe PF1, based on a caged fluorescein, displays the remarkable ability to selectively visualize exogenously added H_2O_2 in living cells [10]. This strategy has been generalized to develop boronate-based H_2O_2 probes with a wide range of optical properties [11, 12]. One general challenge in the use of a small-molecule fluorescent probe is to control its subcellular localization so that it accumulates in the region of interest within a cell. Mitochondrial localization of a small molecule can be achieved by appending an alkylated triphenyl phosphonium tag as first described by Murphy and Smith [13]. This methodology utilizes the fact that lipophilic cations tend to accumulate in the mitochondria due to the highly negative mitochondrial membrane potential [14]. We show here that a mitochondrially localizing H_2O_2 probe can be designed based on the PF dye series by incorporation of a triphenyl phosphonium tag to provide the probe MitoPY1, which can be used to selectively image H_2O_2 in the mitochondria of living cells [15, 16].

The protocol outlined in this chapter provides a detailed description of the synthesis of MitoPY1 and its application to imaging H_2O_2 in the mitochondria of living cells. As a working example, we use MitoPY1 to image mitochondrial H_2O_2 in a cellular model of Parkinson's disease.

2 Materials

2.1 Synthetic Chemistry Components

2.1.1 Step 1: Synthesis of Fmoc-Piperazine Rhodol

1. 2-(2,4-Dihydroxybenzoyl)benzoic acid [17] (*see* **Note 1**).
2. 1-(2-Hydroxyphenyl)-piperazine.
3. Trifluoroacetic acid (TFA).
4. Diethyl ether.
5. Methanol.
6. 9-Fluorenylmethyl chloroformate (Fmoc-Cl).
7. Sodium bicarbonate (NaHCO$_3$).
8. Rotary evaporator.
9. Acetonitrile.
10. Ethyl acetate.
11. Hexanes.
12. Silica gel.
13. Sodium sulfate.

2.1.2 Step 2: Synthesis of Fmoc-Piperazine Rhodol Triflate	1. *N*-Phenyl bis(trifluoromethanesulfonamide). 2. Sodium carbonate (Na$_2$CO$_3$). 3. Dimethylformamide (DMF).
2.1.3 Step 3: Synthesis of Fmoc-Piperazine Rhodol Boronate	1. Pd(dppf)Cl$_2$·CH$_2$Cl$_2$. 2. Potassium acetate (KOAc). 3. Toluene. 4. Microwave reactor.
2.1.4 Step 4: Synthesis of MitoPY1	1. Piperidine. 2. (4-Iodobutyl)triphenylphosphonium iodide [18] (*see* **Note 2**). 3. Glove box. 4. Dichloromethane.
2.2 Cellular Imaging Components (see Note 3)	1. Eppendorf tubes. 2. Fluorescence spectrometer (*see* **Note 4**). 3. Human embryonic kidney cells (HEK293T). 4. Dulbecco's modified Eagle's medium (DMEM). 5. Fetal bovine serum (FBS). 6. Trypsin. 7. Cell culture dishes appropriate for microscopy (*see* **Note 5**). 8. Dimethyl sulfoxide (DMSO). 9. Dulbecco's phosphate-buffered saline (DPBS). 10. Hydrogen peroxide (H$_2$O$_2$). 11. Paraquat. 12. Glutamine. 13. Poly-L-lysine. 14. Inverted confocal microscope (*see* **Note 6**).

3 Methods

3.1 Synthetic Chemistry (Fig. 1) *3.1.1 Synthesis of Fmoc-Piperazine Rhodol*	1. Add 2-(2,4-dihydroxybenzoyl)benzoic acid (1.0 equiv., *see* **Note 1**) and 1-(3-hydroxyphenyl)-piperazine (1.0 equiv.) to a heavy-walled pressure flask (*see* **Note 7**). 2. Add 16 mL TFA per 1 g 2-(2,4-dihydroxybenzoyl)benzoic acid to the flask. Seal the flask and stir for 3 h in an oil bath held at 95 °C (*see* **Note 8**). 3. Cool the reaction to room temperature and add 240 mL diethyl ether per 1 g 2-(2,4-dihydroxybenzoyl)benzoic acid. A solid red precipitate should be observed.

Step 1:

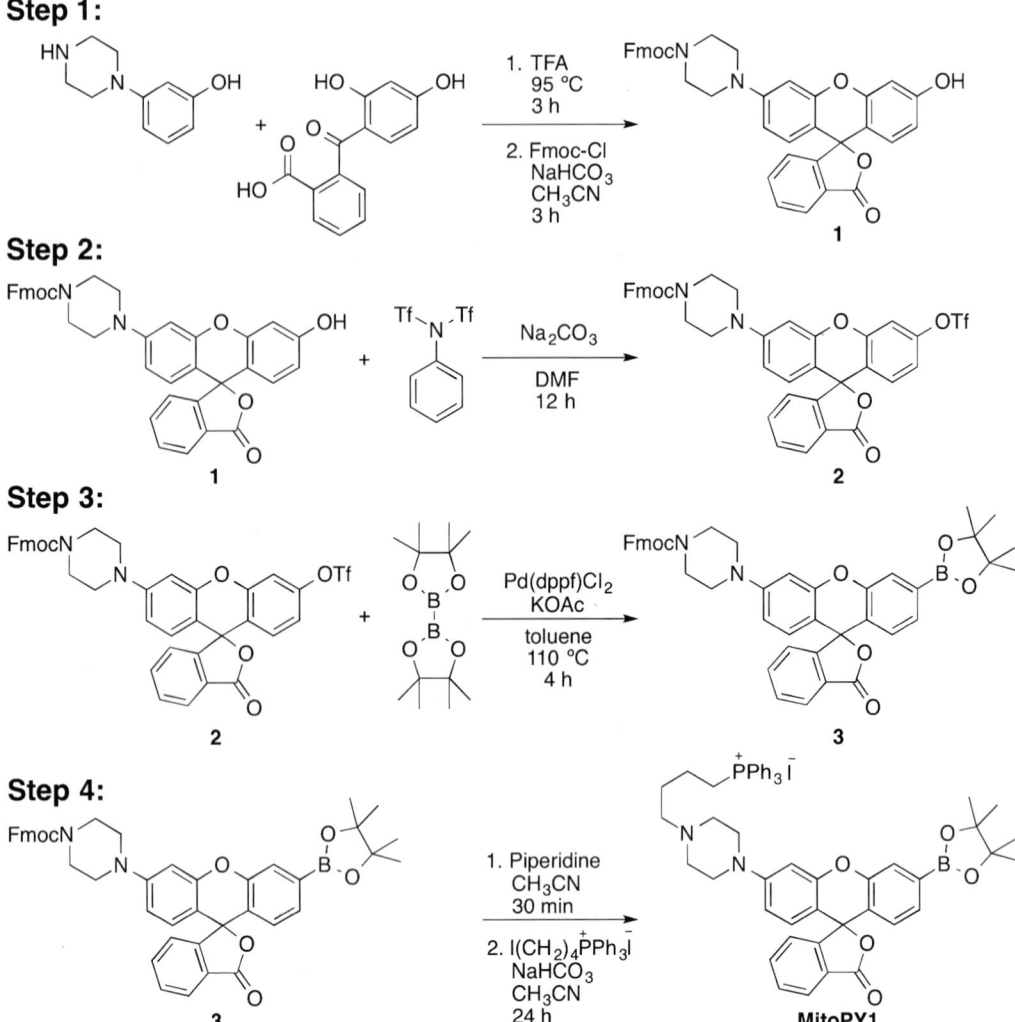

Fig. 1 Stepwise synthetic scheme for the preparation of MitoPY1

Step 2:

Step 3:

Step 4:

4. Collect the precipitate by filtration and immediately dissolve in methanol (*see* **Note 9**).

5. Transfer to a round-bottom flask and dry under high vacuum. Add this unpurified product to a dry Schlenk tube (*see* **Note 10**).

6. Add Fmoc-Cl (0.68 equiv.), NaHCO₃ (1.7 equiv.), and approximately 8 mL anhydrous acetonitrile per 1 g 2-(2,4-dihydroxybenzoyl)benzoic acid, enough to dissolve the material. Stir under ambient temperature for 3 h.

7. Dilute the reaction in ethyl acetate and then wash with water (*see* **Note 11**).

8. Evaporate the organic layer using a rotary evaporator. This crude product can be purified by silica column chromatography, eluting with 1:1 hexanes:ethyl acetate (*see* **Note 12**) to provide Fmoc-piperazine rhodol as a red solid in 39 % yield.

3.1.2 Synthesis of Fmoc-Piperazine Rhodol Triflate

1. Add the Fmoc-piperazine rhodol (1 equiv.), *N*-phenyl bis(trifluoromethanesulfonamide) (2 equiv.), and Na$_2$CO$_3$ (10 equiv.) to a dry Schlenk tube (*see* **Note 10**).

2. Add 2 mL anhydrous DMF per 100 mg Fmoc-piperazine. Fill the flask with nitrogen and stir at ambient temperature for 12 h. After this time, dilute reaction in ethyl acetate and wash with water (*see* **Note 11**).

3. Evaporate the organic layer using a rotary evaporator.

4. Purify using silica column chromatography (*see* **Note 12**), eluting with 1:1 hexanes:ethyl acetate to give Fmoc-piperazine rhodol triflate as a white solid in 46 % yield.

3.1.3 Synthesis of Fmoc-Piperazine Rhodol Boronate

1. Add the Fmoc-piperazine rhodol triflate (1 equiv.), Pd(dppf) Cl$_2$·CH$_2$Cl$_2$ (0.3 equiv.), bis(pinacolato)diboron (1.0 equiv.), KOAc (10 equiv.), and 4.5 mL toluene per 100 mg Fmoc-piperazine rhodol triflate to a dry pressure flask (*see* **Note 7**) inside an inert atmosphere glove box. Remove the tube from the box and heat at 110 °C in a microwave reactor for 4 h.

2. Cool to ambient temperature and transfer to a round-bottom flask using dichloromethane. Concentrate using a rotary evaporator.

3. Purify by silica column chromatography (*see* **Note 12**), eluting with 1:1 hexanes:ethyl acetate to yield Fmoc-piperazine rhodol boronate as a white solid in 74 % yield.

3.1.4 Synthesis of MitoPY1

1. Add the Fmoc-piperazine rhodol boronate (1 equiv.) to a round-bottom flask, and dissolve in 15 mL 15 % piperidine in acetonitrile per 100 mg Fmoc-piperazine rhodol boronate. Stir under ambient temperature for 30 min.

2. Evaporate all solvent using a rotary evaporator. Transfer the reaction vessel into an inert atmosphere glove box. Add (4-iodobutyl)triphenylphosphonium iodide (2 equiv., *see* **Note 2**), NaHCO$_3$ (5 equiv.), and 1.5 mL acetonitrile per 10 mg Fmoc-piperazine rhodol boronate.

3. Stir the reaction for 24 h under ambient temperature inside the glove box.

4. Remove the reaction from the glove box. Filter any solid material and evaporate to dryness using a rotary evaporator.

5. Purify by silica column chromatography (*see* **Note 12**), eluting with 9:9:1 dichloromethane:ethyl acetate:methanol to yield MitoPY1 as a light pink solid in 76 % yield.

3.2 Cellular Imaging with MitoPY1

1. Dissolve MitoPY1 at a concentration of 4.3 mg/mL methanol for a final concentration of 5 mM (*see* **Note 13**).

3.2.1 Preparation of Aliquots

2. Aliquot 20 μL portions of the MitoPY1 solution into Eppendorf tubes. Allow the methanol to evaporate (*see* **Note 14**).

3. Store dry stocks at –20 °C. On the day of the imaging experiment, dissolve dry stocks in 20 μL DMSO to make 5 mM stock solutions.

3.2.2 Validation of Probe Response to H_2O_2 In Vitro

1. Add 2 μL of the 5 mM stock solution of MitoPY1 in DMSO to 1.998 mL of DPBS to make a 5 μM solution of MitoPY1 in DPBS.

2. Separate this solution into two 1 mL aliquots.

3. Prepare a 100 mM solution of H_2O_2 by adding 11 μL of 30 % (wt/vol) H_2O_2 to 989 μL H_2O.

4. Treat a 1 mL aliquot of MitoPY1 in DPBS with 1 μL of 100 mM H_2O_2 for a final concentration of 100 μM H_2O_2. For the negative control, treat the other 1 mL aliquot of 5 μM MitoPY1 in DPBS with 1 μL H_2O.

5. Collect the fluorescence spectra with a fluorescence spectrometer (*see* **Note 4**), using an excitation wavelength of 503 nm and collecting the spectrum between 513 and 650 nm, with an emission maximum at $\lambda_{em} = 530$ nm. An increase in fluorescence intensity should be observed for the aliquot that was treated with H_2O_2 compared with the aliquot treated with H_2O (Fig. 2a).

3.2.3 Cell Culture

1. Maintain HEK293T cells in exponential growth as a monolayer in DMEM supplemented with 10 % FBS (*see* **Note 15**).

2. One or two days before imaging, seed a 24-well plate (*see* **Note 5**) with cells (*see* **Note 16**). At this stage, cytotoxicity (*see* **Note 17**) and mitochondrial membrane potential (MMP, *see* **Note 18**) can be measured to ensure that the probe is not perturbing normal mitochondrial function.

3.2.4 Validation of Probe Response to Mitochondrial H_2O_2 in Cell Culture

1. Add 12 μL of 5 mM MitoPY1 in DMSO to 6 mL DPBS for a final concentration of 10 μM MitoPY1.

2. Remove the 24-well plate from the incubator and replace the medium in six of the wells with 1 mL of the 10 μM MitoPY1 solution in DPBS (*see* **Note 19**).

3. Add MitoTracker Deep Red at a concentration of 25–100 nM (*see* **Note 20**).

4. After 30 min of incubation at 37 °C to allow dye uptake into cells (*see* **Note 21**), wash the cells with 2 × 1 mL DPBS (*see* **Note 22**).

5. Treat three wells with 100 μM H_2O_2 in 1 mL DPBS and the other three wells with 1 mL DPBS as a control.

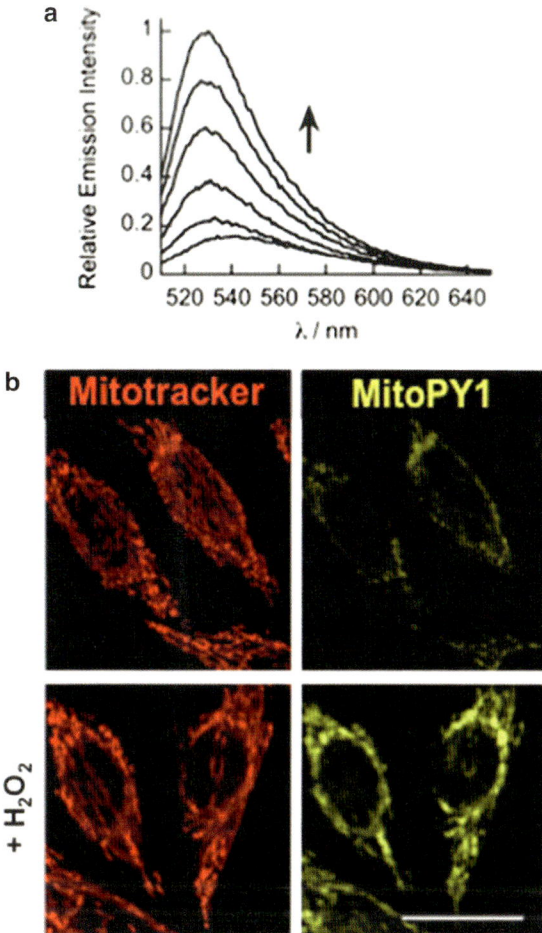

Fig. 2 (**a**) Fluorescence turn-on response of 5 μM MitoPY1 to H$_2$O$_2$. Time points represent 0, 5, 15, 30, 45, and 60 min after the addition of 100 μM H$_2$O$_2$. λ_{ex} = 503 nm. (**b**) Imaging mitochondrial H$_2$O$_2$ in live cells with MitoPY1. HeLa cells loaded with 5 μM MitoPY1 and 50 nM MitoTracker Deep Red for 20 min at 37 °C and then stimulated with either H$_2$O control or 100 μM H$_2$O$_2$ for 40 min. Scale bar = 20 μm. Adapted with permission from Dickinson, B. C. and Chang, C. J. (2008) A targetable fluorescent probe for imaging hydrogen peroxide in the mitochondria of living cells. J Am Chem Soc 130:9638–9639. Copyright 2008 American Chemical Society

6. Image the cells using a fluorescence microscope (*see* **Note 6**). Either a 488 or 510 nm excitation wavelength may be used with a collection window between 527 and 580 nm. MitoTracker Deep Red can then be imaged using a 633 nm excitation wavelength with a collection window between 666 and 698 nm. The mean fluorescence intensity of the H$_2$O$_2$-treated cells should be significantly higher than the control cells (*see* **Note 23**). An example of these results is shown in Fig. 2b.

3.2.5 Imaging
of Mitochondrial H₂O₂
in a Cellular Model
of Parkinson's Disease

Once the synthesized MitoPY1 has been validated in vitro and in cells, biological imaging experiments may be performed. As an example, an imaging experiment that uses an overnight exposure to 1 mM paraquat as a cellular model of Parkinson's disease will be described:

1. The day before imaging, prepare a 1 mM solution of paraquat in DMEM containing 10 % FBS and 2 mM glutamine.

2. Remove the medium from three wells and replace with 1 mL of the 1 mM paraquat solution.

3. Incubate for 24 h at 37 °C.

4. Remove the 24-well plate from the incubator and replace the medium of the three paraquat-treated wells and three untreated control wells (*see* **Note 24**).

5. Replace with 1 mL of the 10 µM MitoPY1 solution in DPBS.

6. Add MitoTracker Deep Red at a concentration of 25–100 nM (*see* **Note 20**).

7. After 30 min to allow dye uptake into cells (*see* **Note 21**), wash the cells with 2×1 mL DPBS (*see* **Note 22**).

8. Image the cells using a fluorescence microscope (*see* **Note 6**). Either a 488 or 510 nm excitation wavelength may be used, with a collection window between 527 and 580 nm. MitoTracker Deep Red can then be imaged using a 633 nm excitation wavelength with a collection window between 666 and 698 nm. Relative quantification of mitochondrial H_2O_2 can be evaluated by measuring the mean pixel intensity of the fluorescence images (*see* **Note 23**).

4 Notes

1. 2-(2,4-Dihydroxybenzoyl)benzoic acid can be prepared by an adapted literature procedure [17]. Briefly, dissolve resorcinol (1.1 equiv.) and phthalic anhydride (1.0 equiv.) in 30 mL dry nitrobenzene per 1 g of phthalic anhydride under an inert atmosphere. Cool the reaction mixture to 0 °C and add aluminum (III) chloride (2.3 equiv.). The reaction mixture should turn dark green. Stir the suspension for 16 h and then pour into a rapidly stirring mixture of 2:1 hexanes:1 M HCl. Filter the precipitate. The solid can be recrystallized from methanol/water.

2. (4-Iodobutyl)triphenylphosphonium iodide can be prepared from a literature procedure [18]. Briefly, add 1,4-diiodobutane (5 equiv.) to triphenylphosphine (1 equiv.) in a screw-top pressure flask. Cover the flask with aluminum foil to block out any light and heat at 100 °C for 1.5 h. After cooling to ambient

temperature, wash the resulting yellow solid three times with diethyl ether. Dissolve the solids in a small amount of dichloromethane, and then add diethyl ether. A yellow solid should precipitate. Carefully decant the solvent by slowly pouring the liquid into another flask or gently pipetting the solvent without removing any solids. Repeat this process three times, combine all solids, and remove any residual solvent under high vacuum. The synthesized (4-iodobutyl)triphenylphosphonium iodide should be used immediately or stored in the dark at –20 °C.

3. Given components are for HEK293T cells, but these can be adapted to other cell lines as needed.

4. A plate reader with fluorescence mode can also be used to validate the probes in vitro.

5. When using an inverted microscope, cells must be imaged through a thin layer of glass to minimize refraction. Cells can be grown in chamber slides (e.g., Nunc makes Lab-Tek chambered coverglass with 1, 2, 4, or 8 wells) or in glass-bottom multi-well plates (MatTek Corporation supplies such plates). Alternatively, cells can be grown on sterile coverslips in plastic multi-well plates and mounted onto glass slides immediately before imaging. Experiments described here involve use of a 24-well glass-bottom plate.

6. Imaging may be performed with an inverted or an upright fluorescence microscope. In addition, these dyes can be analyzed by flow cytometry. If an upright microscope is used, the cells should be grown on sterile coverslips. 12 mm round coverslips fit easily into 24-well plates, so cells can be grown as described below (*see* **Note 15**). Immediately before the imaging experiment, wash the cells with DPBS, pick up the coverslip with forceps, and transfer it to a 4 cm diameter Petri dish containing 2 mL phenol red-free DMEM and any dyes of interest. This Petri dish can be placed on the microscope stage, and the appropriate objective can be dipped into the medium.

7. The flask should have a recommended working pressure of at least 60 psi. These flasks are available from Chemglass (CG-1880-10). Ensure that there are no cracks or deformities in the glass and use a blast shield in case of explosion.

8. A silicone oil bath on a hotplate stirrer equipped with temperature control can be used.

9. After the filtration, the crude product tends to turn into an oil, most likely due to residual TFA. Immediate dissolution into methanol followed by evaporation usually results in a solid product.

10. Schlenk tubes are available from Chemglass (AF-0537). The Schlenk tube should be pre-dried in an oven for 4 h at 130 °C

and then attached to a Schlenk line using a gas inlet adapter. With the stopcock closed, open the flask to the high vacuum. Vacuum for 10–15 min and then fill with dry N_2 using the Schlenk line. Attach the stopcock side of the Schlenk tube to the N_2 line and begin a positive flow of N_2. Open the stopcock and replace the gas inlet adaptor with a septum. A round-bottom flask equipped with a septum can alternatively be used in this step.

11. This operation is typically performed with a separatory funnel. Pour the reaction mixture into the separatory funnel. Next, add the water or other aqueous solution to the funnel. Shake the separatory funnel, being sure to release pressure at regular intervals by pointing the vent away from the face or body. The mixture will separate into two phases, with the water composing the bottom layer and the ethyl acetate composing the top layer. If the organic solvent used is dichloromethane, it will fall to the bottom layer because dichloromethane is denser than water. Remove the stopper, open the stopcock, and collect the desired layer.

12. Silica chromatography can be performed by the method of Still [19]. Add a small piece of cotton to a chromatography column and attach the tip to a house vacuum line. Push the cotton down with a wooden rod, using the house vacuum line to hold it in place if necessary. Add a small layer of sand (~1–2 cm), and then add silica gel (~15–25 cm, depending on the amount of material and width of the column). Tap gently to make a flat layer and then turn on the vacuum to pack the silica gel. Keep under house vacuum for 10–15 min. Add another layer of sand (~1–2 cm). Prepare eluent (~1 L) and then carefully pour into the column with the vacuum on, being careful not to upset the silica bed. For best results, the top of the silica layer should be flat. Close the stopcock and turn off the vacuum before any solvent enters the vacuum line. Remove the vacuum line and clean the column tip with acetone. Fill the column with eluent and then push through using air pressure. Allow the solvent to empty until just above the sand layer. Close the stopcock. Dissolve the crude product in the minimum amount of dichloromethane, or other low-polarity solvent that dissolves the crude material, and then gently add it to the column using a pipet to drip it slowly down the side of the column. Open the stopcock and allow the crude product to be adsorbed to the top of the silica layer, washing with the minimum amount of eluent or dichloromethane. Once the product has been adsorbed, fill the reservoir with eluent and collect the eluent that comes off of the column into different test tubes. Analyze the test tubes using thin-layer chromatography, and then combine and evaporate the fractions that contain the pure product.

13. The concentration of MitoPY1 can be confirmed by using the extinction coefficient of 14,200 M^{-1} cm^{-1} at 510 nm.

14. The evaporation can be accomplished by placing the tubes in a fume hood overnight, by blowing air or nitrogen gas over the tubes, taking care to avoid splashing, or by placing in a vacuum desiccator.

15. Cells should be maintained between 10 and 90 % confluence in a 37 °C, 5 % CO$_2$ tissue culture incubator. HEK293T cells typically double every 36 h, so maintaining exponential growth will require passage of the culture twice a week. Culture cells in a T75 culture flask. At 100 % confluence, the flask will hold approximately 1×10^7 cells. All manipulations that expose cells should be performed in a sterile laminar flow hood. For passaging, remove the medium from the flask by aspirating or carefully decanting. Wash with 2 mL DPBS. Add 0.05 % trypsin in DMEM at 37 °C and incubate for 2 min. After confirming that cells have detached from the plastic, add 10 mL DMEM and determine cell density using a hemocytometer. Pellet the cells at $500 \times g$ for 5 min, remove the supernatant by aspiration or decanting, and resuspend the cells in DMEM. These cells can then be added to new flasks containing medium. For propagating cells, seed 1×10^6 cells into 10 mL DMEM in a T75 culture flask.

16. Cells should be plated in phenol red-free medium to avoid fluorescence background from the media and an appropriate cell number plated for 50–70 % confluence on the day of imaging. 100 % confluence in a 24-well plate corresponds to approximately 5×10^5 cells. Seeding 2×10^5 cells overnight will typically give an appropriate confluence. Use between 0.5 and 1.0 mL medium per well.

17. A number of mitochondrially localizing compounds disrupt mitochondrial function, causing cell death. It is therefore valuable to confirm cell viability upon treatment with MitoPY1 or similar cellular probes. Cell viability can be confirmed by a number of standard assays. Simple methods include counting of cell populations treated with trypan blue or propidium iodide, which only stain dead or dying cells. For more robust statistics, plate readers or flow cytometers enable collection of data from much larger numbers of cells. The colorimetric MTT or WST-1 assays can be performed in a multi-well plate and analyzed on a plate reader. Flow cytometric methods include concurrent use of a live stain (such as calcein AM) with a dead stain (such as propidium iodide).

18. Molecules that localize to the mitochondria can interact with the mitochondrial membrane, thus perturbing the membrane potential and hence overall mitochondrial function.

Measurement of MMP in the presence of MitoPY1 can confirm that this sensor does not act in such a way. MMP can be measured by flow cytometric assays that utilize two different mitochondrial dyes, such as in the method described by Pendergrass et al. [20].

19. For short-term experiments in robust cell lines, a laminar flow hood is usually not necessary. For longer-term experiments and/or sensitive cell lines, it is best to perform all cellular operations in a laminar flow hood with sterile solutions.

20. MitoTracker Deep Red is available from Life Technologies and should be used according to the manufacturer's instructions. This dye will confirm mitochondrial localization and serves as an important control.

21. Depending on the cell type, the optimal time for cellular uptake ranges from 15 to 40 min.

22. Wash cells by removing the medium and adding 1 mL of DPBS warmed to 37 °C and repeating this step with an additional 1 mL of DPBS at 37 °C.

23. Mean fluorescence intensity is typically quantified in ImageJ (National Institutes of Health) by setting identical thresholds in both wells, such that the intracellular fluorescence in both images is selected. Other quantification methods can be used, but these should be clearly reported to ensure lab-to-lab reproducibility.

24. In general, it is best to have at least three biological replicates for every experiment to ensure reproducibility and significance.

References

1. Desagher S, Martinou JC (2000) Mitochondria as the central control point of apoptosis. Trends Cell Biol 10:369–377

2. McBride HM, Neuspiel M, Wasiak S (2006) Mitochondria: more than just a powerhouse. Curr Biol 16:R551–R560

3. Ernstner L, Schatz G (1981) Mitochondria: a historical review. J Cell Biol 91:227s–255s

4. Petros JA, Baumann AK, Ruiz-Pesini E, Amin MB, Sun CQ, Hall J, Lim SD, Issa MM, Flanders WD, Hosseini SH, Marshall FF, Wallace DC (2005) mtDNA mutations increase tumorigenicity in prostate cancer. Proc Natl Acad Sci U S A 102:719–724

5. Fato R, Bergamini C, Leoni S, Strocchi P, Lenaz G (2008) Generation of reactive oxygen species by mitochondrial complex I: implications in neurodegeneration. Neurochem Res 33:2487–2501

6. Lenaz G, Bovina C, D'Aurelio M, Fato R, Formiggini G, Genova ML, Giuliano G, Pich MM, Paolucci U, Castelli GP, Ventura B (2002) Role of mitochondria in oxidative stress and aging. Ann N Y Acad Sci 959:199–213

7. Veal EA, Day AM, Morgan BA (2007) Hydrogen peroxide sensing and signaling. Mol Cell 26:1–14

8. Chan J, Dodani SC, Chang CJ (2012) Reaction-based small-molecule fluorescent probes for chemoselective bioimaging. Nat Chem 4:973–984

9. Lippert AR, Van de Bittner GV, Chang CJ (2011) Boronate oxidation as a bioorthogonal reaction approach for studying the chemistry of hydrogen peroxide in living systems. Acc Chem Res 44:293–804

10. Chang MCY, Pralle A, Isacoff EY, Chang CJ (2004) A selective, cell-permeable optical

probe for hydrogen peroxide in living cells. J Am Chem Soc 126:15392–15393

11. Miller EW, Albers AE, Pralle A, Isacoff EY, Chang CJ (2005) Boronate-based fluorescent probes for imaging cellular hydrogen peroxide. J Am Chem Soc 127:16652–16659

12. Dickinson BC, Huynh C, Chang CJ (2010) A palette of fluorescent probes with varying emission colors for imaging hydrogen peroxide signaling in living cells. J Am Chem Soc 132:5906–5915

13. Murphy MP, Smith RAJ (2007) Targeting antioxidants to mitochondria by conjugation to lipophilic cations. Annu Rev Pharmacol Toxicol 47:629–656

14. Ross MF, Kelso GF, Blaike FH, James AM, Cochemé HM, Filipovska A, Da Ros T, Hurd TR, Smith RAJ, Murphy MP (2005) Lipophilic triphenylphosphonium cations as tools in mitochondrial bioenergetics and free radical biology. Biochemistry (Mosc) 70:222–230

15. Dickinson BC, Chang CJ (2008) A targetable fluorescent probe for imaging hydrogen peroxide in the mitochondria of living cells. J Am Chem Soc 130:9638–9639

16. Dickinson BC, Lin VS, Chang CJ (2013) Preparation and use of MitoPY1 for imaging hydrogen peroxide in live cells. Nat Protoc 8:1249–1259

17. Chang CJ, Nolan EM, Jaworski J, Okamoto KI, Hayashi Y, Sheng M, Lippard SJ (2004) ZP8, a neuronal zinc sensor with improved dynamic range; imaging zinc in hippocampal slices with two-photon microscopy. Inorg Chem 43:6774–6779

18. Lin TK, Hughes G, Muratovska A, Blaikie FH, Brookes PS, Darley-Usmar V, Smith RAJ, Murphy MP (2002) Specific modification of mitochondrial protein thiols in response to oxidative stress. J Biol Chem 277:17048–17056

19. Still WC, Kahn M, Mitra A (1978) Rapid chromatographic techniques for preparative separation with moderate resolution. J Org Chem 43:2923–2925

20. Pendergrass W, Wolf N, Poot M (2004) Efficacy of MitoTracker Green™ and CMXrosamine to measure changes in mitochondrial membrane potentials in living cells and tissues. Cytometry A 61A:162–169

Chapter 22

Simultaneous High-Resolution Measurement of Mitochondrial Respiration and Hydrogen Peroxide Production

Gerhard Krumschnabel, Mona Fontana-Ayoub, Zuzana Sumbalova, Juliana Heidler, Kathrin Gauper, Mario Fasching, and Erich Gnaiger

Abstract

Mitochondrial respiration is associated with the formation of reactive oxygen species, primarily in the form of superoxide ($O_2^{\bullet-}$) and particularly hydrogen peroxide (H_2O_2). Since H_2O_2 plays important roles in physiology and pathology, measurement of hydrogen peroxide has received considerable attention over many years. Here we describe how the well-established Amplex Red assay can be used to detect H_2O_2 production in combination with the simultaneous assessment of mitochondrial bioenergetics by high-resolution respirometry. Fundamental instrumental and methodological parameters were optimized for analysis of the effects of various substrate, uncoupler, and inhibitor titrations (SUIT) on respiration versus H_2O_2 production. The sensitivity of the H_2O_2 assay was strongly influenced by compounds contained in different mitochondrial respiration media, which also exerted significant effects on chemical background fluorescence changes. Near linearity of the fluorescence signal was restricted to narrow ranges of accumulating resorufin concentrations independent of the nature of mitochondrial respiration media. Finally, we show an application example using isolated mouse brain mitochondria as an experimental model for the simultaneous measurement of mitochondrial respiration and H_2O_2 production in SUIT protocols.

Key words Amplex Red, Resorufin, High-resolution respirometry, Oxygraph, Respiration media, Substrate–uncoupler–inhibitor titration, Mouse brain mitochondria

1 Introduction

Mitochondria are the primary source of aerobic ATP production and at the same time main producers of reactive oxygen species (ROS) in the cell [1, 2]. ROS are a by-product of respiratory activity, act as signaling molecules and as toxic factors for cell function when physiological ranges are exceeded, and are thus implicated in many physiological and pathological pathways. Accordingly, there is considerable interest in the evaluation of the interplay between mitochondrial metabolism and ROS formation. The mitochondrial electron transfer complexes I and III are considered as the main

Volkmar Weissig and Marvin Edeas (eds.), *Mitochondrial Medicine: Volume I, Probing Mitochondrial Function*,
Methods in Molecular Biology, vol. 1264, DOI 10.1007/978-1-4939-2257-4_22, © Springer Science+Business Media New York 2015

producers of ROS, primarily in the form of superoxide anions ($O_2^{\cdot-}$) (reviewed in ref. 3). Most of the superoxide radicals produced are immediately converted to hydrogen peroxide (H_2O_2) in a reaction catalyzed by mitochondrial superoxide dismutase (SOD). In contrast to the superoxide anion, H_2O_2 passes biological membranes easily, and therefore, for theoretical reasons and practical purposes, measurement of H_2O_2 emission is the method of choice for estimation of mitochondrial ROS formation. A widely applied method to detect H_2O_2 emission is the Amplex Red (AmR) assay which combines relative stability of the assay components, i.e., the substrate AmR and the product resorufin, and a high fluorescence output [4–6]. This makes it a primary choice for the detection of H_2O_2 production at high sensitivity and stability as required for the simultaneous measurement of ROS formation and mitochondrial respiration. The AmR assay has been intensively described and discussed in the literature focusing not only on applicability but also on limitations of the method [7–10]. So far, however, with one exception [11], most applications were designed to detect H_2O_2 production over rather short experimental periods of time. Here, we present a basal evaluation of the use of the AmR assay in prolonged experiments aiming to simultaneously measure mitochondrial respiration by high-resolution respirometry (HRR) and the production of H_2O_2. We address the impact of the composition of the respiration medium on sensitivity, on linearity and stability, and specifically on the time-dependent changes of chemical background fluorescence. Potential sources of error were analyzed and avoided by optimized strategies for calibrations carried out repeatedly during the time course of substrate–uncoupler–inhibitor titration (SUIT) protocols.

2   Materials

2.1   System for High-Resolution Respirometry (HRR)

1. The Oxygraph-2k for HRR.
   The Oxygraph-2k (O2k, OROBOROS INSTRUMENTS, Austria; http://www.oroboros.at) for measurement of respiration is described in detail elsewhere [12, 13]. The O2k is a closed-chamber respirometer for high-resolution respirometry (HRR), monitoring oxygen concentration, c_{O2} [μM], in the incubation medium over time. Supported by an application-dedicated and user-friendly software (DatLab), it allows to follow aerobic respiration of biological samples (J_{O2}) in real time, to repeatedly reoxygenate the respiration medium in order to prevent undesired hypoxic conditions, and to add and titrate in unlimited steps various substrates, uncouplers, inhibitors, and other substances during prolonged experimental runs.

2. The O2k-Fluorescence LED2-Module.

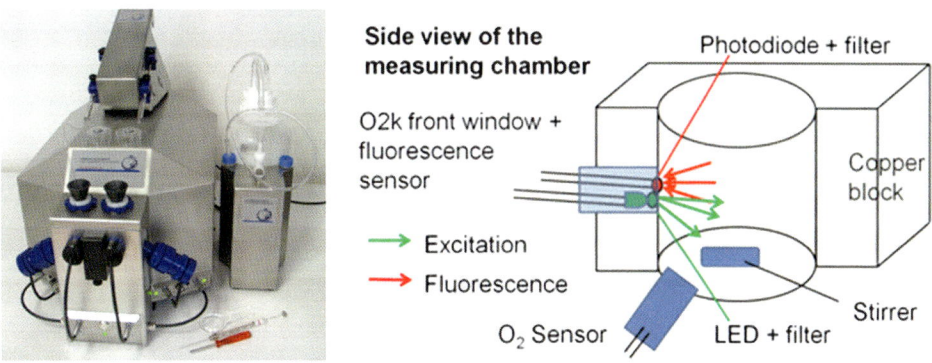

Fig. 1 OROBOROS Oxygraph-2k with O2k-Fluorescence LED2-Module. Excitation: green LED (520 nm) or blue LED (465 nm) with short-pass filter. Detector: photodiode and long-pass filter. Recording: amperometric channel of Oxygraph-2k and DatLab

The O2k-Fluorescence LED2-Module is an add-on component of the O2k with optical sensors including a light-emitting diode, a photodiode, and specific optical filters. These sensors are inserted through the front window of the O2k chambers (Fig. 1), using the Fluorescence-Sensor Green (525 nm) with Amplex Red (H_2O_2, 11) or TMRM (mt-membrane potential) or the Fluorescence-Sensor Blue (465 nm) with Magnesium Green™ (ATP production, 14), safranin (mt-membrane potential, 15), Calcium Green™ (Ca^{2+}), and numerous other applications open for O2k-user innovation.

Excitation and emission filters were optimized for applications with Amplex® UltraRed to ensure that only emission (linear relationship with analyte concentration, positive signal) but not absorption phenomena (logarithmic relationship with analyte concentration, negative signal) contribute to the signal change detected by the photodiode (Fig. 2). The photodiode (Fig. 2a) was replaced by a 600-µm light guide, connected to a mini spectrometer. Fluorescence was induced by the addition of resorufin, which is the reaction product of H_2O_2 formed with Amplex® UltraRed in the presence of horseradish peroxidase (HRP) (Fig. 2b). Difference spectra for 0 and 3 µM resorufin in 100 mM potassium phosphate buffer (pH = 7) were obtained for different filter configurations (Fig. 2b–d). The difference spectrum obtained initially using only a long-pass emission filter (50 % transmission at approximately 600 nm) indicated a distinct absorption (negative) contribution below approximately 580 nm to the output signal. Examining a multitude of filter combinations, we ultimately selected a short-pass excitation filter with 50 % transmission at 445 nm and a long-pass emission filter with 50 % transmission at 620 nm (Fig. 2d) resulting in an optimized signal output specific for changes in resorufin concentration. Further information is available in the O2k-Fluorescence LED2-Module manual [16].

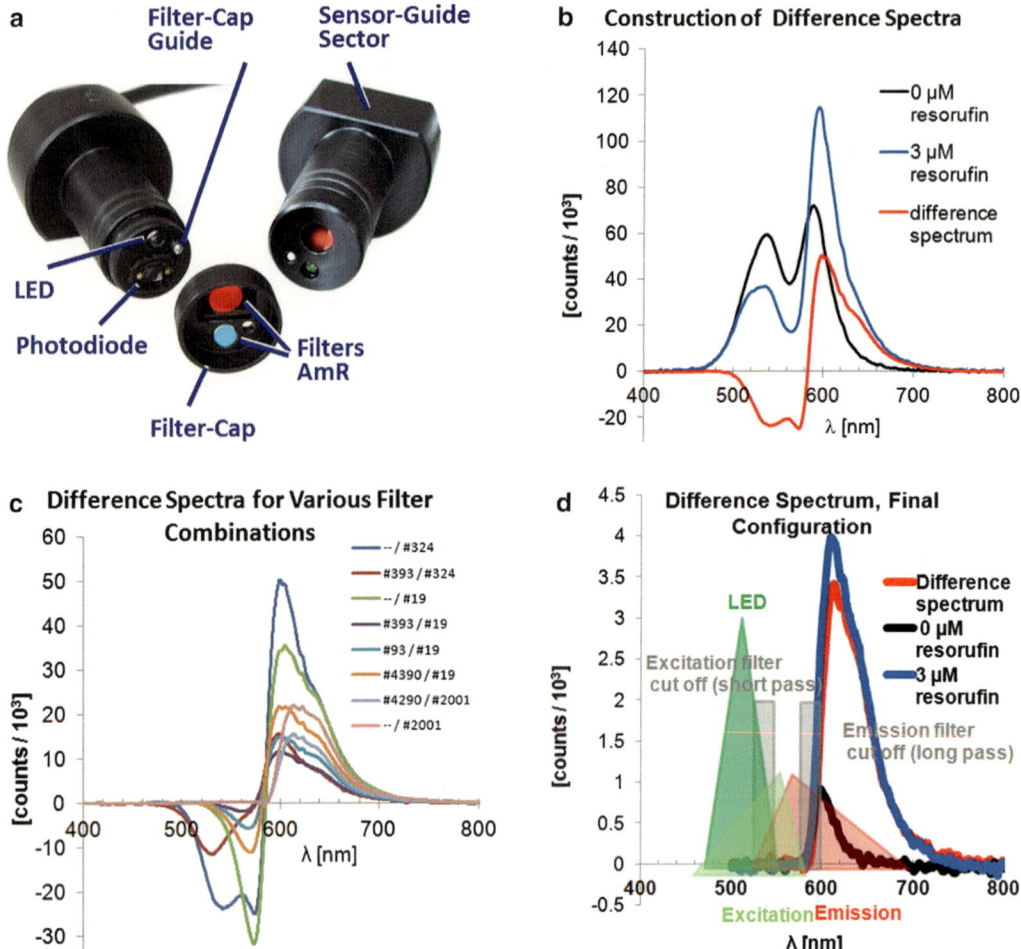

Fig. 2 Optimization of fluorescence-sensor filters for detection of H$_2$O$_2$. (**a**) Fluorescence-Sensor Green with light source (LED with *green* excitation filter), detector (photodiode with *red* emission filter). (**b**) The difference spectrum obtained with an initial filter combination reveals a distinct absorption (negative) contribution. (**c**) Difference spectra with different filter combinations. (**d**) Optimized filter configuration: The difference spectrum (*red line*) shows no negative contribution. Triangular areas show schematically the LED spectrum (*dark green*) and excitation and emission spectra of resorufin (*light green* and *red triangles*). Approximate cutoff regions of the excitation and emission filters are shown by *gray bars*

2.2 Media and Reagents

2.2.1 Mitochondrial Respiration Media

1. MiR05: 110 mM sucrose, 60 mM K$^+$-lactobionate, 0.5 mM EGTA, 3 mM MgCl$_2$, 20 mM taurine, 10 mM KH$_2$PO$_4$, 20 mM HEPES adjusted to pH 7.1 with KOH at 30 °C, and 1 g/l BSA essentially fatty acid-free [17].

2. MiR06: MiR06 is MiR05 plus 280 U/ml catalase.

3. MiRK03: 130 mM KCl, 20 mM HEPES free acid, 10 mM KH$_2$PO$_4$, 3 mM MgCl$_2$, 0.5 mM EGTA adjust pH to 7 with KOH, and 0.1 % BSA [18]. Creatine (Cr; 20 mM) was added to both media (MiR05Cr, MiRK03Cr).

4. MiR05Cr: add 20 mM creatine (Cr) to MiR05.

5. MiRK03Cr: add 20 mM creatine to MiRK03.

2.2.2 Mitochondrial
Isolation Medium

1. Isolation medium for mouse brain mitochondria: 320 mM sucrose, 1 mM EDTA, and 10 mM TRIS, pH adjusted to 7.4 with HCl (25 °C).

2.2.3 Substrates,
Uncouplers, and Inhibitors
for Titrations (SUIT
Chemicals)

1. ADP (D) stock solution: 0.5 M ADP in distilled water, store at –20 °C.

2. Glutamate (G) stock solution: 2 M glutamate in distilled water, adjust to pH 7 with KOH, store at –20 °C.

3. Malate (M) stock solution: 400 mM malate in distilled water, adjust to pH 7 with KOH, store at –20 °C.

4. FCCP or CCCP: 1 mM FCCP or CCCP in ethanol (*see* **Note 1**).

5. Rotenone (Rot):1 mM rotenone in ethanol.

6. Antimycin A (Ama): 5 mM antimycin A in ethanol.

7. Oligomycin (Omy): 4 mg/ml oligomycin in ethanol.

8. Pyruvate (P) stock solution: 2 M pyruvate in distilled water, prepare fresh daily.

9. Malonic acid (Mna): 2 M malonic acid in distilled water, adjusted to pH 7 with KOH (25 °C), prepare fresh daily, *see* Fontana-Ayoub et al. [19] for further information.

2.3 Solutions
for Measurement
of Hydrogen Peroxide
Production

1. Amplex® UltraRed: a 10 mM storage solution of Amplex® UltraRed reagent (AmR) is prepared by adding 340 μl of fresh, high-quality DMSO to one commercial vial of Amplex® UltraRed reagent (1 mg) with rigorous vortexing. The storage solution is diluted 1:5 with DMSO to yield a 2 mM stock solution. Aliquots (100 μl) of the 2 mM stock solution are stored at –20 °C in the dark, protected from moisture. When stored properly, the 2 mM stock solution is stable for at least 6 months (*see* **Note 2**).

2. Horseradish peroxidase (HRP): a stock solution containing 500 U HRP/ml in MiR05 or MiR05Cr is prepared for storage at –20 °C. 4 μl of stock solution are titrated into the 2-ml O2k chamber for a final concentration in the chamber of 1 U/ml.

3. Superoxide dismutase (SOD): use the enzyme at a concentration of 5 U/ml (*see* **Notes 3** and **4**).

4. Calibration standards of H_2O_2: prepare fresh by diluting 284 μl of a commercial solution of 3 wt.% H_2O_2 (880 mM) with 10 μM HCl (1 ml of 1 mM HCl filled to 100 ml with distilled water) to a total volume of 25 ml, to obtain a 10 mM H_2O_2 solution A. Dilute 400 μl of solution A with 10 μM HCl to a volume of 50 ml to obtain a stock solution of 80 μM H_2O_2. Store the stock solution in the refrigerator (*see* **Note 5**).

3   Methods

3.1   Isolation of Mouse Brain Mitochondria

1. Dissect the brain from the mouse skull and place immediately in ice-cold MiR05.

2. Determine wet weight (W_w [mg]).

3. Transfer brain to a precooled glass beaker (20 ml) and wash twice with ice-cold MiR05.

4. Mince the tissue into small pieces using a pair of sharp scissors and suspend with five volume W_w ice-cold MiR05.

5. Transfer to a precooled glass/Teflon potter and homogenize the tissue with eight strokes at $1,000 \times g$.

6. Transfer the homogenate to 20 ml of ice-cold mitochondrial isolation medium with 2.5 mg BSA/ml.

7. Centrifuge at $1,000 \times g$ for 10 min at 4 °C.

8. Transfer the supernatant into new tube and centrifuge at $6,200 \times g$ for 10 min at 4 °C.

9. Discard the supernatant and wash mitochondrial pellet with ice-cold mitochondrial isolation medium.

10. Resuspend mitochondria in a small volume of MiR05 (~200 µl/brain).

11. Store isolated mitochondria on ice until use [20].

3.2   Determination of H_2O_2 Production

1. Linearity of the emission signal.

The Amplex method is based on the H_2O_2-dependent oxidation of AmR to the red fluorescent compound resorufin catalyzed by the enzyme HRP. The increase of the resorufin-linked fluorescence signal with time yields a slope which is related to the reaction flux of H_2O_2 production, while the H_2O_2 concentration is maintained practically at (or close to) zero. Thus, the concentration of AmR diminishes during an experiment, and the resorufin concentration increases concurrently, which has to be taken into account particularly in extended SUIT protocols designed for evaluation of mitochondrial energetics [21]. As is generally true for enzyme-catalyzed reactions, the accumulated product concentration must not surpass a certain threshold to prevent product inhibition of the reaction [22]. In addition, accumulation of resorufin may lead to an allosteric inhibition of HRP, which ultimately limits the AmR reaction with H_2O_2 [23]. Therefore, it is critical to evaluate inhibitory thresholds of AmR depletion and resorufin accumulation. In addition, the concentration of accumulated resorufin should ideally be linearly related to the fluorescence emitted, so as to allow for simple one-step calibrations (see **Note 6**).

We determined the relation between the concentration of resorufin and the fluorescence emission signal in a set of pilot experiments. Two basically different respiration media were used, MiR05Cr and MiRK03Cr. The first medium has been specifically developed for the assessment of mitochondrial respiration and preservation of mitochondrial function during prolonged experimental incubation times [17]. The latter medium was based on the respiration buffer used by Komary et al. [18] and modified for the determination of H_2O_2 production with AmR. In these experiments, resorufin was titrated stepwise into the Oxygraph-2k chambers containing 20 μM AmR, 1 U/ml HRP, and 5 U/ml SOD, using the Titration-Injection microPump TIP2k (OROBOROS INSTRUMENTS) programmed to titrate resorufin in 0.1 μM steps up to 5 μM followed by 0.2 μM steps up to 7 μM. As shown in Fig. 3, the fluorescence signal was nonlinearly dependent on resorufin concentration over the entire concentration range, showing considerable deviation from linearity above 3 μM resorufin. This deviation from the ideal behavior may be corrected for with repeated calibrations (see below) and should be taken into account when experiments are planned where a high cumulative H_2O_2 production is expected (*see* **Note 6**). Nonlinearities were very similar in experiments with MiR05Cr and MiRK03Cr (Fig. 3).

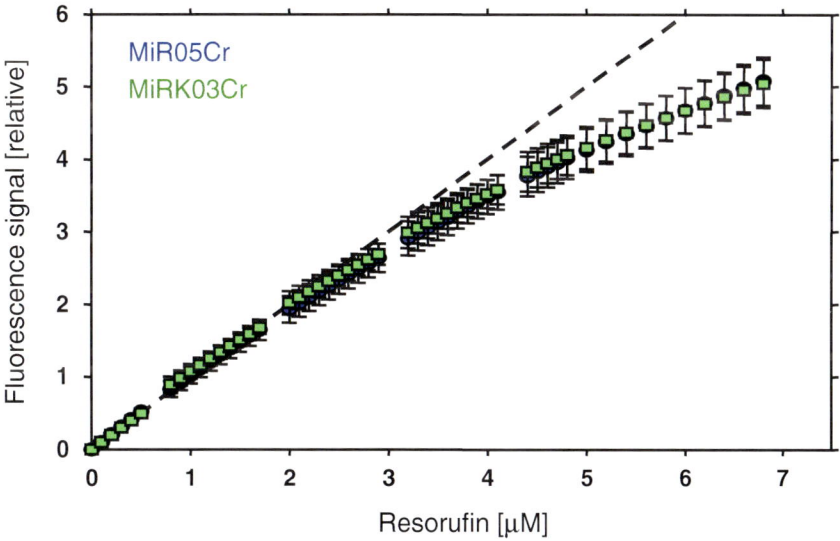

Fig. 3 Relative fluorescence signal in resorufin titrations with MiR05Cr or MirK03Cr, containing AmR (20 μM), HRP, and SOD. Resorufin was titrated automatically into the 2-ml Oxygraph-2k chambers in 0.1 μM steps up to 5 μM and 0.2 μM steps up to 7 μM using the Titration-Injection microPump TIP2k. Fluorescence was normalized to the change observed after the first injection of 0.1 μM resorufin (extrapolated *dashed line*) to account for differences in sensor sensitivities. Means ± SE of 5 experiments

2. Calibrations with H_2O_2 and factors impacting on assay sensitivity.

Theoretically, calibrations of the fluorescence signal can be performed by titrations of H_2O_2 or resorufin [11], as fluorescence of the latter is the experimental readout. However, considering the scavenging power of different media based on their antioxidant properties, it appears that actual mitochondrial production of H_2O_2 is best evaluated based on calibrations by H_2O_2. Here, we investigated the potential use of calibrations with H_2O_2 and resorufin and the necessity of repeated calibrations to account for changes in sensitivity over time. In addition, we studied whether additional calibrations with titration of resorufin can be used to separate effects influencing the direct sensitivity of the assay (see **Note 7**) for detection of resorufin concentration from H_2O_2-scavenging effects of the respiration medium and any experimental substances used in a titration protocol. To this end, we determined the sensitivity to calibrations with H_2O_2 and resorufin in background SUIT protocols (without biological sample) over the duration of a typical extended experiment. Calibrations were performed by first adding resorufin followed by the titration of H_2O_2 within 1–2 min, avoiding time-dependent differences in sensitivity toward addition of H_2O_2 and resorufin. In order to account for the different sensitivities obtained in MiR05Cr and MiRK03Cr (see below), data were normalized to the value obtained in the first calibration step. Figure 4 summarizes the results on the influence of respiration medium (MiR05Cr, blue lines; MiRK03Cr, green lines), experimental time, and AmR concentration (10 µM, straight lines; 20 µM, dashed lines) on the sensitivity. In both media, these AmR concentrations had no significant influence on sensitivity. However, the relative H_2O_2 sensitivity over time did show considerable variations, as we observed an initial 1.1-fold increase in MiR05Cr but a slowly progressing decrease to 0.65-fold in the initial value in MiRK03Cr. In comparison, we observed a relatively steady decrease of sensitivity over time in both media in resorufin calibrations but also found that day-to-day reproducibility was inexplicably bad (not shown). The observed decrease of sensitivity over time seen with both H_2O_2 and resorufin calls for repeated calibrations at different time points of an experimental run. At the same time, total accumulation of resorufin by direct resorufin titrations or after titrations of H_2O_2 should be limited to minimize nonlinearities and avoid depletion of AmR. Taken together, we thus strictly advise against the use of resorufin for calibration and suggest performing calibrations with H_2O_2 (see **Note 8**) at the start and end of an experiment, complemented by repeated one-step calibrations throughout prolonged experiments (see **Note 9**).

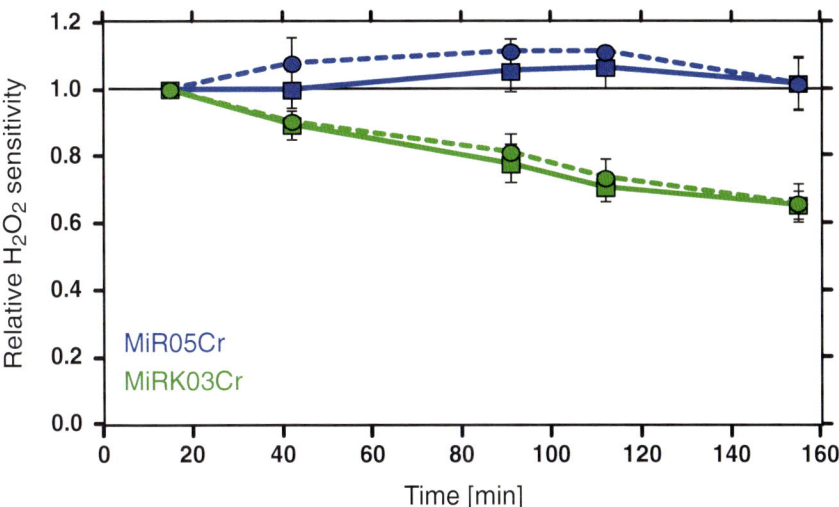

Fig. 4 Changes in relative fluorescence sensitivity over the period of a SUIT protocol in MiR05Cr (*blue lines*) and MiRK03Cr (*green lines*) at initial AmR concentrations of 10 μM and 20 μM (*full* and *dashed lines*). The relative sensitivity for H_2O_2 titrations declined in MiRK03Cr but remained practically constant in MiR05Cr. Three experiments were performed each in respiration media without titrations, titrations of carrier, and SUIT protocol titrations (Fig. 4). These titrations did not exert a significant effect on sensitivity. Means were calculated from the averages ($n = 3$) for the three regimes and plotted ± SD ($N = 3$)

A further important insight derived from these experiments was that changes in medium composition dramatically affected the sensitivity of the reaction between AmR and H_2O_2. A systematic evaluation of this issue comparing assay sensitivities at two different concentrations of AmR, two different calibration standards, and a number of different sensors (*see* **Note 10**) uncovered a generally much lower sensitivity in MiR05Cr as compared to MiRK03Cr (Fig. 5) which was largely independent of the other parameters considered (*see* **Note 11**). Varying medium composition of MiR05, we observed that sensitivity in this medium could be considerably elevated by replacing lactobionate with KCl, whereas it was little affected by omitting either sucrose or taurine in the medium (Fig. 6). Taken together, these data indicate that sensitivity needs to be carefully determined to assess the suitability of any experimental medium, while at the same time the suitability of any such medium for the evaluation of mitochondrial energetics must not be forgotten.

3. Chemical background flux.

Under unfavorable conditions [5], AmR is spontaneously oxidized even in the absence of H_2O_2. Spontaneous oxidation is also observed in the presence of HRP, measured as background flux of H_2O_2 without biological sample and is, therefore, unrelated

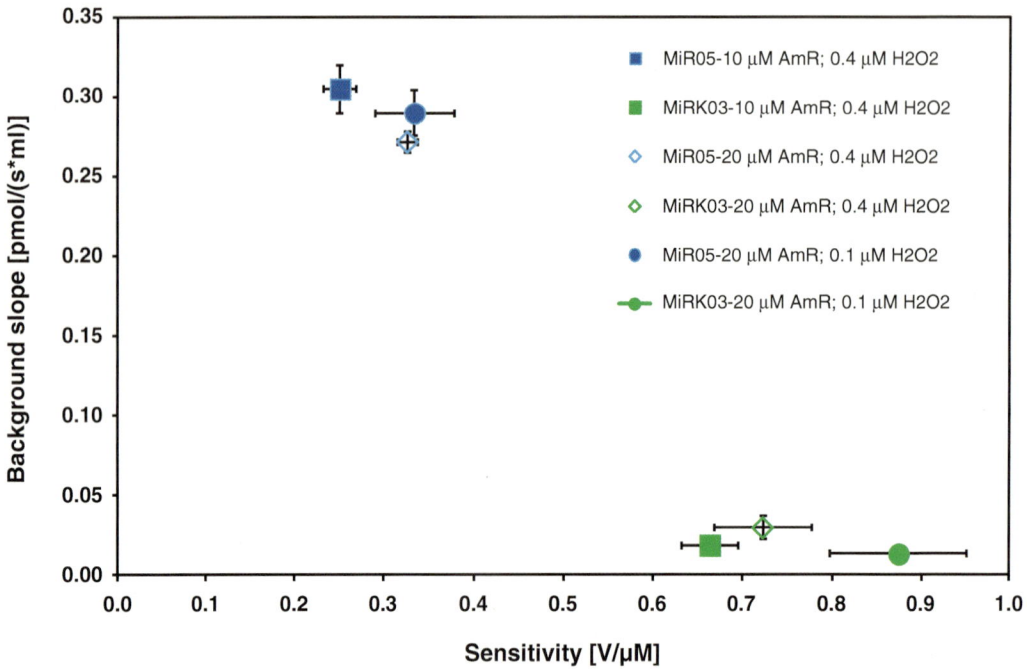

Fig. 5 Effect of Amplex Red and H_2O_2 concentration on background slope [pmol/s/ml] and sensitivity [V/μM H_2O_2] in MiR05Cr and MiRK03Cr. The change in fluorescence was recorded over time in the presence of either 10 μM or 20 μM AmR, and with 1 U/ml HRP and 5 U/ml SOD, at a temperature of 37 °C and a setting of 1 mA for light intensity. Experiments were performed with different sensors of the type Fluorescence-Sensor Green. MiR05Cr (*blue symbols*) and MiRK03Cr (*green symbols*) are compared after the addition of either 0.4 μM or 0.1 μM H_2O_2. Data are means ± SE of averages (*n* = 3 replica) of different sensors (*N* = 3)

to mitochondrial H_2O_2 production. The magnitude of this background flux depends on the light intensity used to excite the fluorescence probe [7, 24] and on components of the respiration medium. Opposing these effects, mitochondrial respiration media that are designed to exert protective antioxidant properties will scavenge a variable fraction of the H_2O_2 that is either produced by mitochondria or added for calibration purposes before it can react with AmR and HRP, thereby diminishing the sensitivity of the H_2O_2 detection assay. To evaluate the impact of different respiration media, we analyzed the background flux based on resorufin fluorescence changes over time in MiR05Cr and MiRK03Cr. These experiments showed a much higher background flux in MiR05Cr (apparent flux of $H_2O_2 \approx 0.28–0.45$ pmol/s/ml) as compared to MiRK03Cr (apparent flux of $H_2O_2 \approx 0.05$ pmol/s/ml) (Fig. 7). When observed over the entire protocol, this resulted in a much higher background accumulation of resorufin in MiR05Cr compared to MiRK03Cr, leading to a final apparent resorufin concentration of $6.9 ± 0.95$ μM resorufin in MiR05Cr as

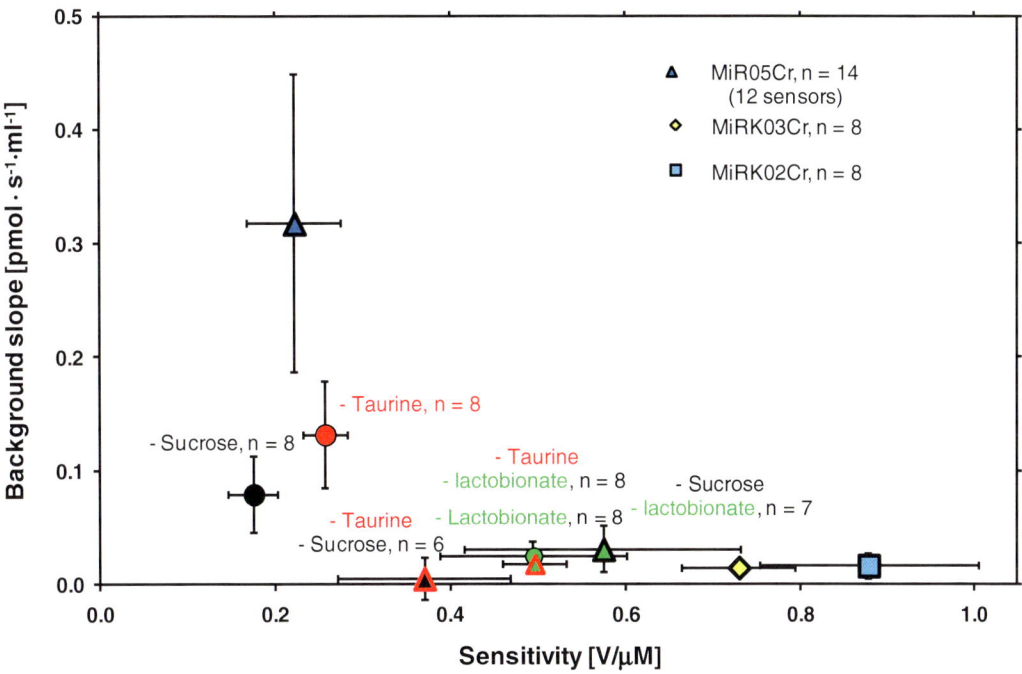

Fig. 6 Effect of components in mitochondrial respiration media on background slope (fluorescence increase) [pmol/s/ml] and sensitivity [V/µM H_2O_2] of the determination of H_2O_2 with Amplex Red. Background slope was estimated from the change in fluorescence over time observed in the absence of biological sample but in the presence of 5 µM AmR and 1 U/ml HRP at 37 °C at a setting of 1 mA for light intensity and a gain of 1,000. Sensitivity was determined by conducting three-point calibrations consisting of two consecutive injections each of 0.04 µM to 0.4 µM H_2O_2. Media used were either MiR05Cr, MiRK03Cr, MiRK02Cr, or alternative formulations of MiR05Cr where lactobionate, sucrose, taurine, or combinations of these compounds were omitted and osmotically replaced by KCl. Elimination of lactobionate increased the sensitivity, whereas taurine and sucrose exerted small effects on sensitivity. The background slope was reduced by elimination of either taurine, sucrose, or lactobionate. Replacing taurine and sucrose by increasing lactobionate from 60 to 130 mM increased the background slope from 0.04 (*red/black triangle*) to 1.2 pmol/s/ml (not shown)

compared to 2.8 ± 0.2 µM in MiRK03Cr within 2 h and 40 min. Interestingly, the background flux was independent of the concentration of AmR originally added to the medium (Fig. 7), suggesting that 20 µM rather than 10 µM AmR may be advantageous to avoid depletion of the probe during prolonged experiments (*see* **Note 12**). Correction for chemical background flux is important, particularly in MiR05Cr at low experimental mitochondrial densities. Based on these observations, we recommend analyzing the background in each experiment. We did not observe any influence of the tested SUIT chemicals on background flux (*see* **Note 13**), indicating that respirometric analyses using various substrates, uncouplers, and inhibitors are compatible with the combined measurement of H_2O_2 production.

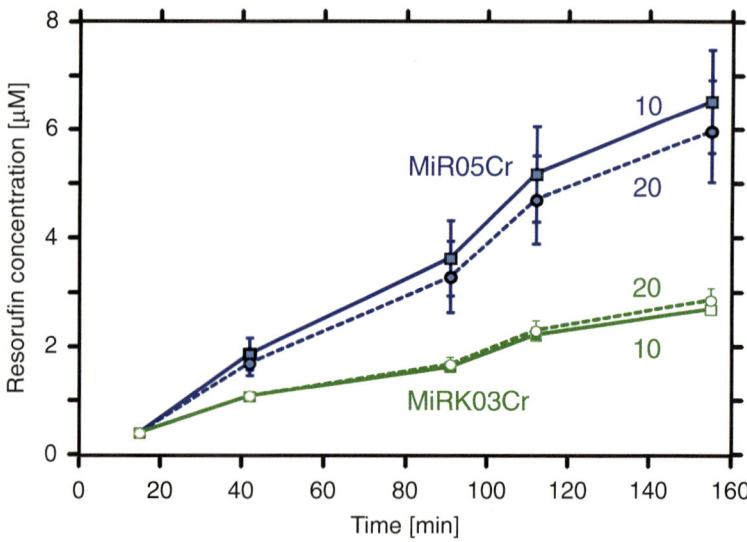

Fig. 7 Background accumulation of resorufin in a SUIT protocol without sample in respiration media MiR05Cr and MiRK03Cr, at initial AmR concentrations shown by numbers (10 and 20 μM). Resorufin concentrations were calculated from the first H_2O_2 calibration. Means ± SD of three experiments. Mitochondrial isolation medium was injected to mimic the addition of isolated mitochondria followed by the addition of 2.5 mM ADP, 10 mM succinate, and three uncoupler titrations each with 0.5 μM CCCP, 0.5 μM complex I inhibitor rotenone, and 5 mM complex II inhibitor malonic acid. Data points are shown for specific time points where resorufin and H_2O_2 were added in steps of 0.4 μM to assess changes in sensitivity

In order to analyze which components of MiR05Cr are responsible for the high background flux, MiR05Cr-based media were prepared with different compositions (Fig. 6). Analysis of background flux showed that not only sucrose but also the two compounds with antioxidant activity, taurine and lactobionate, are responsible for elevated resorufin production. When these compounds were omitted in MiR05Cr (and isotonically replaced with KCl), the background flux was decreased to values similar to MiRK03Cr.

3.3 Experiment with Isolated Mouse Brain Mitochondria

1. Prepare Oxygraph-2k chambers and settings as described above.

2. Inject 10 (or 20) μl of the AmR stock solution into each respiration chamber, obtaining an f.c. of 10 (or 20) μM, and add 4 μl of HRP stock (f.c. 1 U/ml) and 5 U/ml SOD.

3. Let the system stabilize for 15 min to get a stability of oxygen and H_2O_2 flux.

4. Titrate 20 μl of isolated mouse brain mitochondria into the chambers (Fig. 8).

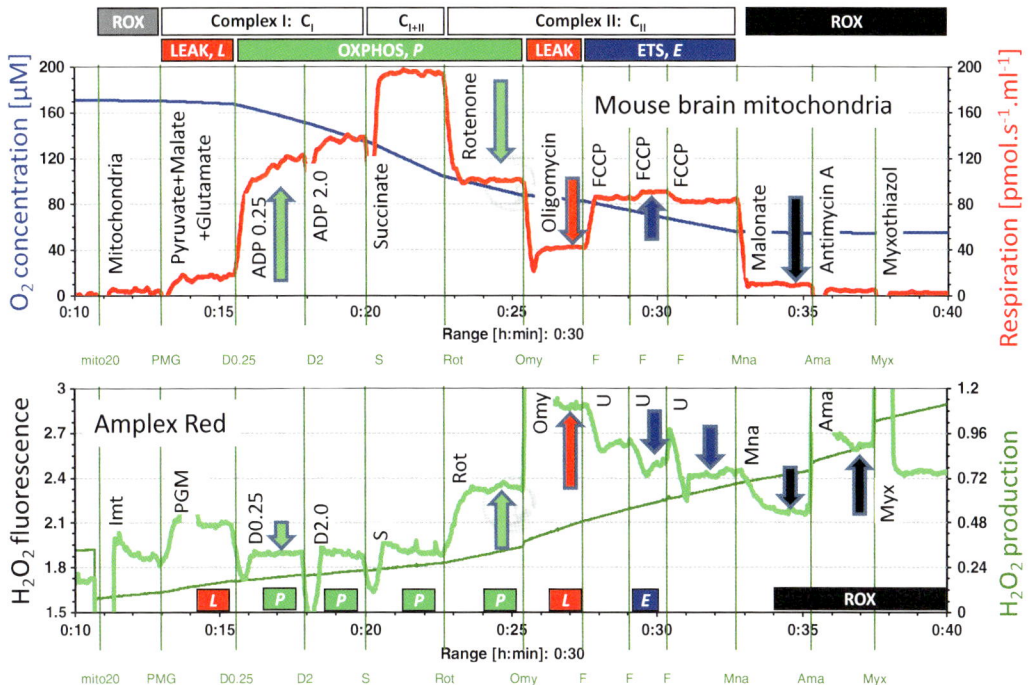

Fig. 8 Simultaneous measurement of respiration (*top*) and of H_2O_2 production (*bottom*) of isolated mouse brain mitochondria: C_I-linked in the LEAK state followed by ADP titration, and substrate control in the OXPHOS state from C_I-, C_{I+II}-, to C_{II}-linked states, sequential titration of oligomycin and FCCP, and inhibition to the ROX state. H_2O_2 production (*bottom*) is independent of respiratory rate and is a function of metabolic state, decreasing with stimulation by ADP and uncoupling (FCCP) but increasing with inhibition by rotenone, oligomycin, and antimycin A

5. Add C_I-linked substrates glutamate (10 mM f.c.), malate (2 mM f.c.), and pyruvate (5 mM f.c.) to obtain C_I-linked LEAK respiration and H_2O_2 production.

6. Add ADP (2.5 mM f.c.) to induce C_I-linked OXPHOS, followed by succinate (10 mM f.c.) to observe C_{I+II}-linked OXPHOS.

7. Add rotenone (0.5 μM f.c.) to inhibit complex I and measure C_{II}-linked OXPHOS.

8. Add oligomycin (2 μg/ml f.c.) to inhibit ATP synthase and measure LEAK respiration in the presence of adenylates.

9. Titrate FCCP to obtain maximum noncoupled respiration fueled by complex II.

10. Add malonate (5 mM f.c.) to inhibit C_{II} and observe residual oxygen consumption (ROX).

11. Add antimycin A (2.5 μM f.c.) to inhibit C_{III} at the Qi site and observe ROX.

12. Add myxothiazol (0.5 μM f.c.) to inhibit C_{III} at the Qo site and observe ROX.

4 Notes

1. CCCP: A 1 mM stock solution (ethanol) is used. Other preparations such as intact yeast may require much higher concentrations of uncoupler compared to mammalian cells or mitochondrial preparations.

2. The preparation of the 10 mM storage solution follows the procedure suggested by the manufacturer. The manufacturer states only an approximate molecular weight for the Amplex® UltraRed formulation and does not publish details how the Amplex® UltraRed formulation deviates from the substance 10-acetyl-3,7-dihydroxyphenoxazine (CAS# 119171-73-2), known as Amplex® Red.

3. Superoxide dismutase is optionally included to generate H_2O_2 from superoxide. Results may be compared with and without SOD for evaluation of the contribution of superoxide not endogenously dismutated with the formation of peroxide.

4. Different batches of SOD preparation may contain a specific activity of 2,000–6,000 U/mg protein so the final volume to be added to the respiratory chamber has thus to be adjusted accordingly.

5. Different calibration concentrations are used dependent on the expected amounts of H_2O_2 produced. We have tested standard calibrations from 0.1 to 0.4 μM. The commercial 880 mM H_2O_2 stock is stabilized with approx. 200 ppm acetanilide, which is diluted by a factor of 11,000 and should thus not affect the measurements.

6. It should be considered that H_2O_2 calibration steps also consume AmR and produce resorufin, which affects the total amount of resorufin production in an experiment. Therefore we recommend the use of 0.1 μM H_2O_2 calibration steps when performing experiments where a high H_2O_2 production is expected.

7. Sensitivity is defined here as the signal change expressed as voltage change (or fluorescence change) of the sensor per unit sample [V/μM] after addition of H_2O_2 or resorufin.

8. Stocks of H_2O_2 appeared fairly instable at room temperature, and therefore the 80 μM H_2O_2 stock had to be prepared freshly before the experiment and was immediately divided into several Eppendorf tubes which were then kept on ice or in the refrigerator during the experiment. For each calibration, a fresh cooled aliquot was taken for immediate use.

9. A comparison of one- and two-step additions of H_2O_2 standards for calibration indicated that sensitivity does not differ between these procedures. Sensitivity [V/μM] amounted to

0.219 ± 0.048, 0.226 ± 0.061, and 0.222 ± 0.054 ($n = 14$ assays in each case) when either the emission signal before and after the first calibration standard, before and after the second addition, or before the first and after the second addition were taken for calibration, respectively. Thus, we recommend single-step calibrations as sufficiently accurate while at the same time minimizing accumulation of resorufin.

10. Differences in sensitivity between sensors may be caused by different luminous intensities of LEDs, slightly different geometrical alignments of the LED in respect to measuring chamber and photodiode, different photosensitivities of photodiodes, and different absorption characteristics of filters. We observed strong effects on the observed signal when the orientation (rotation, angle) of an LED was slightly changed. Further, we measured the light intensities recorded by the photodiode of each sensor ($n = 16$) using either the sensors' integrated LED as light source (experiment A) or using an external (always the same) LED as light source (experiment B). Although the geometric reproducibility of experiment B was problematic, the standard deviation was reduced from 7 % in experiment A to 2 % in experiment B. This indicates that the variation of LED luminosity and small variations in LED alignment may be the main contributing factors to the total variance in sensor sensitivity. In fact, the variance of LED luminosity is a well-known problem, partially addressed by LED manufacturers in a presorting process ("binning" or "ranking"). Even within such presorted "bins" (that are not always available), manufacturers state typical luminosity variations of ±15 to ±20 %, if any statement is given at all.

11. In a set of preliminary experiments, we also tested for the effect of medium oxygenation on assay sensitivity, as a classical study by Boveris and Chance [25] indicated significant oxygen dependence of mitochondrial H_2O_2 production from the hypoxic to the hyperoxic range. We did not detect any significant impact of hyperoxic conditions but observed potential differences in hypoxia, the underlying mechanism of which still awaits further clarification.

12. The assay concentration of Amplex® UltraRed suggested by the commercial supplier is 50 μM, whereas concentrations as low as 1 μM were used in very short experimental assays [26]. In experiments with 20 μM Amplex® UltraRed, we observed that oxygen fluxes of 293T human embryonic kidney cells and specifically complex I-linked OXPHOS may be significantly inhibited by the probe (unpublished observation). In contrast, Hickey et al. [11] observed robust complex I-linked OXPHOS at 50 μM Amplex® UltraRed in permeabilized ventricle fibers of the epaulette shark. Respiratory fluxes in yeast were also not

affected by 20 µM Amplex® UltraRed (unpublished observation). We thus recommend that the final concentration of the probe be tested for potential impacts on respiratory fluxes in each experimental system.

13. A number of substances and classes of substances are strictly incompatible with the AmR method: inhibitors of horseradish peroxidase (cyanide, azide), enzymes and substances consuming or scavenging H_2O_2 (catalase, pyruvate at high concentrations), and strongly redox active substances (cytochrome c, TMPD/ascorbate). The effect of substances in the medium that consume H_2O_2 slowly is taken into account by H_2O_2 calibration and is indicated by a decline of the H_2O_2 sensitivity. In preliminary experiments, we evaluated the applicability of typical SUIT chemicals which can be used in conjunction with the AmR method: DMSO, ethanol, malate, glutamate, pyruvate (a strong scavenger of H_2O_2), succinate, (ADP + Mg^{2+}), (ATP + Mg^{2+}), rotenone, FCCP, oligomycin, antimycin A, malonate, and myxothiazol.

Acknowledgment

The study was funded by the Tyrolean government (K-Regio project MitoCom Tyrol) and the European Regional Development Fund.

References

1. Tahara EB, Navarete FD, Kowaltowski AJ (2009) Tissue-, substrate-, and site-specific characteristics of mitochondrial reactive oxygen species generation. Free Radic Biol Med 46:1283–1297

2. Kowaltowski AJ, de Souza-Pinto NC, Castilho RF, Vercesi AE (2009) Mitochondria and reactive oxygen species. Free Radic Biol Med 47:333–343

3. Dröse S, Brandt U (2012) Molecular mechanisms of superoxide production by the mitochondrial respiratory chain. Adv Exp Med Biol 748:145–169

4. Mohanty JG, Jaffe JS, Schulman ES, Raible DG (1997) A highly sensitive fluorescent microassay of H2O2 release from activated human leukocytes using a dihydroxyphenoxazine derivative. J Immunol Methods 202:133–141

5. Zhou M, Diwu Z, Panchuk-Voloschina N et al (1997) A stable nonfluorescent derivative of resorufin for the fluorometric determination of trace hydrogen peroxide: applications in detecting the activity of phagocyte NADPH oxidase and other oxidases. Anal Biochem 253(2):162–168

6. Towne V, Will M, Oswald B, Zhao Q (2004) Complexities in horseradish peroxidase-catalyzed oxidation of dihydroxyphenoxazine derivatives: appropriate ranges for pH values and hydrogen peroxide concentrations in quantitative analysis. Anal Biochem 334:290–296

7. Summers FA, Zhao B, Ganini D et al (2013) Photooxidation of Amplex Red to resorufin: implications of exposing the Amplex Red assay to light. Methods Enzymol 526:1–17

8. Starkov AA (2010) Measurement of mitochondrial ROS production. Methods Mol Biol 648:245–255

9. Malinska D, Kudin AP, Debska-Vielhaber G et al (2009) Quantification of superoxide production by mouse brain and skeletal muscle mitochondria. Methods Enzymol 456:419–437

10. Votyakova TV, Reynolds IJ (2004) Detection of hydrogen peroxide with Amplex Red: interference by NADH and reduced glutathione auto-oxidation. Arch Biochem Biophys 431(1):138–144

11. Hickey AJ, Renshaw GM, Speers-Roesch B, Richards JG, Wang Y, Farrell AP, Brauner CJ

(2012) A radical approach to beating hypoxia: depressed free radical release from heart fibres of the hypoxia-tolerant epaulette shark (Hemiscyllum ocellatum). J Comp Physiol B 182:91–100

12. Gnaiger E (2001) Bioenergetics at low oxygen: dependence of respiration and phosphorylation on oxygen and adenosine diphosphate supply. Respir Physiol 128:277–297

13. Pesta D, Gnaiger E (2012) High-resolution respirometry. OXPHOS protocols for human cells and permeabilized fibres from small biopsies of human muscle. Methods Mol Biol 810:25–58

14. Chinopoulos C, Vajda S, Csanady L, Mandi M, Mathe K, Adam-Vizi V (2009) A novel kinetic assay of mitochondrial ATP-ADP exchange rate mediated by the ANT. Biophys J 96:2490–2504

15. Krumschnabel G, Eigentler A, Fasching M, Gnaiger E (2014) Use of safranin for the assessment of mitochondrial membrane potential by high-resolution respirometry and fluorometry. Methods Enzymol 542:163–181

16. Fasching M, Gradl P, Gnaiger E (2014) O2k-fluorescence LED2-module. Mitochondr Physiol Network 17.05(05):1–7. http://www.bioblast.at/index.php/MiPNet17.05

17. Gnaiger E, Kuznetsov AV, Schneeberger S, Seiler R, Brandacher G, Steurer W, Margreiter R (2000) Mitochondria in the cold. In: Klingenspor M, Heldmaier G (eds) Life in the cold. Springer, Heiderlberg, pp 431–442

18. Komary Z, Tretter L, Adam-Vizi V (2010) Membrane potential-related effect of calcium on reactive oxygen species generation in isolated brain mitochondria. Biochim Biophys Acta 1797:922–928

19. Fontana-Ayoub M, Fasching M, Gnaiger E (2014) Selected media and chemicals for respirometry with mitochondrial preparations. Mitochondr Physiol Network 03.02(17):1–9. http://bioblast.at/index.php/MiPNet03.02

20. Sumbalová Z, Kucharská J, Kristek F (2010) Losartan improved respiratory function and coenzyme Q content in brain mitochondria of young spontaneously hypertensive rats. Cell Mol Neurobiol 30:751–758

21. Gnaiger E (2012) Mitochondrial pathways and respiratory control. An introduction to OXPHOS analysis. Mitochondr physiol network 17.18. OROBOROS MiPNet Publications, Innsbruck, p 64

22. Frieden E, Walter C (1963) Prevalence and significance of the product inhibition of enzymes. Nature 198:834–837

23. Piwonski HM, Goomanovsk M, Bensimon D et al (2012) Allosteric inhibition of individual enzyme molecules trapped in lipid vesicles. Proc Natl Acad Sci U S A 109(22): E1437–E1443

24. Cooper WJ, Zika RG, Petasne RG, Plane JM (1988) Photochemical formation of hydrogen peroxide in natural waters exposed to sunlight. Environ Sci Technol 22(10):1156–1160

25. Boveris A, Chance B (1973) The mitochondrial generation of hydrogen peroxide. General properties and effect of hyperbaric oxygen. Biochem J 134:707–716

26. Tretter L, Adam-Vizi V (2012) High Ca^{2+} load promotes hydrogen peroxide generation via activation of α-glycerophosphate dehydrogenase in brain mitochondria. Free Radic Biol Med 53:2119–21130

Chapter 23

Measurement of Mitochondrial NADH and FAD Autofluorescence in Live Cells

Fernando Bartolomé and Andrey Y. Abramov

Abstract

In the process of energy production, mitochondrial networks are key elements to allow metabolism of substrates into ATP. Many pathological conditions have been associated with mitochondrial dysfunction as mitochondria are associated with a wide range of cellular processes. Therefore, any disruption in the energy production induces devastating effects that can ultimately lead to cell death due to chemical ischemia. To address the mitochondrial health and function, there are several bioenergetic parameters reflecting either whole mitochondrial functionality or individual mitochondrial complexes. Particularly, metabolism of nutrients in the tricarboxylic acid cycle provides substrates used to generate electron carriers (nicotinamide adenine dinucleotide [NADH] and flavin adenine dinucleotide [$FADH_2$]) which ultimately donate electrons to the mitochondrial electron transport chain. The levels of NADH and $FADH_2$ can be estimated through imaging of NADH/NAD(P)H or FAD autofluorescence. This report demonstrates how to perform and analyze NADH/NAD(P)H and FAD autofluorescence in a time-course-dependent manner and provides information about NADH and FAD redox indexes both reflecting the activity of the mitochondrial electron transport chain (ETC). Furthermore, total pools of NADH and FAD can be estimated providing information about the rate of substrate supply into the ETC. Finally, the analysis of NADH autofluorescence after induction of maximal respiration can offer information about the pentose phosphate pathway activity where glucose can be alternatively oxidized instead of pyruvate.

Key words Mitochondria, NADH, NAD(P)H, FAD

1 Introduction

Mitochondrial health and function are key elements in the maintenance of cellular homeostasis and are responsible for numerous vital processes including energy production (ATP). As a result, it is not surprising that mitochondrial dysfunction is related to disruptions in cellular activities which can ultimately lead to cell death [1, 2]. Many human pathological conditions affecting a variety of organs [3] have been linked to malfunctioning mitochondria [4–7]. In fact, a majority of neurodegenerative disorders have been associated with altered mitochondrial physiology which has been directly linked to neuronal death. Therefore, it is of high importance to investigate

Volkmar Weissig and Marvin Edeas (eds.), *Mitochondrial Medicine: Volume I, Probing Mitochondrial Function*,
Methods in Molecular Biology, vol. 1264, DOI 10.1007/978-1-4939-2257-4_23, © Springer Science+Business Media New York 2015

mitochondrial health with powerful tools to allow a leap forward in the understanding of mitochondrial health and disease.

ATP production in mitochondria requires functional coupling between the electron transport chain (ETC) and oxidative phosphorylation. The tricarboxylic acid cycle (TCA, also known as Krebs cycle) provides electron carriers in the form of nicotinamide adenine dinucleotide (NADH) and flavin adenine dinucleotide ($FADH_2$) that donate their electrons to complex I and II, respectively, enabling the respiratory chain to function. Commercially available biochemical assays for assessments of mitochondrial complexes, their functional efficiency, and possible interactions between mitochondrial and non-mitochondrial proteins require the use of isolated mitochondria which render very important information but bearing some limitations.

Fluorescent imaging of mitochondria in living cells provides a helpful tool to monitor mitochondrial function in more "native" conditions [8–10]. Here, live cell imaging can be employed to capture the autofluorescence of NADH/NAD(P)H and the oxidized form of $FADH_2$ (FAD) which allows the quantification of NADH as well as $FADH_2$. The assessment of mitochondrial activity through NADH and FAD autofluorescence by live cell microscopy gives a range of outputs reflecting the activity of the ETC as well as substrate supply which is conceptually and practically appealing.

This report provides a guide on how to measure NADH and FAD autofluorescence through the employment of live cell imaging in a time-course-dependent manner. Furthermore, it is shown how to analyze these bioenergetic intermediates in order to monitor the ETC activity and substrate supply in the mitochondria through NADH and FAD redox indexes, NADH and FAD pools, and NADH production rates. Moreover, monitoring of non-mitochondrial autofluorescence of NADH and NAD(P)H can provide information about the activity of the pentose phosphate pathway.

2   Materials

1. Primary cells, cell lines, or tissue slices (*see* **Note 1**).

2. HBSS: 156 mM NaCl, 3 mM KCl, 2 mM $MgSO_4$, 1.25 mM KH_2PO_4, 2 mM $CaCl_2$, 10 mM glucose, and 10 mM HEPES; pH adjusted to 7.35 with NaOH (*see* **Note 2**).

3. Phospho-trifluoromethoxy carbonyl cyanide phenylhydrazone (FCCP) (*see* **Note 3**).

4. Sodium cyanide (NaCN) (*see* **Note 4**).

5. Fluorescent imaging setting (*see* **Note 5**).

3   Methods

The autofluorescence of mitochondrial NADH and FAD (from complex II flavoprotein) can be differentiated from cytosolic NADH/NAD(P)H and FAD as well as other sources (including mitochondrial monoamine oxidase's flavoprotein) through the maximization of substrate consumption followed by a total inhibition of electron donation to the ETC. The redox state of NADH or FAD reflects the balance between the mitochondrial ETC activity and the rate of substrate supply. Here it is shown how to analyze the NADH (substrate for the ETC complex I) and FAD autofluorescence to obtain information about the respiration process and other possible pathways affected in cells.

Once the basal autofluorescence levels are obtained through live acquisition (2 min), the maximally oxidized signal through the stimulation of maximal respiration (FCCP) and the maximally reduced signal by full inhibition of respiration are obtained. The maximally oxidized signal is defined as response to the uncoupler FCCP (1 µM) that stimulates maximal respiration. Conversely, the maximally reduced signal is defined as the response to NaCN (1 mM) that fully inhibits respiration. Finally, the "NADH and FAD redox indexes" will be generated by expressing the basal NADH or FAD levels respectively as a percentage of the difference between the maximally oxidized and maximally reduced signals. The NADH and FAD redox indexes along the NADH and FAD pools provide information about ETC activity as well as any mitochondrial complex impairment. Additionally, obtaining the NAD(P)H levels through the NADH autofluorescence analysis allows the assessment of pentose phosphate pathway activity where glucose can be alternatively oxidized.

3.1   NADH Autofluorescence

1. Place the coverslip (containing cells or tissue slices) into the microscope chamber (see **Note 6**).

2. Transfer the coverslip containing the attached cells to the imaging chamber.

3. Wash the cells with HBSS buffer and replace it with fresh HBSS buffer for all imaging experiments.

4. Place the chamber onto the microscope. For NADH autofluorescence measurements, ensure that the UV-compatible quartz objective is used.

5. Once visualized the cells, change the excitation light to 360 nm wavelength which allows visualization of the NADH/NAD(P)H autofluorescence. Adjustment of the focus may be required.

6. Set up time intervals for the recording of fluorescence (every 5–10 s).

7. Start the acquisition.

8. Once the signal is stabilized (usually after 1 min of acquisition), add 1 μM FCCP to maximize mitochondrial respiration. Here, the fluorescence signal represents the minimum as the cells oxidize NADH due to the uncoupling effect exerted from FCCP.

9. Upon fluorescence signal stabilization (but lower than basal levels), add 1 mM NaCN which fully blocks respiration. Here, the fluorescence signal will be at its maximum as the ETC is not working leading to NADH accumulation (record for a few minutes to ensure capture of the maximum fluorescence).

10. Analyze the data representing the traces of fluorescence in a time-dependent manner as Fig. 1 shows. The difference between the average fluorescence values after application of NaCN and the average fluorescence values after addition of FCCP provides the total NADH pool value in these cells (Fig. 1). The averaged maximum fluorescence values obtained after NaCN addition and the averaged minimum fluorescence values obtained after FCCP addition represent 100 % and 0 % respectively (Fig. 1). The values within the 0–100 % scale represent the NADH redox index (Fig. 1). The averaged fluorescence values after the addition of FCCP correspond to the NAD(P)H values as all mitochondrial NADH is oxidized (Fig. 1) (*see* **Note 7**). Further, a rate of NADH production can be monitored through the analysis of the fluorescence signal increase after NaCN addition and the NaCN-induced fluorescence peak (Fig. 1b). The rate of NADH production is a direct reflection of the TCA efficiency since NaCN blocks all respiration taking place.

3.2 FAD Autofluorescence

1. Follow **steps 1–3** from Subheading 3.1 (above) for FAD measurements and ensure that an UV-compatible quartz objective is used.

2. Using the confocal microscope software, start the continuous scan. Choose the appropriate imaging optics, excite at 454 nm, and use a band-pass filter between 505 and 550 nm to collect the emitted light. To allow localization of FAD inside the cells, a regulation of the pinhole may be required to achieve an optical slice of ~2 μm. The laser power should again not exceed 0.2 % to prevent cell damage. However, an increase in the gain may be required to visualize FAD autofluorescence (allowing a reasonable signal-to-noise ratio).

3. Open the settings panel in the software and adjust the laser power and gain to avoid saturated fluorescence.

4. Stop the continuous scanning again and increase the image averaging reducing the scan speed if you wish to have higher-quality acquired images with better resolution of your FAD autofluorescence (*see* **Note 8**).

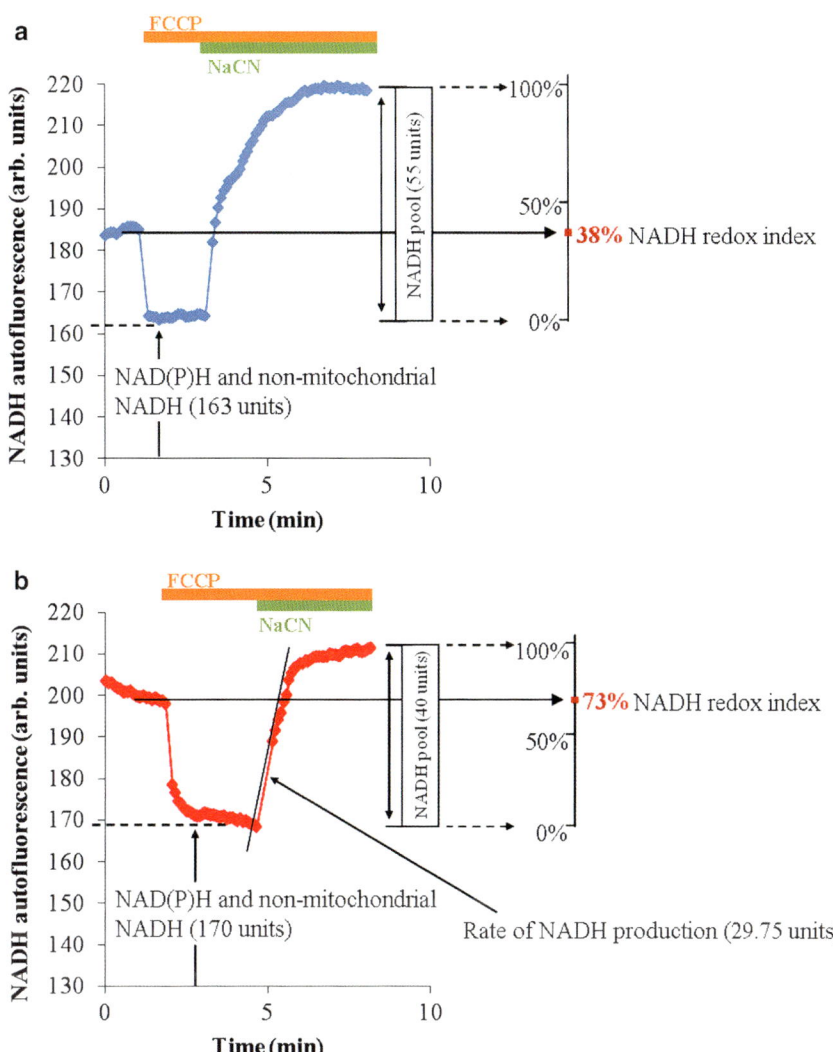

Fig. 1 Representative traces of NADH in healthy SH-SY5Y cells (**a**) and SH-SY5Y cells deficient in p62 (**b**). NADH pool, NADH redox index, and NAD(P)H levels can be obtained through the analysis of NADH autofluorescence. (**a** and **b**) The basal autofluorescence is taken for the first 1–2 min. Once the signal is stable, 1 μM of the uncoupler FCCP is added to maximize the respiration, oxidizing all the NADH leading to the lowest fluorescence signal. After stabilization of the signal, respiration is blocked through addition of 1 mM NaCN obtaining the highest fluorescence signal. The NADH pool is calculated by subtracting the lowest fluorescence value (after FCCP) from the highest (after NaCN) (55 arbitrary units in panel (**a**) and 40 arbitrary units in panel (**b**)). Thereafter, the minimum NADH autofluorescence (after FCCP) is normalized to 0 % and the maximum auto-fluorescence (NaCN) is normalized to 100 %. The NADH redox index is represented in the basal autofluores-cence before the addition of any inhibitors and is expressed as a percentage (38 % in panel (**a**) and 73 % in panel (**b**)). Finally, the NAD(P)H value is calculated by subtracting the background from the minimum fluores-cence (163 arbitrary units in panel (**a**) and 170 arbitrary units in panel (**b**)). This analysis enabled the identifica-tion of a higher NADH redox index in p62-deficient SH-SY5Y cells when compared to its healthy counterpart

5. Set up the time intervals for recording the fluorescence (every 5–10 s).

6. Start the acquisition.

7. Once the signal is stabilized (usually after 1 min of acquisition), add 1 μM FCCP to maximize the respiration. The induced fluorescence signal will be the maximum as the cells reduce the FAD due to the uncoupling effect exerted from FCCP.

8. Upon stabilization of the fluorescence (higher than basal levels), add 1 mM NaCN. The fluorescence signal will reach its minimum as all respiration including complex II is blocked. Ensure that the recording is continued until a stable signal is recorded.

9. Analyze the data representing the traces of fluorescence in a time-dependent manner as Fig. 2 shows. The difference between the recorded average fluorescence values after application of NaCN and the average fluorescence values after addition of FCCP provides the total FAD pool value in these cells (Fig. 2). The averaged maximum fluorescence values obtained after NaCN addition and the averaged minimum fluorescence values obtained after FCCP addition represent 100 % and 0 % respectively (Fig. 2). The basal fluorescence values within the 0–100 % scale represent the FAD redox index (Fig. 2).

4   Notes

1. In all cases, cells are grown on glass coverslips in 6-well plates, whereas tissue slices are placed on glass coverslips just before the experiment. While imaging, cells and tissue should be buffered using HEPES-buffered salt solution (HBSS medium).

2. This buffer allows the imaging of cells or slices in a static chamber avoiding the need for continuous CO_2-equilibrated buffering.

3. FCCP is a lipid-soluble weak acid used as a mitochondrial uncoupling agent. FCCP is negatively charged allowing the anions to diffuse freely through nonpolar media, such as phospholipid membranes. It abolishes the obligatory linkage between the respiratory chain and the oxidative phosphorylation system which occurs in intact mitochondria.

4. NaCN is a respiratory chain inhibitor and it blocks respiration in the presence of either ADP or uncouplers such as FCCP. It specifically blocks the cytochrome oxidase (complex 4) and prevents both coupled and uncoupled respirations despite the presence of substrates, including NADH, and succinate.

5. NADH and NAD(P)H autofluorescence can be measured using an epifluorescence inverted microscope equipped with a ×20 fluorite objective. Excitation light should be fixed at 360 nm and the emitted fluorescence light must be registered at 455 nm. NADH/NAD(P)H can also be measured using a

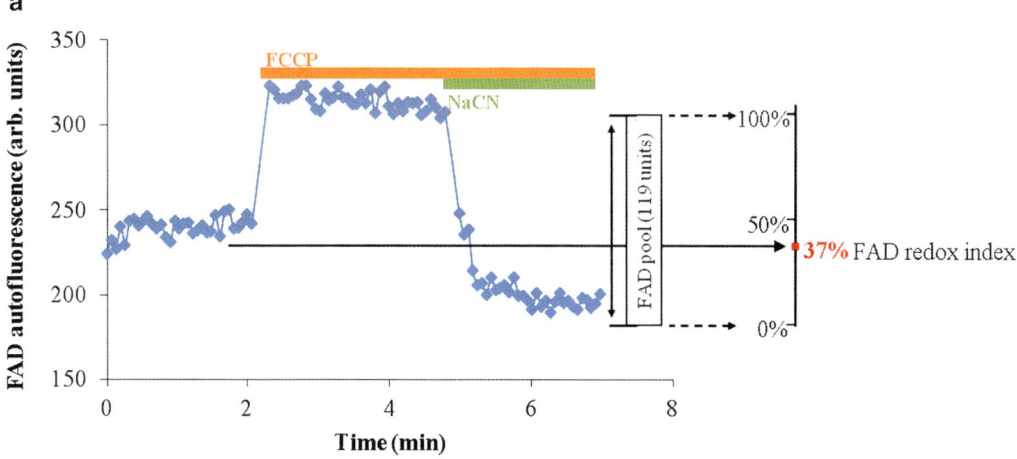

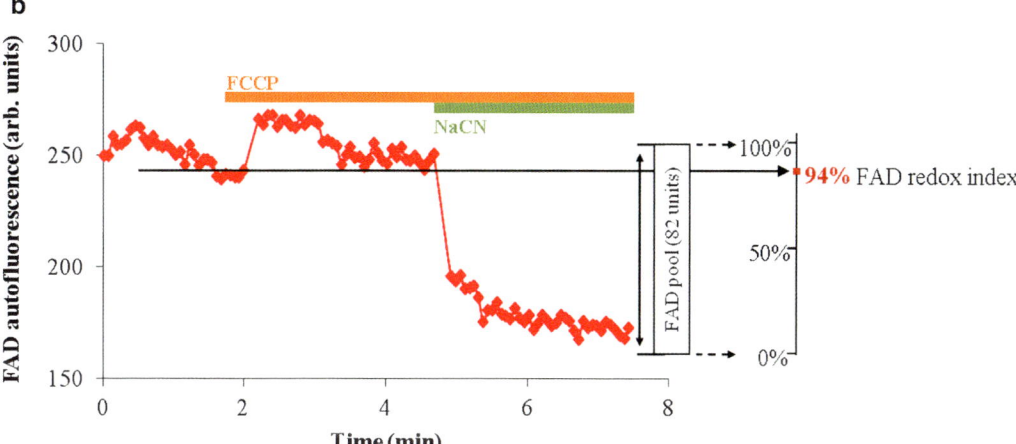

Fig. 2 Representative traces of FAD in healthy SH-SY5Y cells (**a**) and SH-SY5Y cells deficient in p62 (**b**). FAD pool and FAD redox index can be obtained through the analysis of FAD autofluorescence. (**a** and **b**) The basal autofluorescence is taken for the first 1–2 min. Once the signal is stable, 1 μM of the uncoupler FCCP is added to maximize the respiration, reducing all the FAD leading to the highest fluorescence signal. After stabilization, respiration is blocked through addition of 1 mM NaCN obtaining the lowest fluorescence signal. The FAD pool is calculated by subtracting the highest fluorescence value (after FCCP) from the lowest (after NaCN) (119 arbitrary units in panel (**a**) and 82 arbitrary units in panel (**b**)). Thereafter, the maximum FAD autofluorescence (after FCCP) is normalized to 100 %, and the minimum autofluorescence (NaCN) is normalized to 0 %. The FAD redox index is represented in the basal autofluorescence before the addition of any inhibitors and is expressed as a percentage (37 % in panel (**a**) and 94 % in panel (**b**)). This analysis enabled the identification of a higher FAD redox index in p62-deficient SH-SY5Y cells when compared to its healthy counterpart

confocal microscope equipped with a UV lasers. The excitation spectrum of NADH/NAD(P)H autofluorescence is covering 400 nm as we recently demonstrated when analyzing the mitochondrial NADH using the 405 nm laser from the Zeiss 710 VIS CLSM system [11]. The convenience of using a confocal microscope equipped with the UV laser also provides additional advantages as it allows simultaneous measurements of NADH and FAD autofluorescence. FAD autofluorescence should be monitored using a confocal microscope. Cells should be excited by 454 nm (Argon laser line) and the emission fluorescence light must be registered from 505 to 550 nm. Illumination intensity must be kept to a minimum (at 0.1–0.2 % of laser output) to avoid phototoxicity and the pinhole set to give an optical slice of ~2 μm.

6. Here it should be noted that ~70 % of cell confluence is required to allow NADH or the FAD autofluorescence measurements.

7. The registered NADH fluorescence upon addition of FCCP represents both the non-mitochondrial NADH autofluorescence and the NAD(P)H autofluorescence. To ensure that only NAD(P)H autofluorescence is recorded and analyzed, background fluorescence is deducted from the total fluorescence output.

8. The settings for the FAD autofluorescence may not record solely FAD autofluorescence but also autofluorescence signal from the mitochondrial monoamine oxidase's flavoprotein which may be increased due to cellular stress. To distinguish between these signals, FCCP and NaCN are added during the experiment. Here, the uncoupler FCCP allows only mitochondrial FAD to increase, whereas NaCN triggers only for mitochondrial FAD to decrease due to the inhibition of respiration.

References

1. Lopez-Otin C, Blasco MA, Partridge L, Serrano M, Kroemer G (2013) The hallmarks of aging. Cell 153(6):1194–1217

2. Edeas M, Weissig V (2013) Targeting mitochondria: strategies, innovations and challenges: the future of medicine will come through mitochondria. Mitochondrion 13(5):389–390

3. Monsalve M, Borniquel S, Valle I, Lamas S (2007) Mitochondrial dysfunction in human pathologies. Front Biosci 12:1131–1153

4. Burchell VS, Gandhi S, Deas E, Wood NW, Abramov AY, Plun-Favreau H (2010) Targeting mitochondrial dysfunction in neurodegenerative disease: part I. Expert Opin Ther Targets 14(4):369–385

5. Burchell VS, Gandhi S, Deas E, Wood NW, Abramov AY, Plun-Favreau H (2010) Targeting mitochondrial dysfunction in neurodegenerative disease: Part II. Expert Opin Ther Targets 14(5):497–511

6. Gandhi S, Abramov AY (2012) Mechanism of oxidative stress in neurodegeneration. Oxid Med Cell Longev 2012:428010

7. Lin MT, Beal MF (2006) Mitochondrial dysfunction and oxidative stress in neurodegenerative diseases. Nature 443(7113):787–795

8. Bartolome F, Wu HC, Burchell VS, Preza E, Wray S, Mahoney CJ, Fox NC, Calvo A, Canosa A, Moglia C, Mandrioli J, Chio A, Orrell RW, Houlden H, Hardy J, Abramov AY, Plun-Favreau H (2013) Pathogenic VCP mutations induce mitochondrial uncoupling and reduced ATP levels. Neuron 78(1):57–64

9. Moncada S, Bolanos JP (2006) Nitric oxide, cell bioenergetics and neurodegeneration. J Neurochem 97(6):1676–1689

10. Brand MD, Nicholls DG (2011) Assessing mitochondrial dysfunction in cells. Biochem J 435(2):297–312

11. Ludtmann MH, Angelova PR, Zhang Y, Abramov AY, Dinkova-Kostova AT (2014) Nrf2 affects the efficiency of mitochondrial fatty acid oxidation. Biochem J 457(3):415–424

Chapter 24

Mitochondrial Coenzyme Q10 Determination via Isotope Dilution Liquid Chromatography Tandem Mass Spectrometry

Outi Itkonen and Ursula Turpeinen

Abstract

Coenzyme Q10 (CoQ10) is an essential part of the mitochondrial respiratory chain. Here, we describe an accurate and sensitive liquid chromatography tandem mass spectrometry (LC-MS/MS) method for determination of mitochondrial CoQ10 in isolated mitochondria. In the assay, mitochondrial suspensions are spiked with CoQ10-[^{2}H$_6$] internal standard, extracted with organic solvents, and CoQ10 quantified by LC-MS/MS using multiple reaction monitoring (MRM).

Key words Ubiquinone, Coenzyme Q10, LC-MS/MS, Mitochondrial disease

1 Introduction

Coenzyme Q10 (CoQ10), also called ubiquinone, is an essential part of the mitochondrial respiratory chain. The high transfer energy-containing electrons of NADH and FADH$_2$ formed in glycolysis, fatty acid oxidation, and the citric acid cycle are transferred to O$_2$ by the respiratory chain in a process called oxidative phosphorylation (OXPHOS). In this way, the energy in the diet is transformed to adenosine triphosphate (ATP) for cellular functions [1]. In the respiratory chain, CoQ10 transfers reducing equivalents from respiratory complexes I and II to complex III [2]. Apart from its role in the respiratory chain, CoQ10 also allows protons to be extruded from mitochondrial matrix to the intermembrane space [3], acts as pro- or antioxidant [4], has a role in pyrimidine biosynthesis [5], and modulates apoptosis [6].

Disorders in OXPHOS chain cause lack of energy in cells and tissues resulting in heterogeneous symptoms in the patient. Patients with primary and secondary CoQ10 deficiency have been reported [7]. There are no curative treatments for OXPHOS disorders. However, some patients with CoQ10 deficiency benefit

Volkmar Weissig and Marvin Edeas (eds.), *Mitochondrial Medicine: Volume I, Probing Mitochondrial Function*,
Methods in Molecular Biology, vol. 1264, DOI 10.1007/978-1-4939-2257-4_24, © Springer Science+Business Media New York 2015

from CoQ10 substitution [8]; the treatment is relatively cheap and lacks severe side effects. CoQ10 quantification may be used to aid in diagnosis of mitochondrial CoQ10 deficiency and to identify patients for potentially effective substitution therapy.

The reported assays for CoQ10 in biological samples are based on liquid chromatography (LC) with ultraviolet (UV) [9–12], electrochemical (EC) [13–16], mass spectrometric (MS) [17], and tandem mass spectrometric (MS/MS) detection [18–23]. The highest sensitivity and specificity is reached by LC-MS/MS technology employing an isotopically labeled internal standard [18].

2 Materials

Always use reagents of the highest analytical quality and deionized water with a resistance of 18 MΩ at 25 °C. The performance of the analytical balance, spectrophotometer, and the LC-MS/MS instrument should be assured either by an external quality assurance scheme or by the procedures of the respective instrument manufacturer.

2.1 Mitochondrial Isolation

1. Homogenizing buffer: 100 mM KCl, 50 mM potassium dihydrogen phosphate, 50 mM tris(hydroxymethyl)aminomethane, 5 mM $MgCl_2$, 1.8 mM ATP, and 1 mM EDTA, pH 7.2. Weigh 1.86 g potassium chloride (KCl, p.a.), 1.70 g potassium dihydrogen phosphate, 1.51 g tris(hydroxymethyl)aminomethane, 0.254 g magnesium chloride hexahydrate, 0.101 g ethylenediaminetetraacetic acid, dipotassium salt, and 0.248 g adenosine-5′-triphosphate, disodium salt. Add 200 mL of water into a 250 mL beaker. Under mixing with a magnetic stirrer, add the weighed reagents. Adjust the pH to 7.2 with 1 mM HCl. Transfer the solution into a volumetric flask and make up to 250 mL with water. Mix well. Store in 3 mL aliquots at –20 °C in glass vials. Use within a year.

2. Resuspension solution: 250 mM sucrose, 15 mM K_2HPO_4, 2 mM $MgAc_2$, 0.5 mM EDTA, and 0.5 g/L albumin, pH 7.2. Weigh 0.25 g human albumin (essentially fatty acid-free) and put aside (see Note 1). Weigh 42.8 g sucrose, 1.02 g KH_2PO_4, 0.215 g magnesium acetate tetrahydrate, and 0.101 g EDTA×2H$_2$O. Add 400 mL of water into a 500 mL beaker. Under stirring with a magnetic stirrer, add the weighed reagents except albumin (see Note 1). Adjust the pH to 7.2 with 5 M KOH. Add albumin to the solution and stir until dissolved. Transfer the solution into a volumetric flask and make up to 500 mL with water. Mix well. Store in 4 mL aliquots at –20 °C in plastic vials. Use within a year.

3. Cooled (+4 °C) centrifuge for Eppendorf vials with a speed up to $15,000 \times g$.

4. Homogenizer (60 rpm) and Potter-Elvehjem homogenizer tube (inner diameter 8.0 mm, frosted walls) with a Teflon pestle (diameter 7.8 mm).

2.2 Assay

1. Spectrophotometer and quartz cuvettes.

2. CoQ10 calibrator stock solution, 400–700 μM. Prepare a stock solution by weighing 20–30 mg of CoQ10 (Sigma). Make up to 50 mL in a volumetric flask with chloroform (p.a.) and mix thoroughly (*see* **Note 2**). The actual CoQ10 concentration of the stock solution is calculated based on the molar absorptivity of CoQ10. Set the spectrophotometer at 275 nm. Blank with chloroform. Read the absorbance of the CoQ10 solution and calculate the concentration (*see* **Note 3**). Divide into 1.5 mL aliquots into glass (borosilicate) vials and store at –20 °C.

3. Working calibrators. Prepare a 1,000 nM working calibrator by diluting the stock solution with mobile phase A in a 50 mL volumetric flask (*see* **Note 4**). Mix well. Divide into 1.5 mL aliquots into glass (borosilicate) vials and store at –20 °C. Other working calibrators are prepared freshly into Eppendorf tubes as follows:

   500 nM: mix 250 μL of 1,000 nM solution and 250 μL of mobile phase A.

   250 nM: mix 250 μL of 1,000 nM solution and 750 μL of mobile phase A.

   100 nM: mix 100 μL of 1,000 nM solution and 900 μL of mobile phase A.

   10 nM: mix 10 μL of 1,000 nM solution and 990 μL of mobile phase A.

   1 nM: mix 100 μL of 10 nM solution and 900 μL of mobile phase A.

4. Stable isotope-labeled internal standard (IS). Prepare a 1.2 mM (1 g/L) stock solution of IS ($[^2H_6]$-CoQ10, IsoSciences, King of Prussia, PA, USA, http://isosciences.com/) in chloroform. Mix well (*see* **Note 2**). Store as 1.5 mL aliquots in glass vials at –80 °C. Prepare a 0.5 μM IS working solution in 1-propanol by diluting 10.4 μL of IS stock solution to 25 mL in a volumetric flask. Store at +4 °C.

5. Ethanol-hexane (2 + 5). Mix 80 mL of ethanol and 200 mL of hexane.

6. Multi-tube vortexer (Lab-Tek International, Christchurch, New Zealand).

7. Sample evaporator TurboVap® LV (Caliper Life Sciences, PerkinElmer, Hopkinton, MA, USA).

8. High-pressure liquid chromatograph with a binary pump, degasser, column oven (+40 °C), and autosampler.

9. Triple quadrupole mass spectrometer with an electrospray ion source (we use an AB Sciex 4000 triple quadrupole MS with a Turbo V electrospray ionization ion source, AB Sciex, Toronto, Canada).

10. SunFire C18 analytical column (2.1 × 50 mm, 3.5 μm particle size, Waters Corporation, Milford, MA, USA).

11. Mobile phase A: 82 % (v/v) methanol and 18 % 1-propanol containing 500 μM ammonium acetate. Mix 205 mL of methanol (MS grade), 45 mL of 1-propanol, and 125 μL of 1 mol/L ammonium acetate (*see* **Note 5**).

12. Mobile phase B: 50 % methanol and 50 % 1-propanol containing 500 μM ammonium acetate. Mix 125 mL of methanol (MS grade), 125 mL of 1-propanol, and 125 μL of 1 M ammonium acetate.

3 Methods

3.1 Isolation of Mitochondria

For isolation of mitochondria, use published standard procedures like that described for muscle samples in 24 (*see* **Note 6**).

1. Homogenize the muscle sample (roughly 100 mg) for 4 min in a homogenizer fitted with a Teflon pestle. Use a volume (μL) of homogenizing buffer that corresponds to muscle weight (mg) × 20 (min 700 μL) (*see* **Note 7**).

2. Centrifuge first the homogenate for 3 min at $650 \times g$ and then the supernatant for 3 min at $15,000 \times g$.

3. Wash the pelleted mitochondria with 300 μL of homogenizing buffer, centrifuge for 3 min at $15,000 \times g$, and resuspend the pellet in resuspension solution in a volume (μL) corresponding to muscle weight (mg) × 4 (min 200 μL) (*see* **Notes 8** and **9**). This suspension corresponds to 0.25 mg tissue wet weight/μL. The suspension can be stored at −80 or at −20 °C until analysis (*see* **Note 10**).

3.2 Sample and Calibrator Preparation

1. Dilute 10 μL of the mitochondrial suspension (0.25 mg tissue wet weight/μL) with 240 μL of water to obtain a solution corresponding to 0.01 mg wet weight/μL (*see* **Note 11**).

2. Transfer 100 μL of calibrators and diluted mitochondrial solution into Eppendorf tubes (*see* **Note 12**).

3. Add 20 μL of IS working solution and vortex-mix. Calibrators are then ready for analysis.

4. To QA and patient samples, add 300 μL of 1-propanol and mix.

5. Centrifuge at $10,000 \times g$ for 2 min at room temperature.

6. Transfer 350 μL of the upper phase into 10 mL glass tubes.

7. Add 3 mL of ethanol-hexane $(2+5)$ and vortex-mix for 3 min in a multi-tube vortexer.

8. Add 0.5 mL of water and mix for further 3 min. Let the tubes stand for 15 min at room temperature.

9. Transfer the upper phase into clean glass tubes. Evaporate to dryness under a flow of nitrogen (*see* **Note 13**).

10. Dissolve the residue into 100 μL of mobile phase A and transfer into autosampler vials.

3.3 LC-MS/MS Analysis

1. Set up an HPLC method with a flow rate 300 μL/min and the following gradient:
   (a) $t=0$ min, $B=0$ %
   (b) $t=0.2$ min, $B=0$ %
   (c) $t=0.21$ min, $B=100$ %
   (d) $t=2$ min, $B=100$ %
   (e) $t=2.1$ min, $B=0$ %
   (f) $t=10$ min, $B=0$ %

2. Set the column oven at +40 °C and the injection volume at 5 μL. The MS is operated in positive ion mode with the ion source spray voltage at +5,500 V, declustering potential at 70 V, temperature at 275 °C, and collision energy at 27 V. The curtain gas setting is 20 L/min; gas 1, 45 L/min; gas 2, 30 L/min; and collision gas setting 4 (*see* **Note 14**). For MS/MS detection, follow the transitions of m/z $880.8 \rightarrow 197.2$ (dwell time 500 ms) corresponding the ammonium adduct of CoQ10 and m/z $886.8 \rightarrow 203.2$ (250 ms) for ammonium adduct of $[^2H_6]$-CoQ10, respectively.

3. Equilibrate the HPLC system with mobile phase A and keep the column at 40 °C for 20 min (*see* **Note 15**). Let the ion source and MS equilibrate before connecting the column outlet to the ion source and then let equilibrate for further 15 min (*see* **Note 16**).

4. Inject 5 μL of the calibrators and extracted samples to the LC-MS/MS. CoQ10 and the IS elute at retention time of about 4.7 min.

3.4 Calculation of the Results

Review peak integration and construct a calibration curve based on the ratio of the peak areas of CoQ10 and IS using $1/x^2$ weighted linear least-squares regression by the instrument software (*see* **Note 17**). The automatically calculated results are expressed as nmol/L. To express the result as nmol/g wet weight, the obtained result (nM) must be divided by 10 (if the suspension was 0.01 mg wet weight/μL) (*see* **Note 18**).

4 Notes

1. The pH of the solution is low before adjustment. To prevent albumin denaturation, first adjust the pH to 7.2 and then add albumin.

2. We found it important to dissolve the stock calibrator into chloroform because CoQ10 showed poor solubility in 1-propanol. Especially at −20 °C, CoQ10 tends to precipitate in 1-propanol but not in chloroform.

3. Calculate the concentration of CoQ10 according to the equation $c\,(M) = A_{275\,nm}/(\varepsilon \times l)$, where $A_{275\,nm}$ is the absorbance, ε is 14,250 L/mol cm (the molar absorptivity of CoQ10), and l is the sample path length (cm).

4. The volume of the stock solution to be diluted is calculated according to the equation: stock solution volume (mL) = 1 μmol/L × working calibrator volume (mL)/stock solution concentration (μM).

5. To ensure efficient mixing of the mobile phases, it is important that also mobile phase A contains 1-propanol.

6. Enriched mitochondria from any tissue or cell line can be used for the CoQ10 assay. The present procedure has been optimized for isolation of muscle mitochondria.

7. Avoid foaming and high pressure during homogenization.

8. All reagents and materials were cooled and kept ice cold during the entire isolation procedure.

9. Practically all mitochondrial CoQ10 is oxidized during sample pretreatment, and no ubiquinol oxidation is needed for quantification of total CoQ10 in isolated mitochondria [18].

10. Mitochondrial CoQ10 is stable for several years at −80 °C and at least 4 weeks at −20 °C when stored in concentrated (0.25 mg/μL) rather than in dilute suspensions [18].

11. A mitochondrial suspension corresponding to 0.01 mg wet weight/μL is optimal for analysis of CoQ10 in human muscle mitochondria. Different cells and tissues contain varying amounts of mitochondria, and therefore, optimal dilution depends on the samples used and needs to be tested.

12. It is highly recommended to prepare own quality assurance (QA) samples to ascertain assay quality. Treat QA samples identically to unknown samples.

13. For rapid evaporation of the solvents, we recommend the TurboVap® evaporator (Caliper Life Sciences) with the temperature set at 37 °C.

14. The MS settings are instrument dependent and must be optimized.

15. Proper equilibration of the HPLC column prior to analysis and a thorough cleanup procedure after analysis are essential for repeated high-quality analysis.

16. For accurate and repeatable analysis, always inspect the ion source spray prior to analysis, run a blank sample and a test solution, e.g., a calibrator, and record the signal-to-noise ratio for each assay series.

17. The method linearity depends on the instrumentation and should be validated in each laboratory. In our laboratory, the linear range was 0.5–1,000 nM (0.432–863 ng/mL) with an LOD of 0.06 nmol/L (52 pg/mL) and LOQ of 0.5 nM (432 pg/mL) [18].

18. The activity of citrate synthase (CS) is generally accepted as a matrix enzyme and a marker for mitochondrial abundance in a sample. Clinically, the most important application of CoQ10 assay is detection of CoQ10 deficiency of the respiratory chain. For this purpose, normalization of the CoQ10 concentration in relation to CS activity, i.e., mitochondrial abundance, is likely to reflect the available CoQ10 content most accurately [18]. CS activity in the sample can be determined by following the reduction of 5,5′-dithiobis(2-nitrobenzoic acid) (DTNB) at 412 nm coupled to the reduction of coenzyme A in the presence of oxaloacetate [25] in a solution containing 0.12 % n-dodecyl-β-d-maltopyranoside (Anatrace). The mean mitochondrial CoQ10 concentration in *quadriceps* (*vastus lateralis*) muscle is 9.6 nmol/g wet weight (95 % CI 8.6–10.5 nmol/g wet weight) or CoQ10/CS 1.7 nmol/U (95 % CI 1.6–1.7 nmol/U) [18].

References

1. Kroger A, Klingenberg M (1973) The kinetics of the redox reactions of ubiquinone related to the electron-transport activity in the respiratory chain. Eur J Biochem 34:358–368

2. Sun IL, Sun EE, Crane FL, Morre DJ, Lindgren A, Low H (1992) Requirement for coenzyme Q in plasma membrane electron transport. Proc Natl Acad Sci U S A 89:11126–11130

3. Trumpower BL (1990) The protonmotive Q cycle. Energy transduction by coupling of proton translocation to electron transfer by the cytochrome bc1 complex. J Biol Chem 265:11409–11412

4. Ernster L, Forsmark-Andree P (1993) Ubiquinol: an endogenous antioxidant in aerobic organisms. Clin Investig 71:S60–S65

5. Lopez-Martin JM, Salviati L, Trevisson E, Montini G, DiMauro S, Quinzii C et al (2007) Missense mutation of the COQ2 gene causes defects of bioenergetics and de novo pyrimidine synthesis. Hum Mol Genet 16:1091–1097

6. Papucci L, Schiavone N, Witort E, Donnini M, Lapucci A, Tempestini A et al (2003) Coenzyme q10 prevents apoptosis by inhibiting mitochondrial depolarization independently of its free radical scavenging property. J Biol Chem 278:28220–28228

7. Emmanuele V, Lopez LC, Berardo A, Naini A, Tadesse S, Wen B et al (2012) Heterogeneity of coenzyme Q10 deficiency: patient study and literature review. Arch Neurol 69:978–983

8. Hirano M, Garone C, Quinzii CM (2012) CoQ(10) deficiencies and MNGIE: two treatable mitochondrial disorders. Biochim Biophys Acta 1820:625–631

9. Rotig A, Appelkvist EL, Geromel V, Chretien D, Kadhom N, Edery P et al (2000) Quinone-responsive multiple respiratory-chain dysfunction

due to widespread coenzyme Q10 deficiency. Lancet 356:391–395

10. Ogasahara S, Engel AG, Frens D, Mack D (1989) Muscle coenzyme Q deficiency in familial mitochondrial encephalomyopathy. Proc Natl Acad Sci U S A 86:2379–2382

11. Benoist JF, Rigal O, Nivoche Y, Martin C, Biou D, Lombes A (2003) Differences in coenzyme Q10 content in deltoid and quadriceps muscles. Clin Chim Acta 329:147–148

12. Paiva H, Thelen KM, Van Coster R, Smet J, De Paepe B, Mattila KM et al (2005) High-dose statins and skeletal muscle metabolism in humans: a randomized, controlled trial. Clin Pharmacol Ther 78:60–68

13. Tang PH, Miles MV, DeGrauw A, Hershey A, Pesce A (2001) HPLC analysis of reduced and oxidized coenzyme Q(10) in human plasma. Clin Chem 47:256–265

14. Pastore A, Giovamberardino GD, Bertini E, Tozzi G, Gaeta LM, Federici G et al (2005) Simultaneous determination of ubiquinol and ubiquinone in skeletal muscle of pediatric patients. Anal Biochem 342:352–355

15. Menke T, Niklowitz P, Adam S, Weber M, Schluter B, Andler W (2000) Simultaneous detection of ubiquinol-10, ubiquinone-10, and tocopherols in human plasma microsamples and macrosamples as a marker of oxidative damage in neonates and infants. Anal Biochem 282:209–217

16. Miles MV, Tang PH, Miles L, Steele PE, Moye MJ, Horn PS (2008) Validation and application of an HPLC-EC method for analysis of coenzyme Q10 in blood platelets. Biomed Chromatogr 22:1403–1408

17. Li L, Pabbisetty D, Carvalho P, Avery MA, Avery BA (2008) Analysis of CoQ10 in rat serum by ultra-performance liquid chromatography mass spectrometry after oral administration. J Pharm Biomed Anal 46:137–142

18. Itkonen O, Suomalainen A, Turpeinen U (2013) Mitochondrial coenzyme q10 determination by isotope-dilution liquid chromatography-tandem mass spectrometry. Clin Chem 59:1260–1267

19. Teshima K, Kondo T (2005) Analytical method for ubiquinone-9 and ubiquinone-10 in rat tissues by liquid chromatography/turbo ion spray tandem mass spectrometry with 1-alkylamine as an additive to the mobile phase. Anal Biochem 338:12–19

20. Ruiz-Jimenez J, Priego-Capote F, Mata-Granados JM, Quesada JM, Luque de Castro MD (2007) Determination of the ubiquinol-10 and ubiquinone-10 (coenzyme Q10) in human serum by liquid chromatography tandem mass spectrometry to evaluate the oxidative stress. J Chromatogr A 1175:242–248

21. Schaefer WH, Lawrence JW, Loughlin AF, Stoffregen DA, Mixson LA, Dean DC et al (2004) Evaluation of ubiquinone concentration and mitochondrial function relative to cerivastatin-induced skeletal myopathy in rats. Toxicol Appl Pharmacol 194:10–23

22. Arias A, Garcia-Villoria J, Rojo A, Bujan N, Briones P, Ribes A (2012) Analysis of coenzyme Q(10) in lymphocytes by HPLC-MS/MS. J Chromatogr B Analyt Technol Biomed Life Sci 908:23–26

23. Duberley KE, Hargreaves IP, Chaiwatanasirikul KA, Heales SJ, Land JM, Rahman S et al (2013) Coenzyme Q10 quantification in muscle, fibroblasts and cerebrospinal fluid by liquid chromatography/tandem mass spectrometry using a novel deuterated internal standard. Rapid Commun Mass Spectrom 27:924–930

24. Wibom R, Hagenfeldt L, von Dobeln U (2002) Measurement of ATP production and respiratory chain enzyme activities in mitochondria isolated from small muscle biopsy samples. Anal Biochem 311:139–151

25. Trounce IA, Kim YL, Jun AS, Wallace DC (1996) Assessment of mitochondrial oxidative phosphorylation in patient muscle biopsies, lymphoblasts, and transmitochondrial cell lines. Methods Enzymol 264:484–509

Chapter 25

Assessing the Bioenergetic Profile of Human Pluripotent Stem Cells

Vanessa Pfiffer and Alessandro Prigione

Abstract

Assessing the bioenergetics of human pluripotent stem cells (hPSCs), including embryonic stem cells (ESCs) and induced pluripotent stem cells (iPSCs), provides considerable insight into their mitochondrial functions and cellular properties. This might allow exposing potential energetic defects caused by mitochondrial diseases. However, certain challenges have to be met due to unique growth conditions in highly specialized and costly culture media. Here, we describe a method that facilitates the assessment of the bioenergetic profiles of hPSCs in a noninvasive fashion, while requiring only small sample sizes and allowing for several replicates. Basal respiratory and glycolytic capacities are assessed using a XF24 Extracellular Flux Analyzer by simultaneous measurements of the oxygen consumption rate (OCR) and extracellular acidification rate (ECAR), respectively. In addition, bioenergetic parameters are estimated by monitoring OCR and ECAR values upon metabolic perturbations via the consecutive introduction of mitochondria-specific inhibitors.

Key words Pluripotent stem cells, iPS cells, Mitochondria, Bioenergetics, Extracellular Flux Analyzer

1 Introduction

Pluripotent stem cells (PSCs) are defined by their unique features: indefinite propagation and ability to differentiate into any given cell type of the body. Embryonic stem cells (ESCs) were first isolated from the inner cell mass of the mouse blastocyst by Martin Evans in 1981 [1], and later in humans, by James Thomson in 1998 [2]. Additionally, a groundbreaking discovery of Shinya Yamanka in 2006 demonstrated that PSCs can also be established from somatic cells following the introduction of ESC-specific transcription factors [3].

The mitochondrial and metabolic features of human PSCs (hPSCs) have been recently started to be investigated [4–7]. Understanding the mitochondrial and bioenergetic features of PSCs, and how to modulate them, may shed light on the general properties of stem cells and also contribute to the development of

Volkmar Weissig and Marvin Edeas (eds.), *Mitochondrial Medicine: Volume I, Probing Mitochondrial Function*, Methods in Molecular Biology, vol. 1264, DOI 10.1007/978-1-4939-2257-4_25, © Springer Science+Business Media New York 2015

innovative cellular modeling systems for mitochondrial disorders. Indeed, iPSCs have been lately generated from somatic cells derived from patients carrying mitochondrial DNA mutations with the aim of addressing the related disease mechanisms (*see* Review Chap. 24, Volume II, in this book).

Valuable insights into the energy metabolism of hPSCs can be acquired through measuring the oxygen consumption rate (OCR), an indicator for mitochondrial respiration, and the extracellular acidification rate (ECAR), suggestive of glycolytic activity. By using the Extracellular Flux Analyzer (Seahorse Bioscience), these two key parameters can be determined simultaneously within the same cell population and without the need of laborious mitochondrial isolation, thereby allowing bioenergetic investigations of hPSCs within their natural culture environment. In order to provide accurate measurements, adherent cells need to be grown in uniform monolayers. Since hPSCs grow as colonies under standard culture conditions, the bioenergetic analysis of this cell type harbors unique challenges [8].

The OCR/ECAR ratio related to basal respiration can serve as an indicator to assess cellular preference for oxidative phosphorylation (OXPHOS) or glycolysis to respond to their energy demand. Additionally, by monitoring OCR values upon metabolic perturbation caused by consecutive addition of different mitochondrial inhibitors, the bioenergetic profiles can be analyzed in more details and different parameters can thus be estimated, including basal respiration, ATP turnover, maximal respiration, and spare respiratory capacity. In particular, oligomycin is first added after measuring basal OCR and ECAR values, resulting in the inhibition of ATP synthase. Since protons are prevented from moving through the ATP synthase, the proton gradient across the mitochondrial inner membrane increases while the consumption of oxygen drops, as shown by OCR reduction. The injection of the uncoupling agent FCCP reestablishes proton flux nonspecifically across the inner membrane resulting in complete uncoupling of electron transport and ATP generation and thereby giving rise to the maximal consumption of oxygen. Finally, concurrent addition of the electron transfer chain (ETC) inhibitors rotenone and antimycin A shuts down oxygen consumption completely by preventing the transfer of protons in the ETC (Fig. 1).

2 Materials

2.1 Equipment

1. CO_2 incubator (set to 37 °C, 5 % CO_2).
2. Non-CO_2 incubator (set to 37 °C).
3. Neubauer hemocytometer.
4. pH meter.

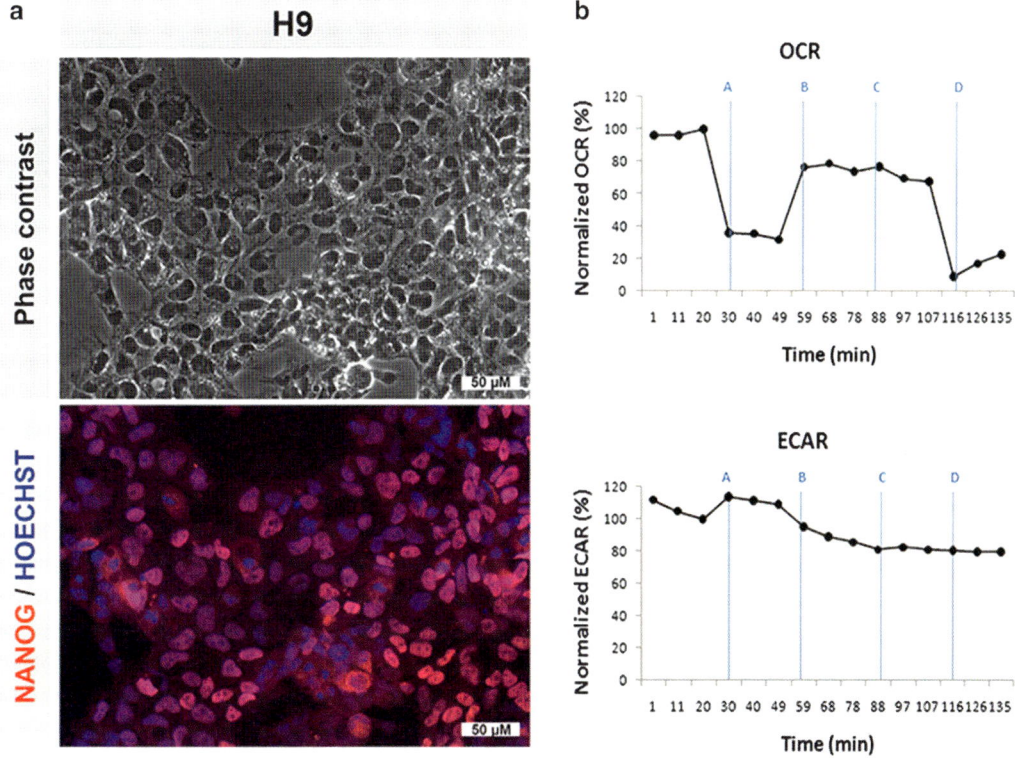

Fig. 1 (**a**) The day of the assay, hPSCs appear as a cellular monolayer but still maintain the expression of pluripotency-specific protein markers, such as NANOG. The human ESC line H9 is shown here as an example. (**b**) Typical OCR and ECAR profiles of hPSCs. The human ESC line H9 is reported here as an example. The four injection times in the specific four ports (*A–D*) are reported in *blue*. The bioenergetic profile of PSCs is quite remarkable. In the *upper graph*, it is noticeable the low OCR response to FCCP (port *B*), suggestive of reduced spare respiratory capacity. In the *lower graph*, the low ECAR response to oligomycin (port *A*) and FCCP (port *B*) can be appreciated, indicative that the cells are functioning with maximal glycolytic capacity that cannot be further increased upon mitochondrial inhibition. The OCR and ECAR values are first normalized to the DNA content using the CyQUANT kit and then presented in the graphs as the percentage of the basal state (rate number 3, before the injection of oligomycin in port *A*)

5. Plate reader (Tecan, Infiniti M200).

6. "Rapid"-Filtermax.

7. Six-well tissue culture plates.

8. Water bath (set to 37 °C).

9. XF24 Extracellular Flux Analyzer (Seahorse Bioscience).

10. XF24 FluxPak (Seahorse Bioscience).

11. XF24 V7 cell culture microplate (Seahorse Bioscience).

2.2 Reagents

1. Antimycin A.
2. B-27 (50×).
3. Basic fibroblast growth factor (bFGF).
4. BSA (30 %).
5. Carbonyl cyanide-4-(trifluoromethoxy)phenylhydrazone (FCCP).
6. CyQUANT® Cell Proliferation Assay Kit (Invitrogen).
7. DMSO.
8. Dulbecco's Modified Eagle's Medium Base (8.3 g/L).
9. Dulbecco's Modified Eagle Medium: Nutrient Mixture F-12 (DMEM/F-12).
10. Glucose.
11. L-glutamine (200 mM).
12. Matrigel Basement Membrane Matrix (BD Bioscience).
13. MEM nonessential amino acids (NEAA, 100×).
14. N-2 supplement (100×).
15. NaCl.
16. Nuclease-free distilled water (dH$_2$O).
17. Oligomycin.
18. PBS (1×, pH 7.4).
19. Penicillin–Streptomycin (P/S, 10 mg/ml).
20. Phenol Red.
21. Rotenone.
22. Sodium pyruvate (100 mM).
23. Thiazovivin.
24. Trypsin-EDTA (0.05 %).
25. Trypsin inhibitor.
26. XF Assay Medium (pH 7.4, Seahorse Bioscience).
27. XF Calibrant Solution (pH 7.4, Seahorse Bioscience).

2.3 Stem Cell Medium for Feeder-Free Conditions Prior to the Assay

In a "rapid"-FilterMax (0.22 μm pore size), supplement DMEM/F-12 medium with final concentrations of 1 mM sodium pyruvate, 2 mM L-glutamine, 0.1 mg/ml P/S, 1× NEAA, 1× N-2, 1× B-27 and 0.05 % BSA. Filter-sterilize and store the solution at 4 °C. Example: supplement 500 ml DMEM/F-12 with 5 ml 100 mM sodium pyruvate, 5 ml 200 mM L-glutamine, 5 ml 10 mg/ml P/S, 5 ml 100× NEAA, 5 ml 100× N-2, 10 ml 50× B-27, and 800 μl 30 % BSA. Filter-sterilize the medium. Before usage, add bFGF to a final concentration of 10 ng/ml.

2.4 Stem Cell Medium for Feeder-Free Conditions During the Assay in XF24 V7 Cell Culture Microplates

Supplement DMEM/F-12 medium as described above and filter-sterilize the resulting solution. Before usage, add bFGF and the ROCK inhibitor thiazovivin to final concentrations of 20 ng/ml and 0.5 μM, respectively.

2.5 XF Assay Medium

Supplement commercially available XF Assay Medium with sodium pyruvate (1 mM final concentration) and glucose (25 mM final concentration). Prepare this medium freshly with each use. Warm medium to 37 °C and adjust a 7.4 pH using NaOH. Filter-sterilize the solution.

Alternatively, XF Assay Medium can be prepared as follows: separately, dissolve DMEM Base in 500 ml dH$_2$O and dissolve 1.85 g NaCl in 500 ml dH$_2$O. Combine both solutions and remove 20 ml from the mixture. To the remaining 980 ml, add 10 ml 200 mM L-glutamine, 10 ml 100 mM sodium pyruvate and 15 mg Phenol Red. Add 4.5 g glucose for a final concentration of 25 mM. Warm medium to 37 °C and adjust a 7.4 pH using NaOH. Filter-sterilize the solution and store 50 ml aliquots at –20 °C.

2.6 Mitochondrial Inhibitors

Store all inhibitors at –20 °C for up to 6 weeks.

1. Oligomycin: 2.5 mM in DMSO.
2. FCCP: 2.5 mM in DMSO.
3. Rotenone: 2.5 mM in DMSO.
4. Antimycin A: 2.5 mM in DMSO.

3 Methods

All working steps including viable human PSCs need to be carried out under sterile conditions on clean benches.

3.1 Growth of PSCs Under Feeder-Free Conditions on Matrigel-Coated Plates Prior to the Assay

1. Coat a six-well plate with Matrigel and prewarm the plate at 37 °C. Depending on cell density and growth, prepare 2–3 six-wells per PSC line.

2. Prewarm stem cell medium for feeder-free conditions supplemented with 10 ng/ml bFGF (*see* Subheading 2.3) in a 37 °C water bath.

3. Split PSCs from MEF feeder layer to Matrigel-coated six-wells (cut-and-paste technique): split selected colonies with a needle into equal pieces. Replace old culture medium with 1 ml fresh prewarmed stem cell medium for feeder-free conditions supplemented with 10 ng/ml bFGF. Detach the colony pieces

with a pipette tip, carefully transfer the cell suspension onto the Matrigel-coated six-wells prepared in **step 1** containing prewarmed stem cell medium for feeder-free conditions supplemented with 10 ng/ml bFGF and incubate in a CO_2 incubator (37 °C, 5 % CO_2). The total volume for each six-well is 2.5 ml culture medium, which should be changed daily (*see* **Note 1**).

3.2 Day 1: Seeding PSCs onto a XF24 V7 Cell Culture Microplate and Preparation of XF24 FluxPak

1. Coat a XF24 V7 cell culture microplate with Matrigel and prewarm the plate at 37 °C.

2. Prewarm 1× PBS and stem cell medium for feeder-free conditions supplemented with 20 ng/ml bFGF and 0.5 μM thiazovivin (*see* Subheading 2.3 and 2.4) in a 37 °C water bath.

3. Aspirate the culture medium from the PSCs grown under feeder-free conditions (*see* Subheading 3.1) and wash each six-well with 1 ml 1× PBS.

4. Aspirate the PBS, add 1 ml of 0.05 % trypsin-EDTA and incubate cells for 7 min at 37 °C. Add 1 ml trypsin inhibitor and carefully transfer the cell suspension into a 15 ml Falcon tube (*see* **Note 2**). Collect cells by centrifuging for 4 min at 120 rcf at room temperature (RT), remove the supernatant, and resuspend the cell pellet in PBS (*see* **Note 3**).

5. Count the cells (e.g. with a Neubauer hemocytometer, *see* **Note 3**) and calculate the required amount of culture medium to obtain a concentration of 50,000 cells per 100 μl. After collecting cells by centrifugation (4 min, 120 rcf, RT), resuspend the cell pellet in the determined amount of prewarmed stem cell medium for feeder-free conditions supplemented with 20 ng/ml bFGF and 0.5 μM thiazovivin (*see* **Note 4**).

6. Seed 50,000 cells per well of a Matrigel-coated XF24 V7 cell culture microplate prepared in **step 1** by pipetting 100 μl PSC suspension per well (*see* **Note 5**). Leave 2–4 wells of the XF24 microplate empty and fill these wells with 100 μl medium only to serve as background controls. Incubate the seeded XF24 V7 cell culture microplate in a CO_2 incubator (37 °C, 5 % CO_2) for 1–2 h until cells have adhered to the well's bottom.

7. Add 150 μl of prewarmed stem cell medium for feeder-free conditions supplemented with 20 ng/ml bFGF and 0.5 μM thiazovivin into each well to a final volume of 250 μl (*see* **Note 6**).

8. Incubate the XF24 V7 cell culture microplate in a CO_2 incubator (37 °C, 5 % CO_2) overnight. Proceed to **steps 9** and **10** before overnight incubation.

9. Add 1 ml XF Calibrant Solution into each well of the lower part (24-well microplate) of a XF24 FluxPak and place the

upper part (green XF24 sensor cartridge) onto the hydrated microplate (*see* **Note 7**). Incubate the hydrated XF24 FluxPak in a non-CO_2 incubator at 37 °C overnight (*see* **Note 8**). Proceed to **step 10** before overnight incubation.

10. Turn on the XF24 Extracellular Flux Analyzer and start the XF24 software to allow the instrument to stabilize at 37 °C overnight.

3.3 Day 2: Exchange of Assay Medium, Loading of XF24 Sensor Cartridge with Injection Compounds and Assay Run

1. Prewarm XF Assay Medium in a 37 °C water bath (~50 ml per XF24 V7 cell culture microplate, *see* **Note 9**).

2. Inspect PSCs on the XF24 V7 cell culture microplate, which has been incubated overnight, under a microscope to ensure confluent cell monolayers. Remove the stem cell medium from all wells without disturbing the cells (*see* **Note 10**).

3. Wash the cells once with 1 ml prewarmed XF Assay Medium before adding XF Assay Medium to a final volume of 500 μl per well. Incubate the plate in a non-CO_2 incubator at 37 °C for 1 h.

4. Meanwhile, prepare 10× solutions of mitochondrial inhibitors in XF Assay Medium (*see* **Note 11**). Prepare 10 μM oligomycin by adding 8 μl of 2.5 mM oligomycin stock solution to 2 ml XF Assay Medium, 10 μM FCCP by adding 12 μl of 2.5 mM FCCP stock solution to 3 ml XF Assay Medium and 10 μM rotenone +10 μM antimycin A by adding 8 μl of both 2.5 mM rotenone and antimycin A stock solutions to 2 ml XF Assay Medium.

5. Load XF24 sensor cartridge placed in the hydrated microplate with 10× solutions of mitochondrial inhibitors prepared in **step 4**. Load 50 μl of 10 μM oligomycin solution into port A, 55 μl of 10 μM FCCP solution into port B, 61 μl of 10 μM FCCP solution into port C and 67 μl of 10 μM rotenone + antimycin A solution into port D (*see* **Note 12**). The final concentrations of the inhibitors will be 1 μM after injection during the assay (*see* **Note 13**). Place the loaded cartridge in the hydrated microplate in a non-CO_2 incubator at 37 °C until starting the assay.

6. Set up a protocol in the XF24 Extracellular Flux Analyzer software.

7. Press 'START', take out the loaded cartridge placed in the hydrated microplate from the non-CO_2 incubator and place it into the XF24 Extracellular Flux Analyzer for calibration (*see* **Note 14**).

8. When calibration is finished, remove the calibrant plate, take out the seeded XF24 V7 cell culture microplate from the non-CO2 incubator and place it into the XF24 Extracellular Flux Analyzer. Click 'CONTINUE' to start measurements (*see* **Note 14**).

9. When the run is finished, remove the XF24 V7 cell culture microplate from the XF24 Extracellular Flux Analyzer and remove the medium from all wells without disturbing the cells. Wash each well with 0.5 ml 1× PBS (*see* **Note 15**), remove the PBS, and freeze the plate at –20 °C (*see* **Note 16**).

3.4 Normalization by Measuring Cellular DNA Content and Data Analysis

1. Per XF24 V7 cell culture microplate, prepare 5 ml 1× cell-lysis buffer by diluting 0.25 ml of 20× cell-lysis buffer in 4.75 ml nuclease-free dH$_2$O and thaw the frozen XF24 V7 cell culture microplate containing the PSCs at RT.

2. Add 25 μl of the 400× CyQUANT® GR dye into 5 ml of 1× cell-lysis buffer prepared in **step 1** and mix thoroughly (*see* **Note 17**).

3. Add 200 μl of the cell-lysis buffer/CyQUANT® GR dye mixture into each well and incubate the plate for 5 min at RT protected from light. Measure the fluorescence at 480 nm excitation and 520 nm emission wave lengths.

4. Analyze normalized data.

4 Notes

1. In order to avoid contamination with mouse embryonic fibroblasts (MEFs), usually required for culturing of human PSCs, cells should be grown under feeder-free conditions on Matrigel-coated plates for 1–2 passages before seeding onto XF24 V7 cell culture plates.

2. It is crucial to dissociate PSCs into single cells using trypsin to facilitate the formation of a cell monolayer when seeding onto a XF24 V7 cell culture microplate, which is required for consistent measurements of OCR and ECAR. In case large cellular aggregates remain after 7 min of trypsinization, do not exceed incubation time to more than 10 min, since cells might suffer resulting in a decreased cellular yield. Add trypsin inhibitor to stop the reaction and dissociate cells mechanically by careful up- and down-pipetting before transferring the cells into a 15 ml Falcon tube.

3. PSCs are very small cells. Therefore, uniformly resuspend the cell pellet in ~3 ml and dilute cells further if the cell density is too high for accurate counting. If resuspending cells in PBS, work quickly to avoid additional stress. Alternatively, cells can be resuspended in culture medium before determining cell number.

4. The increased amount of bFGF as well as the addition of the ROCK inhibitor thiazovivin is needed to minimize cell death and maintain the survival of human PSCs as single cells.

Both components can be stored at −20 °C. Thiazovivin should be stored as small aliquots and be used freshly, since its activity decreases upon repeated thawing and freezing steps.

5. When seeding, rest the pipette tip against the lower circular wall of the well and carefully release the cell suspension to produce a homogenous monolayer of cells.

6. Rest the pipette tip against the upper rectangular wall of the well and add the medium gently without disturbing the only just attached cells.

7. Make sure that the sensor cartridge is placed on top of the microplate in the correct orientation (notch on the bottom left corner) and that all sensors are covered in the XF Calibrant Solution completely.

8. Incubation in a non-CO_2 incubator is crucial, since CO_2 affects the pH of the XF Calibrant Solution and therefore results in distorted ECAR measurements.

9. It is crucial that the pH of the media is 7.4 at 37 °C.

10. To remove the medium, carefully invert the whole plate and then dry the plate on clean paper, as we have not found cellular loss during this procedure. Alternatively, remove the medium with a pipette, while avoiding contact of the pipette tip with the wells' bottoms. Since measurements will only occur within the area confined by three dots in each XF24 V7 well, rest the pipette tip at the well's border outside this area. Work quickly to avoid cell drying.

11. We found that mitochondrial inhibitors' activity decreases after a while. Therefore, only use stock solutions, which are stored at −20 °C no longer than 6 weeks.

12. Before loading the XF24 sensor cartridge, check its orientation (notch on the bottom left corner) and the order of the injection ports A, B, C, and D (starting at the bottom right). The second injection of FCCP provides additional information about maximal respiration and spare respiratory capacity of the investigated PSCs, but can be omitted if required. In that case load 10 μM rotenone + antimycin A solution into port C and leave port D empty. Alternatively, port A can be employed for the injection of other drugs (such as 2DG), and the other ports can be used for oligomycin, FCCP, and rotenone and antimycin A.

13. The ideal concentrations of mitochondrial inhibitors may differ among cell lines and should be titrated before.

14. Make sure the plates are inserted into the XF24 Extracellular Flux Analyzer in the correct orientation and without the lid.

15. A washing step with PBS is recommended to clear cells from the Phenol Red inside the culture medium, which might interfere with adjacent determination of cellular DNA content by CyQUANT® GR dye.

16. The freezing step is required for efficient cell lysis in the adjacent CyQUANT® assay. The plates should not be stored longer than 4 weeks.

17. We find that 2× CyQUANT® GR dye dilution, which provides a linear detection range of ~50 to 250,000 cells, allows more accurate results than a 1× CyQUANT® GR dye dilution, which only allows a linear detection range up to 50,000 cells.

Acknowledgments

The authors declare no competing financial or commercial interests and acknowledge support from the Fritz Thyssen Foundation and the Deutsche Forschungsgemeinschaft (DFG).

References

1. Evans MJ, Kaufman MH (1981) Establishment in culture of pluripotential cells from mouse embryos. Nature 292(5819):154–156

2. Thomson JA, Itskovitz-Eldor J, Shapiro SS, Waknitz MA, Swiergiel JJ, Marshall VS, Jones JM (1998) Embryonic stem cell lines derived from human blastocysts. Science 282(5391): 1145–1147

3. Takahashi K, Yamanaka S (2006) Induction of pluripotent stem cells from mouse embryonic and adult fibroblast cultures by defined factors. Cell 126(4):663–676

4. Prigione A, Fauler B, Lurz R, Lehrach H, Adjaye J (2010) The senescence-related mitochondrial/oxidative stress pathway is repressed in human induced pluripotent stem cells. Stem Cells 28(4):721–733

5. Folmes CD, Nelson TJ, Martinez-Fernandez A, Arrell DK, Lindor JZ, Dzeja PP, Ikeda Y, Perez-Terzic C, Terzic A (2011) Somatic oxidative bioenergetics transitions into pluripotency-dependent glycolysis to facilitate nuclear reprogramming. Cell Metab 14(2):264–271

6. Zhang J, Khvorostov I, Hong JS, Oktay Y, Vergnes L, Nuebel E, Wahjudi PN, Setoguchi K, Wang G, Do A, Jung HJ, McCaffery JM, Kurland IJ, Reue K, Lee WN, Koehler CM, Teitell MA (2011) UCP2 regulates energy metabolism and differentiation potential of human pluripotent stem cells. EMBO J 30(24):4860–4873

7. Prigione A, Rohwer N, Hoffman S, Mlody B, Drews K, Bukowiecki R, Blumlein K, Wanker EE, Ralser M, Cramer T, Adjaye J (2014) HIF1alpha modulates cell fate reprogramming through early glycolytic shift and up-regulation of PDK1-3 and PKM2. Stem Cells 32(2):364–376

8. Zhang J, Nuebel E, Wisidagama DR, Setoguchi K, Hong JS, Van Horn CM, Imam SS, Vergnes L, Malone CS, Koehler CM, Teitell MA (2012) Measuring energy metabolism in cultured cells, including human pluripotent stem cells and differentiated cells. Nat Protoc 7(6):1068–1085

Chapter 26

Integrative Methods for Studying Cardiac Energetics

Philippe Diolez, Véronique Deschodt-Arsac, Guillaume Calmettes, Gilles Gouspillou, Laurent Arsac, Pierre dos Santos, Pierre Jais, and Michel Haissaguerre

Abstract

The more recent studies of human pathologies have essentially revealed the complexity of the interactions involved at the different levels of integration in organ physiology. Integrated organ thus reveals functional properties not predictable by underlying molecular events. It is therefore obvious that current fine molecular analyses of pathologies should be fruitfully combined with integrative approaches of whole organ function. It follows an important issue in the comprehension of the link between molecular events in pathologies, and whole organ function/dysfunction is the development of new experimental strategies aimed at the study of the integrated organ physiology. Cardiovascular diseases are a good example as heart submitted to ischemic conditions has to cope both with a decreased supply of nutrients and oxygen, and the necessary increased activity required to sustain whole body—including the heart itself—oxygenation.

By combining the principles of control analysis with noninvasive ^{31}P NMR measurement of the energetic intermediates and simultaneous measurement of heart contractile activity, we developed MoCA (for Modular Control and Regulation Analysis), an integrative approach designed to study in situ control and regulation of cardiac energetics during contraction in intact beating perfused isolated heart (Diolez et al., Am J Physiol Regul Integr Comp Physiol 293(1):R13–R19, 2007). Because it gives real access to integrated organ function, MoCA brings out a new type of information—the "elasticities," referring to internal responses to metabolic changes—that may be a key to the understanding of the processes involved in pathologies. MoCA can potentially be used not only to detect the origin of the defects associated with the pathology, but also to provide the quantitative description of the routes by which these defects—or also drugs—modulate global heart function, therefore opening therapeutic perspectives. This review presents selected examples of the applications to isolated intact beating heart and a wider application to cardiac energetics under clinical conditions with the direct study of heart pathologies.

Key words Modular control analysis, Noninvasive ^{31}P NMR, Calcium, Adrenaline, Levosimendan, Calcium sensitizer, Perfused heart, System's biology

1 Introduction

Being the first cause of mortality worldwide, cardiovascular diseases are responsible for about 700,000 deaths each year in Europe. Half of this mortality is due to heart failure (ineffective contraction)

Volkmar Weissig and Marvin Edeas (eds.), *Mitochondrial Medicine: Volume I, Probing Mitochondrial Function*, Methods in Molecular Biology, vol. 1264, DOI 10.1007/978-1-4939-2257-4_26, © Springer Science+Business Media New York 2015

which is a chronic and complex process. While the impact of cardiac diseases of genetic origin (known molecular origin) is relatively low, prevalence of acquired modifications of the heart (ischemic events, infarction scar, cardiomyopathy or aging) is essential. Cardiac contraction is entirely dependent, downstream of the electrical stimulation, on the adequacy in the response of contractile apparatus and in energy balance, both disrupted during heart failure.

Heart failure is a complex syndrome of numerous dysfunctional components which converge to cause chronic progressive failure of heart contractile function and maintenance of cardiac output demand [1]. Therefore, cardiovascular disease, as well as the cardiovascular effects of drugs, are essentially multifactorial problems involving interactions between many proteins, dependent on highly organized cell, tissue, and organ structures. As a consequence, biochemical modifications of heart function during pathology must reflect this complexity. This is also one reason why the side effects of drugs are often unanticipated, and remain to date difficult to analyze, since integrated organ function reveals fundamental properties that make it different from the sum of the underlying molecular events. It appears difficult to unravel such questions without using a system's approach, i.e. focussing on processes, not just molecular components [2]. It is also more and more obvious that current fine molecular analyses of pathologies should be fruitfully combined with integrative approaches. Therefore, it appears that an important issue in the comprehension of the link between molecular events in pathologies and whole organ function/dysfunction is the development of new experimental and analytical strategies aimed at the study of integrated organ physiology.

The complete description of a biological or physiological function depends on the synergy between adapted experimental and analytical approaches. System's Biology of important biological functions, like heart contraction, is a current issue in physiology as it not only reveals a new type of fundamental information, but also may turn out as new diagnostic and therapeutic strategies [3]. System's Biology starts with what we have learned from the huge progress at the molecular level and goes further towards the understanding of how small scale events integrate into biological functions. These new strategies require both the development of noninvasive techniques allowing investigations at the integrated level (ex vivo and in vivo) and suitable analytical tools. The analytical tools of the Top-down or Modular approaches to Metabolic Control Analysis (MCA) [4–7] were used by us and others to overcome the complexity of intra-cellular regulations [8–12] and to describe heart energetics [13, 14]. In the Top-down (modular) approach, the complexity of the system is reduced by grouping reactions and reactants into large modules connected by a small number of intermediates. While each module may be of any complexity, their interactions should only take place through the identified intermediate(s). The fundamental principle is that the kinetic interactions (elasticities) [4–7] between the intermediates (substrates and products) and the different modules (or single enzymes) determine and maintain steady state conditions (fluxes and concentrations). Therefore regulation of the pathway (i.e. activity and steady state changes) by a drug or a pathology originates in modulations (adaptation or defect) of these kinetic interactions, which can be studied by MCA.

A wider application to the effects of cardiac drugs in conjunction with the study of heart pathologies models may be considered in the near future. MoCA can potentially be used not only to detect the origin of the defects associated with a pathology (elasticity analyses), but also to provide a quantitative description of how these defects influence global heart function (regulation analysis) and therefore open new therapeutic perspectives. Among these perspectives, these approaches are perfectly fitted for testing toxicity of drugs on cardiac mitochondria. Several key examples of the applications to intact isolated beating heart are presented in this short review, and the perspectives for clinical application are considered.

1.1 Energy Homeostasis in Heart

Like muscle and brain, other energy consuming organs, heart tissues contain apart from ATP a second energy-transferring molecule: phosphocreatine (PCr). However, an important difference has been observed in the regulation of the energetics between skeletal muscle and heart. Indeed, the activation of contraction in skeletal muscle is accompanied by the decrease of both PCr and phosphorylation potential (ΔG_p). This decrease indicates that the energy balance in muscle is achieved with different metabolite concentrations and biochemical steady states [15]. By contrast, the energy balance in heart bioenergetics is characterized by an improved energetics homeostasis [15–17], indicating a perfect energy balance whatever heart activity. Indeed, almost no changes in the energetic intermediates (PCr, ATP, and Pi, and therefore ΔG_p) can be observed even for important increases in heart activity, as seen by ^{31}P NMR spectroscopy on isolated perfused heart [13, 18] or in vivo in human heart [19]. How this homeostasis is achieved in heart—but not in muscle—during the so-called excitation-contraction coupling is still a debated fundamental question. Incidentally, while PCr (ΔG_p) recovery after exercise can be used as a clinical index of the energetics capacity of skeletal muscle for detection of mitochondrial defects (or myopathies) [20–22], the absence of change excludes this direct approach to heart energetics dysfunctions. As a major outcome, integrative approaches like MoCA allow to measure and compare the relative importance of the different routes—i.e. supply, demand, and metabolite changes—by which the effect(s) of a drug/hormone

[23–25] or a pathology [26, 27] are transmitted and alter integrated organ function, e.g. how inotropic drugs/action activates heart contraction. These approaches may appear as an alternative to get better insight in heart energetics regulation [19]. The results of MoCa elasticity analysis applied to isolated rat and guinea pig hearts under reference conditions have been described elsewhere [13, 25] and the present review will emphasize examples of the applications of MoCA Regulation analysis to the study of heart energetics regulation following the action of inotropic drugs.

1.2 Parallel Activation of Mitochondria and Myofibrils by Calcium

By using MoCA, we could demonstrate in intact beating heart the essential role of the activation of mitochondrial oxidative phosphorylation by intracellular calcium in the coupling process between electrical excitation of the myocyte and the energetics of contraction in heart [13], previously demonstrated in vitro [16, 17] and in silico [28]. This has been demonstrated by analyzing the response of heart contraction to changes in Ca^{2+} in the perfusate, which established the almost-perfect parallel activation of both Supply and Demand under these conditions, responsible for energy homeostasis. This activation cascade has also been established for the activation of the different processes of mitochondrial oxidative phosphorylation by Ca^{2+} [29–31] (see Fig. 2 of ref. 32).

We also applied MoCA regulation analysis to the response of heart contraction to adrenaline [24]. The inotropic effect of adrenaline involves a complex physiological mechanism involving important modulations in intracytosolic calcium. These effects include changes in both mean cytosolic concentration and transient Ca^{2+} rises. Interestingly, a direct relationship has been demonstrated between contractile activity and cytosolic calcium concentration when heart contraction was modulated either by varying calcium concentration in the perfusion medium or by stimulating ß1-adrenergic receptor [16, 33]. Figure 1 presents the evolution of the ^{31}P NMR spectra of a typical rat-isolated perfused heart in response to adrenaline addition to the perfusate. Surprisingly, we could observe that in striking contrast with skeletal muscle the energetic intermediate (PCr and therefore ΔG_p, not shown) paradoxically increased in response to the inotropic action of adrenaline. The stronger activation of energy supply (oxidative phosphorylation) by adrenaline when compared to external Ca^{2+} addition could be explained by the marked facilitation of Ca^{2+} transients in the presence of adrenaline [34]. This hypothesis could be tested by applying MoCA regulation analysis to the effect of adrenaline, using information from the MoCA elasticity analysis under reference steady state conditions [13] and the changes in steady state flux (contractile activity) and intermediate concentration (PCr) observed after adrenaline addition (cf. Fig. 1). The results of the regulation analysis are presented in Fig. 2. The total inotropic effect corresponding to the

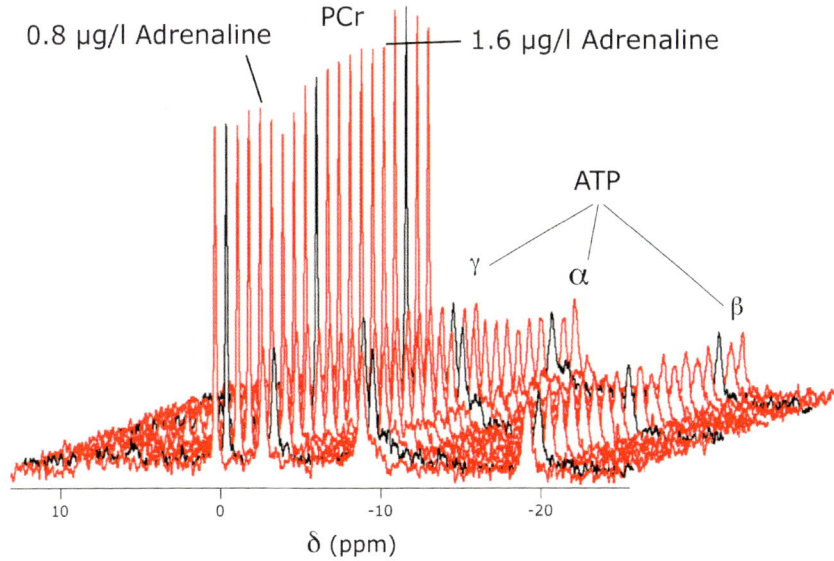

Fig. 1 Typical changes in ^{31}P NMR spectra of perfused rat heart induced by the addition of adrenaline (0.8 and 1.6 µg/l) in the perfusate. On this representative experiment, isolated rat hearts were perfused using pyruvate as substrate and in the presence of 2.0 mM free calcium concentration. Adrenaline increased the contractile activity (Rate Pressure Product, RPP) by 73 or 84 % at the concentration of 0.8 and 1.6 µg/l, respectively. At the same time, the PCr concentration was surprisingly increasing for both adrenaline concentrations (+2 % and +4 % for 0.8 and 1.6 µg/l respectively, $n=6$ different experiments) above the mean value of control spectra (*see* ref. 24 for more details)

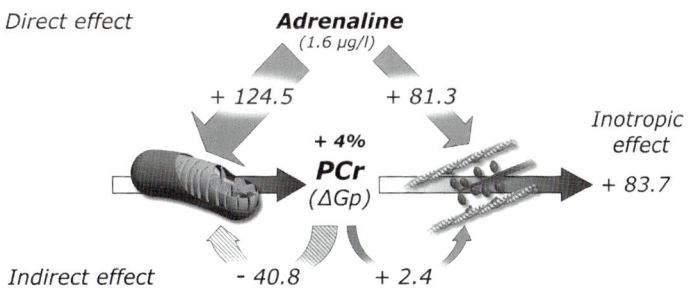

Fig. 2 MoCA regulation analysis of the integrated effect of adrenaline on rat isolated heart energetics. The Direct and Indirect effects of the addition of 1.6 µg/l adrenaline on both supply and demand modules were experimentally determined as described in ref. 13 and experimental values are from ref. 24. Indirect effects are calculated from the observed change in PCr concentration (+4 %) and the responsiveness of the module considered to PCr change (i.e. the elasticity). This allows the determination of the Direct effects of adrenaline on Supply and Demand modules as the integrated elasticity (Inotropic effect – indirect effect) [9]. Since a new steady state is reached after adrenaline addition, the final increase in flux is the same for both modules (83.7 %) and is the sum of the direct effect (by adrenaline) and the indirect effect (through PCr changes). The sizes of the *arrows* evoke the magnitude of the different effects, expressed as the percentage change from the starting steady state condition in the absence of adrenaline

small but significant increase in PCr of 4 % amounted to about 84 % for the contractile activity. The analysis effectively revealed that adrenaline had a much greater direct effect on the supply module (about 50 % higher, 125 % compared to 80 %) than on the demand module [24]. While comparable activations of supply and demand modules were obtained after addition of Ca^{2+} in the perfusate (parallel activation) [13, 32], resulting in an almost perfect homeostasis of the energetic intermediates, the overactivation of energy supply by adrenaline resulted in a paradoxical increase in cell energy status. The analysis also showed that the new steady state of contraction is obtained through a marked back inhibition of oxidative phosphorylation by the increase of PCr (about 40 %). Since Ca^{2+} activates both modules the same way, these results clearly show that physiological activation by adrenaline cannot be reduced to a simple calcium effect and that a specific activation of mitochondrial oxidative phosphorylation occurs. There is still no clear explanation for these results, and the role of cAMP on calcium transients and relaxation (re-uptake of calcium by sarcoplasmic reticulum), and/or mitochondrial integration of Ca^{2+} signal, is currently under study.

1.3 Drug Effects: Levosimendan

We also tested the potential of MoCA regulation analysis to detect the side effects of drugs on the energetics of heart contraction. This study was carried out on Levosimendan, member of a new class of cardiovascular drugs, the calcium-sensitizers. Levosimendan is a positive inotropic drug developed for the treatment of acute de novo or decompensated heart failure [34–36]. Levosimendan acts via selective binding on calcium-saturated cardiac troponin C (TnC) [37, 38] and stabilizes the conformation of the $Ca^{2+}TnC$ complex [34, 39, 40]. As a consequence, Levosimendan increases the sensitivity of myofilaments to calcium, leading to increased myocardial contraction. However, some phosphodiesterase 3 (PDE 3) inhibition has been observed on permeabilized myocytes in the presence of Levosimendan at high concentrations [41]. This side effect may interfere with the inotropic effect mechanisms of levosimendan since PDE 3 inhibition could induce an increase in cyclic AMP (cAMP) and a subsequent increase in intracellular calcium concentration. As demonstrated above, since MoCA regulation analysis may reveal direct mitochondrial activation, it appeared sensible to test the possibility of a side effect of levosimendan besides the Ca^{2+}-sensitization mechanism.

Since it has no effect on rats, the inotropic effects of Levosimendan have been studied on Guinea-pig paced perfused hearts [25]. The results of the MoCA regulation analysis are presented in Fig. 3. By contrast with the analysis of the effects of adrenaline, the inotropic effect of levosimendan (+45 %) is accompanied by a decrease in heart PCr content (about 7 %,

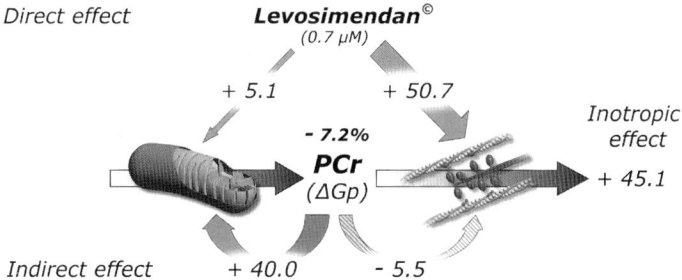

Fig. 3 MoCA regulation analysis of the integrated effect of Levosimendan—a Ca²⁺ sensitizer—on Guinea pig-isolated heart energetics. This figure presents the regulation analysis of the effects of the addition of 0.7 µM Levosimendan in the perfusion medium of Guinea pig-isolated hearts (data are from ref. 25). In the almost total absence of direct effect of Levosimendan on energy Supply (+5.1 %), the activation of oxidative phosphorylation (+40 %) occurs through a significant decrease in PCr (and therefore ΔG_p) in order to match the ATP requirement by the inotropic effect. See methods for details on the calculation

spectra not shown). The regulation analysis presented in Fig. 3 show that this decrease in PCr (ΔG_p) is responsible for the activation of mitochondrial activity to sustain energy demand (40 %) and allow the inotropic effect. Most interestingly, these results also evidenced a very small direct effect of Levosimendan on energy supply (only 5 %), as compared to the 50 % direct activation of demand. Even if it is close to the experimental error, this slight activation of Supply may be correlated to a light inhibition of PDE 3 activity and/or rise in cAMP and the corresponding rise in cytosolic Ca²⁺ and do not allow to rule out completely this side effect. This result evidences that under our conditions, a possible effect of Levosimendan on PDE3 activity and/or intracellular calcium cannot be totally ruled out but remains very low on mitochondrial activity and rather insignificant on integrated cardiac energetics. Thus, it may be concluded that, at least under our ex vivo conditions, Levosimendan inotropic effect on Guinea pig heart depends almost entirely on the Ca²⁺-sensitizing properties leading to myofilaments activation and on the concomitant activation of energy supply by the decrease in PCr (and ΔG_p) [25, 42].

1.4 Conclusion and Perspectives

The results presented here show that regulation of heart contraction energetics can be studied in intact beating heart, as a supply-demand system, by using MoCA elasticity and regulation analyses. MoCA regulation analysis has first been used to demonstrate for the first time in intact beating heart that physiological activation of heart contraction involves a parallel activation of both energy supply and demand, observed during calcium or adrenaline activation in heart, which is crucial for the homeostasis of energetic intermediates and for heart physiological response. The results presented in this paper show how MoCA may be used to describe the internal regulations of energy flux in heart contraction and to describe qualitatively and quantitatively the effects of drugs. Therefore, MoCA may also be used to study pathological deregulations, as well as for the detection of side effects of drugs, with a special access to mitochondrial toxicity.

System's biology analysis delivers the description of healthy heart function through integrated internal regulations which are currently not accessible with available medical tools. One of the challenges for the future is to develop integrated in vivo experimental approaches to evaluate such complex biological processes as intact organ function and to lead to a comprehensive understanding of heart function [19]. The transfer of the analysis under in vivo conditions will require preliminary experiments on large animals in order to determine in vivo elasticities under physiological conditions (oxygenation, multiple substrates, etc.) and adequate modulations of heart activity and energetics.

MoCA will now be tested as a noninvasive clinical tool for human heart contraction energetics. ^{31}P spectroscopy is already available on medical MRI apparatus and specific coils have been developed for human chest. MoCA application also requires the noninvasive assessment of heart contractile activity, which can be measured by a specific 3D MRI protocol that we developed recently for in vivo mouse heart activity [43]. The different steady states of heart activity required for the elasticity analysis will be obtained by allowing patients to exercise inside the magnet. A careful analysis of how these regulations (elasticities) change in response to heart pathologies should be helpful for diagnosis and therapeutic design. We propose that these complementary information may be of some help in the future for diagnosis and therapeutics follow-up in human heart pathologies.

2 Methods and Rationale

2.1 Methods

2.1.1 Ethical Approval

The present study complies with the European Community Guidelines for the Care and Use of Experimental Animals, with the recommendations of the Directive 2010/63/EU of the European Parliament. All the protocols used were also approved by our ethics committee "Comité d'éthique regional d'Aquitaine," France (# 3308010). PD has a permanent license for experimentation on animals (03/17/1999). Surgery was performed under isoflurane anesthesia, and every effort was made during animal manipulation to minimize stress.

2.1.2   Heart Perfusion and ^{31}P NMR Spectroscopy

Male rats (weighing 300–350 g, Janvier, France) for adrenaline protocol or male guinea-pigs (300–350 g, Charles River, France) for levosimendan protocol, received water and food ad libitum. Animals were anesthetized with isoflurane (2–3 %) using a face mask (4 L/min), and 30 mg/kg of sodium pentobarbital were injected intraperitoneally before the thorax was opened and heart quickly excised and immediately cooled in iced Krebs buffer. Hearts were perfused by an aortic canula delivering warmed (37 °C) Krebs Henseleit solution at constant pressure (100 mmHg), containing (in mM: NaCl 118, NaHCO$_3$ 12.5, KCl 5.9, EDTA 0.5, MgSO$_4$ 1.2, glucose 16.7, NaPyruvate 10, CaCl$_2$ 1.75–3.5), continuously gased with a mixture of 95 % O$_2$ and 5 % CO$_2$ (pH 7.35, temperature 37 °C at the level of the aortic cannula). Left ventricle pressure (LVP) of heart was measured under iso-volumic conditions of contraction thanks to a latex balloon inserted into the left ventricle through the left atrial appendage and con-nected to a pressure transducer (Statham Kelvin P23Db). Mechanical performance is expressed as the Rate Pressure Product (RPP in mmHg beat/min) and evaluated as the product between heart rate (HR, beat/min) and developed pressure (ΔP, mmHg).

Perfused heart was placed into a ^{1}H/^{31}P double-tuning 20-mm probe in a 9.4-T super-conducting magnet (Bruker DPX400 Avance). The probe was tuned to the phosphorus resonance fre-quency of 161.94 MHz, and the magnetic field homogeneity was optimized by shimming on the proton signal coming from the heart and the surrounding medium. Partially saturated ^{31}P NMR spectra were obtained by averaging data from 150 free induction decays (5-min acquisition time).

In order to eliminate endogenous substrates and possible remaining anesthetics, 20 min were allowed for stabilization of heart activity before any measurements was carried out. For the elasticity analysis (not presented in this paper), hearts were either submitted to an increase of the left intraventricular pressure by inflating the balloon (from 75 to 150 μl) (rat, $n=6$; Guinea pig, $n=7$) to obtain the elasticity of the Supply module (see below) or to the infusion of NaCN (0.1–0.2 mM final concentration) by using a clinical electric syringe (Carefusion, France) and a 60 time dilution to obtain the elasticity of the Demand module.

For regulation analysis, hearts were treated as control ($n=9$) or subjected to the (1) addition of 0.7 μM Levosimendan (synthe-sized by Abbot Scandinavia, Solna, Sweden) [44] ($n=7$) or (2) addition of 1.6 μg/L adrenaline ($n=8$) [24].

2.1.3   Myocardium Oxygen Consumption

In order to assess mitochondrial oxidative phosphorylation in per-fused heart, MVO$_2$ (μmol O$_2$/min/g dry weight) was measured continuously in hearts under all conditions. Immediately after per-fusion, the pulmonary artery was cannulated with a plastic cannula, which was connected to specific tubing and perfusate was pulled by a peristaltic pump (1 ml/min) through the oxygen sensor (Strathkelvin Instruments, Scotland, UK).

2.2 Modular Control Analysis of Heart Energetics (MoCA)

2.2.1 Rationale and System Definition

We designed MoCA (Modular Control and Regulation Analysis) for the study of the regulation of the energetics of heart contraction. The main concern from the very beginning was the development of a noninvasive integrated description of heart energetics which could be applicable not only to basic research but also for clinical purposes.

By combining the general principles of MCA with the experimental determination of energetic intermediates by noninvasive ^{31}P-NMR measurement and the simultaneous measurement of heart contractile activity, it was possible to design an integrative approach to study control and regulation of cardiac energetics during contraction in intact beating perfused isolated heart [13]. Within MoCA, the energetics of heart contraction is simplified in the framework of modular control analysis [9] in a two-module system (energy Supply and energy Demand) only connected by the energetic intermediate ΔG_p (phosphorylation potential or ΔG for ATP hydrolyis, *see* 45 as reference for bioenergetics). The system of heart contraction energetics considered for modular control analysis is presented in Fig. 4. Concerning the energetic intermediate, since ADP concentration is too low (in the tens of µM range) to be detected by ^{31}P-NMR spectroscopy on intact organ and in the total absence of changes in ATP, Phosphocreatine (PCr) concentration changes are used as representatives of the changes in ΔG_p, the true energetic intermediate [13].

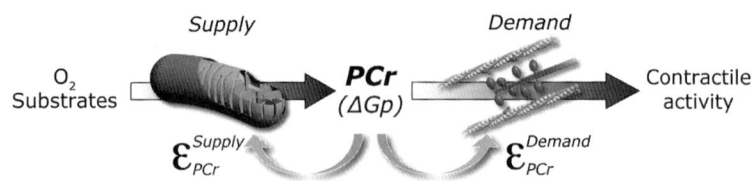

Fig. 4 System Definition for Modular Control Analysis of the energetics of perfused heart. Heart contraction energetics is defined as two modules connected by the energetic intermediate (phosphorylation potential, ΔG_p). The Supply module comprises all the metabolic steps from substrate and oxygen supply to mitochondrial production of ATP/PCr, responsible for ΔG_p generation. The Demand module comprises all the steps consuming ATP/PCr linked to contractile activity; i.e. ATP used by myofilaments for contraction and ATP used for Ca^{2+} recapture during relaxation. The elasticities (responsiveness) of supply ($\varepsilon_{PCr}^{Supply}$) and Demand ($\varepsilon_{PCr}^{Demand}$) module towards changes in the intermediate ΔG_p (represented by changes in PCr measured by ^{31}P NMR) are experimentally determined after induction of slight changes in steady state contractile activity and energetic intermediates by alteration—inhibition or activation—of the other module (*see* ref. 13)

In heart, since energetics is almost completely aerobic, the Supply module thus comprises all the metabolic steps from substrate and oxygen supply to the intact heart down to mitochondrial production of ATP/PCr and ΔG_p generation and poise. On the other hand, the Demand module comprises all the metabolic steps consuming ATP/PCr that are linked to contractile activity. This includes not only the ATP directly used by myosin-ATPases of myofilaments for contraction, but also the ATP used for Ca^{2+} recapture occurring during relaxation (mainly recapture by SERCA in sarcoplasmic reticulum and Ca^{2+} exchange over the sarcolemmal membrane).

The rationale of the modular control analysis is that the internal regulation of the entire system is the consequence of the interactions between the different modules and connecting intermediates. By reducing the complexity of the energetics of heart contraction to two modules and one intermediate (*see* Fig. 4), it follows that the entire system may be described by determining the responsiveness (called "elasticity," ε) of Supply and Demand modules to changes in PCr concentration. Control and regulation of our modular system can then be analyzed using MoCA as long as intermediates concentrations and modules activities (fluxes) can be studied. Continuous ^{31}P NMR spectroscopy gives access to phosphorylated energetic metabolites, including our intermediate: PCr. The energetic flux through the two modules of the system may be concomitantly assessed as heart oxygen consumption (Supply) and contractile activity (Demand) measured by heart rate-ventricular pressure product (RPP).

As this will be described below, these elasticities may be experimentally determined in intact beating heart and have been used to describe normal or pathological control pattern and further perform regulation analysis of pathologies and drugs effects (*see* refs. 13, 32 for a complete description of the analysis).

2.2.2 Experimental Determination of the Elasticities

The principle of the experimental determination of the elasticities of the different modules towards changes in the intermediate has been extensively described by the group of M. Brand [8, 9]. By applying these principles to heart energetics, we could determine the elasticity ($\varepsilon_{PCr}^{Module}$) of each module towards the energetic intermediate from the changes in module's activity (oxygen consumption or contractile activity for supply and demand modules, respectively) induced by a slight change in PCr concentration. This change is typically obtained by a slight modulation of the other module, activation or inhibition, and the obtention of a new steady state. This modulation should only affect the module under study through the changes in the intermediate (*see* refs. 8, 14 for details).

The analysis requires that only small changes in the intermediate should be considered for calculations of the elasticities, which may sometimes be difficult to obtain experimentally. However, the response of the module under study around the reference steady state should be the same for an increase or a decrease in the intermediate, and therefore the best way to ascertain that deviations are not occurring due to too large changes in the intermediate is to carry out both positive and negative modulations of PCr.

Experimentally, in order to measure Demand module elasticity ($\varepsilon_{PCr}^{Demand}$), low concentrations of cyanide (0.1–0.2 mM NaCN) are used to progressively inhibit mitochondrial cytochrome oxidase and thus obtain a new steady state of contractile activity and PCr concentration. Then the evolution in RPP is plotted as a function of [PCr] and $\varepsilon_{PCr}^{Demand}$ calculated from the following equation, where the indices "start" stand for the starting (reference) situation for which the elasticity is determined:

$$\varepsilon_{PCr}^{Module} = \frac{\Delta RPP / RPP_{start}}{\Delta[PCr] / [PCr]_{start}} \tag{1}$$

Supply module elasticity ($\varepsilon_{PCr}^{Supply}$) is usually measured the same way, but changes in [PCr] were induced by inducing a Frank-Starling effect by increasing the internal volume of the balloon placed in the ventricle and utilized for recording heart contractile activity. This inotropic physiological effect induces an increase in myofibrils ATPase activity resulting in a concomitant increase in contractile activity and a decrease in PCr [13]. The elasticity may also be calculated by using Eq. 1.

The control coefficients of both modules (C_{Supply}^{Flux} and C_{Demand}^{Flux}) may be calculated from experimentally measured $\varepsilon_{PCr}^{Supply}$ and $\varepsilon_{PCr}^{Demand}$ according to Summation and Connectivity theorems [6, 7].

2.2.3 Regulation Analysis

The rationale for Regulation Analysis can be found elsewhere [13, 46, 47], and only the specific application to heart energetics is presented here. While Modular Control Analysis (see above) describes how steady states are maintained (flux and metabolites homeostasis), Regulation Analysis [9, 46] has the potential to provide a quantitative description of how regulation by an external or internal effector leads the system to move to a new steady state. This includes the integrated effect of drugs (external effector) and of pathologies (internal modifications of modules' elasticities).

For the complete Regulation Analysis of the effect of any type of effector on the energetics of heart contraction, the only data required are:

– The complete elasticity analysis (see above) under the reference conditions (absence of drug or healthy heart, depending on the effect analyzed).

– The global effect of the effector on both contractile activity and intermediate concentration—expressed as relative changes in RPP and [PCr]—by comparison with reference conditions.

The final effect of any effector on the integrated modular system of heart energetics is complex, since it is the sum of the direct effects on each module and the indirect effects due to reciprocal effects on the intermediate [6–8, 13]. However, knowing the response of every module to the change in the intermediate (the elasticities determined previously), these indirect effects may be easily calculated from the observed change in the intermediate (PCr) in response to the effector. For each module, the Indirect Effect due to this relative change in PCr depends on its responsiveness—its elasticity—and is the product of the relative change in PCr and module's elasticity towards PCr:

$$\text{Indirect effect on supply}: \varepsilon_{PCr}^{Supply} \times \Delta[PCr] / [PCr]_{Start} \qquad (2)$$

$$\text{Indirect effect on demand}: \varepsilon_{PCr}^{Demand} \times \Delta[PCr] / [PCr]_{Start} \qquad (3)$$

Finally, it is possible to decipher the Direct Effect of the effector on each Module (defined as the Integrated Elasticity, IE) [46, 47] as the difference between the total measured effect on contraction ($\Delta RPP/RPP_{Start}$) and the Indirect Effect determined above:

$$\text{Direct effect on supply}: \Delta RPP / RPP_{Start} - \varepsilon_{PCr}^{Supply} \times \Delta[PCr] / [PCr]_{Start} \qquad (4)$$

$$\text{Direct effect on demand}: \Delta RPP / RPP_{Start} - \varepsilon_{PCr}^{demand} \times \Delta[PCr] / [PCr]_{Start} \qquad (5)$$

The direct effect determined experimentally here may be expressed as percent of the activity of the module under reference (starting) conditions and reflects the initial effect(s) of the effector (or pathology) on each module and therefore describes by which route (qualitatively) and how (quantitatively) the effector modulates heart energetics.

Acknowledgments

Part of this work has been supported by the "Association Française contre les Myopathies" (grant #AFM 12338), CNRS (P. Diolez) and IHU-LIRYC (Université de Bordeaux). The author thanks G. Raffard and R. Rouland for their technical help and Y. Chatenet for 3D drawings (Figs. 1, 3, and 4).

References

1. Sheeran FL, Pepe S (2006) Energy deficiency in the failing heart: linking increased reactive oxygen species and disruption of oxidative phosphorylation rate. Biochim Biophys Acta 1757(5–6):543–552

2. Fink M, Noble D (2010) Pharmacodynamic effects in the cardiovascular system: the modeller's view. Basic Clin Pharmacol Toxicol 106(3):243–249

3. Cascante M, Boros LG, Comin-Anduix B et al (2002) Metabolic control analysis in drug discovery and disease. Nat Biotechnol 20(3):243–249

4. Fell DA (1992) Metabolic control analysis: a survey of its theoretical and experimental development. Biochem J 286(Pt 2):313–330

5. Heinrich R, Rapoport TA (1974) A linear steady-state treatment of enzymatic chains. General properties, control and effector strength. Eur J Biochem 42(1):89–95

6. Kacser H, Burns JA (1973) The control of flux. Symp Soc Exp Biol 27:65–104

7. Kacser H, Burns JA (1995) The control of flux. Biochem Soc Trans 23(2):341–366

8. Brand MD (1996) Top down metabolic control analysis. J Theor Biol 182(3):351–360

9. Brand MD, Curtis RK (2002) Simplifying metabolic complexity. Biochem Soc Trans 30(2):25–30

10. Brown GC, Hafner RP, Brand MD (1990) A 'top-down' approach to the determination of control coefficients in metabolic control theory. Eur J Biochem 188(2):321–325

11. Dufour S, Rousse N, Canioni P et al (1996) Top-down control analysis of temperature effect on oxidative phosphorylation. Biochem J 314(Pt 3):743–751

12. Hafner RP, Brown GC, Brand MD (1990) Analysis of the control of respiration rate, phosphorylation rate, proton leak rate and proton-motive force in isolated mitochondria using the 'top-down' approach of metabolic control theory. Eur J Biochem 188(2):313–319

13. Diolez P, Deschodt-Arsac V, Raffard G et al (2007) Modular regulation analysis of heart contraction: application to in situ demonstration of a direct mitochondrial activation by calcium in beating heart. Am J Physiol Regul Integr Comp Physiol 293(1):R13–R19

14. Diolez P, Raffard G, Simon C et al (2002) Mitochondria do not control heart bioenergetics. Mol Biol Rep 29(1–2):193–196

15. Kushmerick MJ (1995) Skeletal muscle: a paradigm for testing principles of bioenergetics. J Bioenerg Biomembr 27(6):555–569

16. Balaban RS (2002) Cardiac energy metabolism homeostasis: role of cytosolic calcium. J Mol Cell Cardiol 34(10):1259–1271

17. Balaban RS, Bose S, French SA et al (2003) Role of calcium in metabolic signaling between cardiac sarcoplasmic reticulum and mitochondria in vitro. Am J Physiol Cell Physiol 284(2):C285–C293

18. Diolez P, Simon C, Leducq N et al (2000) Top down analysis of heart bioenergetics. In: Hofmeyr J-HS, Rohwer M, Snoep JL (eds) BTK2000: animating the cellular map. Stellebosch University Press, Stellenbosch, pp 101–106

19. Balaban RS (2006) Maintenance of the metabolic homeostasis of the heart: developing a systems analysis approach. Ann N Y Acad Sci 1080:140–153

20. Bendahan D, Desnuelle C, Vanuxem D et al (1992) 31P NMR spectroscopy and ergometer exercise test as evidence for muscle oxidative performance improvement with coenzyme Q in mitochondrial myopathies. Neurology 42(6):1203–1208

21. Kemp GJ, Taylor DJ, Radda GK (1993) Control of phosphocreatine resynthesis during recovery from exercise in human skeletal muscle. NMR Biomed 6(1):66–72

22. Kemp GJ, Taylor DJ, Thompson CH et al (1993) Quantitative analysis by 31P magnetic resonance spectroscopy of abnormal mitochondrial oxidation in skeletal muscle during recovery from exercise. NMR Biomed 6(5):302–310

23. Korzeniewski B, Deschodt-Arsac V, Calmettes G et al (2009) Effect of pyruvate, lactate and insulin on ATP supply and demand in unpaced perfused rat heart. Biochem J 423(3):421–428

24. Korzeniewski B, Deschodt-Arsac V, Calmettes G et al (2008) Physiological heart activation by adrenaline involves parallel activation of ATP usage and supply. Biochem J 413(2):343–347

25. Deschodt-Arsac V, Calmettes G, Raffard G et al (2010) Absence of mitochondrial activation during levosimendan inotropic action in perfused paced guinea pig hearts as demonstrated by modular control analysis. Am J Physiol Regul Integr Comp Physiol 299(3):R786–R792

26. Calmettes G, Deschodt-Arsac V, Thiaudiere E et al (2008) Modular control analysis of effects of chronic hypoxia on mouse heart. Am J Physiol Regul Integr Comp Physiol 295(6):R1891–R1897

27. Calmettes G, Vr D-A, Gouspillou G et al (2010) Improved energy supply regulation in chronic hypoxic mouse counteracts hypoxia-induced altered cardiac energetics. PLoS One 5(2):e9306

28. Korzeniewski B, Noma A, Matsuoka S (2005) Regulation of oxidative phosphorylation in intact mammalian heart in vivo. Biophys Chem 116(2):145–157

29. Kavanagh NI, Ainscow EK, Brand MD (2000) Calcium regulation of oxidative phosphorylation in rat skeletal muscle mitochondria. Biochim Biophys Acta 1457(1–2):57–70

30. Johnston JD, Brand MD (1987) Stimulation of the respiration rate of rat liver mitochondria by sub-micromolar concentrations of extramitochondrial Ca2+. Biochem J 245(1):217–222

31. Mildaziene V, Baniene R, Nauciene Z et al (1996) Ca2+ stimulates both the respiratory and phosphorylation subsystems in rat heart mitochondria. Biochem J 320(Pt 1):329–334

32. Deschodt-Arsac V, Calmettes G, Gouspillou G et al (2013) Non-invasive integrative analysis of contraction energetics in intact beating heart. Int J Biochem Cell Biol 45(1):4–10

33. Wu ST, Kojima S, Parmley WW et al (1992) Relationship between cytosolic calcium and oxygen consumption in isolated rat hearts. Cell Calcium 13(4):235–247

34. Endoh M (2006) Signal transduction and Ca2+ signaling in intact myocardium. J Pharmacol Sci 100(5):525–537

35. Delgado JF (2006) Levosimendan in acute heart failure: past, present and future. Rev Esp Cardiol 59(4):309–312

36. Papp Z, Csapo K, Pollesello P et al (2005) Pharmacological mechanisms contributing to the clinical efficacy of levosimendan. Cardiovasc Drug Rev 23(1):71–98

37. Pollesello P, Ovaska M, Kaivola J et al (1994) Binding of a new Ca2+ sensitizer, levosimendan, to recombinant human cardiac troponin C. A molecular modelling, fluorescence probe, and proton nuclear magnetic resonance study. J Biol Chem 269(46):28584–28590

38. Sorsa T, Pollesello P, Solaro RJ (2004) The contractile apparatus as a target for drugs against heart failure: interaction of levosimendan, a calcium sensitiser, with cardiac troponin c. Mol Cell Biochem 266(1–2):87–107

39. Edes I, Kiss E, Kitada Y et al (1995) Effects of Levosimendan, a cardiotonic agent targeted to troponin C, on cardiac function and on phosphorylation and Ca2+ sensitivity of cardiac myofibrils and sarcoplasmic reticulum in guinea pig heart. Circ Res 77(1):107–113

40. Haikala H, Linden IB (1995) Mechanisms of action of calcium-sensitizing drugs. J Cardiovasc Pharmacol 26(Suppl 1):S10–S19

41. Szilagyi S, Pollesello P, Levijoki J et al (2005) Two inotropes with different mechanisms of action: contractile, PDE-inhibitory and direct myofibrillar effects of levosimendan and enoximone. J Cardiovasc Pharmacol 46(3):369–376

42. Deschodt-Arsac V, Calmettes G, Gouspillou G et al (2010) System analysis of the effect of various drugs on cardiac contraction energetics. Biochem Soc Trans 38(5):1319–1321

43. Lefrancois W, Miraux S, Calmettes G et al (2011) A fast black-blood sequence for four-dimensional cardiac manganese-enhanced MRI in mouse. NMR Biomed 24(3):291–298

44. Kaheinen P, Pollesello P, Levijoki J et al (2004) Effects of levosimendan and milrinone on oxygen consumption in isolated guinea-pig heart. J Cardiovasc Pharmacol 43(4):555–561

45. Nicholls DG, Ferguson SJ (2013) Bioenergetics, 4th edn. Academic, London

46. Brand MD (1997) Regulation analysis of energy metabolism. J Exp Biol 200(Pt 2):193–202

47. Ainscow EK, Brand MD (1999) Quantifying elasticity analysis: how external effectors cause changes to metabolic systems. Biosystems 49:151–159

Chapter 27

Computer-Based Prediction of Mitochondria-Targeting Peptides

Pier Luigi Martelli, Castrense Savojardo, Piero Fariselli, Gianluca Tasco, and Rita Casadio

Abstract

Computational methods are invaluable when protein sequences, directly derived from genomic data, need functional and structural annotation. Subcellular localization is a feature necessary for understanding the protein role and the compartment where the mature protein is active and very difficult to characterize experimentally. Mitochondrial proteins encoded on the cytosolic ribosomes carry specific patterns in the precursor sequence from where it is possible to recognize a peptide targeting the protein to its final destination. Here we discuss to which extent it is feasible to develop computational methods for detecting mitochondrial targeting peptides in the precursor sequences and benchmark our and other methods on the human mitochondrial proteins endowed with experimentally characterized targeting peptides. Furthermore, we illustrate our newly implemented web server and its usage on the whole human proteome in order to infer mitochondrial targeting peptides, their cleavage sites, and whether the targeting peptide regions contain or not arginine-rich recurrent motifs. By this, we add some other 2,800 human proteins to the 124 ones already experimentally annotated with a mitochondrial targeting peptide.

Key words Targeting peptide, Prediction of subcellular localization, Arginine motifs, Cleavage site, Machine learning

1 Introduction

1.1 Targeting Peptides

Recent estimates in mammals indicate that mitochondrial proteomes comprise from about 1,500 to 2,500 protein types, all involved in biological processes including to different extent the mitochondrion [1, 2]. Large-scale proteomics experiments, based mainly on mass spectrometry and green fluorescent protein (GFP) tagging, allow characterizing proteins clearly localized in the mitochondrial space, including outer and inner mitochondrial membranes [3]. Consequently, different databases gather the currently available data and therefrom annotate mammalian proteins based on direct assays or upon sequence similarity detection. Among them, MitoMiner (http://mitominer.mrc-mbu.cam.ac.uk) [4] and

Volkmar Weissig and Marvin Edeas (eds.), *Mitochondrial Medicine: Volume I, Probing Mitochondrial Function*,
Methods in Molecular Biology, vol. 1264, DOI 10.1007/978-1-4939-2257-4_27, © Springer Science+Business Media New York 2015

MitoCarta (http://www.broadinstitute.org/pubs/MitoCarta/) [3] currently list 2,241 and 1,013 human mitochondrial proteins, respectively. Targeting peptides have different compositional and structural features that may unravel the different types of translocating protein machineries in the different eukaryotic species. Apparently, the mitochondrial sequence precursors may carry also characteristic arginine-rich motifs at the N-terminal position [6].

In humans, the mitochondrial genome encodes only for 13 proteins that are part of the respiratory complexes labeled as I, III, IV, and V. The nuclear genome encodes for the vast majority of the remaining mitochondrial proteins that are synthesized on the cytosolic ribosomes and therefrom translocated to the different submitochondrial localizations (the outer and the inner membrane, the intermembrane space, and the matrix). Different protein machineries that recognize protein precursors harboring either N-terminal cleavable signals or internal non-cleavable signals control the import into the mitochondrion and the sorting toward submitochondrial compartments [5, 6]. The most characterized and widespread mechanism involves the detection of an N-terminal pre-sequence, called targeting peptide, which directs the protein across the translocase complexes located on the outer and the inner mitochondrial membranes and is then proteolytically cleaved when the protein reaches the final destination. In the genomic era, the detection of targeting peptides starting from the coding protein sequences is then an important step in order to characterize the function, the localization, and the sequence of the mature protein.

1.2 Available Computational Methods for Targeting Peptide Detection

The dataset of experimentally characterized targeting peptides is still very small, and its analysis reveals that they are very variable in length and in primary sequence. This hampers the adoption of simple alignment methods for the detection of targeting peptides in uncharacterized proteins. Furthermore, the cleavage site is quite heterogeneous, and although common motifs have been recognized [6], they are not sufficient to perform prediction based on pattern matching.

Different computational methods can help the annotation of targeting peptides in proteins when experimental data are not available. Most of the available methods implement statistical or machine-learning approaches, such as neural networks (NNs), support vector machines (SVMs), hidden Markov models (HMMs), and grammatical-restrained hidden conditional random fields (GRHCRFs) [7–10]. Machine learning methods can extract rules of association between the input features derived from the protein sequence and features to be predicted. After training on the few available proteins with experimentally characterized targeting peptides and cleavage sites, a machine learning-based method can infer the presence or not (with an associated likelihood) of the targeting peptide and the site of cleavage, taking as input the protein sequence.

The generalization capability of the different methods routinely depends on the different implementations, training procedures, training datasets, and input encodings. In order to evaluate the generalization power, the predictive performance of a method is firstly statistically validated on the same training dataset with a leave-one-out procedure and secondly on a blind dataset of known data. Based on the number of correct and incorrect evaluations, different scoring indexes can be computed in order to measure and compare the different tool reliabilities.

Table 1 lists the most recent available tools for the prediction of targeting peptides (all of them are publicly available as web servers). They differ in many implementation aspects, such as the method, the adopted training set, and the input encoding. In particular, all but MitoProt [11] and iPSORT [12] are based on machine learning approaches, and all but MitoProt have been trained on datasets containing both mitochondrial and plastidic targeting peptides. All methods take as input the protein residue sequence, as translated from its coding sequence and different physical and chemical features that are again derived directly from the sequence, without necessity of any

Table 1
Computational methods available for the prediction of mitochondrial targeting peptides

Method (references)	Type[a]	Site	Method	Input encoding[b]	Web server
MitoProt [11]	Mito	Yes	Discriminant function	47 physicochemical properties	http://ihg.gsf.de/ihg/mitoprot.html
iPSORT [12]	Mito Plast	No	Rule-based algorithm	434 propensity scales	http://ipsort.hgc.jp/
Predotar [7]	Mito Plast	No	NN	Residue sequence Hydrophobicity Charged residues	https://urgi.versailles.inra.fr/ predotar/predotar.html
PredSL [8]	Mito Plast	Yes	NN, HMM, scoring matrices	Residue sequence	http://hannibal.biol.uoa.gr/PredSL
TargetP [9]	Mito Plast	Yes	NN	Residue sequence Hydrophobicity Charged residues	http://www.cbs.dtu.dk/services/ TargetP/
TPpred [10]	Mito Plast	Yes	GRHCRF	Residue sequence Hydrophobicity Charged residues Hydrophobic moment	http://tppred.biocomp.unibo.it/

Mito mitochondrial, *Plast* plastidic
[a]The targeting peptide type predicted
[b]Features considered by the method

functional or structural information. All the predictors compute a probability for the presence of a targeting peptide by analyzing an N-terminal portion of the precursor protein, ranging from 40 to 160 residues depending on the method. Only MitoProt, PredSL [8], TargetP [9], and TPpred [10] are also able to predict the position of the cleavage site along the protein sequence.

An important issue to be considered in developing and using a predictor is that the proportion of proteins endowed with a targeting peptide with respect to the whole proteome is expected to be quite small ($\approx$2–4 %). In this situation, even a small rate of incorrect predictions in the negative set would lead to a large amount of proteins misclassified as carrying a targeting peptide.

We previously developed BaCelLo [13], a tool discriminating five different classes of subcellular localization (secretory way, cytoplasm, nucleus, mitochondrion, and chloroplast), and more recently TPpred, a method based on labeling procedures and specifically addressing the problem of inferring organelle-targeting peptides. Here we benchmark on the human proteome TPpred for its capability of correctly inferring mitochondrial targeting peptides and two widely adopted tools, TargetP and MitoProt. We also describe the protocol for the usage and the output interpretation of the newly developed TPpred server.

2 Materials

2.1 Benchmark Datasets

The UniProtKB (release Dec 2013) [14] contains 73,697 human sequences, excluding fragments. By searching for the keyword "transit" in the "feature" field, we extracted a dataset of experimentally characterized human mitochondrial targeting peptides. We also retained only mitochondrial targeting peptides with known cleavage site and excluded annotations labeled as "by similarity," "potential," "probable," or "not cleaved." With this procedure, the positive dataset (DB+) consists of 124 proteins (Table 2). The negative dataset (DB−) comprises 6,398 proteins with experimentally known non-mitochondrial localization, as derived

Table 2
The human protein dataset

Total	DB+[a]	DB−[a]	DB*[a]
73,697	124	6,398	67,175

The human protein dataset was downloaded from UniProtKB (release Dec 2013) as described in Subheading 2.1

[a]DB+: mitochondrial proteins with an experimentally detected targeting peptide, DB−: proteins with an experimentally known non-mitochondrial localization; DB*: the remaining human protein set

from the subcellular location section of the comment fields of UniProtKB and from the gene ontology (GO) annotations for cellular component. The remaining 67,175 proteins form the DB* dataset, containing: (a) mitochondrial proteins without experimentally annotated targeting peptide, (b) proteins annotated with a subcellular localization that are compatible with the mitochondria (e.g., "cytoplasm," "membrane," "ribosome"), (c) proteins annotated with a non-mitochondrial localization without experimental evidence, and (d) proteins lacking any annotation for the subcellular localization. DB* is also used to predict new potential targeting peptides with TPpred. The datasets are available at our website http://tppred.biocomp.unibo.it/about.

2.2 Length and Residue Composition of Targeting Peptides

The length of experimentally annotated targeting peptides in human proteins (DB+) ranges from 7 to 105 residues, with an average length of 37 residues and a standard deviation of 17 residues (Fig. 1).

The overall residue composition of proteins included in DB+ and DB− datasets shows only minor differences, lower than 2 %, proving that the whole composition is not sufficient to perform a prediction (Fig. 2). On the contrary, the composition of human targeting peptides is quite peculiar, since they are strongly enriched in alanine, leucine, and arginine and are poor in isoleucine, asparagine, aspartic acid, glutamic acid, and lysine. These data are in agreement with previous observations carried out on larger, non-organism-specific datasets [10].

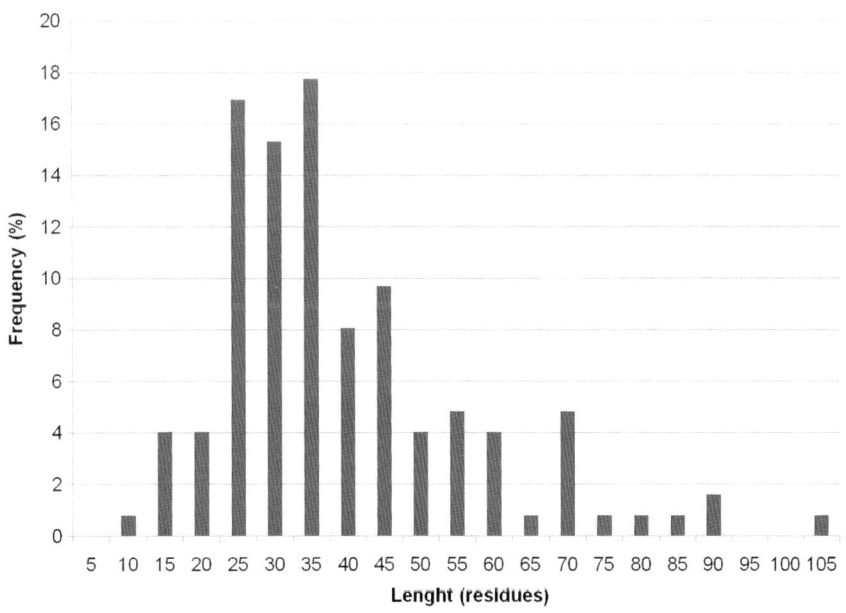

Fig. 1 Length distribution of targeting peptides in the DB+ dataset

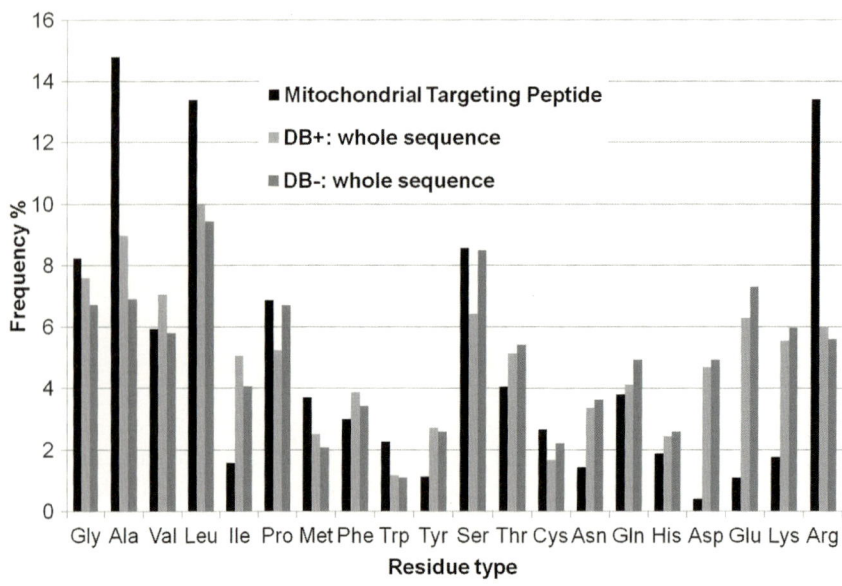

Fig. 2 Residue composition of benchmark datasets. Composition of targeting peptides is reported with *black bars. Light and dark gray bars* represent the composition of the whole sequences included in DB+ and DB− datasets, respectively

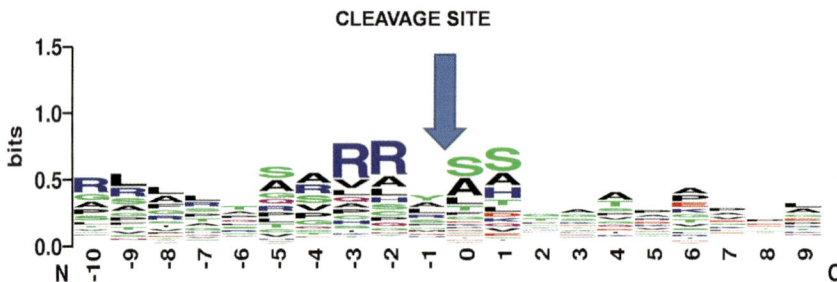

Fig. 3 Sequence logo of the sequence segment surrounding the cleavage site. Sequence logo is computed by the WebLogo server (weblogo.berkeley.edu) [15]. Position "1" is the first residue of the mature protein. Color codes cluster residues in apolar (*black*), polar (*green*), positively charged (*blue*), and negatively charged (*red*). *Height of each letter* indicates the information content of the corresponding residue in the input profile. Information is measured in bits and ranges between 0 and $\log_2(20) \approx 4.3$

2.3 Features of the Cleavage Site

Compositional features of the region surrounding the cleavage site are important for the interaction between the protein precursor and their peptidase complexes, and they can be visualized with a sequence logo obtained by piling up the 20-residue-long segments centered on each one of the cleavage sites of the DB+ dataset (Fig. 3). The logo represents the positional composition of the segments, with letters whose heights are proportional to the information content of each residue [15]. Values range from 0 (low information equivalent to uniform composition) to $\log_2(20) = 4.3$ (high information, corresponding to highly conserved residues). It is evident that the information in the surrounding regions of the

Table 3
Regular expression in cleavage sites of the positive dataset (DB+)

Motif name [6]	Motif expression	Mapped cleavage sites in DB+
R2	RX|X	40
R3a	RX[YFL]|[SA]	24
R3b	RX[YFL]|X	20
R10	RX[FLI]XX[TSG]XXXX|X	19
Rnone	X|XS	31

R arginine, *S* serine, *Y* tyrosine, *F* phenylalanine, *L* leucine, *I* isoleucine, *T* threonine, *G* glycine, *X* wild card (any of the 20 residues), | cleavage site

cleavage site is moderate, although positions −2 and −3 appear significantly richer in arginine (R), positions 0 and 1 are richer in serine (S) and alanine (A), and position 1 is also moderately richer in histidine (H).

These features reflect also the presence of previously recognized motifs around the cleavage site [6]: (a) R2 = RX|X, (b) R3 = RXX|[SAX], (c) R10 = RX[FLI]XX[TSG]XXXX|X, and (d) Rnone = X|X[SX]. In this representation, the character "|" indicates the cleavage position, the wild card character X represents any residue, and the residues enclosed in square brackets indicate multiple choices. Table 3 shows the number of cleavage sites in DB+ matching with each one of the motifs. When the multiple choice contains a wild card character, the patterns have been split, as in the case of R3a (RXX|[SA]) and R3b (RXX|X). The pattern X|XX was not considered. Since the patterns are not mutually exclusive, a single cleavage site can match with more than one of them (Table 4). On the overall, 88 out of 124 (71 %) sites match with a least one pattern, and the most represented are, as expected, the shortest ones: R2 and Rnone.

3 Methods

3.1 TPpred

TPpred [10] is based on grammatical-restrained hidden conditional random fields (GRHCRFs), a recently introduced machine learning paradigm that has been proven effective in addressing sequence annotation problems [16]. Indeed, similarly to hidden Markov models (HMMs), they allow casting into a model the grammatical restraints of the problem at hand, and at the same time, they share with neural networks (NNs) and support vector machines (SVMs) the ability to deal with complex input encodings, consisting of several features besides the residue sequence [16]. GRHCRF model states connected by transitions and the graphical

Table 4
Distribution of motifs around the cleavage site in the positive dataset of human proteins (DB+)

Number of motifs mapping on the cleavage site	Number of proteins
0	36
1	64
2	22
3	2
4 or more	0

DB+contains the human proteins with an experimentally detected targeting peptide

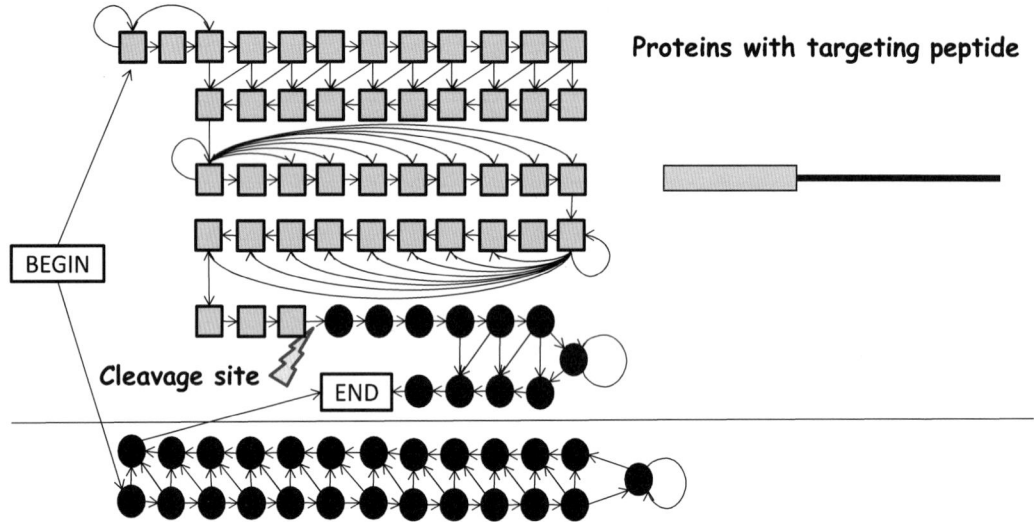

Fig. 4 Graphical model of TPpred. TPpred is based on grammatical-restrained hidden conditional random fields (GRHCRFs, 16). *Gray squares* represent states describing targeting peptides, *while black circles* represent nontargeting peptide states. *The upper sub-model* represents proteins endowed with targeting peptide, and the *lower sub-model* represents proteins devoid of signal peptide

model of TPpred are depicted in Fig. 4, where gray boxes represent targeting peptide positions and black circles represent other positions. The model consists of two major sub-models describing proteins endowed or not with a targeting peptide (the upper and the lower sub-models in Fig. 4, respectively). All states and transitions are associated to potential functions, related to the probability for each residue in the sequence generated by a particular state. Each potential function mixes the different components of the input encoding, namely the residue sequence, the local average hydrophobicity (computed on a 7-residue window centered on each residue), the local average positive and negative charge, and the local hydrophobic moments (computed considering the ideal alpha-helical and beta-strand conformations).

The parameterization of the potential functions was performed during the training phase, considering a positive dataset of 297 non-redundant proteins endowed with a mitochondrial or plastidic targeting peptide and a negative dataset containing 8,010 non-redundant proteins from animal, plants, and fungi [10].

During the prediction phase, parameters are fixed and the input sequence is optimally aligned to the model. A probability value for the sequence putatively endowed with a targeting peptide is computed by averaging the probability of the N-terminal 30 residues to be generated by states representing the targeting peptide. Moreover, if the optimal path associated to the input sequence is lying on the targeting peptide sub-model, it is possible to predict the cleavage site.

3.2 Benchmarking TPpred, TargetP, and MitoProt on the Human Protein Datasets

We evaluate the performance of TPpred, TargetP, and MitoProt on the DB+ and DB– datasets. All the considered methods are able to compute both the probability of carrying a targeting peptide and the position of the cleavage site.

For an overall evaluation of the discriminative power, the receiver-operating characteristic (ROC) curves of the three methods were computed (Fig. 5) by plotting the true-positive rate (TPR or recall) versus the false-positive rate (FPR), defined as:

$$TPR = TP / (TP + FN) \tag{1}$$

$$FPR = FP / (TN + FP) \tag{2}$$

where TP and TN are the numbers of correct (true) predictions in the positive and negative classes, respectively, while FP and FN count the incorrect (false) predictions in the same classes.

The area under the ROC curve (AUC) ranges between 0 and 1, and it measures how far the predictors are from a random predictor that would score with AUC equal to 0.5. All the methods perform well, scoring with AUC values higher than 0.95. TPpred reaches the highest AUC (0.98), and in particular, its curve is higher than that of MitoProt and TargetP in the region of small FPR, respectively, meaning that TPpred can correctly predict a large fraction of positive examples with a still moderate amount of incorrect predictions in the negative class.

Besides the probability estimation, the three methods also perform a binary classification (presence or absence of the targeting peptide) and predict the position of the cleavage site along the

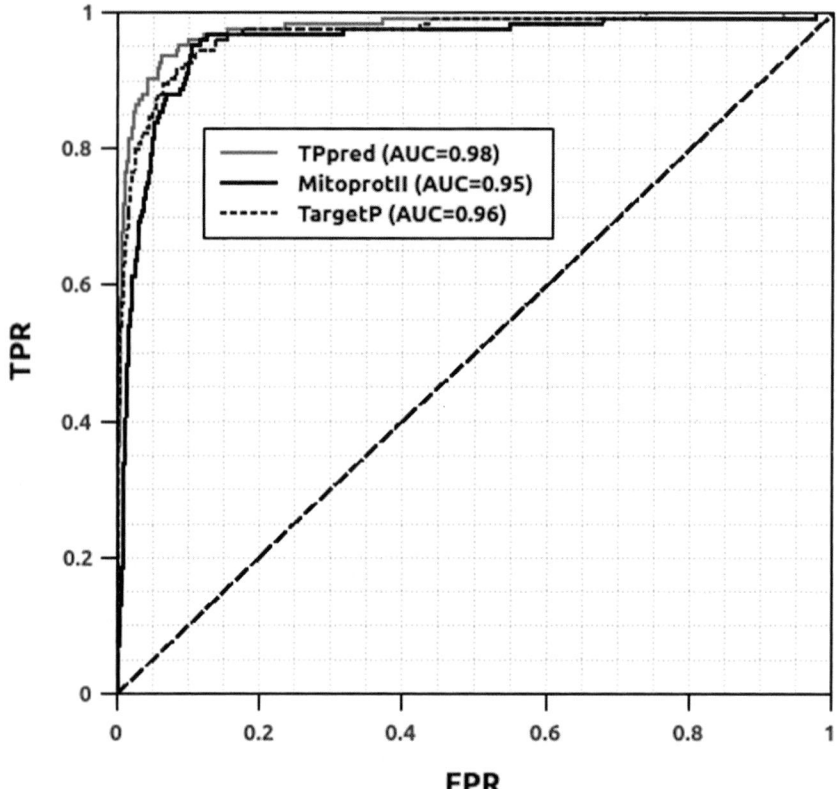

Fig. 5 ROC curves of TPpred, TargetP, and MitoProt. Receiver-operating characteristic (ROC) curves of the three different predictors. *FPR* false-positive rate, *TPR* true-positive rate. For details and definitions, *see* Eqs. 1 and 2 in the text (Subheading 3.2)

Table 5
Prediction scores of MitoProt, TargetP, and TPpred

	TPR	FPR	PPV	ACC	MCC	F1
MitoProt [11]	0.87	0.13	0.06	0.75	0.19	0.12
TargetP [9]	0.94	0.09	0.18	0.92	0.39	0.30
TPpred [10]	0.81	0.02	0.46	0.98	0.60	0.58

TPR true-positive rate, *FPR* false-positive rate, *PPV* positive predictive value, *ACC* accuracy, *MCC* Matthews correlation coefficient, *F1* F1 score. For definitions, *see* Subheading 3.2

sequence in positive cases. Table 5 lists the evaluation of the binary classification. Besides the TPR and FPR, we compute the following indexes:

$$\text{Positive predictive value}\,(\text{precision}) : \text{PPV} = \text{TP} / (\text{TP} + \text{FP}) \quad (3)$$

$$\text{Accuracy} : \text{ACC} = (\text{TP} + \text{TN}) / (\text{TP} + \text{FP} + \text{TN} + \text{FN}) \quad (4)$$

Matthews correlation coefficient:

$$MCC = (TP \times TN - FP \times FN) / \sqrt{((TP + FP) \times (TP + FN) \times (TN + FP) \times (TN + FN))} \quad (5)$$

$$F1\ score : F1 = 2 \times TPR \times PPV / (TPR + PPV) \quad (6)$$

TPpred outperforms the other two methods in terms of ACC, MCC, PPV, and FPR. Only TPR is lower than the values reached by the other two methods, indicating that a larger number of positive examples are not detected in the binary classification (24, in our DB+ dataset, to be compared with 16 and 7, relative to MitoProt and TargetP, respectively). On the other hand, the number of false-positives in the binary prediction of TPpred is 118 out of 6,398 proteins included in DB−, to be compared with 1,636 and 546 that are the values reached by MitoProt and TargetP, respectively.

When the cleavage site position along the sequence is predicted, the rate of perfect prediction ranges between 6.5 % of MitoProt, 21 % of TPpred, and 46 % of TargetP (Table 6). TPpred scores with the lowest average absolute error (8, 9, and 10 residues for TPpred, TargetP, and MitoProt, respectively).

Recurrent motifs in cleavage sites can be used to refine the prediction. Regular expressions for motifs have been scanned in a window of 20 residues centered on the cleavage position predicted by TPpred. The average number of motifs per segment is 4.6 corresponding to an average number of hypothetical cleavage sites equal to 3.8. In 55 % of the cases, at least one of the motifs identifies the correct cleavage site, and the average distance between the real cleavage site and the nearest predicted one is about 4.5 residues. However, no clear rule is presently available for choosing the most correct site among the proposed ones.

3.3 TPpred on the Human Genome

TPpred is adopted to annotate new potential targeting peptides in the human proteome, predicting the DB* dataset. Out of 67,175 sequences, 2,836 are predicted to carry a targeting peptide (Table 7). Given the evaluation of the method presented in the

Table 6
Prediction of the cleavage site

	MitoProt	TargetP	TPpred	TPpred + regular expressions
Average |Err|	9.9 residues	8.8 residues	8.0 residues	4.5 residues
|Err| =0	6.5 %	46 %	21 %	55 %
|Err| ≤5	49 %	62 %	57 %	76 %
|Err| ≤10	66 %	72 %	76 %	83 %

\|Err\| = absolute value in residues (real position − predicted position)

Table 7
TPpred at work: the newly annotated precursor proteins in the human proteome

Current annotation in UniProtKB without targeting peptide	Total	Protein evidence level				
		Protein	Transcript	Homology	Predicted	Uncertain
Mitochondrial	587	346	220	12	6	3
Compatible subcellular localization	192	43	141	3	4	1
Non-experimental non-mitochondrial localization	190	65	108	5	9	3
Non-annotated for subcellular localization	1,867	52	1,492	20	283	20
Total	2,836	506	1,961	40	302	27

See Subheading 3.3 for details

previous sections, we expect that this set contains a limited number of false-positive errors and that, on the other hand, part of the true targeting peptides are missed in the prediction. The large majority of the new predictions (1,867 out of 2,836) involve proteins that are lacking any annotation for the subcellular localization in UniProtKB fields (including GO). Another 779 proteins are annotated to be located either in mitochondria or in compatible localizations. Only 190 proteins are localized in other compartments, although with non-experimental annotations.

The list of the proteins predicted with a targeting peptide and the predicted cleavage sites is available at http://tppred.biocomp.unibo.it/about.

3.4 How to Predict the Targeting Peptide with TPpred

TPpred is available as a web server and it requires in input only the residue sequence of the proteins:

1. Write or download the sequence to predict in FASTA format (*see* **Note 1**).

2. Go to the website http://tppred.biocomp.unibo.it/tppred (Internet Explorer 6 and upper, Firefox, and Google Chrome were tested and support the prediction server). *See* Fig. 6.

3. Copy and paste the sequence in the corresponding field (only one sequence at the time) (Fig. 4).

4. Submit the request and wait for results (approximately 10 s per protein sequence).

3.5 How to Read the Results of TPpred (See Fig. 7)

1. The result page (Fig. 5) will not be stored.

2. The first section of the result page reports: (a) the name of the input sequence, (b) the length of the input sequence, and

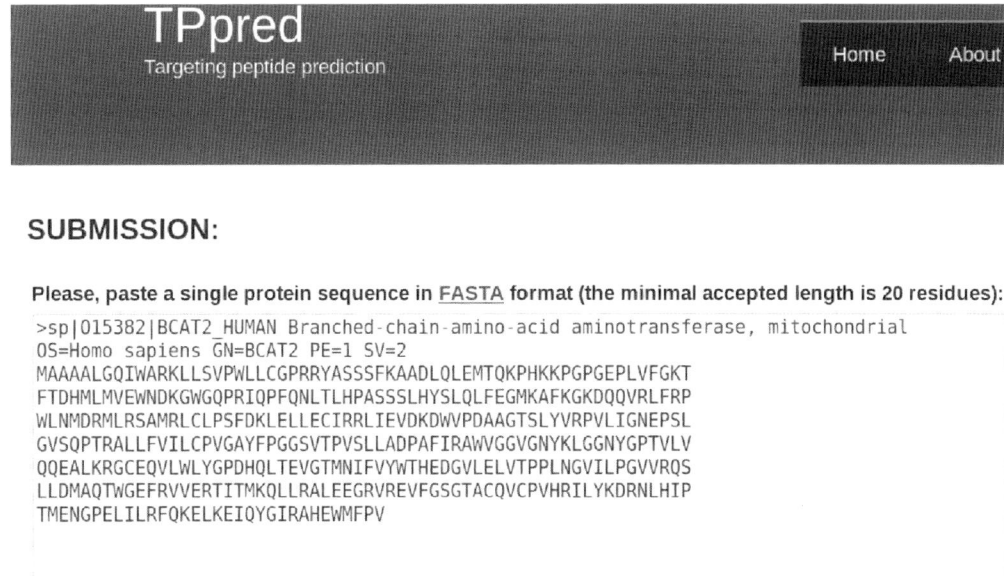

Fig. 6 Input page of TPpred. http://tppred.biocomp.unibo.it

(c) the probability that the input sequence contains an N-terminal mitochondrial targeting peptide, as computed by TPpred (*see* **Note 2**).

3. The second section reports the prediction of the cleavage site, when TPpred recognizes a reliable targeting peptide.

4. The third section reports the sites matching with the regular expression(s) describing the typical cleavage sites in a window of 20 residues centered on the predicted cleavage site. The matches with the motifs can help in better identifying the correct cleavage site, as discussed in Subheading 3.2. However, the procedure usually identifies different hypothetical sites, around the one predicted by TPpred, and the choice is left to the user.

In case TPpred does not predict a cleavage site, this section reports all the matches with the regular expressions occurring in the first 70 residues of the protein.

Results TPpred targeting peptide prediction:

Protein ID	Length	TP probability
sp|O15382|BCAT2_HUMAN	392	0.654911

TPpred **identified a cleavage site at position 27**

Legend	1																																													50

Legend:
- Targeting peptide
- Mature sequence

```
1                                                  50
M A A A A L G Q I W A R K L L S V P W L L C G P R R Y A S S S F K A A D L Q L E M T Q K P H K K P G
P G E P L V F G K T F T D H M L M V E W N D K G W G Q P R I Q P F Q N L T L H P A S S S L H Y S L Q
L F E G M K A F K G K D Q Q V R L F R P W L N M D R M L R S A M R L C L P S F D K L E L L E C I R R
L I E V D K D W V P D A A G T S L Y V R P V L I G N E P S L G V S Q P T R A L L F V I L C P V G A Y
F P G G S V T P V S L L A D P A F I R A W V G G V G N Y K L G G N Y G P T V L V Q Q E A L K R G C E
Q V L W L Y G P D H Q L T E V G T M N I F V Y W T H E D G V L E L V T P P L N G V I L P G V V R Q S
L L D M A Q T W G E F R V V E R T I T M K Q L L R A L E E G R V R E V F G S G T A C Q V C P V H R I
L Y K D R N L H I P T M E N G P E L I L R F Q K E L K E I Q Y G I R A H E W M F P V
```

Results of regular expression pattern matching:

R2-motifs found in the 20-residue long environment of the predicted cleavage site

R2-motif: RX|X	Motif match	Motif-derived cleavage site
	VPWLLCGP **RR	Y** ASSSFKAADL	26
	VPWLLCGPR **RY	A** SSSFKAADL	27

R3a-motifs found in the 20-residue long environment of the predicted cleavage site

R3a-motif: RX[YFL]|[SA]	Motif match	Motif-derived cleavage site
	VPWLLCGP **RRY	A** SSSFKAADL	27

R3b-motifs found in the 20-residue long environment of the predicted cleavage site

R3b-motif: RX[YFL]|X	Motif match	Motif-derived cleavage site
	VPWLLCGP **RRY	A** SSSFKAADL	27

No R10-motif(RX[FLI]XX[TSG]XXXX|X) found

RNone-motifs found in the 20-residue long environment of the predicted cleavage site

RNone-motif: X|XS	Motif match	Motif-derived cleavage site
	VPWLLCGPRR **Y	AS** SSFKAADL	27
	VPWLLCGPRRY **A	SS** SFKAADL	28
	VPWLLCGPRRYA **S	SS** FKAADL	29

Fig. 7 Typical TPpred output. The output page reports the name of the input sequence, its length, and the probability that it contains a mitochondrial targeting peptide, as computed by TPpred. The second section reports the TPpred prediction of the cleavage site, and the third section lists the matches with the regular expressions describing the typical cleavage sites in a window of 20 residues, centered on the predicted cleavage site

4   Notes

1. TPpred prediction is only based on information derived from protein sequence, and it is therefore important to consider the right sequence. In particular, the problem poses when different splicing variants are known. In DB*, 5,832 entries report 11,776 different splicing variants. For 321 entries, TPpred predicts a targeting peptide at least in one of the variants.

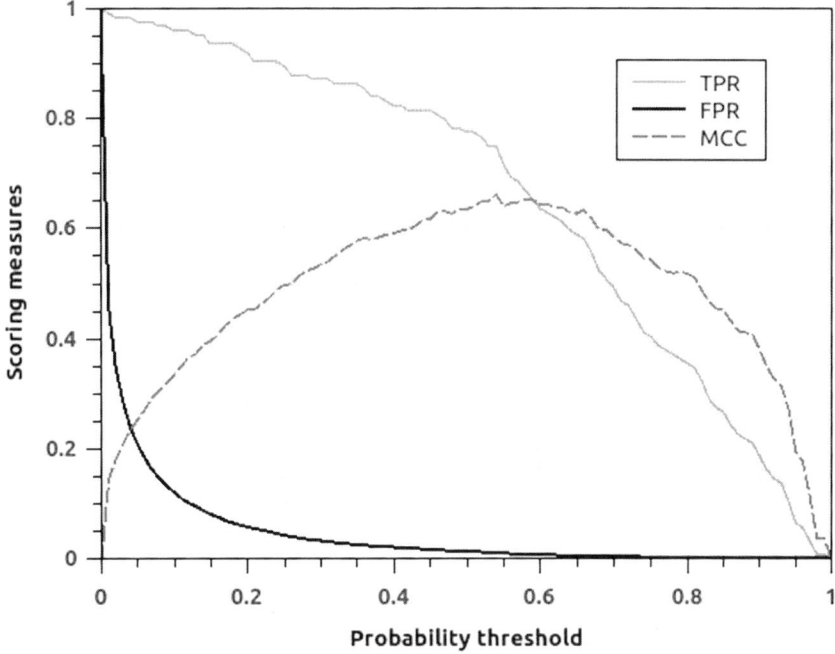

Fig. 8 Calibration curves of TPpred. The most relevant scoring indexes (MCC, FPR, and TPR) are plotted as a function of the probability computed by TPpred of having a mitochondrial targeting peptide in the precursor sequence. *MCC* Matthews correlation coefficient, *FPR* false-positive rate, *TPR* true-positive rate. For details and definitions, *see* Eqs. 1 and 2 in the text (Subheading 3.2)

However, for 47 % of these entries (153), TPpred predicts also variants without targeting peptides.

2. Given the predicted probability value, the user can evaluate the reliability of the prediction using the calibrations reported in Fig. 8, where TPR, FPR, and MCC are plotted as a function of the predicted probability value.

Acknowledgments

This work was supported by the following projects: PRIN 2010–2011 project 20108XYHJS (to P.L.M.) (Italian Ministry for University and Research: MIUR), COST BMBS Action TD1101 (European Union RTD Framework Program, to R.C.), PON projects PON01_02249 and PAN Lab PONa3_00166 (Italian Ministry for University and Research, to R.C. and P.L.M.), and FARB-UNIBO 2012 (to R.C.).

References

1. Goffart S, Martinsson P, Malka F, Rojo M, Spelbrink JN (2007) The mitochondria of cultured mammalian cells: II. Expression and visualization of exogenous proteins in fixed and live cells. Methods Mol Biol 372: 17–32

2. Meisinger C, Sickmann A, Pfanner N (2008) The mitochondrial proteome: from inventory to function. Cell 134:22–24

3. Pagliarini DJ, Calvo SE, Chang B, Sheth SA, Vafai SB, Ong SE, Walford GA, Sugiana C, Boneh A, Chen WK et al (2008) A mitochondrial protein compendium elucidates complex I disease biology. Cell 134:112–123

4. Smith AC, Blackshaw JA, Robinson AJ (2012) MitoMiner: a data warehouse for mitochondrial proteomics data. Nucleic Acids Res 40: D1160–D1167

5. Schmidt O, Pfanner N, Meisinger C (2010) Mitochondrial protein import: from proteomics to functional mechanisms. Nat Rev Mol Cell Biol 11:655–667

6. Mossmann D, Meisinger C, Vögtle FN (2012) Processing of mitochondrial presequences. Biochim Biophys Acta 1819:1098–1106

7. Small I, Peeters N, Legeai F, Lurin C (2004) Predotar: a tool for rapidly screening proteomes for N-terminal targeting sequences. Proteomics 4:1581–1590

8. Petsalaki EI, Bagos PG, Litou ZI, Hamodrakas SJ (2006) PredSL: a tool for the N-terminal sequence-based prediction of protein subcellular localisation. Genomics Proteomics Bioinformatics 4:48–55

9. Emanuelsson O, Brunak S, von Heijne G, Nielsen H (2007) Locating proteins in the cell using TargetP, SignalP and related tools. Nat Protoc 2:953–971

10. Indio V, Martelli PL, Savojardo C, Fariselli P, Casadio R (2013) The prediction of organelle-targeting peptides in eukaryotic proteins with Grammatical-Restrained Hidden Conditional Random Fields. Bioinformatics 29:981–988

11. Claros MG, Vincens P (1996) Computational method to predict mitochondrially imported proteins and their targeting sequences. Eur J Biochem 241:779–786

12. Bannai H, Tamada Y, Maruyama O, Nakai K, Miyano S (2002) Extensive feature detection of N-terminal protein sorting signals. Bioinformatics 18:298–305

13. Pierleoni A, Martelli PL, Fariselli P, Casadio R (2006) BaCelLo: a balanced subcellular localisation predictor. Bioinformatics 22:e408–e416

14. UniProt Consortium (2014) Activities at the Universal Protein Resource (UniProt). Nucleic Acids Res 42:D191–D198

15. Crooks GE, Hon G, Chandonia JM, Brenner SE (2004) WebLogo: a sequence logo generator. Genome Res 14:1188–1190

16. Fariselli P, Savojardo C, Martelli PL, Casadio R (2009) Grammatical-Restrained Hidden Conditional Random Fields for Bioinformatics applications. Algorithms Mol Biol 4:13

Chapter 28

Prediction of Mitochondrial Protein Function by Comparative Physiology and Phylogenetic Profiling

Yiming Cheng and Fabiana Perocchi

Abstract

According to the endosymbiotic theory, mitochondria originate from a free-living alpha-proteobacteria that established an intracellular symbiosis with the ancestor of present-day eukaryotic cells. During the bacterium-to-organelle transformation, the proto-mitochondrial proteome has undergone a massive turnover, whereby less than 20 % of modern mitochondrial proteomes can be traced back to the bacterial ancestor. Moreover, mitochondrial proteomes from several eukaryotic organisms, for example, yeast and human, show a rather modest overlap, reflecting differences in mitochondrial physiology. Those differences may result from the combination of differential gain and loss of genes and retargeting processes among lineages. Therefore, an evolutionary signature, also called "phylogenetic profile", could be generated for every mitochondrial protein. Here, we present two evolutionary biology approaches to study mitochondrial physiology: the first strategy, which we refer to as "comparative physiology," allows the de novo identification of mitochondrial proteins involved in a physiological function; the second, known as "phylogenetic profiling," allows to predict protein functions and functional interactions by comparing phylogenetic profiles of uncharacterized and known components.

Key words Mitochondrial evolution, Comparative genomics, Comparative physiology, Phylogenetic profiling, Orthology

1 Introduction

Given the central role of mitochondria in basic cell biology and human diseases, an urgent next goal is to systematically annotate the function of all mitochondrial proteins. Although tremendous progress has been made toward defining the complete repertoire of proteins targeted to the mitochondria of several organisms, the functional characterization of several hundred mitochondrial proteins remains a challenge [1, 2].

A recent development in high-throughput experimental techniques allows a survey of thousands of genes or proteins simultaneously. However, the experimental characterization of protein function through top-down approaches often requires a dedicated infrastructure such as screening core facilities, therefore remaining

Volkmar Weissig and Marvin Edeas (eds.), *Mitochondrial Medicine: Volume I, Probing Mitochondrial Function*, Methods in Molecular Biology, vol. 1264, DOI 10.1007/978-1-4939-2257-4_28, © Springer Science+Business Media New York 2015

a time-consuming and expensive task that many laboratories cannot afford. Alternatively, several bioinformatics tools and computational approaches have been developed to select among hundreds of mitochondrial components the ones involved in a specific biological process or pathway [1, 3]. A handful of top-scoring candidate proteins can then be tested by targeted genetic screens and biochemical assays [4].

A widely adopted strategy is to infer biological function through the identification of homologous proteins of known function [5]. For example, several mitochondrial proteins in humans have been identified because their yeast homologs with known molecular function encode a product that is targeted to mitochondria [6]. This kind of information transfer is possible for all the biological processes that are conserved between yeast and human [1, 7]. Remarkably, systematic all-against-all comparison of proteins from human and yeast mitochondria has proved instrumental to the prediction of several hundred new components [1, 2, 8]. However, for more than 40 % of the human mitochondrial proteome, an ortholog in yeast cannot be found [1, 9], which limits the use of orthology-based prediction methods.

A recent progress in sequencing and annotating whole genomes of hundreds of species from different taxonomic groups has catalyzed the development of novel computational strategies for the prediction of protein function and protein-protein interaction [10]. Given the complex evolutionary history of modern mitochondrial proteomes, a powerful strategy consists in comparing the pattern of presence/absence of uncharacterized and known mitochondrial proteins in a set of complete proteomes [11]. This phylogenetic profiling approach aims at predicting protein function on the assumption that proteins involved in the same pathway or interacting within the same complex will show similar evolutionary histories. Successful case stories of phylogenetic profiling applied to mitochondrial biology and diseases include the functional annotation of the frataxin gene [12] and the discovery of a novel assembly factor of the NADH-ubiquinone oxidoreductase complex, involved in complex I deficiency [13]. In the first case, the function of an uncharacterized protein is inferred based on the biological process in which phylogenetic neighboring proteins play a role. In the second case, new players in a pathway or complex of interest are identified by looking for phylogenetic neighbors of known components. However, in both cases a prior knowledge of a protein or its function is required.

Recently, we have developed a "comparative physiology" strategy, which allows phenotype-to-genotype predictions in the absence of any previously known components of the biological process under study [4] (Fig. 1). We successfully applied this approach to discover long-sought proteins of the mitochondrial

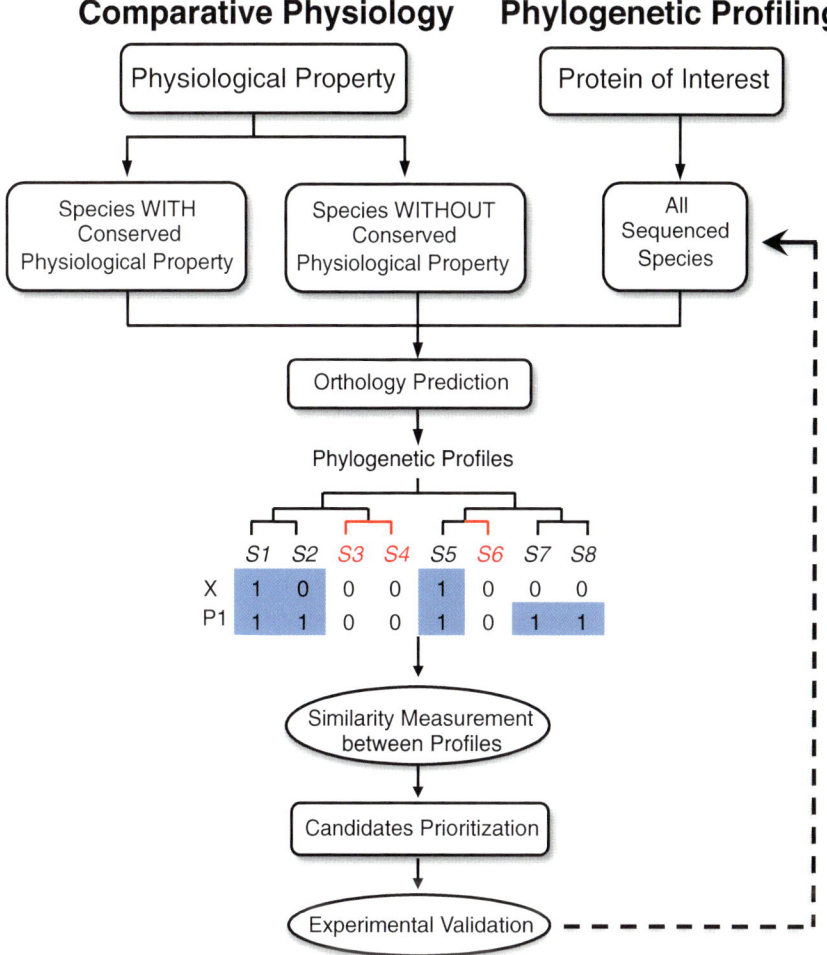

Fig. 1 Workflow for the comparative physiology and phylogenetic profiling methods

calcium uniporter [4, 14] (Fig. 2). This strategy requires the identification of species that show differences in the physiology of the mitochondrial process under study. Such organisms, without the conserved physiological property, represent "negative predictors," whereby, if found, orthologous proteins should not be functionally associated to the physiological property. Together, this defines what we call a "physiological signature" across taxa, which is likely to match to a similar phylogenetic signature (Figs. 1 and 2). Phylogenetic neighbors of the physiological signature may represent candidate proteins involved in the observed physiological property or function.

Here, we describe comparative physiology and phylogenetic profiling methods to address respectively: (1) what are the proteins

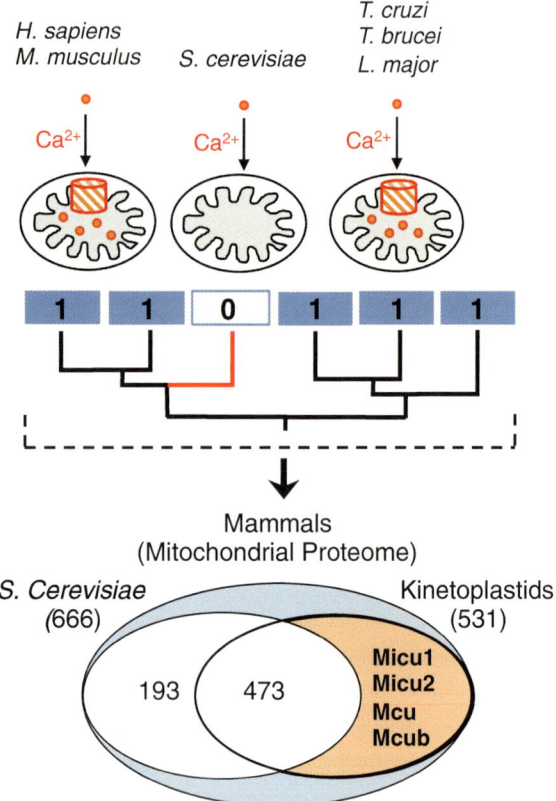

Fig. 2 Comparative physiology strategy applied by Perocchi et al. [4] to identify the founding components of the mitochondrial calcium uniporter. First, a "physiological calcium signature" of mitochondria was defined as being high capacity, membrane potential dependent and ruthenium red sensitive. This signature is common to all mammalian tissues and to unicellular protozoa such as kinetoplastids, yet not measurable in the yeast *Saccharomyces cerevisiae*. Mammalian genes encoding for the mitochondrial calcium uniporter should therefore exhibit a phylogenetic signature that matches the physiological profile across taxa, namely, present in mouse and in kinetoplastids, but absent in yeast. Of a proteomic inventory of 1,098 mouse mitochondrial proteins, 58 proteins fulfilled the above criteria. The 58 candidate proteins include the uniporter gatekeeper MICU1 [4], the pore-forming subunit MCU [14], and their respective paralogs MICU2 [23] and MCUB [24]

responsible for a given biological process or an observed phenotype? (2) what is the function of an uncharacterized protein or what are the additional components of a pathway or a protein complex? Both approaches rely on the availability of genome sequences from evolutionary diverse taxa and on sensitive methods to systematically define orthologs.

2   Materials

Download the following databases and bioinformatics tools.

2.1   The Universal Protein Resource Knowledgebase (UniProtKB) Database

1. Use the link below to download the complete dataset of fully annotated curated protein entries from UniprotKB/Swiss-Prot Protein Knowledgebase database:
   ftp://ftp.uniprot.org/pub/databases/uniprot/current_release/knowledgebase/complete/uniprot_sprot.fasta.gz.

2. Use the link below to download the complete dataset of computer-generated protein entries awaiting full manual annotation from UniprotKB/TrEMBL Protein Database:
   ftp://ftp.uniprot.org/pub/databases/uniprot/current_release/knowledgebase/complete/uniprot_trembl.fasta.gz.

2.2   The Basic Local Alignment Search Tool (BLAST)

Use the link below to download the appropriate BLAST toolbox based on the operating system: ftp://ftp.ncbi.nlm.nih.gov/blast/executables/blast+.

2.3   Database Identifier Mapping

Use the link below to download the identifier mapping data files to map the protein entries in UniProtKB to other databases (e.g., EntrezGene, RefSeq, GI): ftp://ftp.uniprot.org/pub/databases/uniprot/current_release/knowledgebase/idmapping/idmapping.dat.gz.

3   Methods

The following procedures apply to both comparative physiology and phylogenetic profiling methods (Fig. 1). Basic skills in computer programing are required.

3.1   Compiling a List of Informative Species

1. When applying the phylogenetic profiling method, it is recommended to use a comprehensive set of species to cover numerous taxonomic groups. The number of species included in the following analyses will ultimately depend on the computational resources. It is recommended to use the primary proteome sets provided at EBI Reference Proteomes (http://www.ebi.ac.uk/reference_proteomes) (*see* **Note 1**).

2. When applying the comparative physiology approach, choose a set of species based on both the presence and absence of the physiological property under study. This choice should be solely based on conclusive experimental evidence. For example, in the absence of any previously known components of the mitochondrial calcium uniporter, was defined a "physiological calcium signature" of mitochondria based on previous

observations that mitochondrial calcium uniporter activity is evolutionarily conserved in mouse and kinetoplastids, yet not measurable in *Saccharomyces cerevisiae* (Fig. 2) [4].

3.2 Generating Phylogenetic Profiles

1. Based on the selected species, retrieve all protein sequences from UniprotKB.

2. Filter out protein sequences of less than 10 amino acids. Short peptide sequences may confound the search for orthologs as they usually represent partially sequenced proteins.

3. Map protein entries in UniProtKB to EntrezGene. When multiple protein entries are found for the same EntrezGene identifier, choose the longest sequence isoform. This step can substantially reduce computation time (*see* **Note 2**).

4. Use the BLASTP program to perform all-against-all sequence similarity searches. It is recommended to set 1e-5 as *e*-value cutoff and 6 as output format option (-outfmt), which gives a tab-separated output. Also, specify the number of processors used for the parallel computing (use –dbsize option). For all the other options, use default setting (*see* **Note 3**).

5. Use the reciprocal best hit (RBH) to define orthologs. First, for each protein from one species, define the best hit as the orthologous protein with the lowest *e*-value in another species. Two proteins from two different species are defined as an ortholog pair if they find each other as the best hit (*see* **Note 4**). The output file consists of an ortholog pair database, i.e., with each line being a pair of proteins from two different species (*see* **Note 5**).

6. Generate a phylogenetic profile for each protein of a species of interest. The phylogenetic profile of a protein is defined as a binary numerical vector with the size equal to the number of species considered, while each element represents the presence (1) or absence (0) of its ortholog pair within that species (Fig. 1).

7. When applying the comparative physiology approach, generate a profile for the physiological property similarly to the phylogenetic profile, except that each element represents the presence (1) or absence (0) of the conserved physiological property within that species.

3.3 Search for Phylogenetic Neighbors

1. Several methods are available to measure the distance between two phylogenetic profiles. Here, we use the Hamming distance that is calculated as the number of species for which the corresponding elements are different.

2. When applying the phylogenetic profiling method, the Hamming distance is defined between the phylogenetic profiles of the protein of interest and any other protein of the same species.

3. When applying the comparative physiology approach, the Hamming distance is defined between the physiological signature and the phylogenetic profile of any protein of the species of interest.

4. Sort the protein list based on the Hamming distance. Proteins with the lowest Hamming distance represent close phylogenetic neighbors of the protein or the physiological property of interest.

5. Prioritize proteins with the lowest Hamming distance based on further criteria (Table 1), for example, mitochondrial localization [2, 15, 16], tissue-specific expression [2, 16], transmembrane domain, protein domains [18], protein-protein interactions [19], and other relevant proteomics and gene expression dataset.

Table 1
Databases for prioritization of candidate proteins

Database	Organisms	Description	Website
MitoP2 [15]	Human, mouse, *A. thaliana*, *S. cerevisiae*, *N. crassa*	Predicted mitochondrial protein localization, orthology, mutant phenotype, gene expression, and protein-protein interaction	http://ihg.gsf.de/mitop2
MitoMiner [16]	Human, rat, mouse, cow, fruit fly, yeast, *A. thaliana*, *T. thermophila*, *G. lamblia*, *P. falciparum*	Mitochondrial proteomics data and GFP protein localization	http://mitominer.mrc-mbu.cam.ac.uk
MitoCarta [2]	Mouse	Mitochondrial protein distribution across 14 tissues and predicted protein localization	www.broadinstitute.org/pubs/MitoCarta
Mito-Phenome [17]	Human	Manually annotated clinical disease phenotypes for genes involved in mitochondrial biogenesis and function	http://mitophenome.org
HMPDb	Human	Mitochondrial and human nuclear-encoded proteins involved in mitochondrial biogenesis and function	http://bioinfo.nist.gov
String [19]	>1,100 organisms	Known and predicted functional associations	http://string-db.org

4   Notes

1. When compiling a list of species, pay attention to the scientific name of each species. If a species cannot be found in the database, it is possible that different scientific names have been used. A taxonomy database could then be used as a reference to find a synonym of the species. This can be downloaded from the following link: ftp://ftp.ncib.nih.gov/pub/taxonomy/. In addition, some species include numerous strains; it is recommended to use the strain with the largest number of protein sequences available in the database.

2. For some species, most of protein entries in UniProtKB cannot be mapped to an EntrezGene identifier, thus all protein sequences should be used. When multiple protein sequences map to the same EntrezGene identifier, retrieve the longest protein isoform from Swiss-Prot otherwise from TrEMBL.

3. When performing all-vs.-all BLASTP, it is recommended to specify the database size. When additional species have to be included in the analysis, perform one-vs.-all BLASTP with the same database size in order to maintain the e-value comparable with the original all-vs.-all BLASTP. An example of BLASTP command line would be blastp –query input.fasta –db db_name –evalue 1e-5 –outfmt 6 –num_threads num_processors –dbsize 1,000,000 > blast_output. It is noted that to run the all-vs.-all BLASTP on a 32-core computer cluster (Intel Xeon 2.2GHz) takes about 1 day for 11 species consisting of ~60,000 sequences.

4. In a few cases, you may find alignments that have the same e-values. The highest high-scoring segment pair (HSP) could then be used to select the best hit.

5. Several databases of orthologous groups are also available and could be directly used to define ortholog pairs, for example, Clusters of Orthologous Groups (COG) [20], InParanoid [21], and PhyloFacts orthology (PHOG) [22].

Acknowledgment

This work was supported by Deutsche Forschungsgemeinschaft Emmy Noether Programme Grant PE 2053/1-1 and the Bavarian State Ministry of Education, Science and the Arts.

References

1. Perocchi F, Jensen LJ, Gagneur J et al (2006) Assessing systems properties of yeast mitochondria through an interaction map of the organelle. PLoS Genet 2:e170

2. Pagliarini DJ, Calvo SE, Chang BA et al (2008) A mitochondrial protein compendium elucidates complex I disease biology. Cell 134:112–123

3. Gabaldón T (2006) Computational approaches for the prediction of protein function in the mitochondrion. Am J Physiol Cell Physiol 291:C1121–C1128

4. Perocchi F, Gohil VM, Girgis HS (2010) MICU1 encodes a mitochondrial EF hand protein required for Ca(2+) uptake. Nature 467:291–296

5. Rost B, Liu J, Nair R et al (2003) Automatic prediction of protein function. Cell Mol Life Sci 60:2637–2650

6. Barrientos A (2003) Yeast models of human mitochondrial diseases. IUBMB Life 55:83–95

7. Perocchi F, Mancera E, Steinmetz LM (2008) Systematic screens for human disease genes, from yeast to human and back. Mol Biosyst 4:18–29

8. Prokisch H, Scharfe C, Camp DG (2004) Integrative analysis of the mitochondrial proteome in yeast. PLoS Biol 2:e160

9. Gabaldón T, Huynen MA (2007) From endosymbiont to host-controlled organelle: the hijacking of mitochondrial protein synthesis and metabolism. PLoS Comput Biol 3:e219

10. Kihara D (2011) Protein function prediction for omics era. Springer, New York

11. Gabaldón T, Huynen MA (2005) Lineage-specific gene loss following mitochondrial endosymbiosis and its potential for function prediction in eukaryotes. Bioinformatics 21:144–150

12. Huynen MA, Snel B, Bork P et al (2001) The phylogenetic distribution of frataxin indicates a role in iron-sulfur cluster protein assembly. Hum Mol Genet 10:2463–2468

13. Gabaldón T, Rainey D, Huynen MA (2005) Tracing the evolution of a large protein complex in the eukaryotes, NADH:ubiquinone oxidoreductase (Complex I). J Mol Biol 348:857–870

14. Baughman JM, Perocchi F, Girgis HS (2011) Integrative genomics identifies MCU as an essential component of the mitochondrial calcium uniporter. Nature 476:341–345

15. Prokisch H, Andreoli C, Ahting U et al (2006) MitoP2: the mitochondrial proteome database—now including mouse data. Nucleic Acids Res 34:D705–D711

16. Smith AC, Blackshaw JA, Robinson AJ (2012) MitoMiner: a data warehouse for mitochondrial proteomics data. Nucleic Acids Res 40:D1160–D1167

17. Scharfe C, Lu HH, Neuenburg JK et al (2009) Mapping gene associations in human mitochondria using clinical disease phenotypes. PLoS Comput Biol 5:e1000374

18. Sigrist CJ, de Castro E, Cerutti L et al (2013) New and continuing developments at PROSITE. Nucleic Acids Res 41:D344–D347

19. Franceschini A, Szklarczyk D, Frankild S et al (2013) STRING v9.1: protein-protein interaction networks, with increased coverage and integration. Nucleic Acids Res 41:D808–D815

20. Tatusov RL, Koonin EV, Lipman DJ (1997) A genomic perspective on protein families. Science 278:631–7

21. Ostlund G, Schmitt T, Forslund K et al (2009) InParanoid 7: new algorithms and tools for eukaryotic orthology analysis. Nucleic Acids Res 38:D196–D203

22. Datta RS, Meacham C, Samad B et al (2009) Berkeley PHOG: PhyloFacts orthology group prediction web server. Nucleic Acids Res 37:W84–W89

23. Plovanich M, Bogorad RL, Sancak Y et al (2013) MICU2, a paralog of MICU1, resides within the mitochondrial uniporter complex to regulate calcium handling. PLoS One 8:e55785

24. Raffaello A, De Stefani D, Sabbadin D et al (2013) The mitochondrial calcium uniporter is a multimer that can include a dominant-negative pore-forming subunit. EMBO J 32:2362–2376

Chapter 29

Assessment of Posttranslational Modification of Mitochondrial Proteins

Sudharsana R. Ande, G. Pauline Padilla-Meier, and Suresh Mishra

Abstract

Mitochondria play vital roles in the maintenance of cellular homeostasis. They are a storehouse of cellular energy and antioxidative enzymes. Because of its immense role and function in the development of an organism, this organelle is required for the survival. Defects in mitochondrial proteins lead to complex mitochondrial disorders and heterogeneous diseases such as cancer, type 2 diabetes, and cardiovascular and neurodegenerative diseases. It is widely known in the literature that some of the mitochondrial proteins are regulated by posttranslational modifications. Hence, designing methods to assess these modifications in mitochondria will be an important way to study the regulatory roles of mitochondrial proteins in greater detail. In this chapter, we outlined procedures to isolate mitochondria from cells and separate the mitochondrial proteins by two-dimensional gel electrophoresis and identify the different posttranslational modifications in them by using antibodies specific to each posttranslational modification.

Key words Mitochondrial proteins, Isolation of mitochondria, Posttranslational modification, 2D gel electrophoresis, PROTEAN IEF cell, Immunoblotting, Phosphorylation, O-GlcNAc modification, Nitrocellulose membrane

1 Introduction

Mitochondria are the powerhouse of the cell. They are characterized by two membranes called outer and inner membrane. The invaginations of the inner membrane are called as cristae. The membrane system divides the organelle into compartments consisting of matrix, the intercristal space, and the intermembrane space between inner and outer membrane [1]. Each of these compartments has distinct set of enzymes and performs specific functions. Mitochondria produce energy for proper maintenance of the cellular activities [1]. For example, synthesis of ATP (adenosine triphosphate) which is a currency for cellular energy occurs via the process of oxidative phosphorylation through electron transport chain (ETC) [2]. ETC is present at the inner mitochondrial membrane and

Volkmar Weissig and Marvin Edeas (eds.), *Mitochondrial Medicine: Volume I, Probing Mitochondrial Function*, Methods in Molecular Biology, vol. 1264, DOI 10.1007/978-1-4939-2257-4_29, © Springer Science+Business Media New York 2015

consists of protein complexes I to V. The proton gradient that is generated across these complexes creates mitochondrial membrane potential [2]. This membrane potential allows ATP synthase to use the flow of proton gradient to generate ATP. So mitochondria have an efficient system to generate energy in the form of ATP by utilizing the metabolic intermediates. However, defects in proteins or enzymes associated with mitochondria lead to mitochondrial dysfunction. Mitochondrial dysfunction is implicated in various diseases such as cancer, type 2 diabetes, diabetes-associated complications, and various other neurodegenerative diseases. However, the underlying mechanisms between mitochondria and disease state are not well understood. In recent years, it is becoming increasingly evident that mitochondrial proteins undergo various posttranslational modifications such as phosphorylation, acetylation, ubiquitination, and O-GlcNAc modification. Phosphorylation is the most common posttranslational modification present in mitochondrial proteins. Furthermore, majority of the mitochondrial proteins are phosphorylated at multiple sites. For example, one of the well-known mitochondrial outer membrane proteins, voltage-dependent anion channel (VDAC), was reported to be phosphorylated at Ser12 and Ser103 residues [3]. Phosphorylation at these residues is responsible for endostatin-induced apoptotic cell death in endothelial cells. However, mutation of VDAC protein at Ser12 and Ser103 residues prevents the cells from undergoing apoptosis mediated by endostatin [3]. These results provide evidence for the regulatory role of phosphorylation in VDAC. Apart from that, there are many other mitochondrial proteins that are known to be phosphorylated including Bcl-2, BAD, Bax, and TOM70 [4]. Several other mitochondrial enzymes that are involved in fatty acid transport and metabolism such as long-chain acyl-CoA synthase (ACS), carnitine palmitoyltransferase (CPT1), and glycerol-3-phosphate acyl transferase (GPAT1) undergo phosphorylation [4]. Another posttranslational modification that can modulate the activities of mitochondrial proteins is O-GlcNAc modification. For example, we have recently shown that mitochondrial proteins from C2C12 mouse myoblast cells upon exposure to high glucose levels undergo changes in their O-GlcNAc and phosphorylation modification status [5]. Prohibitin (PHB), one of the important mitochondrial chaperones, undergoes elevated levels of tyrosine phosphorylation and O-GlcNAc modification under hyperglycemic conditions indicating the potential role of posttranslational modifications of PHB in modulating its mitochondrial function [5]. Another well-studied mitochondrial protein is heat shock protein 60 (HSP 60). HSP 60 has been shown to be modified by O-GlcNAc under hyperglycemic conditions [6]. Under normal conditions, HSP 60 is bound to Bax protein in the extramitochondrial cytosol. However, under hyperglycemic conditions, HSP 60 undergoes posttranslational modification by O-GlcNAc, and this modification helps in the

6. Place the focusing tray into the PROTEAN IEF cell and close the cover.

7. Program the PROTEAN IEF cell using the appropriate protocol. Focusing is performed at 20 °C.

8. Press start to initiate the electrophoresis run.

3.2.3 Staining IPG Strips with Bio-Safe Coomassie Blue Stain

1. Transfer 2 of the IPG strips to the dry Whatman filter paper by keeping the gel side facing up.

2. Take another filter paper, wet it with nanopure water, and carefully place it on the IPG strips. Gently press over the length of the IPG strips. When finished, carefully remove the top filter paper. This step removes the mineral oil present on the strips.

3. Transfer these IPG strips to a tray containing 50 ml of Bio-Safe Coomassie blue stain. Place the tray on rocking platform for 1 h.

4. Destain the IPG strips using 20 mM Tris–HCl pH 8.8. Complete destaining may take several hours (*see* **Note 4**).

5. Add adequate amount of equilibration buffer I to each channel containing an IPG strip.

6. Place the tray on an orbital shaker for 10 min with shaking at minimum speed (*see* **Note 5**).

7. At the end of the incubation, discard the equilibration buffer I by carefully decanting the liquid from the tray (*see* **Note 6**).

8. Add adequate volume of equilibration buffer II to each strip, return the tray to the orbital shaker, and shake for 10 min.

9. While the IPG strips are incubating, prepare the overlay agarose solution. Microwave on high for 40–60 s until the agarose liquefies. It is best to stop the microwave after 30 s and then swirl the bottle to mix the solution and then microwave it for another 30 s (*see* **Note 7**).

10. Discard the equilibration buffer II by decanting the liquid (*see* **Note 6**).

3.2.4 SDS-PAGE

1. Fill the 100 ml graduated cylinder or tube that is longer than or equal to the length of the IPG strip with 1× Tris-glycine SDS running buffer.

2. Finish preparing SDS gels by blotting away any excess amount of liquid in the IPG strip well by using Whatman 3 MM blotting paper.

3. Remove an IPG strip from the rehydration/equilibration tray and dip into the 100 ml graduated cylinder containing 1× Tris-glycine SDS running buffer. Afterwards lay the IPG strip with gel side facing up on the bottom plate of the SDS gel above the IPG well.

4. Take the first SDS gel with an IPG strip placed on the bottom plate, and let it stand vertically with the small plate facing toward you. With the help of a Pasteur pipette, overlay the agarose solution on top of the IPG well of the gel.

5. Using forceps carefully push the strip into the well and make sure no air bubbles lie between the IPG strip and the gel (*see* **Note 8**).

6. Keep the gel vertical and allow the agarose to solidify for 5 min before proceeding to the next step.

7. Mount the gel in the gel box and fill the reservoir tank with 1× Tris-glycine SDS running buffer and begin the electrophoresis. The migration of the bromophenol blue present in the overlay agarose solution is used to monitor the progression of the electrophoresis.

8. After running the samples in the SDS-PAGE, open the gel cassette and place the gel in a tray containing water.

9. Wash the gel with water for 5 min, and repeat 3 times. Use fresh water to wash the gel every time.

10. Place the gel in a tray containing 50 ml of Bio-Safe Coomassie blue stain.

11. Place the tray containing the gel on rocking platform for 1 h.

12. Discard the stain and wash the gel for 15–30 min with water. The gel can be stored in water for several days. Take a picture of the stained gel using a Gel Doc system (Fig. 1).

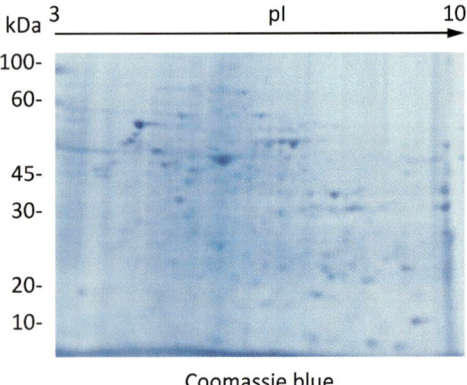

Coomassie blue

Fig. 1 Staining of mitochondrial proteins. Mitochondria were isolated from culture C2C12 mouse myoblast cells. Mitochondrial cell lysates were separated by 2D gel electrophoresis and subsequently stained with Coomassie blue staining

3.3 Transfer of Protein Samples to the Nitrocellulose Membrane

1. Take another unstained gel for Western blotting (do not stain the gel prior to Western blotting).

2. Rinse the gel with distilled water to remove excess of SDS.

3. Place Whatman no. 3 filter paper on the transfer cassette.

4. Place the nitrocellulose membrane on the Whatman no. 3 filter paper. Place the gels on top of the nitrocellulose membrane.

5. Place another Whatman no. 3 filter paper above the gel. Cover the cassette and place it in the transfer apparatus.

6. Fill the transfer apparatus with immunoblot transfer buffer.

7. Connect the transfer apparatus to the power system and run at 100 V for 1 h.

3.4 Immunoblotting

1. After transfer is done, place the membrane in a plastic container and wash it with distilled water.

2. Place the nitrocellulose membrane in blocking buffer (5 % milk in 1× TBST) and incubate it for 1 h at room temperature.

3. Dilute the primary antibody according to the manufacturer's instructions and incubate the membrane in diluted primary antibody for 2 h at room temperature (*see* **Note 9**).

4. Wash the membrane 3 times with 1× TBST for 10 min each.

5. Place the membrane in blocking buffer with diluted HRP-conjugated secondary antibody for 1 h at room temperature.

6. Wash the membrane 3 times with 1× TBST for 10 min each.

7. Develop the membrane using ECL kit by mixing 1 ml of solution A and 1 ml of solution B and pour the mixture over the membrane.

8. Expose the membrane to X-ray film using X-ray cassette or the membrane can be assessed by ChemiDoc system (Fig. 2).

4 Notes

1. Ensure that disruption buffer has been supplemented with protease inhibitor solution.

2. The pellet is very soft. When removing the supernatant, take care that the pellet is not lost or disrupted.

3. It is important that the clear pellet (consisting of density gradient medium) may be at the side of the tube is not disturbed. Due to its transparency, the pellet might be difficult to see.

4. Equilibration buffers should be made 15 min before use. If the IPG strips are frozen at −70 °C, they must be removed and thawed on the lab bench for 10–15 min.

5. While the IPG strips are incubating in equilibrium buffer I, start preparing equilibrium buffer II.

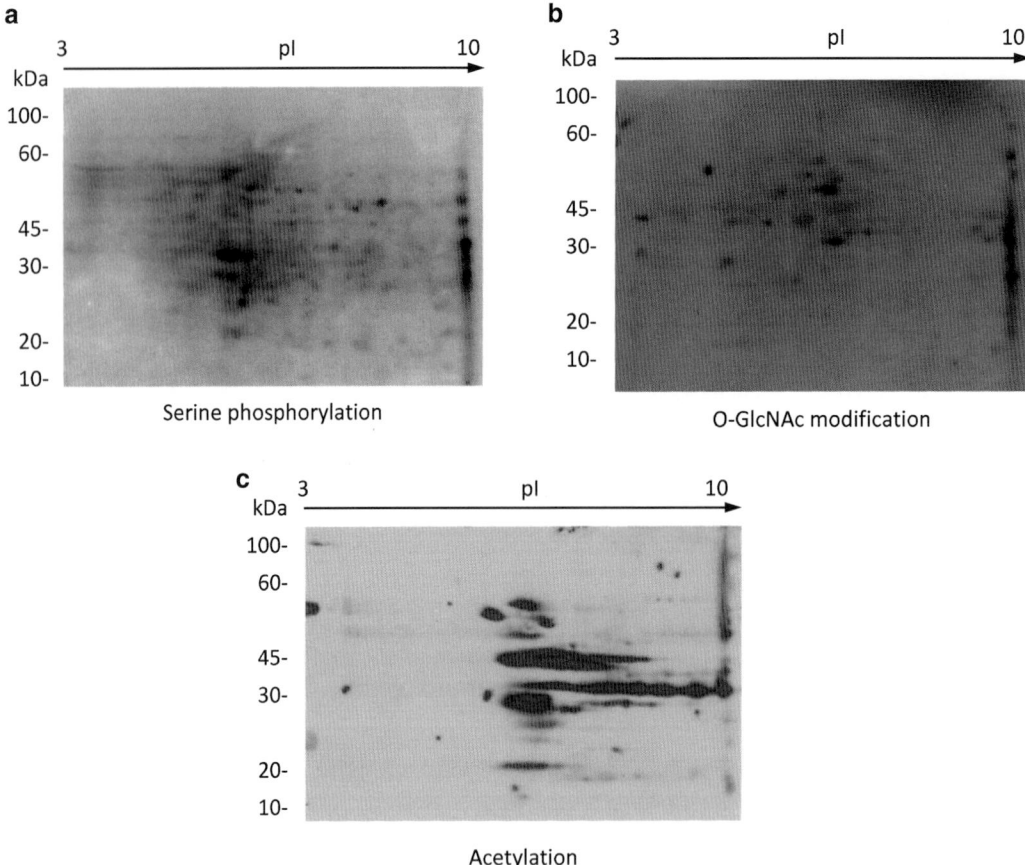

Fig. 2 Analysis of posttranslational modifications in mitochondrial proteins. Mitochondrial proteins were separated by 2D gel electrophoresis and processed for immunoblotting using specific antibodies for each modification. (**a**) Serine phosphorylation, (**b**) O-GlcNAc modification, and (**c**) acetylation

6. Decanting is best carried out by pouring the liquid from the corner of the rehydration/equilibration tray until the tray is positioned vertically. Take care not to pour out the liquid quickly as the strips can slide out of the tray.

7. Avoid overheating the overlay agarose solution to prevent overflow. The SDS in the solution readily forms bubbles when heated.

8. When pushing the IPG strip using forceps, be certain that you are touching the plastic side of the strip and not the gel matrix.

9. Some primary antibodies need to be incubated at 4 °C overnight.

Acknowledgments

Research supported in part by funds from Natural Sciences and Engineering Research Council of Canada.

References

1. Kerner J, Lee K, Hoppel CL (2011) Post-translational modifications of mitochondrial outer membrane proteins. Free Radic Res 45(1):16–28

2. Zapico SC, Ubelaker DH (2013) mtDNA Mutations and Their Role in Aging, Diseases and Forensic Sciences. Aging Dis 4(6): 364–380

3. Yuan S et al (2008) Voltage-dependent anion channel 1 is involved in endostatin-induced endothelial cell apoptosis. FASEB J 22(8): 2809–2820

4. O'Rourke B, Van Eyk JE, Foster DB (2011) Mitochondrial protein phosphorylation as a regulatory modality: implications for mitochondrial dysfunction in heart failure. Congest Heart Fail 17(6):269–282

5. Gu Y, Ande SR, Mishra S (2011) Altered O-GlcNAc modification and phosphorylation of mitochondrial proteins in myoblast cells exposed to high glucose. Arch Biochem Biophys 505(1):98–104

6. Kim HS et al (2006) Heat shock protein 60 modified with O-linked N-acetylglucosamine is involved in pancreatic beta-cell death under hyperglycemic conditions. FEBS Lett 580(9): 2311–2316

7. Anderson KA, Hirschey MD (2012) Mitochondrial protein acetylation regulates metabolism. Essays Biochem 52:23–35

8. Ande SR, Chen J, Maddika S (2009) The ubiquitin pathway: an emerging drug target in cancer therapy. Eur J Pharmacol 625(1–3): 199–205

9. Nijhawan D et al (2003) Elimination of Mcl-1 is required for the initiation of apoptosis following ultraviolet irradiation. Genes Dev 17(12):1475–1486

10. Thompson WE, Ramalho-Santos J, Sutovsky P (2003) Ubiquitination of prohibitin in mammalian sperm mitochondria: possible roles in the regulation of mitochondrial inheritance and sperm quality control. Biol Reprod 69(1):254–260

Chapter 30

Assessment of Mitochondrial Protein Glutathionylation as Signaling for CO Pathway

Ana S. Almeida and Helena L.A. Vieira

Abstract

Protein glutathionylation is a posttranslational process that regulates protein function in response to redox cellular changes. Furthermore, carbon monoxide-induced cellular pathways involve reactive oxygen species (ROS) signaling and mitochondrial protein glutathionylation. Herein, it is described a technique to assess mitochondrial glutathionylation due to low concentrations of CO exposure. Mitochondria are isolated from cell culture or tissue, followed by an immunoprecipitation assay, which allows the capture of any glutathionylated mitochondrial protein using a specific antibody coupled to a solid matrix that binds to glutathione antigen. The precipitated protein is further identified and quantified by immunoblotting analysis.

Key words Glutathionylation, Carbon monoxide, Mitochondria, Glutathione, Immunoprecipitation

1 Introduction

Protein glutathionylation is a posttranslational mechanism involved in redox response, which consists of the regulated formation of mixed disulfides between protein thiol and glutathione disulfide (GSSG) due to glutathione redox changes [1, 2]. The progressive glutathionylation of key proteins can be a molecular switch by which cells respond in an immediate and reversible fashion to oxidative stress by protecting cysteine residues [1]. Still, it can alter protein activity, presenting a physiological signaling function, in the same way as the phosphorylation process. Mitochondria are key organelles for reactive oxygen species (ROS) generation; thus, protein glutathionylation can be crucial for protecting mitochondria from this source of oxidative damage. Furthermore, changes in the redox state of mitochondrial proteins through thiol modifications can transduce redox signals and modulate mitochondrial activity [3].

Several examples of mitochondrial protein glutathionylation or de-glutathionylation are described in the literature: (1) glutathionylation of complex II decreases after myocardial ischemia,

Volkmar Weissig and Marvin Edeas (eds.), *Mitochondrial Medicine: Volume I, Probing Mitochondrial Function*,
Methods in Molecular Biology, vol. 1264, DOI 10.1007/978-1-4939-2257-4_30, © Springer Science+Business Media New York 2015

limiting its electron transfer activity [4]; (2) glutathionylation of specific cysteine residues (C136 and C155) regulates the activity of carnitine/acylcarnitine carrier [5]; (3) glutathionylation of complex I protects against oxidative stress and is mediated by thiyl radical [6]; (4) ANT (ATP/ADP translocator) glutathionylation prevents cell death by improving its activity and limiting mitochondrial membrane permeabilization [7]; and (5) degradation of mitochondrial thymidine kinase 2 is modulated by glutathionylation [8].

Carbon monoxide (CO) is endogenously produced through the cleavage of heme group by heme oxygenase activity (HO), presenting several biological properties: anti-inflammatory, antiproliferative, and antiapoptotic, for review [9]. Cell redox responses, such as ROS signaling, appear to be tightly involved in CO-induced pathways [10], namely, anti-inflammation in macrophages [11]; cytoprotection in cardiomyocytes [12]; antiapoptosis in hepatocytes [13], in neurons [14], or in astrocytes [7]; and cardioprotection [15] and antiproliferation in airway smooth muscle cells [16]. CO also modulates levels of oxidized glutathione, signaling through mitochondrial protein glutathionylation [7].

Herein, a method for assessing CO-induced mitochondrial protein glutathionylation is described, in particular glutathionylation of the mitochondrial inner membrane protein ANT (ATP/ADP translocator). ANT presents critical thiol groups in cysteine residues (cysteines 56, 159, and 256), which can be oxidized and/or derivatized in order to modulate the pore-forming activity of ANT and cell death control [17, 18]. The described protocol can be used with other mitochondrial proteins. For assessing CO-induced mitochondrial protein glutathionylation, two different sources of mitochondria are used: from cell culture (cell lines or primary cultures of astrocytes) and the brain cortex.

2 Materials

Prepare all solutions using ultrapure water (prepared by purifying deionized water to attain a sensitivity of 18 MΩ cm at 25 °C) and analytical grade reagents. Prepare and store all the reagents at 4 °C (unless indicated otherwise).

2.1 Mitochondria Isolation from Cell Culture

1. Phosphate buffer saline (PBS): 1.54 M NaCl, 34 mM Na$_2$HPO$_4$, 20 mM KH$_2$PO$_4$, pH 9.4. In 900 mL of water, dissolve 90 g of NaCl, 4.83 g of Na$_2$PO$_4$, and 2.72 g of KH$_2$PO$_4$. Mix and adjust pH. Make up to 1 L with water.

2. Hypotonic buffer: 0.15 mM MgCl$_2$, 10 mM KCl, 10 mM Tris–HCl, pH 7.6. Weigh 1.43 mg of MgCl$_2$, 74.56 mg of KCl, and 156.6 mg of Tris–HCl. Add water to a volume of 90 mL. Mix and adjust pH. Make up to 100 mL with water. Store at 4 °C.

3. Homogenate buffer (2×): 0.6 M sucrose, 10 mM TES, 0.4 mM EGTA, pH 7.2. Weigh 41.07 g of sucrose, 458.5 mg of TES, and 30.43 mg of EGTA. Add water to a volume of 190 mL. Mix and adjust pH. Make up to 200 mL with water. Store at 4 °C.

4. Homogenate buffer (1×): Dilute 1:2 homogenate buffer (2×) in water.

2.2 Mitochondria Isolation from the Brain Cortex

1. MIB buffer: 225 mM manitol, 75 mM sucrose, 1 mM EGTA, 5 mM HEPES, pH 7.4. Weigh 10.25 g of manitol, 6.42 g of sucrose, 75.09 mg of EGTA, and 279.88 mg of HEPES. Add water to a volume of 220 mL. Mix and increase pH until 8 in order to dissolve EGTA. Adjust pH to 7.4, make up to 250 mL, and store at 4 °C.

2. Brain mitochondrial buffer (complex I): 125 mM KCl, 2 mM K_2HPO_4, 1 mM $MgCl_2$, 1 µM EGTA, 20 mM Tris–HCl, 5 mM glutamate, 5 mM malate, pH 7.2. Add to 220 mL of water 2.33 g of KCl, 73.6 mg of K_2HPO_4, 50.8 mg of $MgCl_2$, 25 µL of an EGTA 10 mM stock solution, 788 mg of Tris–HCl, 233.92 mg of glutamate, and 167.63 mg of malate. Mix and adjust pH to 7.2 at 37 °C. Make up to 250 mL with water and store at 4 °C (*see* **Note 1**).

3. Percoll gradient: Dilute stock solution of Percoll in MIB buffer to obtain final Percoll concentrations of 15 %, 24 %, and 40 % (v/v). Mix 0.75 mL of Percoll stock solution with 4.25 mL of MIB buffer, for 15 % concentrated solution. In order to prepare the 24 % and the 40 % concentrated solutions, pipette 1.2 mL of Percoll and 3.8 mL of MIB and 2 mL of Percoll and 3 mL of MIB, respectively.

2.3 CO Treatment

1. CORM-A1 solution: Prepare a 5 mM solution of CORM-A1 in water. Filtrate the solution with 0.22 µM filter, aliquot, and store at –20 °C. For each use, an aliquot should be thawed and rapidly added into the culture.

2. CO gas solution: Saturate PBS by bubbling 100 % of CO gas for 30 min to produce 10^{-3} M stock solution. 100 % CO was purchased as compressed gas. Fresh stock solutions of CO gas should be prepared each day and sealed carefully (*see* **Note 2**).

2.4 Immunoprecipitation

1. 10 % Triton X-100: Dilute 10 µL of Triton X-100 in 90 µL of water.

2. PBS (*see* Subheading 2.1, **item 1**).

3. Loading buffer: 10 % (v/v) glycerol; 10 mM DTT; 0,005 % (w/v) bromophenol blue. To prepare 20 mL, weigh 30 mg of DTT and 0.001 mg of bromophenol blue. Solubilize both in 18 mL of water. Add 2 mL of glycerol, mix, and store at 4 °C.

2.5 Immunoblotting

1. T-TBS buffer: 0.25 M Tris–HCl; 0.75 M NaCl. Weigh 7.88 g of Tris–HCl and 8.76 g of NaCl. Solubilize in 1 L of water.

2. Blocking buffer: T-TBS with 5 % (w/v) milk. Weigh 5 g of milk and dilute in 100 mL of T-TBS buffer.

3. Running buffer (10×): 0.25 M Tris–HCl, 1.92 M glycine, 35 mM SDS. Weigh 30 g of Tris–HCl, 144 g of glycine, and 10 g of SDS. Solubilize all the components in 1 L of water.

4. Running buffer (1×): Dilute 100 mL of running buffer (10×) in 900 mL of water.

5. Transfer buffer: Running buffer with 20 % (v/v) of methanol. Add 200 mL of methanol to 800 mL of running buffer (1×).

3 Methods

All the steps should be carried out at 4 °C, unless indicated otherwise.

3.1 CO Treatment

1. Cell culture: Add CO gas solution to culture medium, to a final concentration of 50–100 μM. If you are using CORM-A1 solution, add it to the culture medium to a final concentration of 12.5–25 μM (*see* **Note 3**). At the required time point after CO exposure, proceed to mitochondrial isolation from cell culture (Subheading 3.2).

2. Tissue: After isolation from tissue, add CO gas solution (or CORM-A1 solution) directly to isolated mitochondria. Incubate mitochondria at 37 °C and proceed to immunoprecipitation at the different time points after CO exposure (Subheading 3.4).

3.2 Mitochondria Isolation from Cell Culture

Adapted from Vieira et al. [19].

1. Inoculate 175 cm² T-flasks with primary cell culture of astrocytes or a cell line culture.

2. Maintain cells in culture until you achieve the confluence.

3. Wash cell culture (175 cm² T-flask) with 5 mL PBS at 4 °C, in order to eliminate any serum.

4. Trypsinize the cell by adding 5 mL of trypsin, followed by 5 min incubation at 37 °C.

5. Collect the cells in 10 mL of culture medium and centrifuge at $200 \times g$ for 10 min at 4 °C.

6. Discard supernatant, wash the cells with 10 mL of PBS, and centrifuge $200 \times g$ for 10 min at 4 °C.

7. Discard supernatant, add 3.5 mL of hypotonic buffer, and incubate at 4 °C for 5 min.

8. Add an equal volume (3.5 mL) of homogenization buffer twice concentrated (2×) to a final volume of 7 mL.

9. Homogenize samples with a *Dounce* glass homogenizer at 4 °C (*see* **Note 4**).

10. Remove the sample to a 50 mL tube, wash glass homogenizer with homogenization buffer (1×), and add it to the sample.

11. Centrifuge cell extracts at $900 \times g$ for 10 min at 4 °C (to remove nuclei and unbroken cells).

12. Remove supernatant to a clean tube and centrifuge at $10,000 \times g$ for 10 min at 4 °C.

13. Resuspend mitochondrial pellet in 100 μL of homogenization buffer (1×) and quantify the total amount of protein.

3.3 Mitochondria Isolation from Brain Tissue

The non-synaptic mitochondria isolation protocol was adapted by Queiroga et al. [7] from Kristián and colleagues and Sims [20–22].

1. Sacrifice one male Wistar rat (*see* **Notes 5** and **6**) by cervical dislocation.

2. Remove the cerebellum and underlying structures (only the cortex is used). Isolate the cortex 1 min after death.

3. Wash the cortex in MIB in a Petri dish and cut it in small pieces.

4. Homogenize the cortex manually 10 times with tissue homogenizer and centrifuge the tissue extract at $1,300 \times g$ for 3 min at 4 °C.

5. Keep the supernatant, resuspend the pellet, and recentrifuge at $1,300 \times g$ for 3 min at 4 °C.

6. Pool together the two supernatants and centrifuge at $21,000 \times g$ for 10 min at 4 °C in ultracentrifugation tubes.

7. Resuspend the pellet in 3.5 mL of 15 % Percoll solution and add it over to the gradient (Fig. 1).

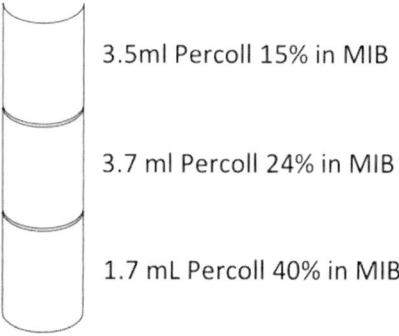

3.5ml Percoll 15% in MIB

3.7 ml Percoll 24% in MIB

1.7 mL Percoll 40% in MIB

Fig. 1 Percoll gradient schema. Pipette 1.7 mL of 40 % Percoll solution, followed by 3.7 mL of 24 % Percoll, and, finally, 3.5 mL of 15 % Percoll. Mitochondrial content will be found between 24 % and 40 % fractions of the gradient after centrifugation

8. Centrifuge the gradient at $31,700 \times g$ for 8 min at 4 °C.

9. Remove mitochondrial fraction from layer 24 % and 40 % with a syringe. Add MIB buffer to mitochondrial fraction to wash Percoll out by centrifugation at $16,700 \times g$ for 10 min at 4 °C.

10. Resuspend pellet in 10 mL of MIB buffer supplemented with 5 mg/mL BSA and centrifuge at $6,800 \times g$ for 10 min at 4 °C.

11. Remove supernatant and resuspend mitochondria in 100 μL of MIB without EGTA.

12. Quantify the total amount of protein.

3.4 Immuno-precipitation of Proteins in Isolated Mitochondria

1. Prepare microtubes with 50–100 μg of isolated mitochondria into 100 μL of homogenization buffer (1×).

2. Permeabilize mitochondria by adding 5 μL of Triton X-100 at 10 % (final concentration 0.5 %).

3. Incubate mitochondria with 20 μL of anti-GSH for 1 h 30 min at 37 °C.

4. Perform the immunoprecipitation by adding 15 μL of Protein A/G PLUS-Agarose beads and incubate them for 30 min at 37 °C with extremely gentle shaking.

5. Discard the supernatant after 10 min of centrifugation at $10,000 \times g$. Wash the pellet with PBS, followed by centrifugation at $500 \times g$ for 10 min, 5 times.

6. Resuspend the pellet (proteins attached to the beads) in 40 μL of loading buffer and freeze at –20 °C for further immunoblot analysis.

3.5 Immunoblotting

1. Load the samples on a 12 % SDS-PAGE gel in order to separate the proteins under reducing electrophoresis conditions. Run the electrophoresis at fixed voltage of 135–150 V for 30 min.

2. Electrically transfer the proteins to a nitrocellulose membrane (fixed 500 mA for 1 h).

3. Incubate the membrane at RT for 1 h with blocking buffer.

4. Dilute 1:1,000 the primary antibody (anti-ANT) in blocking buffer and incubate the membrane for 2 h at RT.

5. Wash the membrane with T-TBS three times for 10 min.

6. Incubate the blot with HRP-labeled anti-mouse IgG antibody, 1:5,000 diluted in blocking buffer, for 1 h at RT.

7. Wash the membrane with T-TBS three times for 10 min.

8. Develop the blot using ECL (enhanced chemiluminescence) detection system (Fig. 2).

9. The area and intensity of bands (Fig. 2) can be quantified by densitometry analysis and presented as a percentage relative to

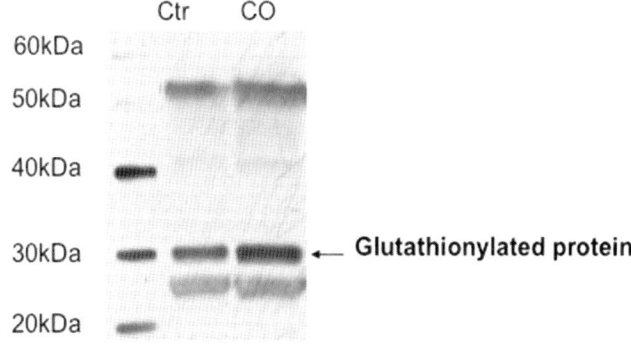

Fig. 2 Example of immunoblotting film image obtained after immunoprecipitation of glutathionylated proteins (α-GSH) of a mitochondrial isolate of astrocytes. The blot was incubated with primary antibody against ANT, which is the target glutathionylated mitochondrial protein

control (100 %) without any treatment. In this example, it can be observed that CO treatment increased the amount of glutathionylated ANT (30 kDa), compared to control.

4   Notes

1. EGTA's concentration can go up to 15 μM, in order to obtain more consistent results.

2. The concentration of CO in solution was determined spectrophotometrically by measuring the conversion of deoxymyoglobin to carbon monoxymyoglobin [23].

3. Homogenize samples with the *Dounce* glass homogenizer 25 times with the loose-fitted pestle and then another 25 times with the tight-fitted pestle at 4 °C.

4. Use CO-saturated solution immediately after opening the vial, about 2 or 3 min. CO releases very easily, changing its final concentration.

5. Animals are allowed water and food ad libitum for 24 h before death.

6. From one male Wistar rat (300–350 g), one might obtain 5 mg of non-synaptic mitochondria.

Acknowledgments

This work was supported by the Portuguese Fundação para a Ciência e a Tecnologia (FCT-ANR/NEU-NMC/0022/2012) and the Portuguese Fundação para a Ciência e a Tecnologia for ASA's SFRH/BD/78440/2011 fellowship.

References

1. Gallogly MM, Mieyal JJ (2007) Mechanisms of reversible protein glutathionylation in redox signaling and oxidative stress. Curr Opin Pharmacol 7(4):381–391

2. Mieyal JJ, Chock PB (2012) Posttranslational modification of cysteine in redox signaling and oxidative stress: focus on s-glutathionylation. Antioxid Redox Signal 16(6):471–475

3. Mp M (2012) Mitochondrial thiols in antioxidant protection and redox signaling: distinct roles for glutathionylation and other thiol modifications. Antioxid Redox Signal 16(6):476–495

4. Chen YR, Chen CL, Pfeiffer DR, Zweier JL (2007) Mitochondrial complex II in the postischemic heart. J Biol Chem 282(45):32640

5. Giangregorio N, Palmieri F, Indiveri C (2013) Glutathione controls the redox state of the mitochondrial carnitine/acylcarnitine carrier Cys residues by glutathionylation. Biochim Biophys Acta 1830:5299–5304

6. Kang PT, Zhang L, Chen C, Green-church KB, Chen R (2012) Protein thiyl radical mediates S-glutathionylation of complex I. Free Radic Biol Med 53(4):962–973

7. Queiroga CSF, Almeida AS, Martel C, Brenner C, Alves PM, Vieira HLA (2010) Glutathionylation of adenine nucleotide translocase induced by carbon monoxide prevents mitochondrial membrane permeabilization and apoptosis. J Biol Chem 285(22):17077–17088

8. Sun R, Eriksson S, Wang L (2012) Oxidative stress induced S-glutathionylation and proteolytic degradation of mitochondrial thymidine kinase 2. J Biol Chem 287(29):24304–24312

9. Motterlini R, Otterbein LE (2010) The therapeutic potential of carbon monoxide. Nat Rev Drug Discov 9(9):728–743

10. Bilban M, Haschemi A, Wegiel B, Chin BY, Wagner O, Otterbein LE (2008) Heme oxygenase and carbon monoxide initiate homeostatic signaling. J Mol Med 86(3):267–279

11. Zuckerbraun BS, Chin BY, Bilban M et al (2007) Carbon monoxide signals via inhibition of cytochrome c oxidase and generation of mitochondrial reactive oxygen species. FASEB J 21(4):1099–1106

12. Suliman HB, Carraway MS, Ali AS, Reynolds CM, Welty-wolf KE, Piantadosi CA (2007) The CO/HO system reverses inhibition of mitochondrial biogenesis and prevents murine doxorubicin cardiomyopathy. J Clin Invest 117(12):3730–3741

13. Kim HS, Loughran PA, Billiar TR (2008) Carbon monoxide decreases the level of iNOS protein and active dimer in IL-1b-stimulated hepatocytes. Nitric Oxide 18:256–265

14. Vieira HLA, Queiroga CSF, Alves PM (2008) Pre-conditioning induced by carbon monoxide provides neuronal protection against apoptosis. J Neurochem 107(2):375–384

15. Scragg JL, Dallas ML, Wilkinson JA, Varadi G, Peers C (2008) Carbon monoxide inhibits L-type Ca2+ channels via redox modulation of key cysteine residues by mitochondrial reactive oxygen species. J Biol Chem 283(36):24412–24419

16. Taillé C, El-Benna J, Lanone S, Boczkowski J, Motterlini R (2005) Mitochondrial respiratory chain and NAD(P)H oxidase are targets for the antiproliferative effect of carbon monoxide in human airway smooth muscle. J Biol Chem 280(27):25350–25360

17. Costantini P, Chernyak BV, Petronilli V, Bernardi P (1996) Modulation of the mitochondrial permeability transition pore by pyridine nucleotides and dithiol oxidation at two separate sites. J Biol Chem 271(12):6746–6751

18. Costantini P, Belzacq AS, Vieira HL et al (2000) Oxidation of a critical thiol residue of the adenine nucleotide translocator enforces Bcl-2-independent permeability transition pore opening and apoptosis. Oncogene 19(2):307–314

19. Vieira HLA, Boya P, Cohen I et al (2002) Cell permeable BH3-peptides overcome the cytoprotective effect of Bcl-2 and Bcl-X(L). Oncogene 21:1963–1977

20. Kristian T, Fiskum G (2004) A fluorescence-based technique for screening compounds that protect against damage to brain mitochondria. Brain Res Brain Res Protoc 13(3):176–182

21. Kristián T, Gertsch J, Bates TE, Siesjö BK (2000) Characteristics of the calcium-triggered mitochondrial permeability transition in nonsynaptic brain mitochondria: effect of cyclosporin A and ubiquinone O. J Neurochem 74(5):1999–2009

22. Sims NR (1990) Rapid isolation of metabolically active mitochondria from rat brain and subregions using Percoll density gradient centrifugation. J Neurochem 55(2):698–707

23. Motterlini R, Clark JE, Foresti R, Sarathchandra P, Mann BE, Green CJ (2002) Carbon monoxide-releasing molecules: characterization of biochemical and vascular activities. Circ Res 90(2):E17–E24

Chapter 31

High-Resolution Melting Analysis for Identifying Sequence Variations in Nuclear Genes for Assembly Factors and Structural Subunits of Cytochrome C Oxidase

Alžběta Vondráčková, Kateřina Veselá, Jiří Zeman, and Markéta Tesařová

Abstract

High-resolution melting (HRM) analysis is a simple, sensitive, and cost-effective screening method. HRM enables the detection of homozygous or heterozygous point sequence variants and small deletions within specific PCR products by observing temperature and shape changes in melting curve profiles using fluorescent dyes. Herein, an updated protocol for routine variant screening of nuclear genes encoding assembly factors and structural subunits of cytochrome c oxidase (COX) is described. Nonetheless, the general recommendations given for HRM analysis can be applicable for examining any genetic region of interest.

Key words COX, HRM analysis, Screening of sequence variants, Melting dye, LCGreen Plus, LightScanner

1 Introduction

The mammalian cytochrome c oxidase (COX, Complex IV, EC 1.9.3.1) is a multimeric copper–heme A metalloenzyme embedded in the inner mitochondrial membrane; its function is to transport electrons from cytochrome c to molecular oxygen, which is then reduced to water. COX likely consists of 14 polypeptide subunits, of which the 3 largest subunits are encoded by the mitochondrial genome (MT-CO1, MT-CO2, and MT-CO3); the other 11 small peripheral subunits are encoded by the nuclear genome [1, 2]. In humans, four of the nuclear-encoded subunits, COX4I, COX6A, COX6B, and COX7A, have tissue-specific isoforms that reflect differences in the energetic demands of the particular tissues [3]. COX deficiency is a clinically heterogeneous group of disorders that predominantly affect tissues with high energy demand. The disorders range from isolated myopathy to severe multisystem disease and exhibit onset from infancy to adulthood; they are caused

Volkmar Weissig and Marvin Edeas (eds.), *Mitochondrial Medicine: Volume I, Probing Mitochondrial Function*, Methods in Molecular Biology, vol. 1264, DOI 10.1007/978-1-4939-2257-4_31, © Springer Science+Business Media New York 2015

by mutations located both in mitochondrial DNA and in the nuclear genes required for mitochondrial function. Rare disease-related mutations have been described for all three of the mitochondrial DNA-encoded COX subunits [4]. The majority of COX defects originate from mutations in nuclear genes involved in the assembly and maintenance of the COX holoenzyme complex. The first mutations were only recently characterized in nuclear genes coding for the structural subunits COX4I2, COX6B1, COX7B, and NDUFA4 [5–8]. Despite advances in the identification of an increasing number of mutations and genes involved in the disease phenotype, the molecular basis of COX deficiency remains elusive in many patients, and this leads to difficulties in genetic counseling.

High-resolution melting (HRM) analysis is a simple, sensitive, and cost-effective method to detect homoduplexes and/or heteroduplexes that form in the presence of a fluorescent dye, while the temperature of amplified PCR products is increasing. HRM analysis is commonly applied for mutation scanning of blindly screened samples from patient's and healthy control's unknown genotypes and for mutation genotyping using control samples of known genotypes. Up to now, HRM analysis has been successfully used for mutation screening and genotyping of various human, animal, plant, or microbial genomic DNA [9], and it has been employed in general population, epigenetics, or cancer studies [10]. As documented by others, HRM techniques can also be fruitfully applied in a targeted mutation analysis of highly polymorphic human mitochondrial DNA [11]. Importantly, HRM screening of genes with greatly variable SNPs is generally not recommended.

The high sensitivity and specificity of HRM have been affirmed for amplicons of up to 1,000 bp (although preferentially less than 500 bp), and HRM has proved to be better than many conventional mutation detection methods, including even Sanger sequencing [12–14]. Shorter PCR amplimers are better for heterozygote detection, whereas longer amplimers seem to be more suitable for homozygote detection [15]. Although the position of a base-pair variant within the PCR product does not affect the accuracy of HRM, SNPs of the third and fourth class and GC-rich regions can be a source of difficulties when optimizing HRM experiments [16–18]. Nevertheless, sequence analysis of GC-rich regions and/or a gene with many related pseudogenes can be challenging even for the currently available next-generation sequencing technology [19–21].

Generally, changes in the shapes of melting curves are more apparent than differences in melting temperatures (Tm) [15]. However, nearly any type of sequence variant can be distinguished by cluster analysis of the melting curves on a specific PCR product when appropriate modification of PCR and HRM design, such as the use of short amplicons (approximately 50–150 bp) [22], internal control temperature calibrators [23, 24], snapback primers [25], masking [26] and unlabeled probes (probes approximately

20–40 bp, amplicons smaller than approximately 250 bp) [27, 28], mixing of reference and patient PCR products, asymmetric PCR (a 1:5–1:10 primer ratio is suggested) [28], and nested PCR, COLD PCR, is applied. All unlabeled probes should be suitably blocked at the 3' end to prevent their extension during PCR [27].

With regard to the accuracy of instruments used for HRM, several parameters need to be observed such as melting rates (°C/s), temperature data acquisitions/°C, and temperature homogeneity over the analyzed set of samples or multiwell plate [29, 30]. At present, several useful fluorescent melting dyes are available for use in HRM: LCGreen Plus (BioFire Defense), EvaGreen (Biotium, Inc.), ResoLight (Roche), Chromofy (TATAA Biocenter), and SYTO9 (Invitrogen), and these are coupled with matching master mixes. SYBR Green I Dye is not recommended for use in HRM because of the effect of dye redistribution and potential PCR inhibition [31, 30].

HRM methodology is especially valuable for large-scale mutation screening studies [32] in which systematic resequencing of the whole gene in all samples would be too laborious and expensive. The excellent screening properties and adaptability of HRM methodology prompted us to implement this procedure in mutation screening of genes assumed to be possible causes of COX deficiency, including ubiquitous and tissue-specific COX subunits and COX assembly factors [33]. Herein, an updated and simplified protocol for routine variant, probe-free, HRM screening of COX-related nuclear genes is provided.

2   Materials

2.1   Genomic DNA (gDNA) Isolation

1. Types of analyzed patient samples: EDTA anticoagulated human blood, cultivated patient fibroblasts, and biopsy tissue (from muscle, heart, and liver) [34, 35] (*see* **Notes 1** and **2**).

2. QIAamp DNA Mini Kit (QIAGEN).

3. 1× RBC lysis solution: 8.3 g NH_4Cl, 0.1 g $KHCO_3$, 2 mL 0.5 M EDTA.

4. Digestion buffer: 5.85 g NaCl, 10 mL 1 M Tris–HCl at pH 8.0, 50 mL 0.5 M EDTA, 50 mL 10 % SDS.

5. Proteinase K.

6. Phenol–chloroform–isoamyl alcohol (25:24:1).

7. 7.5 M ammonium acetate.

8. 1× TE buffer (100 ml): 1 mL 1 M Tris–HCl at pH 8.0, 0.2 mL 0.5 M EDTA).

9. Ethanol: 70 %; 96 %.

2.2 Polymerase Chain Reaction and Melting Analysis

1. Primers for amplifying the coding regions of *COX4I1, COX4I2, COX5A, COX5B, COX6A1, COX6A2, COX6B1, COX6C, COX7A1, COX7A2, COX7B, COX7C, COX8A, COX10*, and *COX15* (*see* Table 1 and **Note 3**).

2. PCR Ultra H_2O (Top-Bio).

3. Plain PP and Plain Combi PP Master Mixes (Top-Bio).

4. 25 mM $MgCl_2$.

5. DMSO.

6. 8-Strip PCR tubes.

7. 10× LCGreen® Plus⁺ melting dye (BioFire Defense).

8. Mineral oil.

9. Low-profile, thin-wall, 96-well, skirted PCR plates.

2.3 Specificity of PCR Products and HRM Results

1. Agarose for preparation of 1.5 % gel.

2. 10× TBE buffer for agarose gel electrophoresis (1 L): 108 g Tris base, 55 g boric acid, 40 mL 0.5 M EDTA at pH 8.0.

3. Ethidium bromide diluted with double-distilled water to a concentration of 0.5 μg/mL.

4. Wizard® SV Gel and PCR Clean-Up System (Promega) for PCR product purification before sequencing.

2.4 Instrumentation

1. NanoDrop ND-1000 UV–Vis Spectrophotometer (NanoDrop Technologies).

2. Thermal cycler with a gradient unit for PCR optimization.

3. Equipment and accessories for DNA agarose gel electrophoresis.

4. UV transilluminator for the visualization of PCR products on DNA agarose gels.

5. Centrifuges and adapters for 96-well PCR plates, tubes, and strips.

6. LightScanner™ 96-well system (BioFire Defense).

7. Genetic analyzer for sequencing.

3 Methods

3.1 Isolation of Genomic DNA (gDNA)

1. Isolation of patient DNA from whole, EDTA anticoagulated blood is performed via a standardized protocol using phenol–chloroform–isoamyl alcohol extraction as per previously published recommendations [36] with slight modifications according to [37]. Store the isolated DNA at 4 °C.

2. Biopsies consisting of up to 25 mg tissue (muscle, heart, or liver) are processed as per the instructions in the QIAamp DNA Mini Kit Handbook. Store the isolated DNA at 4 °C.

Table 1
Primers and PCR conditions for HRM analysis of COX-related nuclear genes

Gene	Amplicon covering exon	Forward (F) primer (5′ → 3′) Reverse (R) primer (5′ → 3′)	Length of amplicon	Cycles	PCR mixture[a]
COX10[b,c]	1	F: AGACACCACGCTCTCCTTTC R: GAGAAGAATTCCCCCAAGG	240 bp	35	1× PM, 4 % DMSO, 800 nM primers
	2	F: TCTGGGGAGGTGTAGTCATCA R: GCAGAAAGTAACAGAGTAAGAATGGTC	249 bp	35	1× PCM, 4 % DMSO, 800 nM primers
	3	F: AACCATTTGAGAGCATTTGG R: GCTTTTTGTTTCACTAAAGATAGAGTT	441 bp	30	1× PCM, 4 % DMSO, 0.5 mM MgCl$_2$, 600 nM primers
	4	F: TGGTAACAGTGTGTCTGCTCTGT R: ACAGCCATCTAGGAAAAAGTGA	218 bp	35	1× PCM, 2 % DMSO, 800 nM primers
	5	F: AATGTTCAGTACTAAAGCGGAAGA R: TCGCAGCTCAAAGCAAAATA	240 bp	30	1× PCM, 4 % DMSO, 800 nM primers
	6	F: cag gtt ctc tgc tct ttt tcc R: ctc ctt gac cga gtg tgc t	306 bp	30	1× PM, 4 % DMSO, 400 nM primers
	7a[d]	F: TCTGGTGATGACTGCCTTTG R: GTCCACGTAGAAGCGGAAGC	328 bp	35	1× PCM, 0.5 mM MgCl$_2$, 8 % DMSO, 800 nM primers
	7b[d]	F: TTCCCATCAATGCGTACATC R: TTCCAGAATTACCACAACATGC	243 bp	40	1× PCM, 8 % DMSO, 800 nM primers

(continued)

Table 1
(continued)

Gene	Amplicon covering exon	Forward (F) primer (5′→3′) Reverse (R) primer (5′→3′)	Length of amplicon	Cycles	PCR mixture[a]
COX15[b,c]	1	F: gtt gtg gaa gag gtg gct gt R: tat ctt tat ccc ggc cct tt	191 bp	30	1× PCM, 4 % DMSO, 800 nM primers
	2	F: TAGCAGCCATTCCCTGTTTC R: TCCTCAGTCAACCTGTGCTT	300 bp	30	1× PCM, 3 % DMSO, 0.5 MgCl$_2$, 600 nM primers
	3	F: TTTTGTGAGTAATCCAGCCTCA R: CAAAAGATCAAATGGGCCTAC	233 bp	30	1× PCM, 2 % DMSO, 720 nM primers
	4	F: gga tgt ttc ctc ctc ct R: tgg gag cat ttc tgg ttt ct	271 bp	30	1× PCM, 1 mM MgCl$_2$, 4 % DMSO, 240 nM primers
	5	F: CAGATCATTTAACCTTGTTTTGTTT R: CAAAAGCGGGGTCTTGAAC	261 bp	30	1× PCM, 2 % DMSO, 600 nM primers
	6	F: GTCACTTGGGTTTGGCCTTA R: GCAAAGCATTAGGCAAGAGG	242 bp	30	1× PCM, 4 % DMSO, 720 nM primers
	7	F: TTGGTCTCTTCTTCCTCATTTT R: CAGTGTTGCCATCAGTGCTT	294 bp	30	1× PCM, 2 % DMSO, 0.5 mM MgCl$_2$, 700 nM primers
	8	F: gaa gag gat ggt gga aga g R: ttt gta gag atg ggg ttt tg	473 bp	30	1× PCM, 4.6 % DMSO, 400 nM primers
	9a[d]	F: tgt gga ggt ttg tgt gtg R: gtc aca gtc cca gga gg	258 bp	30	1× PCM, 800 nM primers
	9b[d]	F: GGCCTTGATGGTGTAGTGC R: ATCTCGATGGGGTCATTCTG	244 bp	30	1× PCM, 4 % DMSO, 720 nM primers
COX4I1	1	F: AGACTCCAGTCGCGCTTC R: CTGCGGACGTGCAGACTT	331 bp	35	1× PCM, 10 % DMSO, 1 µM primers
	2	F: GCTCTGGGGCAAAAGAAG R: AACTCCAGCACAGGGCTTTA	234 bp	30	1× PCM, 1 mM MgCl$_2$, 800 nM primers
	3	F: CTGTGACCCCCTGAGATGAT R: GAGGCTCTGTCACACACG	364 bp	35	1× PM, 800 nM primers
	4	F: TGGTTGAATGTTGCAGAGGA R: AGCCTCAAGGTATGGAGGTC	298 bp	30	1× PCM, 4 % DMSO, 800 nM primers
	5	F: GAGGGATTGGCCTAGAAACA R: CCCTTGGGAGAAACCTATTG	398 bp	30	1× PM, 4 % DMSO, 800 nM primers

Gene		Primers	bp	Cycles	Conditions
COX4I2	1	F: CTGCCGAAGCAGGACGTT R: AACCCTCTAAGACAGGGACCA	267 bp	35	1× PM, 8 % DMSO, 800 nM primers
	2	F: TGATGTGGGGCAGAACT R: TGGGAAGTGTGTGTAGGAACA	242 bp	35	1× PM, 2 % DMSO, 800 nM primers
	3	F: CCCGGCCACCTTCTTTATTA R: GGCCATTCTTTCCAAAGTCA	313 bp	35	1× PM, 2 % DMSO, 800 nM primers
	4	F: GAAGCCGGGATCACTTAGAG R: GTGACCACAAGGGCATGG	249 bp	35	1× PM, 2 % DMSO, 800 nM primers
	5	F: CCTGGCTGGTGTAGGAAGAC R: TGCCTAATTTTAGGTGCCAAGT	356 bp	35	1× PM, 4 % DMSO, 800 nM primers
COX5A	1	F: GTCACCTGACCAGAGACAAGG R: AGGTCACCGCAAGGACAC	390 bp	30	1× PCM, 10 % DMSO, 1 µM primers
	2	F: TTCAATATTTTGCTGCCACA R: GCAAGTTGCATGAAGTAACCA	354 bp	30	1× PM, 2 % DMSO, 800 nM primers
	3	F: GGAGACCCAGACAGATAAGATCA R: TTCTGATACCTCAGCAATAGCC	298 bp	30	1× PM, 2 % DMSO, 800 nM primers
	4	F: TCTGTCCTACCTGCCTCTGC R: GCTCACGGCCATTACCTCTA	354 bp	30	1× PM, 2 % DMSO, 800 nM primers
	5	F: TCGCTTGTGGGTTGACAGTA R: CAGCAAAACCATGAAACCAA	397 bp	30	1× PM, 2 % DMSO, 800 nM primers
COX5B	1	F: ACTACGCGGTGCAGAAAGAG R: CCACTGGGACCTCGAGAAG	399 bp	30	1× PCM, 8 % DMSO, 1.2 µM primers
	2	F: GCACCATTTTCCTTGATCATT R: CTCCCAGAGAGGAGACCACT	237 bp	30	1× PM, 800 nM primers
	3	F: AACAGTCCCCTGAGCTTCTG R: CAACATGCACTCACACACTGA	291 bp	30	1× PCM, 2 % DMSO, 800 nM primers
	4	F: TCAACCATAGTCTTACTTGTGTATCA R: GGCAAGCTAGCATTAACAGACA	397 bp	30	1× PCM, 2 % DMSO, 800 nM primers

(continued)

Table 1
(continued)

Gene	Amplicon covering exon	Forward (F) primer (5′ → 3′) Reverse (R) primer (5′ → 3′)	Length of amplicon	Cycles	PCR mixture[a]
COX6A1	1	F: GGCGCCCAATAGTAACTTCC R: AGGTCACAGTCCCTCCCTGT	267 bp	35	1× PCM, 0.5 mM MgCl$_2$, 6 % DMSO, 800 nM primers
	2	F: CGGGAGGGAAAGTGAGACC R: CACCCATGCCTTCAGAGAA	297 bp	30	1× PM, 4 % DMSO, 800 nM primers
	3a[d]	F: CACCCCGTTATAAGCAGTTCA R: TAACGGTCCAAACCAGTGCT	222 bp	30	1× PCM, 4 % DMSO, 800 nM primers
	3b[d]	F: CCAACTGGCTACGAAGATGA R: CAGCCTAGACCTTCACTGTGG	354 bp	30	1× PM, 4 % DMSO, 600 nM primers
COX6A2	1	F: TGCCTCCTTGCCAAAATAAG R: AGCAGACGCCAGGTACGAG	400 bp	35	1× PM, 6 % DMSO, 800 nM primers
	2[c]	F: TACCCTGCCCACCTGTTC R: CCCGCAGCACCCCCGTGC	246 bp	48	1× PCM, 14 % DMSO, 3 mM MgCl$_2$, 1 µM primers
	3	F: CTCTCTCCACAGCCCTACCC R: AGGAGCGGCTTACCAAGCTG	228 bp	35	1× PCM, 6 % DMSO, 800 nM primers
COX6B1	1	F: GGCCAGAAGTGAGGATGAAC R: CCTCAGCCCGCTAGACTG	288 bp	30	1× PM, 2 % DMSO, 800 nM primers
	2	F: CTGGGTAGTCTGGCTTGCTC R: GGGTCCCCTAGGAAGAGG	249 bp	30	1× PM, 640 nM primers
	3	F: CAGATTGAGACCCTACCTCAAAA R: CACACTCCCCTCTGCTAAGA	245 bp	30	1× PM, 4 % DMSO, 800 nM primers
	4	F: TAGAGGTTGGCACACAGCAG R: GGTCAGGGCACTGATTCC	378 bp	30	1× PM, 2 % DMSO, 800 nM primers

Gene		Primers	bp	Cycles	Conditions
COX6C	1	F: ATGAACTTCGGCTGTCACCT R: CGACTAAATCCGAGGCAGAG	248 bp	30	1× PM, 2 % DMSO, 800 nM primers
	2	F: AGCTCCAATCAATGCTTCCA R: AAAGATTTTTCAACCAAAAACACA	397 bp	30	1× PM, 4 % DMSO, 800 nM primers
	3	F: AAAACATGTGTTCTACCTTGTCTTTA R: GGGACAGTCACCTGTATTTGC	298 bp	30	1× PCM, 2 % DMSO, 800 nM primers
	4	F: CTCAGTTGATCCTCAAAGATGG R: GCTTCATAAACAGTTAAATCCCAAA	284 bp	30	1× PM, 2 % DMSO, 800 nM primers
COX7A1	1a[d]	F: TAAATACCGTTTTACTCCCAAA R: CTCGGATTCGTCCACCAC	398 bp	35	1× PCM, 6 % DMSO, 600 nM primers
	1b[d]	F: TATTCCCTGGTACCGCTTTG R: GCACTTGGAGAGTCGCGTAT	377 bp	35	1× PCM, 10 % DMSO, 800 nM primers
	2	F: GCTAGGGATGGGGCTGTC R: CTGATGAGAAAGGGGTGCTG	297 bp	35	1× PCM, 8 % DMSO, 800 nM primers
	3	F: CTCTAAGGAGCAGCCAGCAC R: AGTCCTGCCCAGAAACCAG	215 bp	35	1× PM, 8 % DMSO, 800 nM primers
	4	F: GATGTCCAGGGAGGGGATTA R: TCCACAGGGCAGAGATCC	243 bp	35	1× PM, 6 % DMSO, 800 nM primers
COX7A2	1a[d]	F: GTTTTACGCCTTCTCGCTCA R: GCTTGCGCTCCTAACCATAG	257 bp	35	1× PCM, 800 nM primers
	1b[d]	F: CCGTACTGCCGCTCTAGTTT R: CCAGGTGAGGGTTTCTGTC	285 bp	35	1× PCM, 4 % DMSO, 800 nM primers
	2	F: CGACTGAAAATAGTTGGTTTTGAA R: CTATGGTACATGTCCTTGACTTTTT	300 bp	30	1× PM, 2 % DMSO, 800 nM primers
	3	F: TCTCAAATTAACGGTGAAAGAAGA R: TGGCATAGCAAAAGCAATAAA	399 bp	30	1× PM, 2 % DMSO, 800 nM primers
	4	F: CAAACTTACAACTTTTGAACTGGA R: CAAACGGCAAGTTGAGACAG	486 bp	30	1× PM, 2 % DMSO, 800 nM primers

(continued)

Table 1
(continued)

Gene	Amplicon covering exon	Forward (F) primer (5′→3′) Reverse (R) primer (5′→3′)	Length of amplicon	Cycles	PCR mixture[a]
COX7B	1	F: AAGGGATTGCAATTACTATAGGTTT R: CGTAAAAGGAAAGCACACGA	291 bp	30	1× PM, 2 % DMSO, 800 nM primers
	2	F: TTCCTTGGCTTTCCTGATTG R: GACACTTGAATGCATAGACTGAGA	292 bp	30	1× PCM, 4 % DMSO, 800 nM primers
	3	F: TCCCAGGTGAGTTCTGTGT R: AACATAAAGGCTAAAGTGATCAAGC	384 bp	30	1× PCM, 800 nM primers
COX7C	1	F: CCGCAATGGTCTGAACTACAA R: AGCCTGGTTTCTGGCTATCA	394 bp	30	1× PCM, 4 % DMSO, 600 nM primers
	2	F: GCCATGTAGTGTTTTGTGATGAA R: GTGATGGGGAAGAGGCTACT	400 bp	30	1× PM, 2 % DMSO, 800 nM primers
	3	F: TCATGAAACTACATGATTTCTGTTAAA R: CACCATTAAATAAGCTAAATCACAGA	285 bp	30	1× PCM, 2 mM MgCl$_2$, 2 % DMSO, 600 nM primers
COX8A	1	F: GCGGTCATTTCCGAGAGACTT R: TCCAGACATGCCCAAACC	387 bp	30	1× PM, 6 % DMSO, 800 nM primers
	2	F: CTTTCTGCTGCCTGGGAACT R: TCCACACCTCCACCCAGT	476 bp	30	1× PCM, 1 mM MgCl$_2$, 8 % DMSO, 400 nM primers

PM Plain PP Master Mix (Top-Bio s.r.o., Czech Republic), *PCM* Plain Combi PP Master Mix (Top-Bio s.r.o., Czech Republic), *DMSO* dimethyl sulfoxide

[a]Each PCR mixture was prepared in a total volume of 25 µL containing 2.5 µL 1× LCGreen Plus melting dye

[b]For HRM analysis, accessory sequences for the forward (5′-CAG GAA ACA GCT ATG AC-3′) and reverse (5′-AAT ACG ACT CAC TAT AG-3′) primers were used in the amplification of exon 6 of COX10 and exons 1, 4, 8, and 9a of COX15. The last exons of COX10 and COX15 were designed to screen the part of the coding region that flanks the stop codon and which corresponds exactly to the coding regions for the COX10 and COX15 proteins

[c]Primers highlighted in bold were newly designed for the routine HRM analysis of sequence variants

[d]The analyzed exon was separated into two overlapping fragments

3. Determine the concentration and purity of the isolated DNA by measuring the absorbance at 260 nm (A_{260}), 280 nm (A_{280}), and 230 nm (A_{230}) and then checking the absorbance ratios of A_{260}/A_{280} (1.8–2.0) and A_{260}/A_{230} (2.0–2.2) [38] (*see* **Notes 4** and **5**).

4. Dilute the samples using pure PCR-grade water (DNase and RNase free) to obtain sample aliquots, whose concentration can range from 15 to 50 ng of gDNA to provide reproducible results in HRM analysis.

3.2 Primer Design for HRM Analysis

1. It is recommended that the latest versions of the DNA and RNA reference sequences for the analyzed genes be acquired, e.g., via NCBI.

2. As population-specific sequence variants may occur, use publicly accessible genome browsers like UCSC, dbSNP NCBI, Ensembl, 1000 Genomes, or HGMD to determine the occurrence and frequency of previously characterized SNPs and pathological variants [39, 40] (*see* **Note 6**).

3. Primers can be designed using any online software, e.g., Primer3Plus [41] (*see* **Note 7**). The targeted exonic regions of COX-related genes were selected so that there were at least 20–30 intronic bases on both the 5′ and 3′sides to screen for the potential presence of deleterious splice mutations [42, 43].

4. The specificity of the optimal primer pairs should be checked against the human genome reference assembly, e.g., with the use of Primer-BLAST on the NCBI website.

5. Optional step: the selected targeted amplified region can be subjected to in silico analysis to get an idea of its melting domains and approximate Tm. Currently available online programs such as uMELT™, uMelt HETS, POLAND, MeltSim, DINAMelt, and Stitchprofiles can be used [17, 44, 45] (*see* **Note 8**).

3.3 Optimization of PCR Conditions

1. The amplicon-specific annealing temperature can be reliably assessed within a temperature range of 50–70 °C for all analyzed amplicons. The PCR conditions must be optimized so that one specific product is amplified, without the presence of primer dimers; it is possible to confirm this by loading 10 μL of the PCR product on a 1.5 % agarose gel.

2. The PCR reaction is carried out in a total volume of 25 μL. Exon 2 of *COX6A2* was amplified as follows: an initial denaturation step was performed at 95 °C for 2 min, followed by 45 PCR cycles of denaturation at 95 °C for 20 s, annealing for 30 s, and extension for 30 s, with a final extension at 72 °C for 7 min. The PCR conditions for all the remaining amplicons included an initial denaturation step at 95 °C for 2 min, followed by

30–40 cycles of denaturation at 95 °C for 30 s, annealing for 30 s, and extension for 1 min, with a final extension at 72 °C for 7 min (*see* Table 1 and **Notes 9** and **10**).

3.4 Design of an HRM Experiment in a 96-Well Plate

1. Perform optimized PCR reactions for a genomic region of interest for all patient and control samples in PCR tube strips (*see* **Notes 11** and **12**).

2. Denaturation step: heat the samples to 94 °C for 30 s and cool to 25 °C for 30 s in PCR strips (*see* **Note 13**).

3. Pipette all samples in a 96-well plate as follows: (a) 10 μL each of the patient and control samples in one well; (b) mix 5 μL of the patient sample with 5 μL of homozygous wild-type control sample at a 1:1 ratio in one well. The sample mixing will produce a heterozygous melting profile sample as the homozygous control sample interacts with the homozygous pathological variant from the patient (*see* **Note 14**).

4. Add 15 μL mineral oil to each pipetted sample (*see* **Note 15**).

5. Spin the plate for 2 min at $568 \times g$ at 4 °C using a microplate swinging bucket rotor.

6. Examine all wells to make sure there are no bubbles. Otherwise, repeat **step 5** once more.

3.5 Melting Data Acquisition and Analysis on the LightScanner

1. Melting curves are obtained by heating the samples from 55 to 97 °C at 0.1 °C/s on a LightScanner instrument, and analyzed manually according to the instructions from the LightScanner Operator's Manual using the supplied software module Call-IT®, version 1.5.0.972.

2. Compare the melting curve profiles of the known/healthy reference control samples with the patient melting curves. Sequence all samples presenting melting profiles not matching with the reference ones (*see* **Note 16**).

3. Optional step: The Web-based program uAnalyze™ can be used to provide an analysis of HRM data and a comparison of the data with thermodynamic prediction of data from specific melting curves [44].

3.6 Confirmation of an Identified Patient Genotype

1. Sequence identity of a sample (patient genotype) is proved by a perfect match of the temperature and shape of melting curves between a reference and an investigated sample, which is also relevant to samples mixed with known control genotype.

2. All samples producing a unique, unclassifiable melting curve are subjected to sequencing. Purify these PCR products with a purification kit prior to sequencing. Because PCR amplification should give just only a specific product, there is generally no need to prepare the sample on an agarose gel (*see* **Note 17**).

4   Notes

1. Samples of tissues from biopsies were placed on sterile gauze moistened with saline to prevent drying. After freezing the tissues in liquid nitrogen, the samples were moved into a pre-cooled test tube and immediately put on dry ice. Samples were then stored at −80 °C until DNA isolation.

2. The most frequently used anticoagulant is probably EDTA; however, methodologies for processing heparinized and/or even clotted blood have recently been reported. These approaches along with simplified extraction protocols could be especially useful in investigating severely affected infants [46–48].

3. Optionally, dilute the stock concentration of primers with PCR-grade water to make aliquots containing 10 μM of both (the forward and reverse primers).

4. If the purity of the DNA is unsatisfactory, PCR amplify the samples together with a reference control, e.g., using real-time PCR or agarose gel separation of amplimers. All DNA samples yielding PCR by-products (primer dimers, nonspecific products) either have to be repurified and eluted into fresh buffer or excluded from the HRM screening procedure to avoid misinterpretation of the results [49].

5. Variable salt content in a solution is commonly known to influence DNA melting. Although patient and control DNAs are isolated by two different methodologies, we did not observe any adverse effects from this on the accuracy of the HRM analysis.

6. If the screened population seems to have SNPs more frequently in one of the targeted regions, consider optimizing the primer design and/or PCR modifications.

7. If an intended amplicon is longer than 500 bp, split the targeted region into at least two overlapping fragments. Manufacturers of HRM instruments and accessories usually supply their own primer and/or probe design software. In addition, there are also many other different, free, Web-based, primer-design programs that are able to design primers suitable for methylation analysis, degenerate and/or mismatched primers, primers for the analysis of GC-rich genetic regions, and primers for qPCR.

8. The accuracy of the prediction is variable and strongly dependent on thermodynamic parameters, adjustment for ingredients of a PCR reaction (ions, dNTPs, DMSO, melting dye), and fluorescence background removal, as well as on the type of analyzed sequence and the melting instrument used [38, 44].

Unfortunately, contemporary predictive programs do not enable prediction of multiple variants and/or melting domains in one PCR product [17, 50].

9. Analyze 10 μL of the PCR product on a DNA agarose gel (1.5 %) to inspect the specificity of selected PCR conditions. Use the other 10 μL of the PCR product for HRM analysis and ascertain the intensity of the resulting fluorescence signal. If the signal is not sufficiently strong, optimize the PCR by increasing the concentration of primers, DMSO, and/or $MgCl_2$ in the reaction mixture. Alternatively, increase the number of PCR cycles. If nonspecific PCR products are present, lower the concentration of primers, $MgCl_2$, or DMSO in the reaction mixture.

10. It is also possible to perform PCR optimization without the LCGreen Plus melting dye, which has to be added to the PCR reaction prior to the denaturation step. Nonetheless, keep in mind the enormous amount of pipetting that would be required in a large-scale study. Moreover, LCGreen Plus increases the Tm of DNA by approximately 1–3 °C (based on the manufacturer's information), which may sometimes compromise the HRM analysis and require further optimization of PCR conditions. In addition, LCGreen evinced higher sensitivity with low-melting-temperature amplimers [51], which can be used by adjusting the PCR conditions (e.g., primer design, reaction mixture composition).

11. This step is possible to perform in a 96-well PCR plate or using PCR strips. If a 96-well PCR plate is used, the PCR should be optimized for a volume of just 10 μL. In this case, overlay all samples in a PCR plate with 15 μL mineral oil.

12. Importantly, do not forget to amplify a sufficient amount of reference control sample for the reliable detection of homozygotes. When performing a large-scale analysis, amplify a sufficient number of reactions containing the reference genotype intended to be mixed with the unknown patient genotypes; ideally, the volume of the reaction mixture for the reference genotype should be the same as that used for the patient samples. After the PCR has finished, combine the identical reference aliquots together, vortex, spin, and add to the patient samples.

13. It is possible to do this step in a PCR plate. After pipetting the samples from the PCR strips to wells of a PCR plate, do not forget to overlay all the samples with 15 μL mineral oil. Then, spin the plate and implement a denaturation step.

14. Although the selected reference and patient samples could be mixed prior to PCR amplification, the robustness and reproducibility of this act should be carefully considered.

15. If the samples had been overlaid with mineral oil in a previous step, then there is no need to add any more.

16. If the samples for HRM analysis were amplified in a PCR plate, add 5 μL of the reference control sample to the 10 μL of screened patient samples. As reported previously, the addition of only 15 % of a known homozygous genotype makes it possible to distinguish a patient's genotype [17, 52]. After pipetting the control samples into the patient samples, do not forget to spin the plate and carry out the denaturation step. Then perform an HRM analysis.

17. As the PCR reaction provides enough templates for conventional Sanger sequencing, there is no need for additional sample amplification. Notably, LCGreen Plus melting dye does not disrupt the sequencing procedure, but other components of the PCR may do so. For that reason, recombine the pure patient sample pipetted in a well of a PCR plate (10–20 μL) and the rest of the unused PCR sample (5 μL), and purify the pooled sample using a PCR purification kit. The mineral oil should not be a hindrance.

Acknowledgements

This work was supported by grants SVV2014/260022, P24/LF1/3 and UNCE 204011 from Charles University in Prague.

References

1. Stiburek L, Hansikova H, Tesarova M, Cerna L, Zeman J (2006) Biogenesis of eukaryotic cytochrome c oxidase. Physiol Res 55(Suppl 2): S27–S41

2. Balsa E, Marco R, Perales-Clemente E, Szklarczyk R, Calvo E, Landazuri MO, Enriquez JA (2012) NDUFA4 is a subunit of complex IV of the mammalian electron transport chain. Cell Metabol 16(3):378–386

3. Fernandez-Vizarra E, Tiranti V, Zeviani M (2009) Assembly of the oxidative phosphorylation system in humans: what we have learned by studying its defects. Biochim Biophys Acta 1793(1):200–211

4. Menezes MJ, Riley LG, Christodoulou J (2013) Mitochondrial respiratory chain disorders in childhood: insights into diagnosis and management in the new era of genomic medicine. Biochim Biophys Acta 1480:1368–1379

5. Indrieri A, van Rahden VA, Tiranti V, Morleo M, Iaconis D, Tammaro R, D'Amato I, Conte I, Maystadt I, Demuth S, Zvulunov A, Kutsche K, Zeviani M, Franco B (2012) Mutations in COX7B cause microphthalmia with linear skin lesions, an unconventional mitochondrial disease. Am J Hum Genet 91(5):942–949

6. Massa V, Fernandez-Vizarra E, Alshahwan S, Bakhsh E, Goffrini P, Ferrero I, Mereghetti P, D'Adamo P, Gasparini P, Zeviani M (2008) Severe infantile encephalomyopathy caused by a mutation in COX6B1, a nucleus-encoded subunit of cytochrome c oxidase. Am J Hum Genet 82(6):1281–1289

7. Pitceathly RD, Rahman S, Wedatilake Y, Polke JM, Cirak S, Foley AR, Sailer A, Hurles ME, Stalker J, Hargreaves I, Woodward CE, Sweeney MG, Muntoni F, Houlden H, Taanman JW, Hanna MG (2013) NDUFA4 mutations underlie dysfunction of a cytochrome c oxidase subunit linked to human neurological disease. Cell Rep 3(6): 1795–1805

8. Shteyer E, Saada A, Shaag A, Al-Hijawi FA, Kidess R, Revel-Vilk S, Elpeleg O (2009) Exocrine pancreatic insufficiency, dyserythropoeitic anemia, and calvarial hyperostosis are

caused by a mutation in the COX4I2 gene. Am J Hum Genet 84(3):412–417

9. Fraley SI, Hardick J, Jo Masek B, Athamanolap P, Rothman RE, Gaydos CA, Carroll KC, Wakefield T, Wang TH, Yang S (2013) Universal digital high-resolution melt: a novel approach to broad-based profiling of heterogeneous biological samples. Nucleic Acids Res 41(18):e175

10. Hernandez HG, Tse MY, Pang SC, Arboleda H, Forero DA (2013) Optimizing methodologies for PCR-based DNA methylation analysis. Biotechniques 55(4):181–197

11. Dobrowolski SF, Hendrickx AT, van den Bosch BJ, Smeets HJ, Gray J, Miller T, Sears M (2009) Identifying sequence variants in the human mitochondrial genome using high-resolution melt (HRM) profiling. Hum Mutat 30(6):891–898

12. Taylor CF (2009) Mutation scanning using high-resolution melting. Biochem Soc Trans 37(Pt 2):433–437

13. Vossen RH, Aten E, Roos A, den Dunnen JT (2009) High-resolution melting analysis (HRMA): more than just sequence variant screening. Hum Mutat 30(6):860–866

14. Do H, Dobrovic A (2009) Limited copy number – high resolution melting (LCN-HRM) enables the detection and identification by sequencing of low level mutations in cancer biopsies. Mol Cancer 8(1):82

15. Wittwer CT (2009) High-resolution DNA melting analysis: advancements and limitations. Hum Mutat 30(6):857–859

16. Tindall EA, Petersen DC, Woodbridge P, Schipany K, Hayes VM (2009) Assessing high-resolution melt curve analysis for accurate detection of gene variants in complex DNA fragments. Hum Mutat 30(6):876–883

17. Dwight ZL, Palais R, Kent J, Wittwer CT (2014) Heterozygote PCR product melting curve prediction. Hum Mutat 35(3):278–282

18. Liew M, Pryor R, Palais R, Meadows C, Erali M, Lyon E, Wittwer C (2004) Genotyping of single-nucleotide polymorphisms by high-resolution melting of small amplicons. Clin Chem 50(7):1156–1164

19. Petruzzella V, Carrozzo R, Calabrese C, Dell'Aglio R, Trentadue R, Piredda R, Artuso L, Rizza T, Bianchi M, Porcelli AM, Guerriero S, Gasparre G, Attimonelli M (2012) Deep sequencing unearths nuclear mitochondrial sequences under Leber's hereditary optic neuropathy-associated false heteroplasmic mitochondrial DNA variants. Hum Mol Genet 21(17):3753–3764

20. Payne BA, Wilson IJ, Yu-Wai-Man P, Coxhead J, Deehan D, Horvath R, Taylor RW, Samuels DC, Santibanez-Koref M, Chinnery PF (2013) Universal heteroplasmy of human mitochondrial DNA. Hum Mol Genet 22(2):384–390

21. Avital G, Buchshtav M, Zhidkov I, Tuval Feder J, Dadon S, Rubin E, Glass D, Spector TD, Mishmar D (2012) Mitochondrial DNA heteroplasmy in diabetes and normal adults: role of acquired and inherited mutational patterns in twins. Hum Mol Genet 21(19):4214–4224

22. Gundry CN, Dobrowolski SF, Martin YR, Robbins TC, Nay LM, Boyd N, Coyne T, Wall MD, Wittwer CT, Teng DH (2008) Base-pair neutral homozygotes can be discriminated by calibrated high-resolution melting of small amplicons. Nucleic Acids Res 36(10): 3401–3408

23. Seipp MT, Durtschi JD, Liew MA, Williams J, Damjanovich K, Pont-Kingdon G, Lyon E, Voelkerding KV, Wittwer CT (2007) Unlabeled oligonucleotides as internal temperature controls for genotyping by amplicon melting. J Mol Diagn 9(3):284–289

24. Liew M, Seipp M, Durtschi J, Margraf RL, Dames S, Erali M, Voelkerding K, Wittwer C (2007) Closed-tube SNP genotyping without labeled probes/a comparison between unlabeled probe and amplicon melting. Am J Clin Pathol 127(3):341–348

25. Zhou L, Errigo RJ, Lu H, Poritz MA, Seipp MT, Wittwer CT (2008) Snapback primer genotyping with saturating DNA dye and melting analysis. Clin Chem 54(10):1648–1656

26. Margraf RL, Mao R, Wittwer CT (2006) Masking selected sequence variation by incorporating mismatches into melting analysis probes. Hum Mutat 27(3):269–278

27. Dames S, Margraf RL, Pattison DC, Wittwer CT, Voelkerding KV (2007) Characterization of aberrant melting peaks in unlabeled probe assays. J Mol Diagn 9(3):290–296

28. Erali M, Palais R, Wittwer C (2008) SNP genotyping by unlabeled probe melting analysis. Methods Mol Biol 429:199–206

29. Herrmann MG, Durtschi JD, Wittwer CT, Voelkerding KV (2007) Expanded instrument comparison of amplicon DNA melting analysis for mutation scanning and genotyping. Clin Chem 53(8):1544–1548

30. Herrmann MG, Durtschi JD, Bromley LK, Wittwer CT, Voelkerding KV (2006) Amplicon DNA melting analysis for mutation scanning and genotyping: cross-platform comparison of instruments and dyes. Clin Chem 52(3): 494–503

31. Price EP, Smith H, Huygens F, Giffard PM (2007) High-resolution DNA melt curve analysis of the clustered, regularly interspaced short-palindromic-repeat locus of Campylobacter

jejuni. Appl Environ Microbiol 73(10): 3431–3436

32. Dwyer S, Carroll L, Mantripragada KK, Owen MJ, O'Donovan MC, Williams NM (2010) Mutation screening of the DTNBP1 exonic sequence in 669 schizophrenics and 710 controls using high-resolution melting analysis. Am J Med Genet B Neuropsychiatr Genet 153B(3):766–774

33. Vondrackova A, Vesela K, Hansikova H, Docekalova DZ, Rozsypalova E, Zeman J, Tesarova M (2012) High-resolution melting analysis of 15 genes in 60 patients with cytochrome-c oxidase deficiency. J Hum Genet 57(7):442–448

34. Lam NYL, Rainer TH, Chiu RWK, Lo YMD (2004) EDTA is a better anticoagulant than heparin or citrate for delayed blood processing for plasma DNA analysis. Clin Chem 50(1): 256–257

35. Meola G, Bugiardini E, Cardani R (2012) Muscle biopsy. J Neurol 259(4):601–610

36. Moore D, Dowhan D (2002) Purification and concentration of DNA from aqueous solutions. In: Current protocols in molecular biology. 59: I: 2.1A: 2.1.1-2.1.10

37. Mulyatno KC. Manual isolation of human dna from lymphoblasts or whole blood. http://www.itd.unair.ac.id/files/pdf/protocol1/Manual%20Isolation%20of%20Human%20DNA%20from%20Lymphoblasts%20or%20Whole%20Blood.pdf

38. Gallagher S (2001) Quantitation of nucleic acids with absorption spectroscopy. Curr Protoc Protein Sci. Appendix 4:Appendix 4K

39. Wong L-J (2013) Biochemical and molecular methods for the study of mitochondrial disorders. In: Wong L-JC (ed) Mitochondrial disorders caused by nuclear genes. Springer, New York, pp 27–45

40. Barreiro LB, Laval G, Quach H, Patin E, Quintana-Murci L (2008) Natural selection has driven population differentiation in modern humans. Nat Genet 40(3):340–345

41. Untergasser A, Nijveen H, Rao X, Bisseling T, Geurts R, Leunissen JA (2007) Primer3Plus, an enhanced web interface to Primer3. Nucleic Acids Res 35(Web Server issue):W71–W74

42. Krawczak M, Thomas NS, Hundrieser B, Mort M, Wittig M, Hampe J, Cooper DN (2007) Single base-pair substitutions in exon-intron junctions of human genes: nature, distribution, and consequences for mRNA splicing. Hum Mutat 28(2):150–158

43. Lewandowska MA (2013) The missing puzzle piece: splicing mutations. Int J Clin Exp Pathol 6(12):2675–2682

44. Dwight ZL, Palais R, Wittwer CT (2012) uAnalyze: web-based high-resolution DNA melting analysis with comparison to thermodynamic predictions. IEEE/ACM Trans Comput Biol Bioinform 9(6):1805–1811

45. Dwight Z, Palais R, Wittwer CT (2011) uMELT: prediction of high-resolution melting curves and dynamic melting profiles of PCR products in a rich web application. Bioinformatics 27(7):1019–1020

46. Ding M, Bullotta A, Caruso L, Gupta P, Rinaldo CR, Chen Y (2011) An optimized sensitive method for quantitation of DNA/RNA viruses in heparinized and cryopreserved plasma. J Virol Methods 176(1–2):1–8

47. Zakaria Z, Umi SH, Mokhtar SS, Mokhtar U, Zaiharina MZ, Aziz AT, Hoh BP (2013) An alternate method for DNA and RNA extraction from clotted blood. Genet Mol Res 12(1): 302–311

48. Wang TY, Wang L, Zhang JH, Dong WH (2011) A simplified universal genomic DNA extraction protocol suitable for PCR. Genet Mol Res 10(1):519–525

49. Norambuena PA, Copeland JA, Krenkova P, Stambergova A, Macek M Jr (2009) Diagnostic method validation: high resolution melting (HRM) of small amplicons genotyping for the most common variants in the MTHFR gene. Clin Biochem 42(12):1308–1316

50. Abtahi H, Sadeghi MR, Shabani M, Edalatkhah H, Hadavi R, Akhondi MM, Talebi S (2011) Causes of bimodal melting curve: asymmetric guanine-cytosine (GC) distribution causing two peaks in melting curve and affecting their shapes. Afr J Biotechnol 10(50):10196–10203

51. Wittwer CT, Reed GH, Gundry CN, Vandersteen JG, Pryor RJ (2003) High-resolution genotyping by amplicon melting analysis using LCGreen. Clin Chem 49(6 Pt 1): 853–860

52. Palais RA, Liew MA, Wittwer CT (2005) Quantitative heteroduplex analysis for single nucleotide polymorphism genotyping. Anal Biochem 346(1):167–175

Chapter 32

Heterologous Inferential Analysis (HIA) as a Method to Understand the Role of Mitochondrial rRNA Mutations in Pathogenesis

Joanna L. Elson*, Paul M. Smith*, and Antón Vila-Sanjurjo

Abstract

Despite the identification of a large number of potentially pathogenic variants in the mitochondrially encoded rRNA (mt-rRNA) genes, we lack direct methods to firmly establish their pathogenicity. In the absence of such methods, we have devised an indirect approach named heterologous inferential analysis or HIA that can be used to make predictions on the disruptive potential of a large subset of mt-rRNA variants. First, due to the high evolutionary conservation of the rRNA fold, comparison of phylogenetically derived secondary structures of the human mt-rRNAs and those from model organisms allows the location of structurally equivalent residues. Second, visualization of the heterologous equivalent residue in high-resolution structures of the ribosome allows a preliminary structural characterization of the residue and its neighboring region. Third, an exhaustive search for biochemical and genetic information on the residue and its surrounding region is performed to understand their degree of involvement in ribosomal function. Additional rounds of visualization in biochemically relevant high-resolution structures will lead to the structural and functional characterization of the residue's role in ribosomal function and to an assessment of the disruptive potential of mutations at this position. Notably, in the case of certain mitochondrial variants for which sufficient information regarding their genetic and pathological manifestation is available; HIA data alone can be used to predict their pathogenicity. In other cases, HIA will serve to prioritize variants for additional investigation. In the context of a scoring system specifically designed for these variants, HIA could lead to a powerful diagnostic tool.

Key words Mitochondrial rRNA, mtDNA, Mitoribosome, Mitochondrial deafness, mtDNA mutation

1 Introduction

Human mitochondrial DNA (mtDNA) encodes 13 essential polypeptide components of the oxidative phosphorylation system, which produces most of the cell's ATP. In addition to these polypeptides, mtDNA also codes for 22 mitochondrial tRNAs (mt-tRNAs) and 2 mt-rRNAs needed to translate these 13 polypeptides within the organelle.

These authors contributed equally to this work.

Volkmar Weissig and Marvin Edeas (eds.), *Mitochondrial Medicine: Volume I, Probing Mitochondrial Function*, Methods in Molecular Biology, vol. 1264, DOI 10.1007/978-1-4939-2257-4_32, © Springer Science+Business Media New York 2015

While the identification of mutations causing disease in the mt-tRNA genes and the mitochondrial protein-encoding genes has seen vast improvements in the last 10 years, the same is not true of mutations in the mt-rRNA genes. This is largely because distinguishing pathogenetic changes from polymorphic variants remains highly problematic for these genes, as many of the functional tests used to distinguish population variants from disease-causing mutations in the protein-encoding genes and mt-tRNAs are simply not applicable in the context of mt-rRNAs.

The lack of direct biochemical evidence prevents the analysis of mt-rRNA variants in their natural context. As a result, indirect methods must be applied to infer the role of such variants in ribosomal function. To date, comparative analysis of mt-rRNA variations has been rather limited and has often relied on methods questionable in this context. Such methods include the estimation of evolutionary conservation of a variant using a small number of sequences which often represent a narrow section of the phylogenetic tree of life. Even for mt-DNA gene categories where this type of analysis has validity, e.g. mt-tRNA genes, the rigor of application has been subject to scrutiny [1]. Given the complexity of the rRNA structure and with the mt-rRNAs being no exception, a more robust approach is necessary in order to assess the pathogenicity of mt-rRNA mutations.

Another common practice among researchers aiming to characterize mt-rRNA mutations, which is borrowed from those considering mt-tRNA mutations, has been to score mutations supposedly leading to the disruption of canonical Watson-Crick (WC) base pairs as likely pathogenic and to dismiss mutations mapping to predicted single-stranded regions as nonpathogenic [2–4]. An additional complication has been the use of non-phylogeny-driven folding algorithms to predict the mt-rRNA structure [5–7]. Unfortunately, these practices disregarded many well-known facts about the rRNA structure and function learned from heterologous ribosomes. For example, while WC base pairing accounts for up to 60 % of the structure present in bacterial ribosomes there are many other hydrogen-bonding interactions involving both rRNA and ribosomal protein residues are necessary to stabilize such a large molecular structure [8]. So much so that according to Harry Noller [8], "bases that are not involved in either Watson-Crick or some kind of noncanonical interaction are very rare." In addition, many WC base pairs are part of tertiary and quaternary interactions involving other rRNA bases and/or ribosomal proteins. Finally, some of the most important residues in the ribosome, usually involved in the binding of ribosomal ligands, are displayed in single-stranded regions in secondary-structure maps of rRNA. To further complicate the issue, the ribosome is a highly dynamic particle that interacts with a myriad of ligands during translation. Therefore, to elucidate the disruptive potential of an rRNA mutation, one needs to know the whole spectrum of secondary, tertiary, and quaternary interactions in which the mutated base might be involved during the translation cycle.

The universality of the ribosome and of the core of the translation cycle, together with the high conservation of the rRNA fold, provides a framework for the identification of functionally and structurally equivalent rRNA residues in rRNAs from phylogenetically diverse origins [9]. Structural predictions made by comparative phylogenetic methods faced their most stringent test once the first high-resolution structures of the ribosomal subunits emerged at the turn of the century. Strikingly, approximately 97–98 % of the base pairings predicted by a covariation analysis were found to be present in the Small SUbunit (SSU) and Large SUbunit (LSU) crystal structures [10]. In addition, these structures validated the rRNA alignments that had been generated by comparative methods [10]. Thus, such alignments can be used as the basis for inferential analysis aimed at establishing the structural relationship among rRNA residues from organisms or organelles for which no high-resolution structures exist. We propose that the use of phylogenetically derived rRNA alignments, together with the recent explosion of data from high-resolution ribosome structures and with the vast amount of literature bearing on the genetic and biochemical characterization of the ribosome and its cycle, can be used to inferentially study the effects of mt-rRNA variations on the structure and function of the mitoribosome. Our method, named heterologous inferential analysis (HIA), takes advantage of all these tools to elucidate whether mt-rRNA variations suspected to be implicated in mitochondrial failure might do so by the disruption of mitoribosomal function. We believe that this is an obligatory first step toward understanding the true role of these mutations in human disease. We have already applied HIA to the study of rare mitochondrial variations mapping to the human mitochondrial 12S [11] and 16S rRNAs (manuscript in preparation).

Applying HIA requires a deep understanding of RNA structure, the ribosome, and the complex process of translation. A review of the vast collection of literature on these topics is well beyond the scope of this paper. Here, we include a collection of protocols that can be used as an introduction to HIA for researchers unfamiliar with RNA structure and the process of protein synthesis. A schematic description of HIA is provided in Fig. 1.

2 Computational Requirements Necessary for HIA Analysis

HIA is a fully computational method. UCSF Chimera, a program for interactive analysis and visualization of molecular structures, was used for the structural analysis (https://www.cgl.ucsf.edu/chimera/). The program was developed by the Resource for Biocomputing, Visualization, and Informatics and funded by

Heterologous Inferential Analysis (HIA)

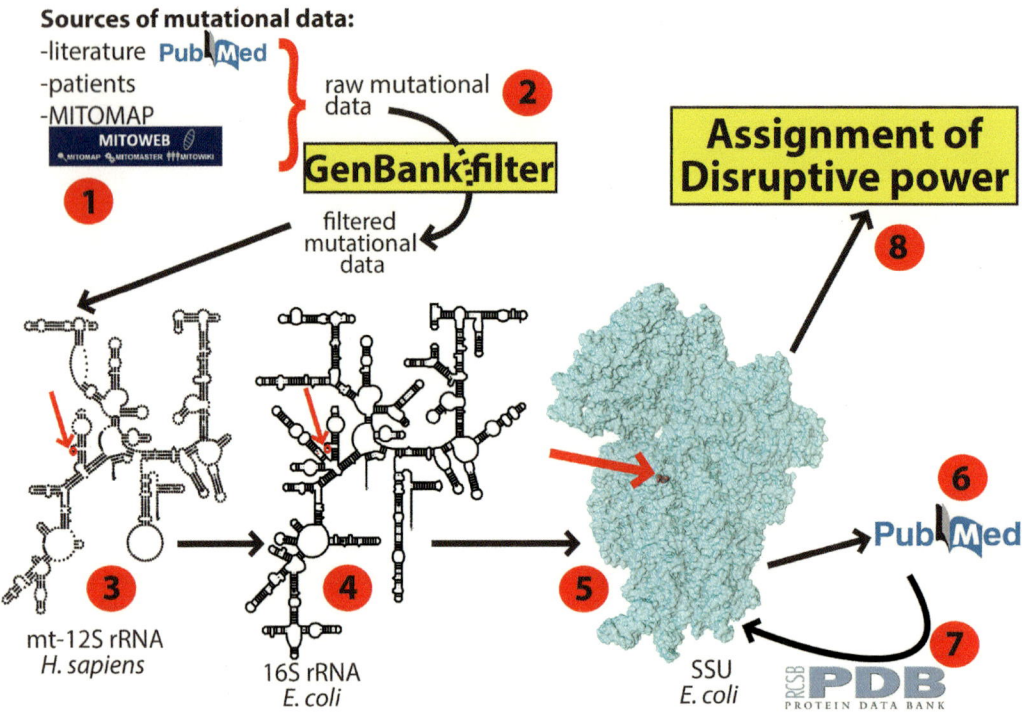

Sources of mutational data:
- literature PubMed
- patients
- MITOMAP

raw mutational data

GenBank filter

Assignment of Disruptive power

filtered mutational data

mt-12S rRNA
H. sapiens

16S rRNA
E. coli

SSU
E. coli

PubMed

PDB
PROTEIN DATA BANK

Fig. 1 HIA scheme. The numbers show the flow of information: (1) sources of mutational data, both experimental and published, (2) the GenBank filter, (4) placement of the site of variation on the secondary-structure map of the *Homo sapiens'* mt-rRNA, (5) assignment of the heterologous equivalent residue on a bacterial rRNA map, (3) visualization of the heterologous equivalent on a reference high-resolution structure, (6) search through relevant literature, (7) visualization in additional, relevant, high-resolution structures (not necessary for all variations), and (8) HIA predictions. The *red arrows* indicate sites of mutation

the National Institutes of Health of the USA (NIGMS P41-GM103311). The analysis of the mt-SSU rRNA, UCSF Chimera was run on an Intel(R) core(TM) i7 CPU Q 720 (1.60 GHz) with 6GB of RAM and an NVIDIA GeForce GT 230M video card, running a Windows 64-bit operating system. PERL scripts were run on the same hardware but with a Linux-based operating system.

While the above hardware specifications were satisfactory for the work on the SSU; the use of a larger number of superimposed heterologous models for the LSU work (up to five entire ribosomes loaded in UCSF Chimera, occupying an average of 3.5–4 GB of RAM) clearly reached the memory limits of the system. This problem should be easily avoided by increasing the system's RAM.

3 Collection and Preparation of Mitochondrial DNA Sequence Data for HIA

3.1 Sources of Mutational Data

The following types of mitochondrial variations have been subjected to HIA:

(a) Variations identified in human subjects with a suspected mitochondrial disease.

(b) Somatic mtDNA mutations that had undergone clonal expansions in cancer patients. To identify the variant as a somatic mutation that had undergone clonal expansion, the existence of a non-mutated and noncancerous control tissue from the same patient was a requirement.

(c) COX-deficient cells located in aged tissues. To identify the variant as a somatic mutation that had undergone clonal expansion, the existence of a COX-proficient surrounding tissue was a requirement.

(d) Variants known to define human haplogroups and thus believed not to play a role in the etiology of disease through the generation of a cellular biochemical defect.

3.2 Analysis of GenBank Mitochondrial Genome Data

At present, HIA has been used only to evaluate the disruptive potential of human mutations with zero appearances in controls, thus increasing the likelihood that the variant plays a role in disease. We have used the ~21,500 sequences of the human mitochondrial genome present in the GenBank database (November 2013) as a filter to "weed out" variations that are present in non-patient sequences. We are aware of the fact that this method has the potential to leave out some truly pathogenic mutations:

1. Retrieve all the GenBank records corresponding to sequenced human mt-genomes. The query "Homo[Organism]+AND+ gene_in_mitochondrion[PROP]+AND+14000:19000[SLEN] +NOT+pseudogene[All Fields]" [12] can be used to download all the available records. All sequences were downloaded with Entrez Programming Utilities and our own PERL scripts [13]. All the data should be easily downloaded by a simple batch retrieval as described under "Retrieving large datasets" (http://www.ncbi.nlm.nih.gov/books/NBK25498/#chapter3. Application_3_Retrieving_large) [13].

We have found that batch retrieval of large datasets is prone to failure. As this problem might have been permanently solved, we recommend to attempt the quick batch method first. In case that some of the records fail to download, sequential download can be attempted. This method is lengthy and requires writing a script that can sequentially post "efetch" queries for the individual missing records while keeping track of the downloaded ones. If need be, our PERL script for the sequential downloading of GenBank's missing records can be made available upon request.

All the sequence data was downloaded in FASTA format and sequentially aligned to the revised Cambridge Reference (NC_012920.1) [14] by using the quick alignment algorithm afforded by the MAFFT package [15]. The process was automated by means of a PERL script, and the generated alignments were subsequently analyzed by means of a second script that recorded the total number of appearances per mitochondrial variation (again, these PERL scripts will be made available upon request). Since the downloaded records follow the International Union of Pure and Applied Chemistry (IUPAC) nucleotide base code (http://www.ncbi.nlm.nih.gov/Class/MLACourse/Modules/MolBioReview/iupac_nt_abbreviations.html), care should be taken to make sure that the correct base change is assigned to individual variations. Mitochondrial genomic positions appearing as "N" were dismissed as uninformative.

For variations with 15 or less appearances in the combined dataset, including the data from both the GenBank and Phylotree databases, the possibility that the individual sequences could be linked to mitochondrial disease was investigated by inspecting the related publication sources. In all cases it was observed that no new patient sequences other than the ones previously found in the literature, were present within GenBank.

4   Assignment of the Disruptive Power Potential by HIA

4.1 Assignment of Variations to mt-rRNA Positions

12S rRNA starts at genomic position 648 and 16S rRNA starts at position 1,671; therefore, mt-rRNA positions are calculated by subtracting either 647 or 1,670 from the corresponding genomic positions. When numerous variations were to be scored, we found it convenient to place the mt-rRNA positions of the variations on a primary-sequence alignment between the homologous human 12S and 16S rRNAs and the heterologous sequences from a model organism (*E. coli* in our case). To do this, we used the program LALIGN [16] due to its simplicity of use and its availability as an online tool (http://www.ch.embnet.org/software/LALIGN_form.html).

Phylogenetically derived secondary-structure maps are available at http://www.rna.icmb.utexas.edu/ [17]. Such maps constitute a very convenient tool to assign heterologous equivalent residues, due to the easy identification of conserved structural features. Once the heterologous equivalent has been found, its position must be located on a high-resolution model of the heterologous ribosome.

4.2 Variations Can Be Investigated Using High-Resolution Ribosomal Models (Selecting Residues and Inspecting Local Structure)

Coordinate files for high-resolution structures of the ribosome are deposited at the RCSB Protein Data Bank (http://www.rcsb.org). At present, only crystal structures obtained by X-ray diffraction afford near atomic resolution structures of the ribosome. Accession codes for some representative high-resolution structures are shown in Table 1. It should be noted that due to crystallographic symmetry, several different structures of the ribosome and/or ribosomal subunits

Table 1
Representative high-resolution structures used in HIA analysis

Accession codes: SSU, *LSU* / RCSB title	Ribosomal composition	Rotation state	mRNA	tRNA	Other ligands	Resolution (Å)	Refinement parameters	Organism	Phylogenic domain	Publication	PubMed ID
2I2P 2I2T 2I2U 2I2V — Crystal structure of ribosome with messenger RNA and the anticodon stem loop of P-site tRNA	SSU/LSU	Non-ratcheted	Y	P(ASL)	None	3.22	R-value: 0.287 (work) R-free: 0.320	*Escherichia coli*	Bacteria	[24]	17038497
2J00 2J01 2J02 2J03 — *Structure of the Thermus thermophilus 70S ribosome complexed with mRNA, tRNA, and paromomycin*	SSU/LSU	Non-ratcheted	Y	A(ASL^a), P, E	Paromomycin	2.3	R-value: (0.271 (obs.) R-free: 0.313	*Thermus thermophilus*	Bacteria	[25]	16959973
4HUB 4I4M — The re-refined crystal structure of the *Haloarcula marismortui* large ribosomal subunit at 2.4 Angstrom resolution: more complete structure of the L7/L125 and L1 stalk, L5, and LX proteins	LSU	NA	N	None	None	2.4	R-value: 0.166 (obs.) R-free: 0.206	*Haloarcula marismortui*	Archaea	[26]	23695244
2XZM 2XZN — Crystal structure of the eukaryotic 40S ribosomal subunit in complex with initiation factor 1	SSU	NA	N	None	eIF1	3.93	R-value: 0.206 (obs.) R-free: 0.243	*Tetrahymena thermophila*	Eucarya	[27]	21205638

(continued)

Table 1
(continued)

Accession codes: SSU, *LSU*	RCSB title	Ribosomal composition	Rotation state	mRNA	tRNA	Other ligands	Resolution (Å)	Refinement parameters	Organism	Phylogenic domain	Publication	PubMed ID
4KIX 4KIY 4KIZ 4KI0 4KJ1 4KI2 4KJ3 4KI4	Control of ribosomal subunit rotation by elongation factor G	SSU/LSU	Ratcheted	N	None	EFG, viomycin	2.9	R-value: 0.223 (work) R-free: 0.271	*Escherichia coli*	Bacteria	[19]	23812721
3U5B 3U5C 3U5D 3U5E 3U5F 3U5G 3U5H 3U5I	The structure of the eukaryotic ribosome at 3.0 a resolution	SSU/LSU	Ratcheted	N	None	None	3.0	R-value: 0.182 (obs.) R-free: 0.228	*Saccharomyces cerevisiae*	Eucarya	[28]	22096102
3D5A 3D5B 3D5C 3D5D	Structural basis for translation termination on the 70S ribosome	SSU/LSU	Non-ratcheted	Y	P, E	RF1	3.21	R-value: 0.292 (obs.) R-free: 0.319	*Thermus thermophilus*	Bacteria	[29]	18596689
3F1E 3F1F 3F1G 3F1H	Crystal structure of a translation termination complex formed with release factor RF2	SSU/LSU	Non-ratcheted	Y	P, E	RF2	3.0	R-value: 0.281 (obs.) R-free: 0.316	*Thermus thermophilus*	Bacteria	[30]	19064930

PDB IDs	Description		Ratcheted		Sites	Ligands	Resolution	R-values	Species		Ref.	PMID
4KBT 4KBU 4KBV 4KBW	70s ribosome translocation intermediate GDPNP-II containing elongation factor EFG/GDPNP, mRNA, and tRNA bound in the PE*/E state	SSU/LSU	Ratcheted	Y	pe*/E[b]	EFG-GDPNP, viomycin	3.86	R-value: 0.264 (obs.) R-free: 0.317	Thermus thermophilus	Bacteria	[31]	23812722
2WRI 2WRJ 2WRK 2WRL	The structure of the ribosome with elongation factor G trapped in the post-translocational state	SSU/LSU	Non-ratcheted	Y	P, E	EFG-fusidic acid	3.60	R-value: 0.227 (obs.) R-free: 0.260	Thermus thermophilus	Bacteria	[32]	19833919
2XQD 2XQE	The structure of EF-TU and aminoacyl-tRNA bound to the 70s ribosome with a GTP analogue	SSU/LSU	Non-ratcheted	Y	A/T[c], P, E	EFTu-GDPCP, paromomycin	3.1	R-value: 0.231 (obs.) R-free: 0.268	Thermus thermophilus	Bacteria	[33]	21051640
2WDK 2WDL 2WDM 2WDN	Structure of the Thermus thermophilus 70S ribosome in complex with mRNA, paromomycin, acylated A- and P-site tRNAs, and E-site tRNA	SSU/LSU	Non-ratcheted	Y	aa-A[d], aa-P[d], E	Paromomycin	3.5	R-value: 0.208 (obs.) R-free: 0.256	Thermus thermophilus	Bacteria	[34]	19363482

(continued)

Table 1
(continued)

Accession codes: SSU, *LSU*	RCSB title	Ribosomal composition	Rotation state	mRNA	tRNA	Other ligands	Resolution (Å)	Refinement parameters	Organism	Phylogenic domain	Publication	PubMed ID
2WDG 2WDH 2WDI 2WDJ	Structure of the *Thermus thermophilus* 70S ribosome in complex with mRNA, paromomycin, acylated A-site tRNA, deacylated P-site tRNA, and E-site tRNA	SSU/LSU	Non-ratcheted	Y	aa-A^d, p, e	Paromomycin	3.3	R-value: 0.223 (obs.) R-free: 0.272	*Thermus thermophilus*	Bacteria	[34]	19363482

RCSB accession codes and relevant crystallographic and phylogenetic information are provided for crystal structures used in HIA analysis (11, manuscript in preparation). SSU accession codes are underlined and LSU accession codes are italicized

[a]Notes from Selmer et al. [25]. "The anticodon stem loop (ASL), comprising nucleotides 26–44, was clearly visible in the decoding center of the 30S subunit, with an orientation that corresponds to the accommodated form in which the acceptor arm is in the PTC. Surprisingly, although full-length aminoacylated tRNAPhe was used in crystallization and its binding was stabilized by paromomycin, the rest of the tRNA was not visible in the density. We believe that this is due to deacylation of tRNAPhe during crystallization, which results in a reduction of affinity for the PTC and a disorder of the acceptor arm"

[b]Notes from Zhou et al. [31]. "Because the ASL is simultaneously bound to P-site elements of the 30S head and to features that lie between the P and E sites of the 30S body and platform, we refer to this new state as a chimeric pe*/E hybrid state. Repositioning of the tRNA is promoted not only by structural changes in the ribosome but also by conformational changes in the tRNA itself. We imagine that the pe*/E state represents an intermediate between the P/E and E/E states"

[c]Notes from Voorhees et al. [33]. "Ribosome binding and codon recognition induce the bent tRNA in the A/T state (where the anticodon interacts with the mRNA in the A site and the acceptor end is bound to EF-Tu)"

[d]*aa* aminoacylated

might be present in the asymmetric crystallographic unit. While in most cases a, single structure from these crystals is sufficient for visualization, which might be not always the case, as on certain occasions the different ribosomes present in the asymmetric unit correspond to different conformational states [18, 19]. In principle, crystal structures of complete ribosomes with phylogenetically conserved ligands are the most informative. The bacterial ribosome has been by far the most thoroughly studied, both biochemically and genetically, as well as structurally. As a result, we currently have several structures of bacterial ribosomes in different structural conformations and in complex with many different conserved ligands that are necessary during translation. The resolution and the R and R-free factors are important parameters to consider, as they inform us about the quality of the model obtained from crystallographic data [20]. Typical R values for near atomic ribosome structures lie between 0.2 and 0.3, with the R-free factor lying slightly above the R factor.

The visualization software UCSF Chimera [21] provides a free, powerful, and user-friendly tool for the structural analysis of ribosomal structures. Online manuals and tutorials are available at https://www.cgl.ucsf.edu/chimera/. Here we will describe some useful instructions to load, visualize, and perform structural analysis with UCSF Chimera (version 1.7). A suggested plan of action for the elucidation of the role of any rRNA residue without any preliminary knowledge of its function goes as follows:

1. Upload the coordinate files for a 70S structure of the unratcheted bacterial ribosome, with the highest possible resolution. The file is downloaded from RCSB Protein Data Bank (choose PDB file) and loaded on UCSF Chimera. Note that RCSB Protein Data Bank provides a list of all the different chains present in the coordinate file.

2. The heterologous residue equivalent to the mitochondrial variant is found by selecting its position on the primary sequence:

   (a) Click **Tools:Sequence** and then select the chain ID corresponding to the rRNA where the residue is located.

   (b) Highlight a residue on the sequence, which selects it for further manipulation.

   (c) Click **Actions:Focus** to make the residue visible.

   (d) Click **Actions:Atoms/Bonds:Show** to display a wire model of the residue.

   (e) Ctrl-click on any atom to select it, and then click **Actions:label:other** to open a window where a label for the residue, placed on the selected atom, can be entered. Additionally, a different color can be used to highlight the residue after clicking on its corresponding ribbon portion and ctrl-clicking it to select all its atoms. A color palette can be found under **Actions:Color**.

(f) Ctrl-click on neighboring rRNA residues to sort out all possible secondary, tertiary, and quaternary interactions. According to Jeffrey [22], the mean donor-acceptor distances for hydrogen bonds in protein and nucleic acids are close to 3.0 Å. Therefore, any donor-acceptor pairs within 3.0 Å likely correspond to a hydrogen bond. In cases where the heterologous equivalent is part of a canonical Watson-Crick base pair, its partner should be easily spotted after displaying its wire representation.

(g) Ctrl-click any two atoms to determine the distance between them. To display the distance between two atoms, use the **Distances** wizard (**Tools:Structure Analysis:Distances**).

(h) Save the Chimera session so that it can be loaded back at a later time (Select **File:Save Session**).

Regardless of how informative 3D representations of the ribosome might be to elucidate the role of individual rRNA residues in the stabilization of local structure; the enormity of the ribosome, relative to the size of a nucleotide, efficiently dilutes down the deleterious contribution of most rRNA mutations to normal ribosomal function. Thus, the conclusions of the preliminary structural analysis must be complemented with a search for biochemical and genetic data that can be used to rationalize the effects of the structural perturbation on ribosome function. A good starting point to initiate this search is to use the Ribosomal Mutation Database (ribosome.fandm.edu) and its associated reference [23] to elucidate whether the heterologous equivalent or a neighboring residue has been targeted by mutagenesis studies of non-mitochondrial ribosomes. Subsequent literature searches with the standard resources (Google, PubMed) will necessarily target papers in which links between the mutated region and specific aspects of ribosome function are established. These investigations should be followed by the visual inspection of such links in high-resolution structures of the ribosome in complex with pertinent ribosomal ligands. In order to best visualize structures of the ribosome in different steps of translation, it is important to first superpose the structures by using the MatchMaker tool of UCSF Chimera. For bacterial ribosomes, we normally use either the 16S or 23S rRNA chains to obtain the best superposition results. To display the result of SSU rotation while comparing differently rotated states of the ribosome, the 23S rRNA must be used for the superposition. The following example will illustrate this process:

1. Let's suppose that two different ribosomes, consisting of four independent PDB files, are opened in UCSF Chimera in the following order: SSU ribosome 1, LSU ribosome 1, SSU ribosome 2, and LSU ribosome 2. UCSF Chimera will automatically assign consecutive numbers to the models, starting at 0.

2. The LSU from ribosome 2 (#3) is superposed onto the LSU from ribosome 1 (#1) by selecting **Tools:Structure Comparison:MatchMaker**. In the **MatchMaker** dialog box, choose the option **Specific chain(s) in reference structure with specific chain(s) in match structure**; make sure that **Matrix** displays the **Nucleic** option and leave all other options as default.

3. To move the SSU from ribosome 2 (#2), the coordinate transformation matrix generated by **MatchMaker** is applied to model #2 by typing **matrixcopy #3 #1 moving #2** in the command line (click **Favorites/Command** line to open the command line dialog box).

4. It is convenient to save the modified PDB files for future use after selecting the corresponding model on the **Model Panel** (**Favorites:Model Panel**) and clicking the **Write PDB** option on the **Model Panel** (select **Favorites:Model Panel**). Make sure that the proper model is chosen on the dialog box, that a new name is given to the file to be saved, and that the **Save relative to model** option is clicked and displays the name of the reference model. Save the Chimera session so that it can be loaded back at a later time.

The same method can be used to superimpose heterologous models from different model organisms.

5 Criteria for the Assignment of Disruptive Power Potential

The final goal of HIA is to establish the disruptive potential of a putative mutation occurring at the heterologous equivalent residue, as a basis for estimating the degree of pathogenicity of the original mitochondrial variation. To do this, we have come up with the following scoring system:

1. Supportive, direct mutagenesis data exists against a deleterious effect of the mutation and/or its base pairing partner on ribosomal function. -->**N**="certainly not disruptive."

2. No direct mutagenesis data exists, but is there substantial indirect data against a deleterious effect of the mutation and/or its base pairing partner on ribosomal function -->**U**="unlikely disruptive."

3. No direct nor indirect data arguing for or against a deleterious effect exists. **NEE**="not enough evidence."

4. Enough indirect data (but not direct mutagenesis data) exists that suggests a deleterious effect associated to the mutation. -->**L**="likely disruptive."

5. Enough direct mutagenesis data demonstrating a deleterious effect associated to the mutation. -->**C**="certainly disruptive."

6. An mtDNA mutation would be considered to have the same disruptive power as its heterologous counterpart as long as the available structural data for the former is not in disagreement with the heterologous evidence, e.g. if the higher-order structural elements taken into consideration are conserved in both the mitochondrial and heterologous ribosomes. Two additional categories were used to classify the mitochondrial mutations:

(a) **Und**="undetermined." Either a heterologous equivalent residue cannot be assigned with enough certainty, or the existing structural differences in the surrounding region are too large to allow the extrapolation of conclusions made in the heterologous case.

(b) **E**="expectedly disruptive." This category contains the mitochondrial equivalents to residues assigned as "certainly disruptive" in the heterologous system. When the mitochondrial mutation is present in homoplasmy, the mutation can be considered pathogenic.

Note added in proof

Recent advances in cryo-electron microscopy have allowed the groups led by R.K. Agrawal, N. Ban, and V. Ramakrishnan to achieve medium-resolution and near-atomic-resolution structures of mammalian mitoribosomal particles. Such advances now permit the placement of sites of mutation directly on mitoribosomal structures, thus dramatically improving the predictive power of the methods described here.

References

1. Yarham JW, McFarland R, Taylor RW, Elson JL (2012) A proposed consensus panel of organisms for determining evolutionary conservation of mt-tRNA point mutations. Mitochondrion 12:533–538

2. Lu J, Li Z, Zhu Y, Yang A, Li R, Zheng J, Cai Q, Peng G, Zheng W, Tang X et al (2010) Mitochondrial 12S rRNA variants in 1642 Han Chinese pediatric subjects with aminoglycoside-induced and nonsyndromic hearing loss. Mitochondrion 10:380–390

3. Shen Z, Zheng J, Chen B, Peng G, Zhang T, Gong S, Zhu Y, Zhang C, Li R, Yang L et al (2011) Frequency and spectrum of mitochondrial 12S rRNA variants in 440 Han Chinese hearing impaired pediatric subjects from two otology clinics. J Transl Med 9:4

4. Mutai H, Kouike H, Teruya E, Takahashi-Kodomari I, Kakishima H, Taiji H, Usami S, Okuyama T, Matsunaga T (2011) Systematic analysis of mitochondrial genes associated with hearing loss in the Japanese population: DHPLC reveals a new candidate mutation. BMC Med Genet 12:135

5. Tang J, Qi Y, Bao XH, Wu XR (1997) Mutational analysis of mitochondrial DNA of children with Rett syndrome. Pediatr Neurol 17:327–330

6. Tang HY, Hutcheson E, Neill S, Drummond-Borg M, Speer M, Alford RL (2002) Genetic susceptibility to aminoglycoside ototoxicity: How many are at risk? Genet Med 4:336–345

7. Trifunovic A (2006) Mitochondrial DNA and ageing. Biochim Biophys Acta 1757:611–617

8. Noller HF (2005) RNA structure: reading the ribosome. Science 309:1508–1514

9. Subhankar B, Dhananjaya S (2003) MITOMAP mtDNA sequence data: unpublished variant 20041220003

10. Gutell RR, Lee JC, Cannone JJ (2002) The accuracy of ribosomal RNA comparative structure models. Curr Opin Struct Biol 12:301–310

11. Smith PM, Elson JL, Greaves LC, Wortmann SB, Rodenburg RJ, Lightowlers RN, Chrzanowska-Lightowlers ZM, Taylor RW, Vila-Sanjurjo A (2014) The role of the mitochondrial ribosome in human disease: searching for mutations in 12S mitochondrial rRNA with high disruptive potential. Hum Mol Genet 23:949–956

12. Pereira L, Freitas F, Fernandes V, Pereira JB, Costa MD, Costa S, Maximo V, Macaulay V, Rocha R, Samuels DC (2009) The diversity present in 5140 human mitochondrial genomes. Am J Hum Genet 84:628–640

13. Sayers E (2010) E-utilities quick start. In: Entrez programming utilities help [Internet]. National Center for Biotechnology Information (US), Bethesda (MD)

14. Andrews RM, Kubacka I, Chinnery PF, Lightowlers RN, Turnbull DM, Howell N (1999) Reanalysis and revision of the Cambridge reference sequence for human mitochondrial DNA. Nat Genet 23:147

15. Katoh K, Misawa K, Kuma K, Miyata T (2002) MAFFT: a novel method for rapid multiple sequence alignment based on fast fourier transform. Nucleic Acids Res 30:3059–3066

16. Huang X, Miller W (1991) A time-efficient, linear-space local similarity algorithm. Adv Appl Math 12:337–357

17. Cannone JJ, Subramanian S, Schnare MN, Collett JR, D'Souza LM, Du Y, Feng B, Lin N, Madabusi LV, Muller KM et al (2002) The comparative RNA web (CRW) site: an online database of comparative sequence and structure information for ribosomal, intron, and other RNAs. BMC Bioinformatics 3:2

18. Schuwirth BS, Borovinskaya MA, Hau CW, Zhang W, Vila-Sanjurjo A, Holton JM, Cate JH (2005) Structures of the bacterial ribosome at 3.5 A resolution. Science 310:827–834

19. Pulk A, Cate JH (2013) Control of ribosomal subunit rotation by elongation factor G. Science 340:1235970

20. Kleywegt GJ, Alwyn JT (1997) Model building and refinement practice. Meth Enzymol 277:208–230

21. Pettersen EF, Goddard TD, Huang CC, Couch GS, Greenblatt DM, Meng EC, Ferrin TE (2004) UCSF chimera – a visualization system for exploratory research and analysis. J Comput Chem 25:1605–1612

22. Jeffrey G (1997) An introduction to hydrogen bonding. Oxford University Press, Oxford

23. Triman KL (2007) Mutational analysis of the ribosome. Adv Genet 58:89–119

24. Berk V, Zhang W, Pai RD, Doudna Cate JH (2006) Structural basis for mRNA and tRNA positioning on the ribosome. Proc Natl Acad Sci U S A 103:15830–15834

25. Selmer M, Dunham CM, Murphy FV 4th, Weixlbaumer A, Petry S, Kelley AC, Weir JR, Ramakrishnan V (2006) Structure of the 70S ribosome complexed with mRNA and tRNA. Science 313:1935–1942

26. Gabdulkhakov A, Nikonov S, Garber M (2013) Revisiting the haloarcula marismortui 50S ribosomal subunit model. Acta Crystallogr D Biol Crystallogr 69:997–1004

27. Rabl J, Leibundgut M, Ataide SF, Haag A, Ban N (2011) Crystal structure of the eukaryotic 40S ribosomal subunit in complex with initiation factor 1. Science 331:730–736

28. Ben-Shem A, Garreau de Loubresse N, Melnikov S, Jenner L, Yusupova G, Yusupov M (2011) The structure of the eukaryotic ribosome at 3.0 A resolution. Science 334:1524–1529

29. Laurberg M, Asahara H, Korostelev A, Zhu J, Trakhanov S, Noller HF (2008) Structural basis for translation termination on the 70S ribosome. Nature 454:852–857

30. Korostelev A, Asahara H, Lancaster L, Laurberg M, Hirschi A, Zhu J, Trakhanov S, Scott WG, Noller HF (2008) Crystal structure of a translation termination complex formed with release factor RF2. Proc Natl Acad Sci U S A 105:19684–19689

31. Zhou J, Lancaster L, Donohue JP, Noller HF (2013) Crystal structures of EF-G-ribosome complexes trapped in intermediate states of translocation. Science 340:1236086

32. Gao YG, Selmer M, Dunham CM, Weixlbaumer A, Kelley AC, Ramakrishnan V (2009) The structure of the ribosome with elongation factor G trapped in the posttranslocational state. Science 326:694–699

33. Voorhees RM, Schmeing TM, Kelley AC, Ramakrishnan V (2010) The mechanism for activation of GTP hydrolysis on the ribosome. Science 330:835–838

34. Voorhees RM, Weixlbaumer A, Loakes D, Kelley AC, Ramakrishnan V (2009) Insights into substrate stabilization from snapshots of the peptidyl transferase center of the intact 70S ribosome. Nat Struct Mol Biol 16:528–533

Chapter 33

Analysis of Mitochondrial Dysfunction During Cell Death

Vladimir Gogvadze, Sten Orrenius, and Boris Zhivotovsky

Abstract

Mitochondria play a key role in various modes of cell death. Analysis of mitochondrial dysfunction and the release of proteins from the intermembrane space of mitochondria represent essential tools in cell death investigation. Here we describe how to evaluate release of intermembrane space proteins during apoptosis, alterations in the mitochondrial membrane potential, and oxygen consumption in apoptotic cells.

Key words Mitochondria, Cell death, Permeabilization, Respiration, Membrane potential

1 Introduction

Investigation of various forms of cell death has become an important area of biomedical research. Recently, several cell death modalities in addition to necrosis and apoptosis have been described and characterized based on morphological and biochemical criteria [1]. The interaction between different forms of cell death is complicated and still a matter of debate. Mitochondria play a crucial role in the execution of various modes of cell death, although the precise mechanisms of their involvement are still unclear.

Currently, it is widely accepted that mitochondria are important participants in the regulation of apoptosis, an evolutionarily conserved and genetically regulated process of critical importance for embryonic development and maintenance of tissue homeostasis in the adult organism [2]. Apoptosis is also involved in the spontaneous elimination of potentially malignant cells and therapeutically induced tumor regression, whereas defects in the apoptosis program may contribute to tumor progression and resistance to treatment [3]. The release of different proapoptotic proteins from the mitochondrial intermembrane space has been observed during the early stages of apoptotic cell death [4]. Among these proteins is a component of the mitochondrial respiratory chain, cytochrome *c*. Once in the cytosol, cytochrome *c* interacts with its adaptor molecule, Apaf-1, resulting in the recruitment, processing,

Volkmar Weissig and Marvin Edeas (eds.), *Mitochondrial Medicine: Volume I, Probing Mitochondrial Function*,
Methods in Molecular Biology, vol. 1264, DOI 10.1007/978-1-4939-2257-4_33, © Springer Science+Business Media New York 2015

and activation of pro-caspase-9, a member of the caspase family of cysteine proteases, in the presence of dATP or ATP [5]. Caspase-9, in turn, cleaves and activates pro-caspase-3 and -7; these effector caspases are responsible for the cleavage of various proteins leading to biochemical and morphological features characteristic of apoptosis [6].

The mechanisms regulating cytochrome *c* release remain partly obscure. However, two distinct models for cytochrome *c* release have emerged, and these can be distinguished on the basis of whether Ca^{2+} is required for the event. In one instance, mitochondrial Ca^{2+} overload causes opening of a nonspecific pore in the inner mitochondrial membrane, with subsequent swelling and rupture of the outer membrane followed by the release of cytochrome *c* and other intermembrane space proteins [7]. The Ca^{2+}-independent model asserts that a more selective protein release occurs without changes in mitochondrial volume or dissipation of the mitochondrial membrane potential. This mechanism involves specific pores in the outer mitochondrial membrane that are formed by certain proapoptotic members of the Bcl-2 family of proteins, including Bax [8]. Truncated Bid, generated by caspase-8 and other proteases, induces a conformational change in Bax that allows this protein to insert into the outer membrane, oligomerize, and mediate cytochrome *c* release. In addition, Bax can modulate cytochrome *c* release by facilitation of opening of the permeability transition pore [9].

The following protocols represent basic tools widely used in estimating release of intermembrane space proteins during apoptosis, assessment of the mitochondrial membrane potential in apoptotic cells, and analysis of mitochondrial oxygen consumption.

2 Materials

2.1 Evaluation of Cytochrome c Release from the Mitochondria of Apoptotic Cells

In order to analyze the release of certain proteins from mitochondria during apoptosis, the cellular plasma membrane should be disrupted and the cytosolic fraction separated from membrane material, including mitochondria. This can be achieved by incubation of cells in a solution containing digitonin, a steroid glycoside from *Digitalis purpurea*, which selectively permeabilizes the plasma membrane leaving the outer mitochondrial membrane intact. Plasma membrane permeabilization occurs due to the interaction of digitonin with cholesterol. At digitonin concentrations between 10 and 100 µg/ml, cholesterol-rich plasma membranes are permeabilized, whereas those of intracellular organelles are not [10]. The molecular weight of digitonin is about three times that of cholesterol; thus binding to cholesterol permeabilizes the plasma membrane by disrupting the packing of lipids.

1. Cells of interest in culture (e.g., Jurkat cells, U 937, HeLa).

2. Apoptotic stimuli (e.g., etoposide, cisplatin, staurosporine).

3. Phosphate-buffered saline (PBS).

4. 1 M Tris: dissolve 12.1 g of Tris in 100 ml of distilled water; adjust pH to 7.4 with HCl.

5. 1 M $MgCl_2 \times 6H_2O$: dissolve 203.3 mg of $MgCl_2 \times 6H_2O$ in 1 ml of distilled water.

6. 0.5 M EGTA: dissolve 38.1 g of EGTA in 80 ml of distilled water, adjust pH to 7.4 using KOH, bring the solution to 100 ml, and store at 4 °C.

7. 0.2 % digitonin: dissolve 10 mg of digitonin in 5 ml of distilled water, aliquot, and store at –20 °C.

8. 10 ml Fractionation buffer: Weigh 0.11 g of KCl, transfer into 15 ml tubes, add 50 µl 1 M Tris, 10 µl 1 M f $MgCl_2$, 50 µl 0.5 M EGTA, 0.5 ml 0.2 % digitonin solution, and make up to 10 ml with distilled water. Store at 4 °C for 4–5 days.

9. Low-speed centrifuge.

10. Eppendorf centrifuge.

11. Reagents and equipment for electrophoresis in polyacrylamide gel and subsequent Western blotting.

2.2 Assessment of the Mitochondrial Membrane Potential

Alteration of the mitochondrial membrane potential is one of the first responses of cells to any insult. The mitochondrial membrane potential, which drives oxidative phosphorylation [11] and mitochondrial calcium uptake [12], is generated by the electron-transporting chain. When electron transport ceases, for example during ischemia, the inner membrane potential is built up at the expense of ATP, hydrolyzed by the mitochondrial ATP synthase. The relationship between mitochondrial depolarization and apoptosis remains controversial. Depending on a cell death stimuli some investigators consider a decrease in the mitochondrial membrane potential an early irreversible signal for apoptosis [13], while others describe it as a late event [14].

1. Cells of interest (e.g., Jurkat cells, U 937, HeLa).

2. RPMI-1640 medium supplemented with 5 % (v/v) heat-inactivated fetal bovine serum, 2 mM L-glutamine, penicillin (100 U/ml), and streptomycin (100 µg/ml).

3. 25 mM TMRE stock solution: dissolve 12.8 mg tetramethyl-rhodamine methyl ester (TMRE; Molecular Probes) in 1 ml ethanol; store according to the manufacturer's instructions.

4. HEPES buffer: 10 mM HEPES, 150 mM NaCl, 5 mM KCl, 1 mM $MgCl_2 \times 6H_2O$, adjust pH to 7.4 with NaOH; Store up to 2–3 days at 4 °C.

5. Carbonyl cyanide 3-chlorophenylhydrazone (CCCP) (10 mM): dissolve 4.1 mg CCCP in 2 ml of ethanol, and store at –20 °C. Dilute the stock solution to 1 mM by adding 50 µl of 10 mM CCCP in 450 µl of ethanol.

6. Flow cytometer (e.g., FACS; Becton Dickinson).

2.3 Assessment of Oxygen Consumption in Intact Apoptotic Cells

In many instances, apoptosis-inducing agents can directly affect mitochondria; thus assessment of vital functions of mitochondria, such as respiration, provides important information concerning involvement of mitochondria in cell death process. Analysis of oxygen consumption can be performed using intact cells as well as cells with digitonin-permeabilized plasma membrane.

1. Cells of interest (Jurkat, U 937, HeLa).

2. Medium in which cells were growing.

3. Carbonyl cyanide 3-chlorophenylhydrazone (CCCP) (10 mM): dissolve 4.1 mg CCCP in 2 ml of ethanol, and store at –20 °C. Dilute the stock solution to 1 mM by adding 50 µl of 10 mM CCCP in 450 µl of ethanol.

4. Oxygraph (Hansatech Instruments), or any other Clark-type oxygen electrode connected to computer or chart recorder.

5. Hamilton-type syringes (10 and 25 µl).

2.4 Assessment of Mitochondrial Respiration in Apoptotic Cells with Digitonin-Permeabilized Plasma Membrane

1. Cells of interest.

2. 1 M Tris: dissolve 12.1 g of Tris in 100 ml of distilled water; adjust pH to 7.4 with HCl.

3. 1 M $MgCl_2 \times 6H_2O$: dissolve 203.3 mg of $MgCl_2 \times 6H_2O$ in 1 ml of distilled water.

4. 0.5 M KH_2PO_4: dissolve 340.2 mg of KH_2PO_4 in 5 ml of distilled water, adjust pH to 7.4, aliquot, and keep frozen.

5. 0.2 % digitonin: dissolve 10 mg of digitonin in 5 ml of distilled water, aliquot, and store at –20 °C.

6. 10 ml respiration buffer: Weigh 0.11 g of KCl, transfer into 15 ml tube, add 50 µl 1 M Tris, 100 µl 0.5 M KH_2PO_4, 10 µl 1 M $MgCl_2$, 0.5 ml 0.2 % digitonin, and make up to 10 ml with distilled water. Store at 4 °C for 4–5 days.

7. Sodium succinate (0.5 M): dissolve 135 mg of sodium succinate in 1 ml of distilled water, aliquot, and store frozen at –20 °C up to 3 months.

8. Sodium pyruvate (0.5 M): dissolve 135 mg of sodium pyruvate in 1 ml of distilled water, aliquot, and store frozen at –20 °C up to 3 months.

9. Malate (0.5 M): dissolve 134.1 mg of malic acid in 5 ml of distilled water, adjust pH to 7.4 with KOH, aliquot, and store frozen at –20 °C up to 3 months.

10. CCCP (10 mM): dissolve 4.1 mg CCCP in 2 ml of ethanol, and store at –20 °C. Dilute the stock solution to 1 mM by adding 50 µl of 10 mM CCCP in 450 µl of ethanol.

11. Rotenone (2.5 mM): dissolve 1 mg rotenone in 1 ml ethanol, store at –20 °C up to 3 months.

12. Malonate (0.5 M): dissolve 260.15 mg of malonic acid in 5 ml of distilled water, adjust pH with KOH, aliquot, and store frozen at –20 °C up to 6 months.

13. Ascorbate (0.5 M): dissolve 440 mg of ascorbic acid in 5 ml of distilled water, adjust pH to 7.4 with KOH, aliquot, and store at –20 °C up to 3 months.

14. Tetramethyl phenylenediamine (TMPD) (0.03 mM): dissolve 4.9 mg of TMPD in 1 ml of ethanol, store at –20 °C up to 3 months.

15. Oxygraph (Hansatech Instruments), or any other Clark-type electrode connected to computer or chart recorder.

16. Hamilton-type syringes (10 and 25 µl).

3   Methods

3.1   Evaluation of Cytochrome c Release from the Mitochondria of Apoptotic Cells

1. Incubate cells with apoptotic stimuli (type, concentration, and incubation time determined by cell type).

2. Harvest cells using Trypsin, transfer into 15 ml tubes, count cells, and spin down at $200 \times g$ for 5 min.

3. Gently remove the supernatant, without touching the pellet.

4. Resuspend cells at concentration of 1×10^6 cells in 100 µl of the fractionation buffer and transfer samples into Eppendorf tubes.

5. Take aliquots (2–3 µl) for protein determination.

6. Incubate samples at room temperature for 10–15 min.

7. In the end of the incubation vortex cells briefly and spin samples down using Eppendorf centrifuge for 5 min at $10,000 \times g$.

8. Gently, without touching the pellet, transfer supernatants (approx. 95 µl) from each sample into new Eppendorf tubes (see **Note 1**).

9. Add 95 µl of the fractionation buffer in each tube and resuspend the pellets.

10. Mix supernatant and pellet fractions with Laemmli buffer. Detection of proteins of interest is performed using polyacrylamide gel electrophoresis with subsequent blotting and probing with specific antibodies.

A typical distribution of cytochrome c between mitochondria and cytosol in apoptotic cells is shown in Fig. 1.

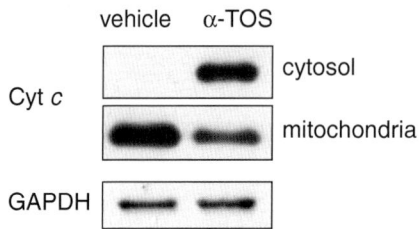

Fig. 1 Release of cytochrome *c* from mitochondria in neuroblastoma Tet21N cells during apoptosis induced by alpha-tocopheryl succinate. GAPDH was used as loading control

3.2 Assessment of the Mitochondrial Membrane Potential

1. Prepare an aliquot of 0.3×10^6 cells in 300 μl of RPMI-1640 medium.

2. Dilute TMRE stock solution 1:1,000 with HEPES buffer (for a concentration of 25 μM).

3. Add an aliquot of the diluted (25 μM) TMRE to cells for a final concentration of 25 nM.

4. Incubate cells with TMRE 20 min at 37 °C.

5. Further dilute the 25 μM TMRE 1:1,000 with HEPES buffer for a final concentration of 25 nM. Centrifuge cells 5 min at $200 \times g$, room temperature, and resuspend in fresh HEPES buffer containing 25 nM TMRE.

6. Analyze membrane potential by flow cytometry according to the manufacturer's instructions for the instrument used (*see* **Note 2**).

7. Use protonophore CCCP (5 μM) to dissipate the mitochondrial membrane potential completely as a positive control.

3.3 Assessment of Oxygen Consumption in Intact Apoptotic Cells

1. Calibrate the oxygen electrode according to the manufacturer protocol.

2. Add 0.3 ml of the medium to the oxygen electrode chamber under conditions of constant stirring. Set the rate of stirring 20–25 rpm.

3. Harvest cells (two to four million per measurement, depending on the cell type), spin them down.

4. Remove the supernatant and start the program. Take 50–60 μl of the medium from the oxygen electrode chamber for resuspending the cell pellet, transfer medium with cells back to the chamber, and close the chamber with the plunger. The plunger has a stoppered central precision bore allowing additions to be made to the reaction mixture using a standard Hamilton-type syringe. Expel all air bubbles through the bore in the plunger (slight twisting of the plunger helps to gather the bubbles). The level of oxygen in the chamber will start decreasing as mitochondria consume oxygen.

5. After 3–4 min add 10 μM CCCP through the bore in the plunger using Hamilton-type syringe.

6. After adding CCCP the rate of respiration will increase as CCCP lowers the mitochondrial potential that stimulates oxygen consumption.

7. Express the rate of respiration as the amount of oxygen consumed in 1 min by one million of cells.

The Protocol allows measuring basal respiration, and respiration stimulated by CCCP (highest activity of the respiratory chain), which is not suppressed by the mitochondrial membrane potential or controlled by the activity of ATP synthase.

3.4   Assessment of Mitochondrial Respiration in Apoptotic Cells with Digitonin- Permeabilized Plasma Membrane

Permeabilization of the plasma membrane allows measurement of mitochondrial activity in situ, without isolation of these organelles and the accompanying potential risk of mitochondrial damage. Selective disruption of the plasma membrane by digitonin makes mitochondria accessible to substrates for various respiratory complexes (*see* **Note 3**). The concentration of digitonin must be chosen carefully and usually should not exceed 0.01 % (w/v), since higher concentrations might affect the outer mitochondrial membrane (*see* **Note 4**).

1. Calibrate the oxygen electrode according to the manufacturer protocol.

2. Add 0.3 ml of the respiration buffer to the oxygen electrode chamber under conditions of constant stirring. Set the rate of stirring 20–25 rpm.

3. Harvest cells (two to four million per measurement, depending on the type of the cells), count them, and spin down in 15 ml tubes.

4. Remove the supernatant and start the program. Take 50–60 μl of the respiration buffer from the oxygen electrode chamber for resuspending the cell pellet, transfer the buffer with cells back to the chamber, and close the chamber with the plunger. Expel all air bubbles through the bore in the plunger (slight twisting of the electrode helps to gather the bubbles at the slot). The level of oxygen in the chamber will start decreasing as mitochondria consume oxygen. The presence of digitonin in respiration buffer will make mitochondria accessible to substrates.

5. Analysis of the activity of the respiratory complexes should be performed after uncoupling oxidation and phosphorylation by adding the protonophore CCCP (10 μM final concentrations) to get maximum rates of respiration. After 2–3 min add succinate, a substrate of Complex II (10 mM final concentration); addition of substrate will stimulate respiration (*see* **Note 5**). After 3–4 min add malonate (10 mM final concentration). The respiration will slow down as succinate dehydrogenase is inhibited.

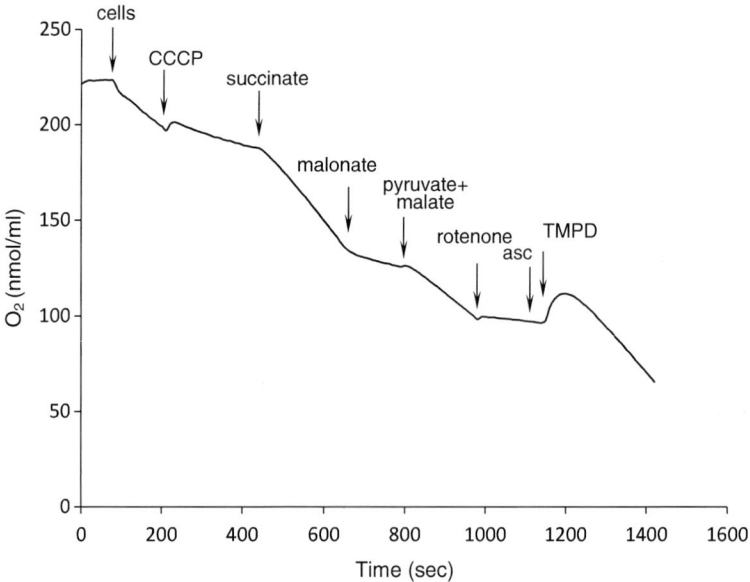

Fig. 2 Assessment of mitochondrial respiration in digitonin-permeabilized cells

6. After 3–4 min add pyruvate and malate, substrates of Complex I (10 mM final concentration each) (*see* **Note 6**). Respiration will be stimulated. After 3–4 min add rotenone, an inhibitor of Complex I (2.5 µM final concentration). This will slow down oxygen consumption.

7. After 3–4 min add 5 mM ascorbate and 0.3 mM (TMPD), an artificial electron donor for cytochrome *c*. Respiration will be stimulated again.

This approach allows assessment of the Complexes I, II, and IV of the mitochondrial respiratory chain. A typical curve of mitochondrial oxygen consumption permeabilized cells is shown in Fig. 2.

4   Notes

1. Separation of the mitochondria from the supernatant should be done thoroughly. Aliquots of supernatant should be taken without disturbing the pellet.

2. The analysis of the membrane potential should be done shortly after staining of the cells. TMRE is only accumulated by mitochondria with high membrane potential and any delay in analysis may negatively affect the functional state of the mitochondria and therefore cause leakage of the dye.

3. The concentration of digitonin can be determined experimentally by staining permeabilized cells with Trypan blue, a vital stain used to selectively color cells with damaged plasma membrane.

The lowest concentration causing permeabilization of 90–95 % cells should be used in the experiment.

4. Digitonin may precipitate after cooling; shake it vigorously before adding to the buffer.

5. Experiment can be started with analysis of Complex I activity instead of Complex II.

6. Pyruvate and malate can be mixed before the experiment and added together.

Acknowledgments

The work was supported by grants from the Swedish Childhood Cancer Foundation, the Swedish Research Council, the Swedish and the Stockholm Cancer Societies, and the Russian Foundation for Basic Research (grant 14-04-00963 A).

References

1. Kroemer G, Galluzzi L, Vandenabeele P, Abrams J, Alnemri ES, Baehrecke EH, Blagosklonny MV, El-Deiry WS, Golstein P, Green DR, Hengartner M, Knight RA, Kumar S, Lipton SA, Malorni W, Nunez G, Peter ME, Tschopp J, Yuan J, Piacentini M, Zhivotovsky B, Melino G (2009) Classification of cell death: recommendations of the Nomenclature Committee on Cell Death 2009. Cell Death Differ 16(1):3–11

2. Kerr JF, Wyllie AH, Currie AR (1972) Apoptosis: a basic biological phenomenon with wide-ranging implications in tissue kinetics. Br J Cancer 26(4):239–257

3. Lowe SW, Lin AW (2000) Apoptosis in cancer. Carcinogenesis 21(3):485–495

4. Gogvadze V, Orrenius S, Zhivotovsky B (2006) Multiple pathways of cytochrome c release from mitochondria in apoptosis. Biochim Biophys Acta 1757(5–6):639–647

5. Zou H, Li Y, Liu X, Wang X (1999) An APAF-1.cytochrome c multimeric complex is a functional apoptosome that activates procaspase-9. J Biol Chem 274(17):11549–11556

6. Robertson JD, Orrenius S, Zhivotovsky B (2000) Review: nuclear events in apoptosis. J Struct Biol 129(2–3):346–358, S1047-8477(00)94254-0 [pii]

7. Crompton M (1999) The mitochondrial permeability transition pore and its role in cell death. Biochem J 341(Pt 2):233–249

8. Martinou JC, Green DR (2001) Breaking the mitochondrial barrier. Nat Rev Mol Cell Biol 2(1):63–67

9. Gogvadze V, Robertson JD, Zhivotovsky B, Orrenius S (2001) Cytochrome c release occurs via Ca2+-dependent and Ca2+-independent mechanisms that are regulated by Bax. J Biol Chem 276(22):19066–19071, M100614200 [pii]

10. Mooney RA (1988) Use of digitonin-permeabilized adipocytes for cAMP studies. Methods Enzymol 159:193–202

11. Mitchell P, Moyle J (1967) Chemiosmotic hypothesis of oxidative phosphorylation. Nature 213(5072):137–139

12. Carafoli E, Crompton M (1978) The regulation of intracellular calcium by mitochondria. Ann N Y Acad Sci 307:269–284

13. Zamzami N, Susin SA, Marchetti P, Hirsch T, Gomez-Monterrey I, Castedo M, Kroemer G (1996) Mitochondrial control of nuclear apoptosis. J Exp Med 183(4):1533–1544

14. Bossy-Wetzel E, Newmeyer DD, Green DR (1998) Mitochondrial cytochrome c release in apoptosis occurs upstream of DEVD-specific caspase activation and independently of mitochondrial transmembrane depolarization. EMBO J 17(1):37–49

Chapter 34

The Use of FLIM-FRET for the Detection of Mitochondria-Associated Protein Interactions

Elizabeth J. Osterlund, Qian Liu, and David W. Andrews

Abstract

Fluorescence lifetime imaging microscopy–Förster resonant energy transfer (FLIM-FRET) is a high-resolution technique for the detection of protein interactions in live cells. As the cost of this technology becomes more competitive and methods are devised to extract more information from the FLIM images, this technique will be increasingly useful for studying protein interactions in live cells. Here we demonstrate the use of the ISS-Alba FLIM/FCS confocal microscope, which was custom-built for supervised automation of FLIM data acquisition. We provide a detailed protocol for collecting and analyzing good FLIM-FRET data. As an example, we use FLIM-FRET to detect the interaction between BclXL and Bad at the mitochondrial outer membrane in live MCF7 breast cancer cells.

Key words FLIM, FRET, Protein–protein interactions, Interactions at membrane, Fluorescence lifetime imaging, mCerulean3, Fluorescence proteins

1 Introduction

1.1 Studying Protein–Protein Interactions at Membranes in Live Cells

Protein–protein interaction studies are central in our effort to further understand biological systems and progress in drug design and development. While there are many available biochemical and biophysical approaches that can be used to detect interactions, no technique is without limitations. Currently co-immunoprecipitation is most practiced; however, this technique is unreliable when it comes to the study of membrane protein interactions due to detergent effects [1]. Many techniques used in live cells, such as spatial co-localization or proximity ligation, offer poor resolution for identifying real interactions and may confuse binding with co-localization. In contrast, fluorescence lifetime imaging microscopy–Förster resonance energy transfer (FLIM-FRET) is exceptionally sensitive for the identification of protein–protein interactions in live cells. Here we will use FLIM-FRET to demonstrate the interaction of BclXL and Bad at the mitochondrial outer membrane [2] and provide helpful insights from our experience.

Volkmar Weissig and Marvin Edeas (eds.), *Mitochondrial Medicine: Volume I, Probing Mitochondrial Function*, Methods in Molecular Biology, vol. 1264, DOI 10.1007/978-1-4939-2257-4_34, © Springer Science+Business Media New York 2015

1.2 Review of Fluorescent Microscopy

Understanding the fundamentals of fluorescent microscopy is essential. The basic principles of fluorescence are best understood with the use of a Jablonski diagram (*see* Fig. 1a). A fluorophore is a molecule that absorbs energy at some wavelength of light (λ_{abs}) causing its electrons to jump to a higher-energy state ($S_1 \ldots S_n$). In time, the electrons will return to the more stable, ground state (S_0), and in so doing energy is released by internal conversion, intersystem crossing, phosphorescence, and *fluorescence* (radiative energy released in the form of light on the nanosecond scale) [3, 4] _ENREF_3. Notably, fluorescence results from the transition of an electron from the baseline of S_1 to ground state S_0; some energy is released before reaching the baseline, and wavelength emitted (λ_{em}) will have lower energy than absorbed (λ_{abs}).

Some pairs of fluorophores have overlapping spectra. If the emission spectrum of one fluorophore, the *donor*, overlaps with the excitation spectrum of another fluorophore, the *acceptor*, then energy may transfer from donor to acceptor without the release of a photon (*see* Fig. 1b). This phenomenon is called Förster resonance energy transfer (FRET). The FRET efficiency (*E*) is inversely

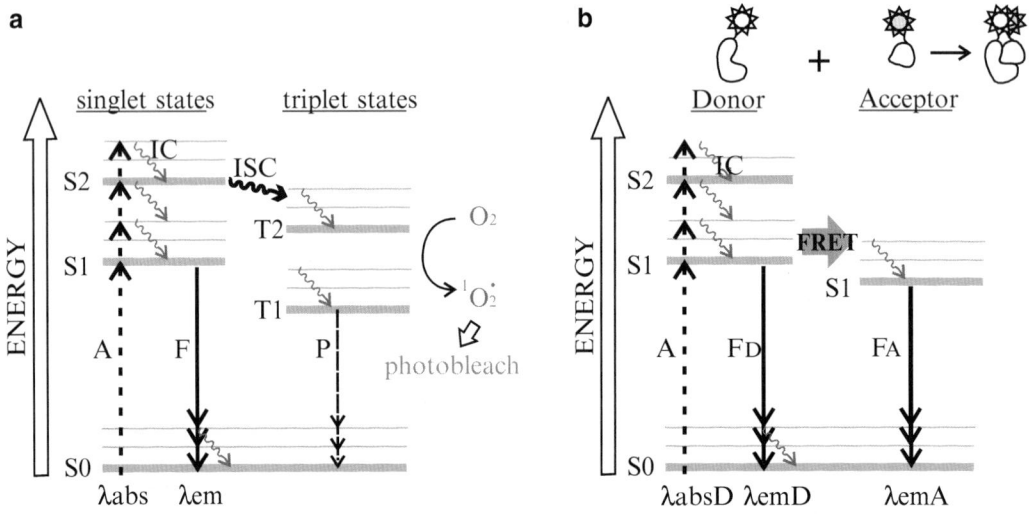

Fig. 1 Basic fluorescence principles. (**a**) Jablonski diagram adapted from [4] illustrating electronic states of a fluorescent molecule. Singlet states are represented by "S" and triplet states by "T." *A* absorbance, *IC* internal conversion/vibrational relaxation, *ISC* intersystem crossing, *F* fluorescence, and *P* phosphorescence. *Arrows* represent transitions between energy states. *Overlapping arrows* represent differences in quantity of energy that may be transferred in the same process (i.e., absorbance), to achieve the same result but occurring at separate events. (**b**) Diagram illustrating the nonradiative energy transfer from a donor to an acceptor fluorophore. Cartoon illustrates the donor and acceptor fluorophores (sun shapes) fused to proteins of interest. Energy levels for donor and acceptor are illustrated below. *FD* fluorescence of the donor, *FA* fluorescence of the acceptor

proportional to the sixth power of the distance (r) between the donor and acceptor, where R_0 is the Förster distance [5, 6]

$$E = \frac{1}{\left(1 - \dfrac{r}{R_0}\right)^6}$$

FRET efficiency also depends on the freedom of the donor and acceptor to orient properly in space for dipole–dipole alignment and other factors, but these effects are generally minor compared to distance. Therefore, in practice for most fluorophores used in biology, FRET can be detected for distances less than ~10 nm [6]. This makes FRET an exceptionally sensitive tool for the detection of protein-binding interactions when a donor is attached to one protein of interest and an acceptor to another protein. Only a direct interaction between the proteins will bring the donor and acceptor fluorophores in close enough proximity for FRET to occur.

1.3 Why Use FLIM to Measure FRET in Live Cells?

As highly specialized equipment is not required, the simplest way to measure FRET is by looking at changes in donor and acceptor intensities. However, intensity-based FRET measurements can only be applied when the quantity of donor and acceptor is known or when the ratio of donor–acceptor is constant in every area measured. The concentration of donor and acceptor cannot be accurately quantified in every pixel (or region of interest) in images of live cells. Additionally, FRET signals are generally small so it is difficult to correct for errors that result from spectral bleed-through, acceptor crosstalk, or changes in laser intensity. Measuring changes in donor lifetime allows more accurate quantification of FRET in live cells since lifetime is concentration independent.

Lifetime is the average time a population of donor fluorophores remains in the excited state before releasing a photon. Both time domain and frequency domain FLIM modes can be used to measure lifetime [5]. The more accurate, although more time-consuming method, time domain FLIM can be measured using the ISS-Alba microscope by time-correlated single photon counting (TCSPC). In brief, the sample is excited with a pulsed laser, and for each pulse the time between the excitation and the first detected emission photon is recorded. Photons collected over time can be used to generate a decay curve and calculate the lifetime of the donor (*see* Fig. 2a). The average donor lifetime will decrease with FRET since those fluorophores that retain their excitation longest in a population are also the most likely to undergo FRET with an acceptor. As a result, FRET efficiency (E) can be calculated by

$$E = \left(1 - \frac{\text{lifetime of Donor + Acceptor}}{\text{lifetime of Donor}}\right).$$

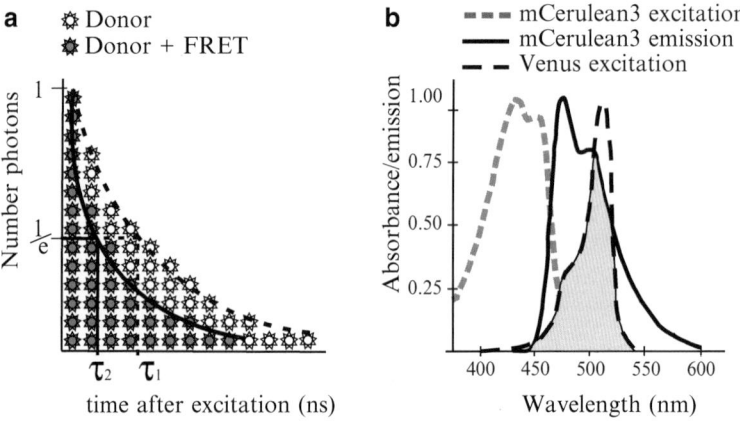

Fig. 2 Lifetime decreases with fluorescence resonance energy transfer. (**a**) Cartoon diagram of a lifetime decay curve. Photons are represented by sun shapes. T1 is the lifetime of the donor alone and T2 is the lifetime of the donor when FRET occurs. *Dashed line* refers to the decay of the donor alone and solid line refers to the decay of the donor when FRET occurs. (**b**) Cartoon representation of the donor mCerulean3 excitation/emission spectra and the overlap (*shaded gray*) with the acceptor Venus excitation [9]

Lifetime is intensity independent and is therefore not affected by small changes in illumination intensity that can arise from changes in laser power or alignment. However, both lifetime and fluorescence intensity are temperature dependent. Therefore, even though time is an absolute measurement, the lifetime of a fluorophore will be consistent from lab to lab only in controlled environmental conditions.

As with all techniques, there are limitations to FLIM-FRET. There is the possibility that addition of the fluorophore will affect the activity of the protein of interest. This is particularly true when FLIM-FRET is used to measure interactions between proteins that are fluorescent due to fusion to one of the many different fluorescence proteins. In these experiments the fluorescence protein may often be as large as the protein being analyzed. In addition, FLIM-FRET requires overexpression of the labeled proteins. Interactions detected by FLIM-FRET in live cells may be impacted by the expression levels of the putative binding partners particularly when they are expressed at defined locations in the cell such as membranes. For these reasons FLIM FRET data need to be validated by additional methods. Nevertheless, FLIM-FRET has been shown to be a powerful technique for the demonstration of distance-dependent protein–protein interactions in live cells [7, 8].

1.4 Overview of the FLIM-FRET Procedure

The first step is to choose an appropriate donor and acceptor pair and generate DNA constructs expressing the necessary fusion proteins. In our case, we chose mCerulean3 as the donor [9] and Venus as the acceptor (*see* **Note 1**). Next, a cell line stably expressing the donor is developed. For example, we created an MCF7 cell line expressing mCerulean3 fused to BclXL. This cell line is then transiently transfected with DNA that will lead to the expression of the acceptor fusion protein(s) and necessary controls. Finally, binding curves are generated for each transfectant, and relative apparent Kd values are used to determine if an interaction has occurred.

2   Materials

Human breast cancer MCF7 cell lines were cultured at 37 °C, 5 % v/v CO_2, and handled aseptically under the fume hood.

2.1   Cell Culture and Transfection Components

1. Complete αMEM: 450 ml minimal essential medium-alpha modification, 50 ml fetal bovine serum (FBS), 5 ml PenStrep.
2. Phosphate-buffered saline (PBS): 10 mM phosphate, pH 7.4, autoclaved.
3. 2× trypsin.
4. FuGENE HD (Promega).
5. MCF7 human breast cancer cell line (or cell line of your choice).

2.2   Microscopy Components

1. Microscope specifications (see Table 1).
2. Lab-Tek eight-well chambered cover glass w/cvr #1 German borosilicate sterile (Thermo Scientific).
3. 0.1 M NaOH: 40 g NaOH, fill to 10 ml with dH_2O.
4. Fluorescein standard: 0.753 g fluorescein disodium salt, fill to 10 ml with 0.1 M NaOH to make 0.2 μM fluorescein stock. For imaging dilute to 10 nM (1 ml 0.2 μM stock in 19 ml 0.1 M NaOH). Store wrapped in tinfoil at 4 °C.

2.3   Data Analysis Components

1. ImageJ version 1.47. Download our custom plug-ins at http://dwalab.ca/software/index.html. See Appendix A for "Macro_Mito.txt".
2. Microsoft Excel and RDBMerge plug-in for Excel (free online).
3. GraphPad Prism® version 6 or a similar biostatistics graphing software.

3   Methods

3.1   Transfecting Donor Cell Line to Prepare Sample for FLIM-FRET on the ISS-Alba

1. Design constructs for the expression of donor fused to one protein of interest and acceptor fused to another (5–10 amino acid linker). For example, here we used the constructs "mCerulean3–BclXL" and "Venus–Bad," respectively (see Note 2).
2. Choose a parent cell line to stably express the donor fusion protein (see Note 3). We used human MCF7 breast cancer cells stably expressing mCerulean3–BclXL.

Table 1
Microscope specifications for the machines used in this text

Microscope	ISS-Alba FLIM/FCS confocal microscope (custom by ISS)	EVOS FL imaging system (AMF4300)
Laser excitation	445 nm, 473 nm, 514 nm, 635 nm laser diodes (by LASOS Lasertechnik GmbH) 588 nm laser (by Coherent)	NA
Laser filter settings	Filter 1 = 442/512/561 nm Filter 2 = 520 nm longpass Channel 1 = 542/27 nm Channel 2 = 479/40 nm	NA
Lamp excitation	Nikon C-HGFIE epifluorescence	LED, CFP light cube 442 nm excitation/510 nm emission, GFP light cube 470 nm excitation, 525 nm emission
Objectives	60× water, NA = 1.20 60× air, NA = 0.70 40× oil, NA = 1.30 20× air, NA = 0.50	10× air, NA = 0.45 20× air, NA = 0.45 40× air, NA = 0.45
Temperature	25 °C	Room temperature
Imaging medium	αMEM complete	αMEM complete
Camera	Nikon DS-Qi1Mc	Sony® ICX445 CCD
Acquisition settings	VistaVision software v4.0, settings adjusted depending on daily laser intensity	50 % light power on 20× objective, find and focus mode

3. Seed the stable donor cell line into an eight-well chambered cover glass for 60–80 % confluency the following day (*see* **Note 4**).

4. After 20–24 h, transfect cells with plasmids expressing controls and protein of interest. BclXL is a well-known antiapoptotic protein that has previously been shown to interact with Bad [2, 10]. Therefore, V–Bad is our positive control for binding BclXL. Our negative controls include Venus, Venus–Bad2A, and Venus–Bad2A–ActA (*see* **Note 5**). For transfecting MCF7 cells, such as our MCF7–mCerulean3–BclXL, follow the steps below.

5. Warm FuGENEHD, αMEM complete, 5 ml autoclaved dH$_2$O, and DNA constructs in 37 °C water bath.

6. Remove FuGENEHD from water bath, vortex, and cool to room temperature. For each transfectant, dilute 0.25 μg of DNA in 11 μl dH$_2$O, label tubes with construct name, and vortex briefly.

7. Add 0.75 μl FuGENE HD, vortex, and quickly centrifuge the reaction to consolidate in the bottom of the tube.

8. Incubate at room temperature for 10–15 min.

9. Add 225 μl αMEM complete to each transfection reaction, mix gently by pipetting, and dispense in a designated well (replacing medium already on the cells). Leave one well untransfected; replace media with fresh αMEM complete.

10. Incubate for 5 h and change media to 300 μl fresh αMEM complete.

11. After 24 h, check transfections (*see* **Note 6**) and again change media to 300 μl fresh αMEM complete before imaging.

3.2 Initializing the ISS-Alba Confocal Spectroscopy and FLIM Workstation (ISS)

1. Turn on components in this order:

   (a) Computer.

   (b) TI-PS100W/A power supply (for DIA pillar illuminator).

   (c) Nikon TI-S-CON motorized stage control.

   (d) Nikon Intensilight C-HGFIE.

   (e) Excitation box, turn keys for the 445 and 514 nm lasers (ext. 20 MHz).

   (f) Detector box.

   (g) Nikon Eclipse T*i* microscope (at the rear of the microscope).

   (h) A320 FastFLIM (91070 2012).

   (i) Load VistaVision 4.0 software.

   (j) A512 3-Axis DAC controller (No. 91014 1001).

   (k) A403 scanning mirror driver (91032 1009).

   (l) Signal conditioning unit (98018 3001).

2. Allow entire apparatus to warm up for 30 min (*see* **Note 7**).

3. Add 300 μl of 10 nM fluorescein in 0.1 M NaOH to one well of an eight-well chambered slide and incubate in the dark for 10 min at room temperature.

3.3 Calibrating the System

Here we use the 10 nM fluorescein reference to check the 514 and 445 nm laser intensities.

1. Lower the water immersion objective lens and add one drop of dH$_2$O.

2. Fix the reference chamber slide onto the stage.

3. Turn on the bright field lamp and move the objective lens up until the water droplet contacts the chamber slide. Move the stage so the objective lens is beneath the edge of the fluorescein sample well. Looking through the eyepiece, continue moving the objective lens up until the bottom of the plate is in focus (*see* Fig. 3).

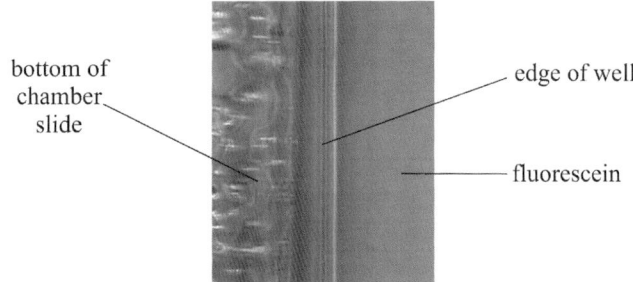

Fig. 3 Example for focus on the bottom of a well. Since there are no structures in the fluorescein sample to determine whether the focus is within the sample or at the *top/bottom* of the plate, we focus on the bottom of the well seen here

4. At this point, reset the z-axis on the microscope, and then move to $z = +100$. Move the stage so the objective lens is in the center of the well and turn off the bright field lamp. Cover sample to block out ambient lighting when imaging.

5. Press L100 to send the information to the computer, and on the ISS VistaVision software, refresh the setup (*see* **Note 8**).

6. Set the 445 and 514 nm lasers to 70 %, click the green circle to turn open shutters, and record the values for Channel 1 (Ch1) and Channel 2 (Ch2). Similarly, record values for each laser individually at 70 % intensity.

7. Finally, set 445 nm to 50 % and 514 nm to 70 %, averaging mode = frames, no. frames = 4, and pixel dwell time = 0.1 ms and click start. In the new window, click "τ," set T1 = 18 and T2 = 80, and calculate lifetime (4 ± 0.1 ns) by region average for Ch1 and Ch2 (*see* Table 2). Save the reference as an ISS file.

Table 2
Example of 10 nM fluorescein reference

	Channel 1	Channel 2
445 nm + 514 nm at 70 %	52 K	203 K
445 nm at 70 %	7 K	202 K
514 nm at 70 %	46 K	0 K
Lifetime (τ)	4.012 ns	4.019 ns

3.4 Choosing the Data Collection Settings

1. Fix the sample chamber slide to the stage. Ensure there is still water on the objective lens; this will require replacing every 2 h.

2. In bright field, focus on the untransfected cells. Press z-reset and then press FOCUS and an orange light will indicate that the autofocus is in use. Press L100, and then change settings to *find focus mode* (pixel dwell time = 0.02 ms, averaging mode = none, repeat scan on).

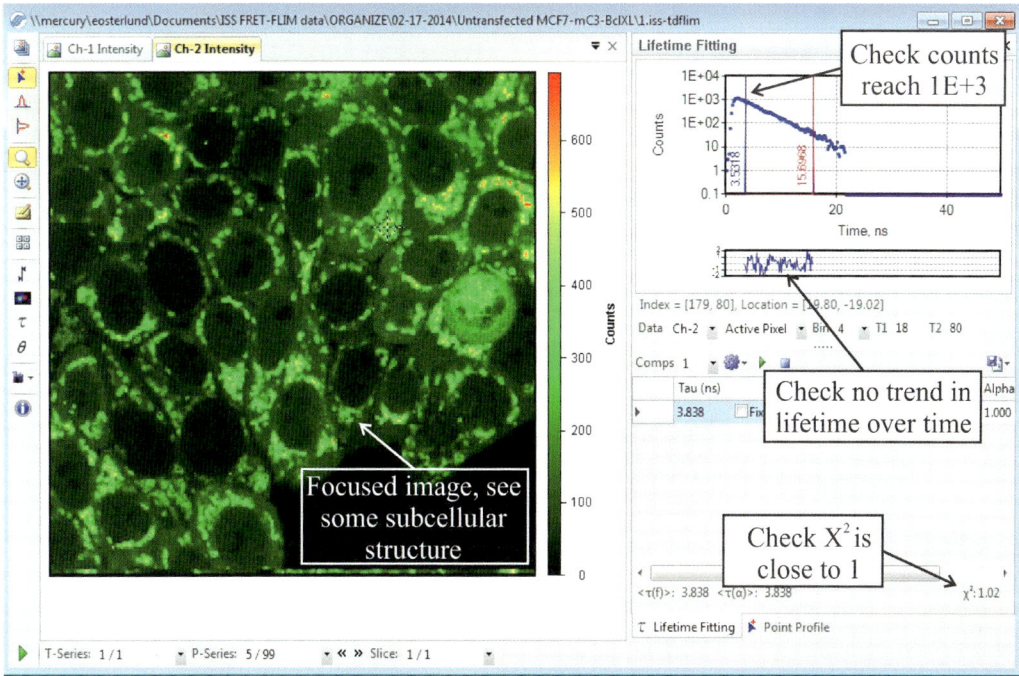

Fig. 4 Print screen of MCF7–mCerulean3–BclXL Ch2 as seen on VistaVision software. MCF7–mCerulean3–BclXL cells untransfected. Image is focused as you can see mCerulean3–BclXL expression in the Ch2, where BclXL is primarily localized to mitochondria (*bright spots*). To the left we have highlighted where to look to acquire good FLIM data: ensure that photon counts (binned by 4) reach 1,000, that there is no trend in lifetime, and that X^2 is ~1 throughout the image

3. Set 445 and 514 nm laser intensities to 70 % and click start. Use the autofocus wheel at coarse setting to gently adjust the focus so the subcellular structures (depending on your donor–protein of interest) are defined and bright as possible in Ch2 (*see* Fig. 4). Once satisfied with the focal plane, zero the *xyz* position on the software, z-reset on the microscope, and then stop the acquisition.

4. Change settings to *FLIM data acquisition mode* (pixel dwell time = 0.1 ms, averaging mode = frames, no. frames = 4, repeat scan off).

5. Click start, set intensity threshold Ch2=50, click lifetime fit "τ," and drag the pointer to an area where there is signal in Ch2.

6. In the lifetime window, set bin number to 4 and observe the accumulated Ch2 photon counts as the number of frames increases. For confident lifetime decay fitting, we require collection of 1,000 photons (binned by 4) (*see* Fig. 4). You may decrease the intensity of the 445 nm laser or decrease the number of frames acquired to meet this requirement.

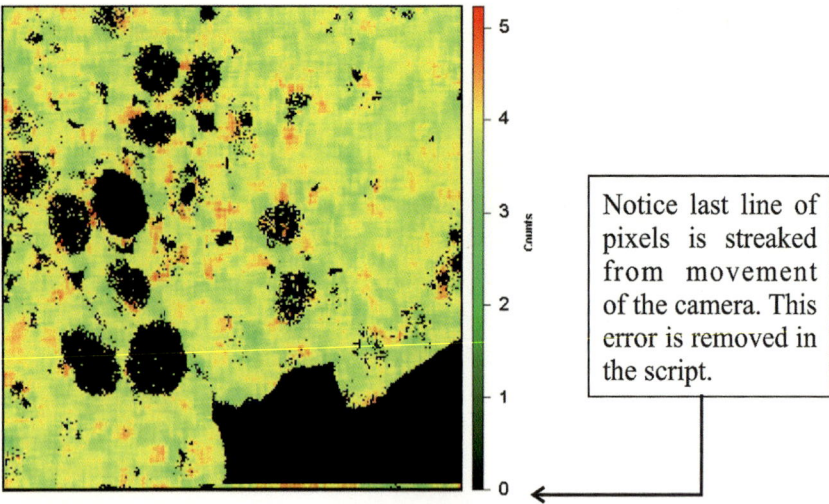

Notice last line of pixels is streaked from movement of the camera. This error is removed in the script.

Fig. 5 MCF7–mCerulean3–BclXL Channel 2 average lifetime. Ch2 average lifetime data processed from the Ch2 image in Fig. 4. Bin number = 4, T1 = 18, T2 = 80, Ch2 intensity cutoff = 50. Lifetime is represented in pseudocolor scale. Note there is little variation across in color across this image, which is desired for these untransfected MCF7–mCerulean3–BclXL. In cells where there is FRET (not shown), the lifetime will drop and color will change correspondingly

Using the untransfected cells, check that bleed-through from Ch2 to Ch1 does not exceed 100 counts after acquisition. If so, decrease the intensity of the 445 nm laser and increase frames acquired. Change "Active pixel" to "Frame" and click the green arrow to process tau values (*see* Fig. 5). Record the final 445 nm laser intensity and the number of frames collected and use for data collection.

7. Move the stage to the first control well (expressing Venus fluorescent protein (FP)). Switch to epifluorescence mode and find cells expressing Venus (*see* **Note 9**).

8. Refocus in *find focus mode* and then switch to *FLIM data acquisition mode*. Click start and observe the Ch1 counts/second across the image. At any point if counts exceed 600K, then decrease the 514 nm laser intensity in intervals of 5 % until this limit is satisfied (*see* **Note 10**). Record the 514 nm laser intensity determined.

3.5 Collecting FRET-FLIM Data on the ISS-Alba

To generate binding curves of publishable quality, we usually acquire 100 images per transfectant. The first image is manually chosen and then set as the (0,0,0) *xyz* reference point. When choosing this point, stay away from the bottom and right edge of the well to avoid scanning outside the well during the position series. For subsequent positions in the series, the stage automatically moves laterally, and Nikon Perfect Focus System (NPFS) continuously adjusts the axial position to maintain a constant distance between the focal plane and the axial reference plane (refractive index boundary at the bottom of the well). The ISS-Alba simultaneously acquires Ch1 and Ch2 intensity images and lifetime data.

1. Start in the untransfected control well and manually choose the first image in the position series.

2. Set *xyz* position "0,0,0" and z-reset on the microscope.

3. Activate the position series and choose the first position. Adjust focus in *find focus mode*. Switch to *FLIM data acquisition mode*, choose the first 20 positions, and click start. Once complete, save the ISS-Alba file as "1" in a file folder named "Untransfected (insert name of cell line)" (*see* **Note 11**). Also save the TIFF image as "1" in the same folder, which will save both Ch1 and Ch2 image stacks.

4. Finally, process the lifetime at Ch2 threshold = 50, bin number=4, T1 = 18, and T2 = 80. Use "active pixel" mode and calculate the lifetime at any region with Ch2 intensity. Then switch to "p-series" mode and click start. Save as "1t" in the untransfected folder (*see* Fig. 6).

5. Similarly, collect a position series for each transfectant. Use the epifluorescence mode to choose a start position with transfected cells in the center of the field of view.

6. Select "add wells Serpentine" in the position series and 100 positions will be added. Data collection takes approximately 1.5 h per transfectant. Create a separate folder for each transfectant (e.g., Venus, Venus–Bad, etc.). Save data with the same naming system used earlier (1.iss, 1.tiff, and 1t.tiff). Finally, organize the TIFF files into three subfolders: "Ch1," "Ch2," and "tau."

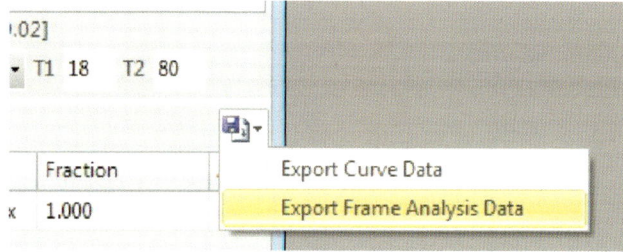

Fig. 6 Print screen of how to save processed lifetime data. The tau data is not saved like all other data (using "File > save as"), instead the data must be exported as seen here. Click "Export Frame Analysis Data," then uncheck "Average Tau," and check "Comp-1-tau." Find the subfolder that the tau dataset came from and save as "stack#t" (stack# is the image stack number for the dataset and t indicates tau for lifetime)

3.6 FLIM-FRET Data Analysis

As examples, we will analyze the untransfected and Venus–Bad sample data (provided in supplemental). Open Macro_Mito file (Appendix) in Microsoft Word and copy all. Open ImageJ, plug-ins, and new macro, then paste, and run script. In the new window, select the untransfected tau subfolder location. Similarly, choose Ch1 and then Ch2 subfolders from the same dataset when prompted. Ch1, Ch2, and tau images are scanned as stacks. A new window opens, and in the untransfected folder location, create a new "analysis" subfolder to designate the saving directory. The macro will then take a short time to run through the dataset (*see* **Note 12**). Depending on the parameters discussed in **Note 12**, for each image Ch2 background is subtracted, an intensity cutoff is applied, and regions of interest (ROIs) are chosen with shape/size constraints appropriate for mitochondrial localization.

Ch1, Ch2, and tau ROI measurements for each image are stored in the analysis folder as separate Excel files. Open Microsoft Excel and use the RDB merge plug-in to compile all untransfected mean values for Ch1, Ch2, and tau (*see* **Note 13**). Save work sheet as "Total_data_Untransfected." Create a scatterplot with Ch2 and lifetime on the *X*- and *Y*-axis, respectively. Estimate the Ch2 intensity cutoff beyond which there is least variation in lifetime measurements (*see* Fig. 7). Also, calculate the *average lifetime donor lifetime alone* by averaging the mean lifetime column (should be approximately 3.8 ns for mCerulean3).

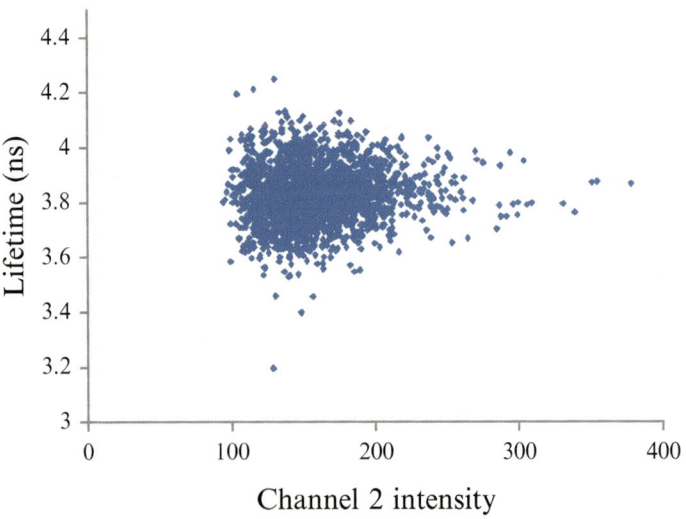

Fig. 7 Example of how to determine Ch2 intensity cutoff. Here we have graphed in Excel the lifetime versus Channel 2 intensity for the untransfected MCF7–mCerulean3–BclXL. At lower values of Ch2 intensity, there is larger variation in lifetime, an effect attributed to cellular auto-fluorescence and insufficient Ch2 counts to calculate tau. From the above graph, we chose a Ch2 intensity cutoff of 200 for further analysis of this dataset. There is less variation in the lifetime calculated for higher Ch2 counts

Table 3
Example of bin sizes

Range		Bin size
0	0.1	0.01
0.1	1	0.1
1	2	0.25
2	5	0.5

Similarly create "Total_data_Venus-Bad" and duplicate the first tab to preserve the merged data prior to analysis. Highlight all columns and sort data from the smallest to largest Ch2 intensity. Delete data below the Ch2 cutoff previously established with the untransfected data. Make a new column with the heading "Acceptor:Donor (A:D) intensity ratio" and apply the formula "=Ch1/Ch2 intensity." Highlight all columns and now sort from the smallest to largest A:D intensity ratio. In a new tab copy the A:D intensity ratio to column A and mean tau to column B. In columns, bin the data by A:D intensity ratio (*see* Table 3 and **Note 14**). Calculate the average for every column. Calculate the standard error in lifetime using the formula "=STDEV(B1:B#last row in bin)/SQRT(COUNT(B1:B#last row in bin)." Drag this formula horizontally to apply to all lifetime data columns. Copy the binned average A:D intensity ratio, lifetime, and standard error of the lifetime to a new tab (*see* Fig. 8).

Calculate

$$\text{"\% FRET efficiency"} = \left(1 - \frac{\text{average lifetime of bin}}{\text{average donor lifetime alone}}\right) \times 100$$

and

$$\text{"Standard Error in \% FRET efficiency"} = \left(\frac{\text{standard error of binned lifetime}}{\text{average donor lifetime alone}}\right) \times 100.$$

Open GraphPad Prism, create a new XY scatterplot, and click "Enter and plot error values already calculated elsewhere> Mean & SEM." Copy the A:D intensity ratio, % FRET efficiency, and standard error in % FRET data to "X," "Mean," and "SEM" columns, respectively. Rename the data table as the transfectant name (i.e., Venus–Bad). Label the *X*- and *Y*-axis "Acceptor:Donor intensity ratio" and "% FRET efficiency," respectively. Click "Analyze" and choose from the panel of saturation binding equations "one site specific binding with Hill Slope." The formula $Y = \dfrac{(\text{Bmax})X^h}{\left(\text{Kd}^h + X^h\right)}$ is

	A	B	C	D	E	F	G	H	I	J
1	0.011324	3.817768	0.001335	0.058436	0.034945			0.011324	0.058436	0.034945
2	0.166405	3.586141	0.007645	6.121976	0.200142			0.166405	6.121976	0.200142
3	0.369507	3.339912	0.008203	12.56776	0.214743			0.369507	12.56776	0.214743
4	0.592966	3.242804	0.009499	15.10984	0.248673			0.592966	15.10984	0.248673
5	0.861122	3.211573	0.014042	15.92741	0.367592			0.861122	15.92741	0.367592
6	0.861122	3.211573	0.014042	15.92741	0.367592			0.861122	15.92741	0.367592
7	1.1995	3.142247	0.016321	17.74224	0.427249			1.1995	17.74224	0.427249
8	1.77059	3.07744	0.021012	19.43874	0.55004			1.77059	19.43874	0.55004
9	2.237364	3.092778	0.024657	19.03723	0.645466			2.237364	19.03723	0.645466
10	2.788959	3.000091	0.030465	21.46359	0.797521			2.788959	21.46359	0.797521
11	3.258499	2.910636	0.030602	23.80533	0.801105			3.258499	23.80533	0.801105

Fig. 8 Example of the final step in lifetime data binning for FLIM-FRET analysis. Here is a print screen of the analysis at the end of binning. Bins are ordered by increasing row number (1–11). Column A = Acceptor:Donor intensity ratio; B = average lifetime; C = standard error in mean lifetime; D = % FRET efficiency [E = (1 − B/average donor lifetime alone) × 100]; E = % standard error in the mean lifetime [%SE = (C/average donor lifetime alone) × 100]. Columns A, D, and E were copied to H, I, and J, respectively (*color coded*), and H–J will be copied to GraphPad Prism software for further analysis

used to calculate the dissociation constant (Kd), and the statistical results are stored in the Prism project section (*see* **Note 15**). Insert a text box on the graph to display Kd ± standard error. Right click any data point, click Format Graph> Error Bars> Direction, and change to "Both" to display error bars. Uncheck "connected line" for the data points.

Remember that in our case, V–Bad is the positive control for binding to mCerulean3–BclXL. The total data fits a binding curve, and the binning performed earlier in Excel improves visual presentation and averages out outliers (*see* Fig. 9a). Importantly, binning should not change the Kd value. To avoid this source of error, ensure that there are binned data points that fall upon the portion of the curve before saturation (around the Kd value calculated from the total data). If the Kd changes from that calculated with the total dataset, reduce binning. We do not manually delete apparent outliers since it becomes much more difficult to discriminate outliers in negative control datasets. If there are many outliers and the binding curve is unclear, return to the beginning of the analysis and try increasing the Channel 2 threshold.

In the same way as the positive control (V–Bad), bin the merged data and use GraphPad Prism to plot all the negative controls (*see* **Note 16**). The negative controls (those that do not bind to BclXL in our example) better fit a straight line (*see* Fig. 9b). FRET efficiency increases linearly because the probability of random acceptor to donor collisions increases with A:D intensity ratio. In addition, when the donor and acceptor are confined in 2D

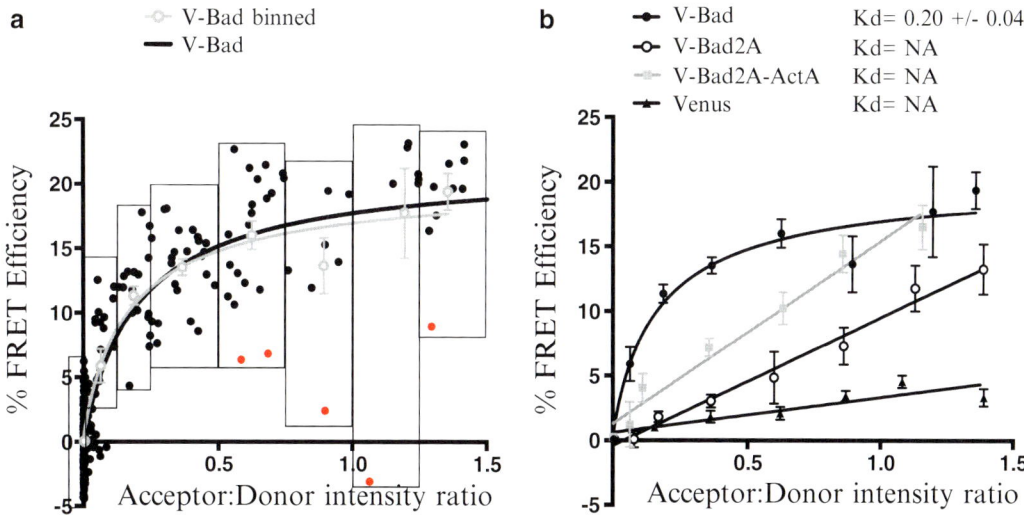

Fig. 9 FRET-FLIM data analysis for MCF7–mCerulean3–BclXL. MCF7–mCerulean3–BclXL cell line was transiently transfected with V–Bad, V–Bad2A, V–Bad2A–ActA, and Venus. FLIM-FRET was collected on the ISS (445 nm 70 %, 514 nm 65 %, 4 frames, 0.1 ms pixel dwell time) and data was plotted using GraphPad Prism software. (**a**) Example binned data for Venus–Bad-positive control binding curve. Venus–Bad total data and the corresponding binding curve from all points are displayed in black. Data points that fall within the rectangles have been binned and the final binned data (*gray circles* and *line*) are overlaid on the total data. Points that may be outliers (*red*) in the dataset were included in the binning. The binding curves of the binned and raw data overlap. (**b**) Total binned FLIM-FRET data. Kd calculated by specific binding with Hill slope equation in GraphPad Prism software. See legend above the graph

space (i.e., the mitochondrial outer membrane), there is an increased chance of collisions. This can account for the increased slope in our V–Bad2A–ActA control (*see* Fig. 9b). In FLIM-FRET experiments, it is important to always control for the effect of localization to detect real interactions at the mitochondrial outer membrane or in any other membrane in the cell. This highlights the sensitivity of FLIM-FRET for real protein–protein interactions and reinforces the fact that co-localization does not equal binding.

Finally, to determine whether an unknown protein binds the donor fusion protein of interest, merge, bin, and plot the data in the same way as the positive control. Try to fit the data both nonlinearly and linearly. If the Kd from the Hill slope equation is ambiguous, the dataset will be better fit linearly in most cases. Again, a linear fit indicates no binding. If the data fits a binding curve and if the Kd is similar to the positive control, this indicates that binding occurs between the two proteins of interest in live cells.

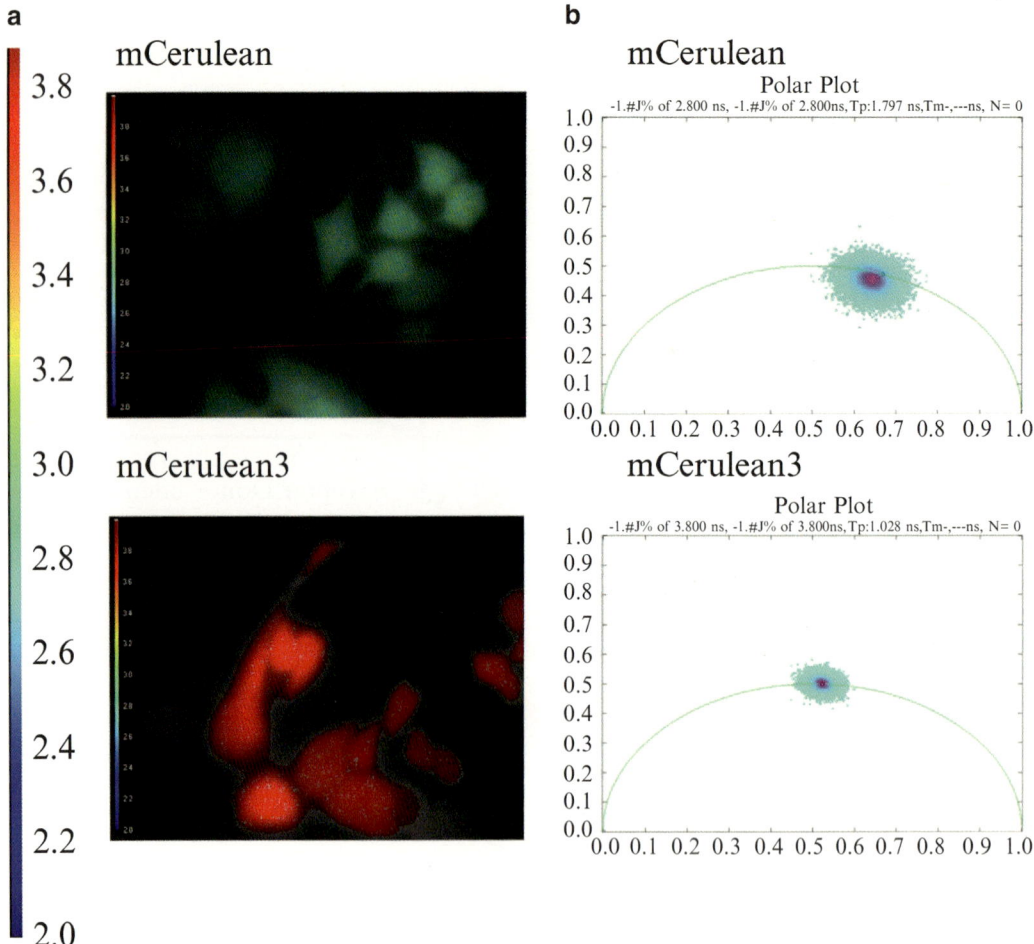

Fig. 10 mCerulean3 is a better donor than mCerulean. MCF7 cells were transfected with mCerulean and mCerulean3. 24 h later, lifetime data was collected on the ISS-Alba in fast FLIM mode at 750 V MCP gain, 30–70 ms exposure time. (**a**) Lifetime images collected are displayed in a pseudocolor scale. The lifetime of mCerulean3 was approximately 3.8 ns, while the lifetime of mCerulean was shorter, at approximately 2.8 ns. (**b**) Polar plots for mCerulean and mCerulean3 data. From the frequency domain FLIM data, there are two models that can be used to calculate lifetime. These are called the "phase" and "modulation" lifetimes, and for a sample that has single exponential decay, these values will be equal (theoretical values indicated by the *green circle*). Since the data for mCerulean3 deviates less from the *green circle*, we conclude mCerulean3 is more mono-exponential than mCerulean

4   Notes

1. mCerulean3 and Venus have good spectral overlap (*see* Fig. 2b). We compared two donor proteins: mCerulean and mCerulean3. mCerulean3 has a longer lifetime (3.8 ns compared to 2.5 ns) giving it a larger dynamic range (*see* Fig. 10a). Monoexponential decay is required for more accurate estimation of changes in lifetime; mCerulean3 is more mono-exponential than mCerulean, as indicated by better alignment with the

single lifetime curve (semicircle on polar plot) (*see* Fig. 10b). Finally, mCerulean3 is brighter than mCerulean making it easier to acquire sufficient photons to generate a decay curve (DNS).

2. Take into account the membrane topology and structural features of your protein of interest when designing the fusion protein. For soluble proteins, it can be beneficial to test fusions to either terminus. The donor and acceptor may be fused to the N or C terminus of the proteins of interest. For example, BclXL is localized to the cytoplasm, the endoplasmic reticulum, or most commonly the outer mitochondrial membrane. The C-terminal transmembrane region of BclXL inserts into membranes and the protein faces the cytoplasm. Consequently we designed mCerulean3 fused to the N terminus of BclXL (mCerulean3–BclXL).

3. Using more than one parent cell line is recommended to ensure the interaction is not cell line specific. Sorting the cell line 3–4 times is necessary to minimize variation in donor–fusion protein expression. This will improve Acceptor:Donor intensity ratio estimation.

4. We find letting the cells sit for ½ an hour at room temperature after seeding reduces the tendency for the cells to cluster on the borders of the well.

5. "Venus" is the acceptor fluorophore alone control. "V–Bad2A" is Venus fused to a mutant of Bad that no longer interacts with BclXL [7]. Adding the ActA sequence to the C terminus of Venus–Bad2A causes it to localize exclusively to the mitochondrial outer membrane. Furthermore, the Venus–Bad2A–ActA control accounts for the increased chance of collision when both donor and acceptor are confined in 2D space on the mitochondrial outer membrane.

6. Using a basic fluorescent microscope such as the EVOS-FL, check the expression of the acceptor (Venus) and visually judge ~20–30 % of the total population of cells are transfected (*see* Fig. 11). If there is low transfection efficiency, then data collection will be inefficient (instead of using the automated imaging features of the ISS, fields of view must be manually chosen to capture transfected cells). If this is the case, then let the cells incubate another 24 h and check expression again. If there is no improvement, optimize transfection before moving on to imaging.

7. The internal temperature of the microscope increases slightly in the first half hour due to contributions of heat from the lasers. Lifetime is sensitive to fluctuating temperatures; allowing the microscope to reach a stable state will result in reproducible measurements.

8. We recommend resetting all the filter positions (*see* Fig. 12) each time you reboot the ISS software.

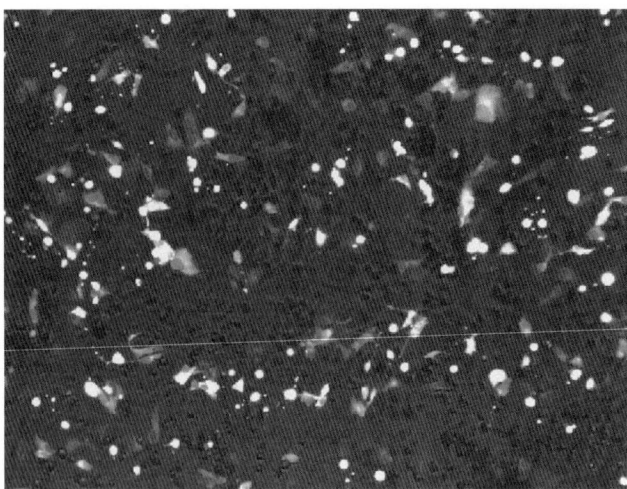

Fig. 11 Example of good transfection efficiency. MCF7 cells were transfected using FuGENEHD at 6:1 reagent to Venus–Bad DNA ratio as recommended by the manufacturers. Cells were imaged on the EVOS FL imaging system (AMF4300) at 20× resolution, GFP light cube. Bright spots are transfected cells. Background cells were imaged in bright field

9. To switch to epifluorescent mode, change the filter on the microscope from "none" to "blue." Then hit "EYE" on the microscope and open the shutter on the Nikon C-HGFIE HG controller. Once transfected cells have been found, remember to close the shutter and reset the filter to "none" before returning to data acquisition.

10. The effect commonly known as detector pileup occurs when the number of photons reaching the detector exceeds the linear range of detection for the single photon counting module (SPCM-AQR series). According to the manufacturer, this detector reaches the nonlinear saturation condition at one million counts/second. We usually apply an upper limit of 600K to Ch1 to safely avoid this source of error.

11. Position series files collected on the ISS are very large in size and may take a few minutes to fully save. Making multiple commands on the same file during the saving process corrupts the saved file. When you save one file, wait until the status at the bottom of the screen is "ready" to process lifetime data or to save other features from that same file.

12. Variables in the script that may be changed according to different expressing levels of donor protein and different laser settings:

    (a) image_intensity: this is the mean intensity of the whole image in Ch2. In default 7 is set to remove images without any cells. It can be adjusted to your needs.

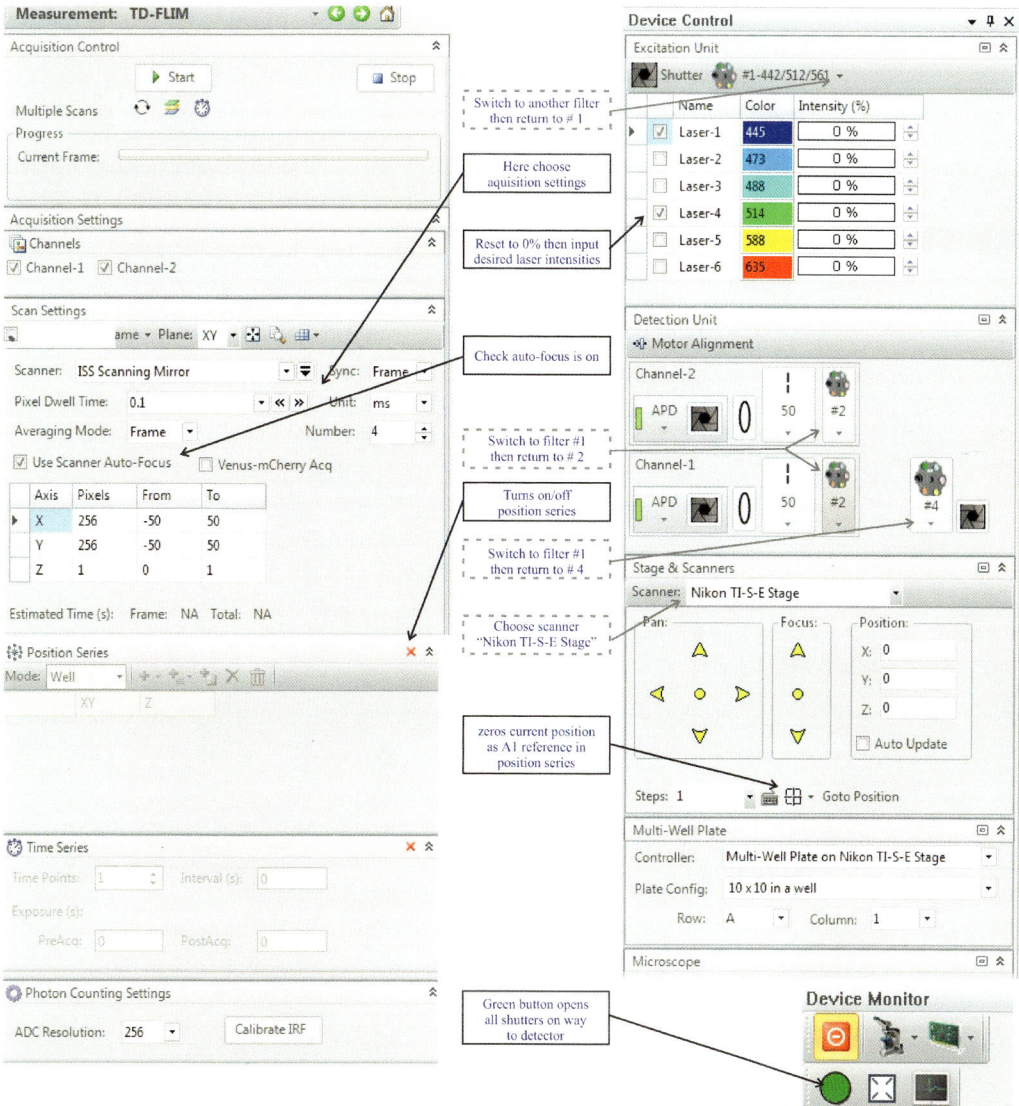

Fig. 12 Print screen layout of VistaVision 4.0 software. This is how the software will look like when it starts up. In TD, FLIM measurement mode, reset all filter positions and choose scanner (*see dashed boxes*). Some commonly used features referred to in the text are also highlighted (*solid boxes*)

(b) In the Canvas Size command, the numbers may be changed according to your instruments' alignment.

(c) new_lower: this is the lower limit we need to set for a better thresholding before converting the image into a mask. The default setting is +50 from auto-thresholding. The bigger the number is set, the better the connected cells will be fragmented and the more dim cells will be removed and vice versa.

(d) Size (in the Analyze Particles command): this is the low limit of the number of pixels in one ROI. It should be adjusted according to different organelles and different resolution of the images.

(e) max_intensity and low_intensity: these are the highest and lowest intensities in one ROI. They are set to remove the unwanted ROIs with too high/low expressions that may affect the lifetime decay fitting. These numbers need to be adjusted according to recorded images.

13. In Microsoft Excel click "RDB Merge" under the data tab. Choose the "analysis" folder as the folder location. Under *Which Files*, select "Merge all files with a name that: **Contains**" Ch1, and under *Which Range*, select "First Cell" and input A2 "till last on Worksheet." In the merged file, column titles are as follows: **A** = file location/image number, **B** = ROI number, **C** = standard deviation of mean, **D** = mean, **E** = lowest value, and **F** = highest Ch1 value.

14. The bin size is variable due to the expression levels of proteins of interest and different laser settings. Number/size of bins can also be reduced or increased depending on the amount of data available. However, keep in mind that when you adjust the bin size of one dataset, you must also adjust the bin size for all controls accordingly.

15. The lower the Kd value, the higher affinity for binding. Note that the Kd calculated here is an apparent value as it does not have units of concentration. Since the Kd is a value of Acceptor:Donor intensity ratio, it will depend on the acquisition settings/laser intensities used during the experiment. As a result, Kd values can only be compared when the same acquisition settings are used and when the fluorescein reference is consistent from day to day. We use the Kd to compare binding instead of change in maximum % FRET efficiency because fusion of the acceptor to different proteins of interest (i.e., V–Bad verses Venus–Bad2A–ActA) may hold the acceptor in a different location/orientation in respect to the donor, which affects FRET efficiency. Comparing the maximum FRET efficiency is only valid when comparing the effect of point mutations in the protein of interest.

16. We have observed that at high A:D intensity ratios, the data may come out of the plateau and continue to increase linearly or there may be points that suddenly drop dramatically in FRET efficiency. Beyond saturation of binding the increased random collisions combined with the lack of sufficient data points at higher intensity ratios could lead to this error. Find the A:D intensity ratio at which there is saturation of the positive control for at least three points. This can be set as the upper limit for the intensity ratio and applied to all datasets before calculating Kd. Apply this cutoff to remaining datasets.

5   Appendix

```
// macro to read in all the files and make them
in a stack
   dir1  =  getDirectory("Choose  tau  Source
Directory");
   list1 = getFileList(dir1);
   for (i=0; i<list1.length; i++) {
       open(dir1+list1[i]);
       nSlices;
       if (nSlices>1) {
           run("Stack to Images");
       }
   }
   run("Images to Stack", "name=tau title=t");
   dir2  =  getDirectory("Choose  Ch1  Source
Directory");
   list2 = getFileList(dir2);
   for (i=0; i<list2.length; i++) {
       open(dir2+list2[i]);
       nSlices;
       if (nSlices>1) {
           run("Stack to Images");
       }
   }
   run("Images to Stack", "name=Ch1 title=Ch1");
   dir3  =  getDirectory("Choose  Ch2  Source
Directory");
   list3 = getFileList(dir3);
   for (i=0; i<list3.length; i++) {
       open(dir3+list3[i]);
       nSlices;
       if (nSlices>1) {
           run("Stack to Images");
       }
   }
   run("Images to Stack", "name=Ch2 title=Ch2");
   // to remove the images that have no cells
   selectWindow("Ch2");
   p=nSlices;
   setBatchMode(true);
   for (i=1; i<=p; i++) {
       setSlice(i);
       run("Measure");
       image_intensity=getResult("Mean");
       if (image_intensity<7) {
       run("Delete Slice");
       selectWindow("Ch1");
```

```
            setSlice(i);
            run("Delete Slice");
            selectWindow("tau");
            setSlice(i);
            run("Delete Slice");
            selectWindow("Ch2");
            i=i-1;
            p=p-1;
            }
    }
    // remove the bottom pixels (due to the prob-
lem of camera aligning) and resave single images
into target folder
    dir4 = getDirectory("Choose saving Directory");
    selectWindow("Ch2");
    n=nSlices;
    for (i=1; i<=n; i++) {
        setSlice(i);
        run("Duplicate…", "title=Ch2-1");
        run("Canvas Size…", "width=256 height=250
position=Top-Center zero");
            saveAs("Tiff", dir4+"Ch2_"+i+".tif");
        close();
        }
    selectWindow("Ch2");
    close();
    selectWindow("Ch1");
    n=nSlices;
    for (i=1; i<=n; i++) {
        setSlice(i);
        run("Duplicate…", "title=Ch1-1");
        run("Canvas Size…", "width=256 height=250
position=Top-Center zero");
            saveAs("Tiff", dir4+"Ch1_"+i+".tif");
        close();
        }
    selectWindow("Ch1");
    close();
    selectWindow("tau");
    n=nSlices;
    for (i=1; i<=n; i++) {
        setSlice(i);
        run("Duplicate…", "title=tau-1");
        run("Canvas Size…", "width=256 height=250
position=Top-Center zero");
            saveAs("Tiff", dir4+"tau_"+i+".tif");
        close();
```

```
        }
   selectWindow("tau");
   close();
   selectWindow("Results");
   run("Clear Results");
   //ROI selection for each image based on Ch2
signals and measure the intensities/lifetimes
accordingly
   for (i=1; i<=n; i++) {
       open(dir4+"Ch2_"+i+".tif");
       run("Duplicate...", "title=temp1.tif");
       selectWindow("temp1.tif");
        run("Subtract Background...", "rolling=50
stack");
       setAutoThreshold("Default dark");
       //run("Threshold...");
       getThreshold(lower, upper);
       new_lower=lower+50;
       setThreshold(new_lower, upper);
       run("Convert to Mask");
       run("Analyze Particles...", "size=8-Infinity
pixel circularity=0.00-1.00 show=Nothing summa-
rize add");
       selectWindow("Ch2_"+i+".tif");
       roiManager("Show None");
       roiManager("Show All");
       roiManager("Measure");
       q=nResults/2;
       o=0;
       for (m=0; m<q; m++) {
           r=m+q;
           max_intensity=getResult("Max", r);
           low_intensity=getResult("Min", r);
           if (max_intensity>1000 || (low_inten-
sity<100 || low_intensity>400)) {
               w=m-o;
               roiManager("Select", w);
               roiManager("Delete");
               o=o+1;
               }
           }
       if (o<q) {
       roiManager("Save", dir4+"ROI"+i+".zip");
       run("Clear Results");
       selectWindow("Ch2_"+i+".tif");
       run("32-bit");
        run("Subtract Background...", "rolling=50
stack");
```

```
        roiManager("Show None");
        roiManager("Show All");
        roiManager("Measure");
        selectWindow("Results");
        saveAs("Results", dir4+"Ch2-"+i+".xls");
        run("Clear Results");
        open(dir4+"Ch1_"+i+".tif");
        run("32-bit");
         run("Subtract Background...", "rolling=50
stack");
        roiManager("Show None");
        roiManager("Show All");
        roiManager("Measure");
        selectWindow("Results");
        saveAs("Results", dir4+"Ch1-"+i+".xls");
        run("Clear Results");
        open(dir4+"tau_"+i+".tif");
        roiManager("Show None");
        roiManager("Show All");
        roiManager("Measure");
        selectWindow("Results");
        saveAs("Results", dir4+"tau-"+i+".xls");
        run("Clear Results");
        roiManager("reset");
        run("Close All");
        }
        else {
        run("Clear Results");
        roiManager("reset");
        run("Close All");
        }
    }
    run("Close All");
    Dialog.create("Analysis is done.");
    Dialog.show();
```

References

1. Hsu YT (1997) Nonionic detergents induce dimerization among members of the Bcl-2 family. J Biol Chem 272(21):13829–13834

2. Yang E, Zha JP, Jockel J, Boise LH, Thompson CB, Korsmeyer SJ (1995) Bad, a heterodimeric partner for Bcl-X(L) and Bcl-2, displaces Bax and promotes cell-death. Cell 80(2):285–291

3. Jaffe HH, Miller AL (1966) The fates of electronic excitation energy. Chem Educ 43(9):469

4. Laptenok SP, Borst JW, Mullen KM, van Stokkum IH, Visser AJ, van Amerongen H (2010) Global analysis of Forster resonance energy transfer in live cells measured by fluorescence lifetime imaging microscopy exploiting the rise time of acceptor fluorescence. Phys Chem Chem Phys 12(27):7593–7602

5. Lakowicz JR (2006) Principles of fluorescence spectroscopy, 3rd edn. Springer, New York

6. Stryer L, Haugland RP (1967) Energy transfer – a spectroscopic ruler. Proc Natl Acad Sci U S A 58(2):719–726

7. Aranovich A, Liu Q, Collins T, Geng F, Dixit S, Leber B, Andrews DW (2012) Differences in the mechanisms of proapoptotic BH3 proteins binding to Bcl-XL and Bcl-2 quantified in live MCF-7 cells. Mol Cell 45(6):754–763

8. Chen Y, Mills JD, Periasamy A (2003) Protein localization in living cells and tissues using FRET and FLIM. Differentiation 71(9–10):528–541

9. Markwardt ML, Kremers GJ, Kraft CA, Ray K, Cranfill PJ, Wilson KA, Day RN, Wachter RM, Davidson MW, Rizzo MA (2011) An improved cerulean fluorescent protein with enhanced brightness and reduced reversible photoswitching. PloS One 6(3):e17896

10. Liu Q, Leber B, Andrews DW (2012) Interactions of pro-apoptotic BH3 proteins with anti-apoptotic Bcl-2 family proteins measured in live MCF-7 cells using FLIM FRET. Cell Cycle 11(19):3536–3542

Chapter 35

Assessment of Mitochondrial Ca²⁺ Uptake

András T. Deak, Claire Jean-Quartier, Alexander I. Bondarenko, Lukas N. Groschner, Roland Malli, Wolfgang F. Graier, and Markus Waldeck-Weiermair

Abstract

Mitochondrial Ca²⁺ uptake regulates mitochondrial function and contributes to cell signaling. Accordingly, quantifying mitochondrial Ca²⁺ signals and elaborating the mechanisms that accomplish mitochondrial Ca²⁺ uptake are essential to gain our understanding of cell biology. Here, we describe the benefits and drawbacks of various established old and new techniques to assess dynamic changes of mitochondrial Ca²⁺ concentration ($[Ca^{2+}]_{mito}$) in a wide range of applications.

Key words Mitochondrial Ca²⁺ uptake, Calcium Green, Fura-2, Rhod-2, FRET, Oxidative phosphorylation, Mitochondrial membrane potential, Mitoplast, Patch-clamp recording, Ca²⁺ imaging

1 Introduction

Ca²⁺ transfer into the mitochondrial matrix is linked to numerous important cellular functions including the regulation of energy metabolism [1], mitochondrial reactive oxygen species production [2], and cell survival and death [3]. The significance and complexity of mitochondrial Ca²⁺ sequestration are highlighted by several recent studies describing key components of the mitochondrial Ca²⁺-uptake machinery [4–12]. In order to study and to visualize the complex process of mitochondrial Ca²⁺ handling, various chemical and genetically encoded Ca²⁺ sensors, as well as electrophysiological techniques, have been developed and utilized over the past decades. Here, we introduce some common methods, indicators applied on isolated mitochondria or permeabilized and intact cells in order to assess and quantify mitochondrial Ca²⁺-uptake dynamics.

Chemical indicators exhibit altered fluorescent properties when bound to Ca²⁺ and are usually loaded into permeabilized or intact cells by passive incubation. The most popular chemical Ca²⁺ indicators used to assess mitochondrial Ca²⁺ uptake are

Volkmar Weissig and Marvin Edeas (eds.), *Mitochondrial Medicine: Volume I, Probing Mitochondrial Function*, Methods in Molecular Biology, vol. 1264, DOI 10.1007/978-1-4939-2257-4_35, © Springer Science+Business Media New York 2015

Rhod-2 and Calcium Green. Rhod-2 accumulates predominantly within the mitochondrial compartment and responds with increased fluorescence upon $[Ca^{2+}]_{mito}$ elevation [13]. On the other hand, Calcium Green is applied as an indirect reporter of mitochondrial Ca^{2+} uptake by measuring the decline of the extramitochondrial Ca^{2+} concentration upon an activation of mitochondrial Ca^{2+} uniport [14].

Genetically encoded mitochondrial Ca^{2+} probes are protein-based fluorescent indicators targeted into mitochondrial matrix (MM) via a specific mitochondrial signal sequence [15], and they are typically incorporated into cells by gene transfer techniques. These probes are usually divided in two classes: the single fluorescent protein (FP) and the Förster resonance energy transfer (FRET)-based Ca^{2+} sensors. Single FP sensors including the pericams [16], the GCamPs [17], and the recently developed GECOs [18] consist of a circularly permuted FP (cpFP) flanked by a genetically modified Ca^{2+}-binding domain, the calmodulin (CaM), and its interacting peptide (M13) from the myosin light chain kinase. A common feature of these sensors is the ability of their integrated cpFPs to modulate their spectral properties in response to changes of Ca^{2+} concentration based on the interaction of CaM with the M13 peptide.

The other large group of fluorescent Ca^{2+} reporters is the FRET-based cameleons, which consist of two different fluorescent proteins possessing overlapping spectral properties. The "prototype" of such sensors contains the previously mentioned CaM and M13 domains, which were inserted in tandem between the cyan and the yellow fluorescent proteins (CFP/YFP) [19]. Cameleons in a Ca^{2+}-bound state undergo a conformational change, whereupon the donor CFP gets in close proximity to the acceptor YFP yielding an enhanced energy transfer between the two fluorophores. Cameleons are thus ratiometric indicators: increase of FRET is coupled with the decrease of CFP fluorescence. Since the introduction of the first cameleon in 1997, several derivates of this Ca^{2+} sensor with improved Ca^{2+} sensitivities [20], higher FRET efficiencies, increased pH stability, and appropriate mitochondrial targeting have been developed [21, 22]. However, none of these mitochondrial-targeted sensors were suitable for a co-imaging with the most popular cytosolic Ca^{2+} indicator, Fura-2, because of a significant overlap in the excitation and emission spectra [23]. To resolve this problem, a novel mitochondrial-targeted, red-shifted FRET-based sensor referred to as mtD1GO-Cam was recently constructed. This genetically encoded Ca^{2+} probe consists of a green and orange fluorescent protein and was successfully used to simultaneously measure $[Ca^{2+}]_{cyto}$ and $[Ca^{2+}]_{mito}$ in a ratiometric manner within same individual cells [12, 24].

Although optical methods are most often used to investigate mitochondrial Ca^{2+} uptake, this approach bears some limitations.

Specifically, it does not allow controlling key parameters which determine driving flux of Ca^{2+} uptake, such as the mitochondrial membrane potential and Ca^{2+} concentration gradient. Additionally, optical methods reflect rather the dynamic changes of Ca^{2+} concentration in a specific cellular compartment, which is always a balance between Ca^{2+} influx, Ca^{2+} extrusion, and Ca^{2+}-buffering capacity, but not the Ca^{2+} flux itself. Patch-clamp approach circumvents the aforementioned drawbacks associated with optical methods and represents a direct method of assessing transmitochondrial ion fluxes and determining channel conductances [10, 25].

Considering the close interdependency of mitochondrial Ca^{2+} uptake and metabolism [1], it is possible to indirectly assess mitochondrial Ca^{2+} signals by measuring cellular O_2 consumption rates (OCRs). In many cases, the analysis of metabolic activity is also an essential tool in estimating mitochondrial Ca^{2+} handling. The availability of analytical instruments for measuring OCRs has led to broad applications of this technique in the field of mitochondrial Ca^{2+} signaling [5, 26]. To this end, we also describe how to use the Seahorse XF96 Extracellular Flux Analyzer in order to measure cellular O_2 consumption in dependence of mitochondrial Ca^{2+}.

2   Materials

Notably, all chemicals and reagents used are suitable for most adherent mammalian cell lines, but need to be adjusted and optimized for every cell type of choice. All buffers should be prepared using deionized water and analytical grade reagents in order to avoid possible ionic contaminants and at room temperature to obtain best solubility options, unless otherwise indicated.

2.1 Chemical Fluorescent Indicator Components

1. Intracellular assay buffer: 110 mM KCl, 0.5 mM KH_2PO_4, 1 mM $MgCl_2$, 0.03 mM HEPES, 5 mM Succinate, and 10 mM Glucose; pH adjusted with KOH to 7.4 (*see* **Notes 1** and **2**). For storage, freeze aliquots at –20 °C.

2. Permeabilization reagent: Prepare 0.1 % or 1 % stock solution of digitonin by dissolving 1–10 mg in 1 ml of H_2O (0.81–8.1 mM). Heat the mixture up to 95 °C until a clear solution is obtained. Let it cool down to RT before use (Table 1) and store at 4 °C. The optimal final concentration of digitonin depends on the cell type or sample material and has to be adjusted accordingly.

3. Chemical sensors: Prepare stock solutions with dimethyl sulfoxide (DMSO), and freeze aliquots corresponding to the manufacturer's protocol (Table 2). Accordingly, on the day of experiment, prepare working dilutions using deionized water or a buffer of choice.

Table 1
Concentration and solvent of relevant compounds in stock and working solution

Compound	Source	Molecular weight	Solvent	Stock (mM)	Working solution/s (µM)
Histamine	Sigma	184.07	H_2O	100	100, 10, 1
ATP	Sigma	507.20	H_2O	100	100, 10, 1
Carbachol	Sigma	182.65	H_2O	100	100, 10, 1
CGP37157	AbCam	324.22	DMSO	100	10, 20
BHQ	Sigma	222.33	DMSO	100	15
Cyclopiazonic acid	AbCam	336.39	H_2O	10	2–10
Thapsigargin	AbCam	650.76	DMSO	1	1
Rotenone	Sigma	394	DMSO	100	50
Ruthenium red	Sigma	790.39	H_2O	1	10
Digitonin	Sigma	1,229.31	H_2O	0.81–8.1	10–100
Oligomycin	AbCam	786.78	DMSO	10	2
Antimycin	Sigma	534.645	$EtOH_{abs}$	100	10
Ionomycin	AbCam	709.01	DMSO	10	1–10
FCCP	AbCam	255.97	$EtOH_{abs}$	10	0.2–4

Table 2
Chemical sensors in stock and working concentrations including corresponding fluorimeter settings

Sensor	Stock (mM)	Final (µM)	K_d (µM)	Excitation λ (nm)	Emission λ (nm)
Calcium Green 5 N	1 (DMSO)	2	14	506 ± 10	532 ± 10
Fura-5 K	1 (DMSO)	1	0.2	$340/380 \pm 5$	510 ± 20
Fura-2-AM	1 (DMSO)	3.3	0.2	$340/380 \pm 10$	510 ± 10
Rhod-2-AM	1 (DMSO)	5	0.6	550 ± 10	590 ± 10

4. Further chemicals for measurement: Prepare 1 mM and 10 mM $CaCl_2$ stock solutions in deionized water, and apply 1–10 µM Ca^{2+} pulses into the cuvette (Fig. 1). Optionally prepare working stocks of inhibitors such as ruthenium red, cyclopiazonic acid, or FCCP (Table 1).

2.2 Components for the Use of Genetically Encoded Ca^{2+} Indicators

1. Vector: All plasmids encoding for mitochondrial-targeted GECIs (mtRP, mtD3cpv, mtD1GO-Cam, and mito-GEM-GECO1) (Table 1) were constructed in a pcDNA3 vector (Invitrogen, Carlsbad, CA, USA). This vector contains a human cytomegalovirus immediate-early (CMV) promoter for high-level expression in a wide range of mammalian cells (*see* **Note 3**).

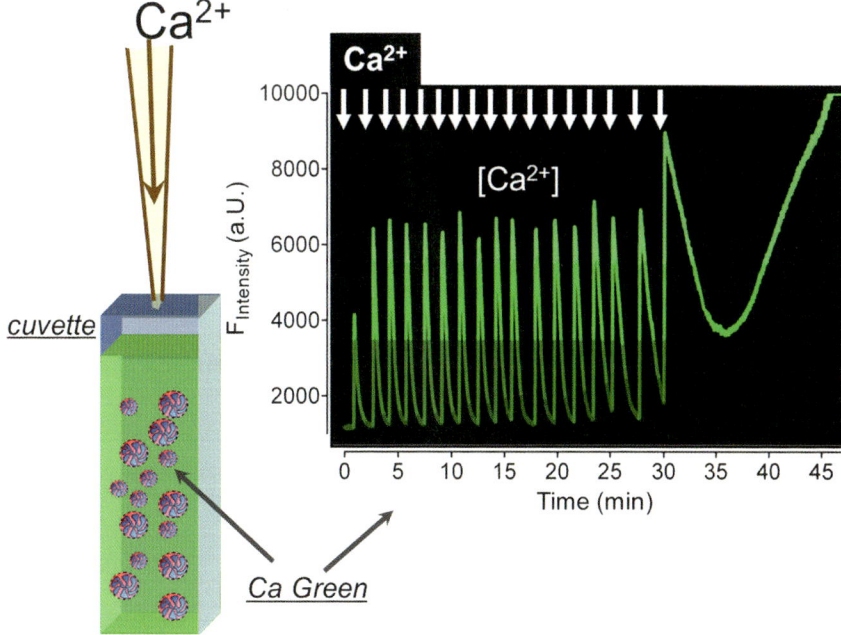

Fig. 1 Indirect assessment of mitochondrial Ca²⁺ uptake by Calcium Green 5 N. Digitonin-treated HeLa cells (10⁸) were exposed to repetitive pulses (100–200 s/pulse) of 50 μM Ca²⁺ added exogenously to the bath, which also contained Calcium Green 5 N. The decline of each Ca²⁺ signal indicates the magnitude of mitochondrial Ca²⁺ uptake

2. Transformation: Many *E. coli* strains are suitable for the transformation of these plasmids including TOP10, TOP10F′, DH5α-T, JM101, or XL1-Blue. Most strains are available as chemically competent or electrocompetent cells. However, the plasmids can be transformed in any method of choice.

3. Antibiotic resistance: Select transformants on Lysogeny broth (LB) plates containing 2 % agar-agar and 50–100 μM/ml ampicillin. For bacterial overnight culture, shake picked colonies in a shaking incubator at 37 °C using 100–250 ml LB (10 g NaCl, 5 g yeast extract, and 10 g tryptone/l) with 50–100 μM/ml ampicillin.

4. Plasmid isolation: Plasmid can be isolated with any plasmid isolation kit. We recommend using the PureYield Plasmid Maxiprep System (Promega, Mannheim, Germany).

5. Transfection medium: Dulbecco's Modified Eagle's Medium—low glucose (Sigma Chemicals, St. Louis, MO, USA) is suitable for most mammalian cell types (*see* **Note 4**).

6. Transfection reagent: We recommend using TransFast™ (Promega, Mannheim, Germany) for transient transfection of the sensor plasmid (*see* **Note 5**).

7. Creation of stable cell lines: Use G418 (Geneticin®) antibiotic resistance of the vector for mammalian cell line (*see* **Note 6**).

8. Storage buffer (SB): 138 mM NaCl, 5 mM KCl, 2 mM $CaCl_2$, 1 mM $MgCl_2$, 10 mM HEPES, 2.6 mM $NaHCO_3$, 0.44 mM KH_2PO_4, and 10 mM D-glucose supplemented with 0.1 % vitamins, 0.2 % essential amino acids, 1 % penicillin/streptomycin, and 1 % fungizone (PAA Laboratories, Linz, Austria); pH adjusted to 7.4 with NaOH (*see* **Notes** 7 and **8**).

9. Experimental buffer (EB): 138 mM NaCl, 5 mM KCl, 2 mM $CaCl_2$, 1 mM $MgCl_2$, 10 mM D-glucose, and 10 mM HEPES; pH adjusted to 7.4 with NaOH (*see* **Note 8**).

10. Stock solution and working solution of compounds: Add agonist(s), mediator(s), or any other compound to EB in a concentration of at least 1:1,000 (Table 1) to mobilize Ca^{2+} in a distinct cell line.

11. Perfusion system: For real-time measurements, it is best to use a perfusion chamber allowing a continuous perfusion of cells with EB ± compound(s).

12. Digital wide-field imaging system (*see* refs. 12, 24, 27, 28): Co-imaging of Fura-2 and mtD1GO-Cam can be performed on an inverted (digital) wide-field microscope using a 40× oil immersion objective (*see* **Note 9**). Use a high-speed polychromator system to allow a fast switching of excitation wavelengths. Select suitable excitation filters according to the sensor(s) needs (Tables 2 and 3), e.g., use an E500spuv and a 495dcxru (Chroma Technology Corp, Rockingham Vermont, USA) for a simultaneous excitation of Fura-2 and mtD1GO-Cam. Collect emission light with proper dichroic filters (Tables 2 and 3); for FRET- and GECO-based indicators, use either a motorized filter wheel or a beam splitter, both equipped with appropriate dichroic filters. Images are best recorded with a thermoelectric-cooled charge-coupled device (CCD) camera. For data acquisition, device controlling, or post-acquisition image analysis, we recommend to use an up-to-date software of MetaFluor, Visiview (Universal Imaging, Visitron), or the live acquisition software (Till Photonics).

2.3 Components for Mitoplast Isolation and Patch-Clamp Recordings

1. Mitochondrial storage buffer: 10 mM HEPES, 250 mM sucrose, 1 mM ATP, 0.08 mM ADP, 5 mM succinate, 2 mM KH_2PO_4, and 1 mM DTT; pH adjusted to 7.4 with KOH.

2. Hypotonic solution: 5 mM HEPES, 5 mM sucrose, and 1mM ethylen glycol tetraacetic acid (EGTA); pH adjusted to 7.2 with KOH.

3. Hypertonic solution (in mM): 750 KCl, 80 HEPES, and 1 EGTA; pH adjusted to 7.2 with KOH.

4. Isolation of mitochondria: There are various protocols and kits available, mostly based on cell disruption and homogenization followed by differential centrifugation, such as the mitochondria isolation kit for cultured cells (Thermo Scientific 89874, USA). Suspend mitochondrial fraction in mitochondrial storage buffer.

Table 3
Description of established mitochondrial-targeted genetically encoded Ca^{2+}

Sensor [reference]	Incorporated fluorophore(s)	Excitation λ (nm)[a]	Emission λ (nm)[a]	Mitochondrial targeting[b]
mtRP [16]	Single FP: cpYFP	430 ± 10 (Ca^{2+}) 480 ± 5 (H$^+$)[c]	535 ± 10	COX4
mtD3cpV [20]	FRET: ECFP (donor) cpVenus (acceptor)	430 ± 10	480 ± 5 535 ± 10	(2×) COX8
mtD1GO-CaM [24]	FRET: cpEGFP (donor) mKO2 (acceptor)	480 ± 5	510 ± 5 560 ± 10	(2×) COX8
mito-CEM-GECO1 [18]	Single FP: cpGFP[d]	377 ± 50	447 ± 40 520 ± 35	(2×) COX8

Abbreviations: *mtRP* mitochondrial-targeted ratiometric pericam, *cp* circularly permutated, *ECFP/CGFP* enhanced CFP/GFP, *mKO2* monomeric Kusabira-Orange 2

[a]Values refer to major excitation/emission peak ± optimal bandwidth

[b]N-terminal targeting sequence obtained from the mitochondrial cytochrome C oxidase subunit (COX) 4/8

[c]mtRP has two excitation maxima: a Ca^{2+}-dependent and a H$^+$-dependent, thus, can measure [Ca^{2+}]$_{mito}$ and [H$^+$]$_{mito}$ simultaneously

[d]Amino acid substitutions relative to GCamP3 are as follows: L60P/K69E/N77Y/D86G/N98I/K191I/L173Q/T223S/N302S/R377P/K380Q/S404G/E430V

5. Micropipettes: For mitoplast patch-clamp recordings, micropipettes are pulled from borosilicate glass capillaries and fire polished [10]. When filled with KCl-containing solution, the pipettes typically have a resistance of 8–12 MΩ.

6. Microscope: For mitoplast visualization, we used a Zeiss Axiovert 135 M microscope equipped for phase contrast with a 63× objective.

7. Standard bath solution (in mM): 150 KCl, 1 EGTA, and 10 HEPES; pH adjusted to 7.2 with KOH.

8. Pipette solution for single-channel recordings (in mM): 105 CaCl$_2$ and 10 HEPES supplemented with 0.01 CsA and 0.01 CGP37187; pH adjusted to 7.2 with Ca(OH)$_2$.

9. Pipette solution for whole-mitoplast recordings (in mM): 120 Cs methanesulfonate, 30 CsCl, 1 EGTA, 110 sucrose, and 10 HEPES; pH adjusted to 7.2 with CsOH.

10. Ca^{2+}-free bath solution (in mM): 130 Trizma HCl, 50 Trizma base, 1 EGTA, 1 EDTA, 10 HEPES, and pH 7.2.

11. Ca^{2+}-containing bath solution (in mM): 130 Trizma HCl, 50 Trizma base, 2–3 CaCl$_2$, 1 EDTA, 10 HEPES, and pH 7.2. For I$_{Ca}$ recording, we add 2–3 mM CaCl$_2$ instead of 1 mM EGTA. For recording monovalent cationic current carried by Na$^+$, use 150 mM NaCl instead of Trizma and the same Cs-based pipette solution and voltage protocol.

2.4 Components for Measuring Cellular O$_2$ Consumption Rates

1. Instruments: XF96 Extracellular Flux Analyzer from Seahorse Bioscience including the controller; 37 °C non-CO$_2$ incubator (or XF Prep Station); XF96 4-port FluxPak or FluxPak mini (see **Note 10**); pH Meter; phase contrast microscope; pipettes including a multichannel pipette and matching tips.

2. Solution: XF Calibrant Solution (Seahorse Bioscience Part #100840-000).

3. Assay medium: non-buffered DMEM (see **Note 11**).

4. Substrates: D-glucose and sodium pyruvate.

5. Reagents: oligomycin A, carbonyl cyanide 4-(trifluoromethoxy)phenylhydrazone (FCCP), and antimycin A (see **Note 12**).

3 Methods

Carry out all procedures at room temperature unless indicated otherwise.

3.1 Measuring Indirect Ca^{2+} Uptake via Calcium Green

1. For each measurement, use 10^6 cells/ml or isolated mitochondria (see **Note 13**) with a protein content of 2 mg/ml (see **Note 14**). Wash cells or isolated mitochondria in EGTA-containing buffer before starting with procedures (see **Note 15**).

2. Resuspend cells in 2 ml assay buffer and add fresh 25 µl of the 0.1–1 % digitonin stock (final concentration of 10–100 µM) before the measurement, or resuspend isolated mitochondria in 2 ml assay buffer.

3. Transfer to 2 ml fluorimeter cuvette (*see* **Note 16**) including a stirrer for mixing.

4. Load protocol on the fluorimeter and insert the cuvette. Recommended settings for Hitachi fluorimeter are given in Table 2. Blank the cuvette before adding the dye.

5. Start the measurement. Upon stable baseline, add a first pulse of 6 µl 10 mM $CaCl_2$ (=30 µM) to overcome the EGTA (in buffer). Add further pulses of 50 µM $CaCl_2$ (Fig. 1) (*see* **Notes 17** and **18**).

6. Repeat the measurement using a new sample, and add various inhibitors as described above before blanking the cuvette.

3.2 Measuring Direct Mitochondrial Ca²⁺ Uptake via Fura-2 or Rhod-2

1. Grow cells on cover slips to ~70 % confluency, or use isolated mitochondria >2 mg/ml in isolation buffer.

2. Incubate cells in growth medium or isolated mitochondria in mitochondrial storage buffer with 3.3 µM Fura-2-AM for 60 min or with 5 µM Rhod-2-AM for 30 min at room temperature in the dark (*see* **Note 20**).

3. After the incubation period, change growth medium to assay buffer or pellet the sample, wash with EGTA-containing buffer (*see* **Note 15**), and let isolated mitochondria settle down onto the microscopy slides in the assay buffer for at least 20 min (*see* **Note 19**).

4. Transfer slide to the microscope, and proceed with the measurement by real-time monitoring Fura-2 or Rhod-2 using an imaging system as described above with the excitation and emission filter settings displayed in (Table 2).

3.3 Transfection of Genetically Encoded Ca²⁺ Indicators

1. Grow adherent mammalian cells in its optimum culture medium in a humidified incubator (37 °C, 5 % CO_2, and 95 % air) on perfusion chamber slides to 60–80 % confluency.

2. For transient transfection of $\sim 5 \times 10^5$ cells, mix 2 µg of a plasmid encoding a mitochondrial-targeted GECI and an appropriate amount of transfection reagent with 1 ml of serum- and antibiotic-free transfection medium. Incubate cells in the incubator for 16–20 h, and change back to complete culture medium. Experiments can be performed 24–72 h after transfection.

3.4 Real-Time Recordings of [Ca²⁺]_{cyto} and [Ca²⁺]_{mito}

Herein, we will describe the simultaneous measurement of Fura-2 and mtD1GO-Cam in detail. For assessing mitochondrial Ca²⁺ using chemical fluorescent indicators (Fura-2 and Rhod-2) or other mitochondria-targeted genetically encoded Ca²⁺ indicators (mtRP, mtD3cpv, and mito-GEM-GECO1), use appropriate filter settings as described in Tables 2 and 3.

1. Load mtD1GO-Cam transfected cells with Fura-2-AM in a concentration of 3.3 μM dissolved in storage buffer for at least 20 min (*see* **Note 20**).

2. Stop Fura-2-AM loading by washing the cells twice with storage buffer.

3. Keep cells in storage buffer prior to measurements (*see* **Note 21**).

4. Put a drop of immersion oil on top of the objective, and place the perfusion chamber slide with the cells upside onto the droplet.

5. Connect the chamber with the perfusion system and start the perfusion.

6. Set the cells in focus by turning the z-tuner of the microscope table, or use the autofocus of the system in the white light mode.

7. Use the ocular for searching cells that have a high- and good-targeted expression of the mitochondrial cameleon using an excitation wavelength at ~480 nm and emission at ~510 nm (cpEGFP, *see* **Notes 22** and **23**).

8. For simultaneous illumination of Fura-2 and mtD1GO-Cam, use the time-lapse function of the imaging software in a triple-wavelength mode. To gain better fluorescence sensitivity, use binning 2 or higher. Expose excitation of Fura-2 at 340 nm and 380 nm for 150 ms or 50 ms, respectively (*see* **Note 24**), and collect emitted light at 510 nm. The mitochondrial sensor gets excited for 400 ms (*see* **Note 25**) at 477 nm and emits at 510 nm (GFP, donor fluorescence) and 560 nm (FRET, acceptor fluorescence), respectively. Accordingly, the two sensors get recorded in the time-lapse mode within 600 ms or less (*see* **Notes 24** and **25**) by alternately exposing the three excitation wavelengths without any fluorescence interference (Fig. 2).

9. Use an appropriate experimental design for cell stimulation via the perfusion system (e.g., 100 μM histamine in EB).

10. Analyze data of the recorded ratios from the two sensors separately to verify the spatiotemporal correlation of $[Ca^{2+}]_{cyto}$ and $[Ca^{2+}]_{mito}$ (*see* **Note 26**).

3.5 Mitoplast Patch-Clamping Recording

1. Mitochondria from cultured cells (e.g., HeLa) are freshly isolated by differential centrifugation steps (*see* **Note 27**) as previously described [10, 25]

2. Mitoplast formation: Incubate isolated mitochondria kept on ice in 4 volumes of hypotonic solution for 10 min. This results in mitochondria swelling and rupture of the outer membrane. Add with 1 volume of hypertonic solution to equilibrate the tonicity (*see* **Note 28**).

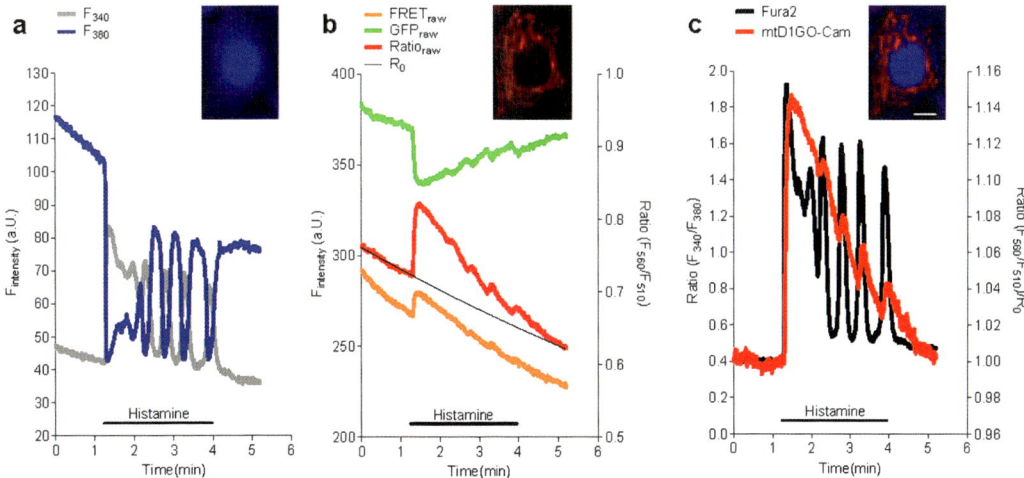

Fig. 2 Simultaneous measurement of $[Ca^{2+}]_{cyto}$ and $[Ca^{2+}]_{mito}$ in the same individual cells. Original traces of (**a**) cytosolic and (**b**) mitochondrial Ca^{2+} signals and (**c**) their respective correlation over time upon cell stimulation with 100 μM histamine in intact mtD1GO-Cam expressing HeLa cells loaded with Fura-2. (**a**) Raw traces of Fura-2 signals at 340 nm (*gray*) and at 380 nm (*blue*) excitation. *Inset image* shows the cytosolic accumulation of Fura-2. (**b**) Raw traces of GFP signal (GFP_raw, *green*) at 510 nm excitation and respective FRET signal (FRET_raw, *orange*) at 560 nm excitation were plotted on the *left y*-axis. *Red curve* (Ratio_raw) indicates the FRET ratio computed from the raw traces (F_{FRET}/F_{GFP} was plotted on the *right y*-axis. *Black curve* represents photobleaching function (R_0) assessed with a one-phase exponential decay function. *Inset image* shows mitochondrial targeting of mtD1GO-Cam. (**c**) Fura-2 (*black*, *left y*-axis) and mtD1GO-CaM signals (*red*, *right y*-axis) were calculated from the raw traces shown in (**a**) and (**b**), respectively. *Inset image* is an overlay of previous *insets*. The scale bar is 10 μM

3. For patch-clamp recordings, place 20–40 μl of mitoplast suspension to the recording chamber (depending on the size of the mitochondrial pellet), and allow mitoplasts to settle down for 10 min prior to experimentation.

4. To form gigaohm contact, position the pipette tip on the chosen mitoplast away from the "cap" region, which represents the attached remnants of the outer membrane. Press the pipette against the mitoplast, and apply negative pressure to the pipette interior. When mitoplast-attached configuration is reached, single-channel openings can be detected during voltage ramps.

5. Before obtaining whole-mitoplast configuration, capacitance transients are compensated, and negative pressure is further applied until the patch is ruptured.

6. Alternatively, voltage steps of 300–600 mV and 10–20 ms duration may be applied.

7. Successful access to the matrix is accompanied by reappearance of capacitance transients. Mitoplast capacitance measured with Membrane Test tool of Clampex is around 1 pF. If membrane rupture is accompanied by a leak current, new mitoplast should be chosen, and the procedure should be repeated again with a new pipette.

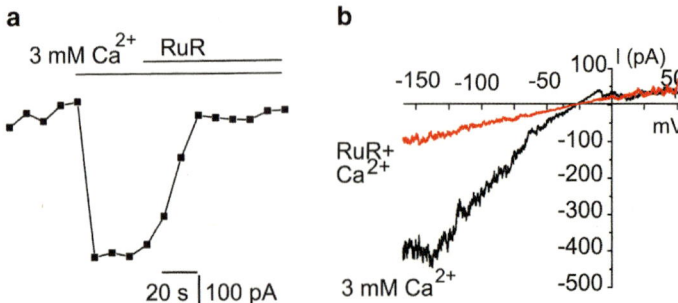

Fig. 3 Patch-clamp recording of transmitochondrial Ca^{2+} flux. (**a**) Time course of the whole-mitoplast current development at −155 mV before and after addition of 3 mM Ca^{2+} followed by addition of 10 μM RuR. (**b**). Corresponding Ca^{2+} current responses to voltage ramps from −160 to 50 mV before and after addition of 10 μM RuR in the presence of 3 mM Ca^{2+}

8. Following a successful membrane rupture, experiments are continued in the same way as they are done with cells.

9. For recordings of Ca^{2+} currents in mitoplast-attached and whole-mitoplast configuration, use respective pipette solutions described above. Signals obtained are sampled at 10 kHz and filtered at 1 kHz.

10. For recordings of Ca^{2+} currents in whole-mitoplast configuration, exchange Ca^{2+}-free bath solution for Ca^{2+}-containing bath solution by bath perfusion. We typically apply voltage ramps from −160 to +60 mV to record whole-mitoplast Ca^{2+} currents (Fig. 3).

11. For recording whole-mitoplast monovalent cationic current carried by Na^+, use 150 mM NaCl instead of Trizma and the same Cs-based pipette solution and voltage protocol.

12. Collect data using the Clampex software of pClamp (Molecular Devices, Sunnyvale, CA, USA). Signals obtained are sampled at 5 kHz and filtered at 1 kHz.

3.6 Assessing Ca^{2+}-Dependent Changes in Mitochondrial Metabolism

1. Harvest, resuspend, and dilute cells in a standard growth medium. Typical cell seeding numbers vary from 5,000 to 100,000 cells per well depending on the cell type, basal metabolic activity, proliferation rate, cell size, and the time of plating and must be determined empirically.

2. Gently seed 100 μl of cell suspension per well in XF96 Polystyrene Cell Culture Microplates (*see* **Note 29**). Wells A1, A12, H1, and H12 are to be left blank for background correction.

3. After seeding, place the microplate with the cells in an incubator at 37 °C, gassed with 5 % CO_2 for ≥12 h to guarantee optimal adherence (*see* **Note 30**).

Table 4
Commonly used protocol template programmed using the Assay Wizard of the Seahorse XF96 Software

Protocol start
1. Calibrate probes
2. Equilibrate
3. Loop 3 times
4. Mix for 3 min 0 s
5. Measure for 3 min 0 s
6. Loop end
7. Inject port A
8. Loop 3 times
9. Mix for 3 min 0 s
10. Measure for 3 min 0 s
11. Loop end
12. Inject port B
13. Loop 3 times
14. Mix for 3 min 0 s
15. Measure for 3 min 0 s
16. Loop end
17. Inject port C
18. Loop 3 times
19. Mix for 3 min 0 s
20. Measure for 3 min 0 s
21. Loop end
Program end

4. Fill each well of the utility plate with 200 μl of XF calibrant solution using a multichannel pipette, and lower the XF96 sensor cartridge onto the plate, fully submerging the biosensors (*see* **Note 31**).

5. Seal both the cartridge and plate, covered by the lid, using parafilm in order to minimize evaporation, and incubate at 37 °C in a non-CO_2 incubator.

6. Switch on the instrument, and open the XF software at least 12 h prior to the assay, in order to allow the system to stabilize at 37 °C. Table 4 summarizes a commonly used protocol

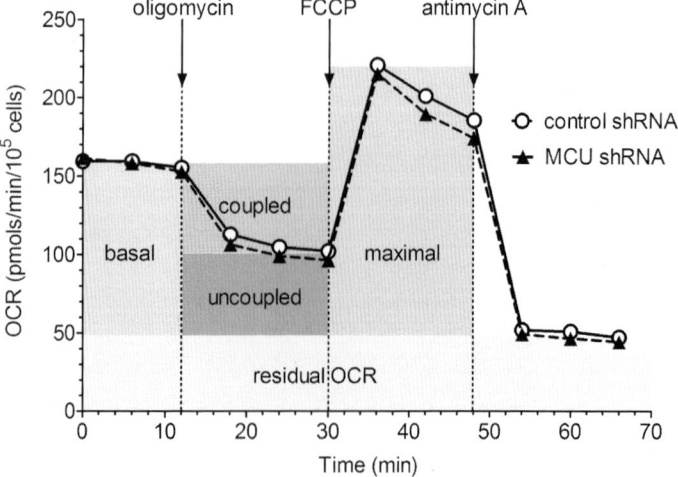

Fig. 4 Mitochondrial respiration assessed by the XF96 Extracellular Flux Analyzer from Seahorse Bioscience: an indirect way to determine mitochondrial Ca^{2+} uptake. O_2 consumption rates (OCRs) of HeLa cells stably expressing control shRNA or shRNA targeting the mitochondrial Ca^{2+} uniporter (MCU shRNA). Cells were treated with 1 μM oligomycin, 500 nM FCCP, and 2.5 μM antimycin A to assess basal, coupled, maximal, and residual OCRs. Data represent means ($n \geq 30$)

template that can be programmed and modified using the Assay Wizard of the Seahorse XF96 Software.

7. On the day of the assay, observe cells under the microscope to assure both sufficient viability and confluency (*see* **Note 29**).

8. Pre-warm non-buffered assay medium containing the desired amount of D-glucose and sodium pyruvate (typically 5.5 and 1 mM, respectively) to 37 °C, and adjust pH to 7.4 using NaOH.

9. Carefully remove the growth medium from the microplate using a multichannel pipette, making sure that ~20 μl of media remain at the bottom of the well at all times. Wash cells two times before replenishing the well with a final volume of 150 μl of assay medium.

10. Incubate the cell plate in a non-CO_2 incubator at 37 °C until ready for use.

11. Prepare compound solutions for injection using assay medium and reconstituted reagents (*see* **Notes 32** and **33**).

12. Open the appropriate assay template and start the assay using the XF96 software.

13. Insert the sensor cartridge and the utility plate (without the lid) into the instrument for calibration.

14. Upon completion of the automated calibration process, replace the utility plate with the microplate containing the cells and click "continue" to start the actual measurement (Fig. 4).

4   Notes

1. Add 8.2 g KCl, 0.114 g KH_2PO_4, 0.203 g $MgCl_2$, 0.766 g HEPES, 1.351 g succinate, and 1.982 g glucose. Dissolve in 800 ml H_2O while stirring for 5 min. Titrate KOH to adjust pH, then add 0.0114 g EGTA, and adjust pH again. While adjusting pH, use low concentrations of acid or base at the end of titrations in order to avoid sudden drops or rises above or below the required pH. Continue stirring for 5 min and check the pH again.

2. For dissolving EGTA properly, adjust pH first to slightly basic conditions. During the addition of the salt, pH will shift to lower values. After complete dissolution, adjust pH again.

3. The vector contains an episomal replication site for cell lines that are latently infected with SV40 or express the SV40 large T antigen (e.g., COS-1, COS-7).

4. For transfection, use the culture medium of the cell line of choice without supplements like FCS, antibiotics, and antimycotics.

5. Transfection reagent and method depend on cell type and equipment of the lab.

6. The pcDNA3.1(−) vector contains a neomycin resistance gene for creating stable mammalian host cell of choice. Transfect the cell line with the sensor plasmid using TransFast™ transfection reagent and transfection medium supplemented with an appropriate concentration of neomycin or G418 (Geneticin®), and feed the cells with selective medium every day. Stable cell colonies can be easily identified on a fluorescence microscope within 3–4 days after the addition of G418. Pick and expand colonies in 96- or 48-well plates.

7. Storage buffer can be stored after sterile filtration in aliquots (of e.g., 30 ml) at 4 °C for at least 3 months.

8. SB and EB are suitable for most non-excitable cell lines (e.g., HeLa, HUVEC, HEK293, and COS), but need to be adjusted to distinct cell type and/or protocol of measurement.

9. Use low-fluorescent immersion oil like a Cargille Immersion Oil Type HF or Type LDF (Optoteam, Vienna, Austria).

10. A FluxPak comprises sensor cartridges containing the fluorescent biosensor as well as the injection ports, utility plates, loading guides, and XF calibrant solution.

11. Non-buffered assay medium is usually based on the formulation of Dulbecco's Modified Eagle Medium including L-glutamine but does not contain any buffering agent (i.e., sodium bicarbonate). Such medium can be obtained from, for example, Seahorse Bioscience or Sigma Aldrich. For measurements under Ca^{2+}-free conditions, Ca^{2+}-free assay medium has to be custom made or replaced by assay solution (6).

12. Dissolve reagents in DMSO at stock concentrations of 5–10 mM and store at –20 °C.

13. Try to use fresh isolated mitochondria, since frozen mitochondria display impaired membrane integrity. Furthermore, optimal number of strokes for Dounce homogenization varies between cell types and homogenizer before impairing the integrity of mitochondrial membranes. Additional to up and down movements, rotation speed can increase homogenization outcome accordingly. Keep all samples during the isolation on ice, and precool centrifuges, all glasswares, flasks, eppis, and tools used during isolation process. Cell samples are handled at RT.

14. Optimal number of cells for each measurement can vary between cell types. Suitable protein content of isolated mitochondria additionally depends on mitochondrial integrity after isolation. If no or weak mitochondrial Ca^{2+} uptake is observed, we recommend using higher amounts of sample material per measurement.

15. Make sure to get rid of any excess Ca^{2+}. In case of harvesting cells, wash away any leftover medium with PBS (at low speed according to cell line ~$700 \times g$). In case of isolated mitochondria, samples are usually prepared in EGTA/EDTA or other Ca^{2+}-chelating agent containing buffers (at high speed for isolated organelles >10,000 g).

16. Fluorimeter cuvettes may vary for volume content. Try to stick to equal concentrations (compare Material section) if shifting down to lower volumes. Furthermore, make sure that the sample is mixed continuously during the measurement; otherwise, both the sensor and the isolated mitochondria or cells will settle down negatively affecting the readout.

17. Further pulses of various Ca^{2+} concentrations may be tried out. The optimal concentration is dependent on the individual cell line and number of mitochondria or cells. Calcium Green 5 N is best used within a concentration of 10–100 μM which also covers the range of MCU. For smaller concentration, one may change to a different sensor with a suitable K_d (Table 1).

18. Observation: The signal goes up upon Ca^{2+} addition and down when taken up by mitochondria. This can be repeated depending on the mitochondrial-uptake capacity and cell number (cave: always normalize to same cell number within samples to overcome laborious calculations afterwards). After a certain addition of Ca^{2+}, the signal will not completely fall down to the last baseline followed by a steady increase of the signal. This is due to permeability pore opening and a release of excess calcium over a particular threshold known as induction of mitochondrial apoptosis.

19. When using a perfusion system on isolated mitochondria, carefully use slow perfusion <0.5 ml/min since mitochondria only settle down and won't stick to the microscopy slides as adherent cells would do. Same procedures should be applied on nonadherent cells.

20. Depending on cell type, loading time is prolongable, and loading concentration of Fura-2-AM is variable between 1 and 10 μM. Keep Fura-2-AM loading solution and Fura-2 loaded cells in dark area at room temperature.

21. Cells are storable in storage buffer for at least 12 h at room temperature.

22. The cpEGFP of the sensor is usually brighter and therefore better visible. Alternatively, the OFP can be excited at ~550 nm with an emission at ~565 nm.

23. No interfering emission light from cytosolic Fura-2 should be visible in this channel.

24. Minimum exposure times for the excitation of Fura-2 are below 20 ms for both wavelengths. Basal ratio levels are around 0.4 and reach above 1 upon stimulation.

25. Minimum exposure time of mtD1GO-Cam is below 100 ms. Basal ratio levels are around 0.8 and reach more than 1 upon stimulation.

26. Usually, mitochondrial Ca²⁺ uptake is delayed to cytosolic Ca²⁺ rises.

27. Some laboratories successfully use French press method [29, 30], which is more advantageous and ensures more gentle mechanical isolation of mitoplasts [31].

28. Mitoplasts (mitochondria devoid of outer membrane) have larger size than mitochondria. When isolated from HeLa cells, mitoplasts are typically around 5 μm in diameter.

29. Even seeding is especially important in order to minimize well-to-well variation. Ideally, cells should reach a confluency of 90 % by the time of the experiment.

30. Nonadherent cells can be immobilized using Corning® Cell-Tak™ cell and tissue adhesive (Corning, Product #354240) according to the protocol supplied by Seahorse Bioscience. Immobilization of cells might also become necessary when measuring under Ca²⁺-free conditions as most cell adhesion molecules are Ca²⁺-dependent, and the mixing steps during the assay do impose considerable strain on cell adhesion.

31. The XF96 sensor cartridge must be hydrated for 16 up to 48 h prior to the start of the assay.

32. Four channels (designated as port A–D) located on the sensor cartridge allow the addition of substrates/inhibitors to the

wells during the measurement in order to resolve different functional states of respiration. Here, we depict the use of three reagents, oligomycin A (an inhibitor of the ATP synthase), FCCP (a protonophore/uncoupling agent), and antimycin A (an inhibitor of complex III) to evaluate basal, coupled, maximal, and residual OCRs. The optimal concentrations of these compounds have to be determined in titration experiments preceding the actual assay. Typical working concentrations are 1–2 μM oligomycin A, 0.1–2.0 μM FCCP, and 2 μM antimycin A. In the course of the assay, 25 μl of compound solution per channel will be injected sequentially. Based on a starting volume of 150 μl per well, 7×, 8×, and 9× stock solutions of the final working concentration will be required for wells A, B, and C, respectively.

33. To ensure complete injection, all 96 channels of one port must be evenly loaded using a multichannel pipette. We do not recommend the use of the supplied loading guide as this is prone to cause air bubbles.

References

1. Jouaville LS, Pinton P, Bastianutto C, Rutter GA, Rizzuto R (1999) Regulation of mitochondrial ATP synthesis by calcium: evidence for a long-term metabolic priming. Proc Natl Acad Sci U S A 96:13807–13812

2. Brookes PS, Yoon Y, Robotham JL, Anders MW, Sheu SS (2004) Calcium, ATP, and ROS: a mitochondrial love-hate triangle. Am J Physiol Cell Physiol 287(4):C817–C833

3. Giorgi C, Romagnoli A, Pinton P, Rizzuto R (2008) Ca^{2+} signaling, mitochondria and cell death. Curr Mol Med 8:119–130

4. De Stefani D, Raffaello A, Teardo E, Szabò I, Rizzuto R (2011) A forty-kilodalton protein of the inner membrane is the mitochondrial calcium uniporter. Nature 476:336–340

5. Mallilankaraman K, Cárdenas C, Doonan PJ, Chandramoorthy HC, Irrinki KM, Golenár T et al (2012) MCUR1 is an essential component of mitochondrial Ca^{2+} uptake that regulates cellular metabolism. Nat Cell Biol 14:1336–1343

6. Baughman JM, Perocchi F, Girgis HS, Plovanich M, Belcher-Timme CA, Sancak Y et al (2011) Integrative genomics identifies MCU as an essential component of the mitochondrial calcium uniporter. Nature 476:341–345

7. Perocchi F, Gohil VM, Girgis HS, Bao XR, McCombs JE, Palmer AE et al (2010) MICU1 encodes a mitochondrial EF hand protein required for Ca^{2+} uptake. Nature 467:291–296

8. Jiang D, Zhao L, Clapham DE (2009) Genome-wide RNAi screen identifies Letm1 as a mitochondrial Ca^{2+}/H^+ antiporter. Science 326:144–147

9. Zecchini E, Siviero R, Giorgi C, Rizzuto R, Pinton P (2007) Mitochondrial calcium signalling: message of life and death. Ital J Biochem 56:235–242

10. Jean-Quartier C, Bondarenko AI, Alam MR, Trenker M, Waldeck-Weiermair M, Malli R et al (2010) Studying mitochondrial Ca^{2+} uptake—a revisit. Mol Cell Endocrinol 353:114–127

11. Sancak Y, Markhard AL, Kitami T, Kovács-Bogdán E, Kamer KJ, Udeshi ND et al (2013) EMRE is an essential component of the mitochondrial calcium uniporter complex. Science 342:1379–1382

12. Waldeck-Weiermair M, Deak AT, Groschner LN, Alam MR, Jean-Quartier C, Malli R et al (2013) Molecularly distinct routes of mitochondrial Ca^{2+} uptake are activated depending on the activity of the sarco/endoplasmic reticulum Ca^{2+} ATPase (SERCA). J Biol Chem 288:15367–15379

13. Bright GR, Fisher GW, Rogowska J, Taylor DL (1989) Fluorescence ratio imaging microscopy. Methods Cell Biol 30:157–192

14. Eberhard M, Erne P (1991) Calcium binding to fluorescent calcium indicators: calcium green, calcium orange and calcium crimson. Biochem Biophys Res Commun 180:209–215

15. Rizzuto R, Simpson AW, Brini M, Pozzan T (1992) Rapid changes of mitochondrial Ca^{2+}

revealed by specifically targeted recombinant aequorin. Nature 358:325–327

16. Nagai T, Sawano A, Park ES, Miyawaki A (2001) Circularly permuted green fluorescent proteins engineered to sense Ca²⁺. Proc Natl Acad Sci U S A 98(6):3197–3202

17. Nakai J, Ohkura M, Imoto KA (2001) A high signal-to-noise Ca²⁺ probe composed of a single green fluorescent protein. Nat Biotechnol 19:137–141

18. Zhao Y, Araki S, Wu J, Teramoto T, Chang YF, Nakano M et al (2011) An expanded palette of genetically encoded Ca²⁺ indicators. Science 333:1888–1891

19. Miyawaki A, Llopis J, Heim R, McCaffery JM, Adams JA, Ikura M et al (1997) Fluorescent indicators for Ca²⁺ based on green fluorescent proteins and calmodulin. Nature 388:882–887

20. Palmer AE, Jin C, Reed JC, Tsien RY (2004) Bcl-2-mediated alterations in endoplasmic reticulum Ca²⁺ analyzed with an improved genetically encoded fluorescent sensor. Proc Natl Acad Sci U S A 101:17404–17409

21. Palmer AE, Giacomello M, Kortemme T, Hires SA, Lev-Ram V, Baker D et al (2006) Ca²⁺ indicators based on computationally redesigned calmodulin-peptide pairs. Chem Biol 13:521–530

22. McCombs JE, Palmer AE (2008) Measuring calcium dynamics in living cells with genetically encodable calcium indicators. Methods 46:152–159

23. Carlson HJ, Campbell RE (2009) Genetically encoded FRET-based biosensors for multiparameter fluorescence imaging. Curr Opin Biotechnol 20:19–27

24. Waldeck-Weiermair M, Alam MR, Khan MJ, Deak AT, Vishnu N, Karsten F et al (2012) Spatiotemporal correlations between cytosolic and mitochondrial Ca²⁺ signals using a novel red-shifted mitochondrial targeted cameleon. PLoS One 7:e45917

25. Bondarenko AI, Jean-Quartier C, Malli R, Graier WF (2013) Characterization of distinct single-channel properties of Ca²⁺ inward currents in mitochondria. Pflugers Arch 465:997–1010

26. Wiederkehr A, Szanda G, Akhmedov D, Mataki C, Heizmann CW, Schoonjans K et al (2011) Mitochondrial matrix calcium is an activating signal for hormone secretion. Cell Metab 13:601–611

27. Vishnu N, Jadoon Khan M, Karsten F, Groschner LN, Waldeck-Weiermair M, Rost R et al (2014) ATP increases within the lumen of the endoplasmic reticulum upon intracellular Ca²⁺-release. Mol Biol Cell 25(3):368–379

28. Nakano M, Imamura H, Nagai T, Noji H (2011) Ca²⁺ regulation of mitochondrial ATP synthesis visualized at the single cell level. ACS Chem Biol 6:709–715

29. Fieni F, Lee SB, Jan YN, Kirichok Y (2012) Activity of the mitochondrial calcium uniporter varies greatly between tissues. Nat Commun 3:1317

30. Fedorenko A, Lishko PV, Kirichok Y (2012) Mechanism of fatty-acid-dependent UCP1 uncoupling in brown fat mitochondria. Cell 151:400–413

31. Decker GL, Greenawalt JW (1977) Ultrastructural and biochemical studies of mitoplasts and outer membranes derived from French-pressed mitochondria. Advances in mitochondrial subfractionation. J Ultrastruct Res 59:44–56

Qualitative Characterization of the Rat Liver Mitochondrial Lipidome Using All Ion Fragmentation on an Exactive Benchtop Orbitrap MS

Susan S. Bird, Irina G. Stavrovskaya, Rose M. Gathungu*, Fateme Tousi*, and Bruce S. Kristal

Abstract

Untargeted lipidomics profiling by liquid chromatography-mass spectrometry (LC-MS) allows researchers to observe the occurrences of lipids in a biological sample without showing intentional bias to any specific class of lipids and allows retrospective reanalysis of data collected. Typically, and in the specific method described, a general extraction method followed by LC separation is used to achieve nonspecific class coverage of the lipidome prior to high-resolution accurate mass (HRAM) MS detection. Here we describe a workflow including the isolation of mitochondria from liver tissue, followed by mitochondrial lipid extraction and the LC-MS conditions used for data acquisition. We also highlight how, in this method, all-ion fragmentation can be used to identify species of lower abundances, often missed by data-dependent fragmentation techniques. Here we describe the isolation of mitochondria from liver tissue, followed by mitochondrial lipid extraction and the LC-MS conditions used for data acquisition.

Key words Mitochondria, Lipidomics, LC-MS, Cardiolipins, HCD

1 Introduction

Mitochondria are intracellular membrane-enclosed organelles that play crucial role in bioenergetics, the biosynthesis of critical cellular constituents, the regulation of cell survival, and the execution of cell death pathways [1, 2]. Lipids are essential to mitochondrial viability, and they are involved in the regulation of the wide range of mitochondrial functions, such as maintenance of membrane structural composition and fluidity, membrane fusion and fission, electron transport and oxidative phosphorylation, signal transduction, interaction with other cellular constituencies such as proteins and glycoproteins, and energy storage [3–5]. The role of lipids in mitochondrial

---

*Author contributed equally with all other contributors.

Volkmar Weissig and Marvin Edeas (eds.), *Mitochondrial Medicine: Volume I, Probing Mitochondrial Function*, Methods in Molecular Biology, vol. 1264, DOI 10.1007/978-1-4939-2257-4_36, © Springer Science+Business Media New York 2015

function is, for example, exemplified by mitochondrial phospholipid cardiolipin (CL). CL, which is almost exclusively found in the inner mitochondrial membrane, plays multiple key roles in the regulation of mitochondrial metabolism, including regulation of essential enzymatic activities involved in electron transport and oxidative phosphorylation, and assembly of respiratory supercomplexes [5–7]. Recent studies suggest a critical role of mitochondrial lipid cardiolipin in apoptotic cell death pathway [8].

Mitochondrial lipids are also both a major target for oxidative damage by reactive oxygen species produced during respiration and a major source of lipid peroxides and peroxidation by-product (e.g., hydroxyalkenals, γ-isoketoaldehydes) production that serve to amplify oxidative damage under pathophysiological conditions [9–13]. Development of analytical methods for accurate quantitative and qualitative analysis of all lipid classes and species that contribute to and reflect mitochondrial function is, therefore, highly important.

Liquid chromatography with high-resolution accurate mass (HRAM) MS is widely used to study the biochemical species (proteins, metabolites, lipids, etc.) that comprise a biological sample [14–16]. These techniques offer the speed, sensitivity, and specificity necessary to determine the biochemical composition of a system by providing both qualitative and relative quantitative results. This is imperative for a successful nontargeted lipidomics analysis, where all lipid species are considered of potential biological importance, and therefore, there can be no deliberate analytical bias given toward any of the 8 lipid categories established and defined by LIPID MAPS [17].

By using HRAM instrumentation, such as the Exactive benchtop Orbitrap MS, for both full scan and all-ion fragmentation (AIF), a broad picture of the lipidome is captured, and the electronic data record obtained can easily be used for both primary and retrospective analyses. LC separation followed by HRAM detection in combination with MS/MS fragmentation allows both known and unknown species to be structurally elucidated and monitored. Using the method described herein, containing both LC and MS with AIF MS/MS, the identification of more than 350 unique lipids in serum, mitochondria, and premature infant fecal samples has been found and reported [18–21].

2   Materials

2.1   Liver Mitochondria Isolation

1. Isolation buffer IB1: 240 mM sucrose, 10 mM HEPES, pH 7.4, 1 mM ethylene glycol-bis(2-aminoethylether)-N,N,N′,N′-tetraacetic acid (EGTA) stock with pH 7.2 adjusted by KOH, 5 g/1 L bovine serum albumin (BSA, Sigma-Aldrich, fatty acid free) (*see* **Notes 1–5**).

2. Petri dish.

3. 50 ml plastic disposable beaker.

4. Potter-Elvehjem glass-Teflon homogenizers 40 ml and 15 ml volume.

5. 50 ml round-bottom centrifuge tubes.

6. Fixed angle JA-20 centrifuge rotor and appropriate centrifuge (both, Beckman, USA), both pre-chilled to 4 °C.

7. Secure the pestle in the chuck of the electrical motor.

8. Isolation buffer IB2: 82.152 g/1 L sucrose, 2.38 g/1 L HEPES, pH 7.4, adjusted by KOH.

9. Bicinchoninic acid kit (Sigma-Aldrich, USA) to determine mitochondrial protein concentration and 1.0 mg/ml BSA as a standard solution (*see* **Note 6**).

10. Spectrophotometer, Uvikon 943, or equivalent.

11. Washing buffer: 125 mM KCl, 10 mM HEPES, pH 7.4 adjusted by KOH.

2.2 Mitochondrial Lipid Extraction

1. 1.5 ml Fisherbrand Siliconized Low-Retention Microcentrifuge Tubes (Fisher Scientific, Pittsburgh, PA).

2. Microcentrifuge floating rack.

3. Mechanical ultrasonic cleaner.

4. 10 ml Falcon tubes.

5. Thermo Scientific Savant SPD111V P1.

6. ACS grade dimethyl sulfoxide (DMSO).

7. ACS grade dichloromethane (DCM).

8. LC-MS grade methanol (MeOH).

9. LC-MS grade isopropanol (IPA).

10. LC-MS grade acetonitrile (ACN).

11. All water (H_2O) was deionized to attain a sensitivity of 18 MΩ cm at 25 °C.

12. Lipid internal standard solution: 1.5 ml, enough for 50 samples; combine 37.50 μl of a 2 mg/ml solution of 1,1′,2,2′-tertraoleoyl cardiolipin (CL(18:1)$_4$) (Avanti Polar Lipids), 30 μl each of a 2.5 mg/ml solution of 1,2-diheptadecanoyl-*sn*-*glycero*-3-phosphocholine (PC(17:0/17:0) and 1,2-dimyristoyl-sn-glycero-3-[phospho-rac-(1-glycerol)] (PG(14:0/14:0) (Avanti Polar Lipids), 15 μl of a 5 mg/ml solution of 11-O-hexadecyl-*sn*-glycero-3-phosphocholine (lysoPC(16:0)) (Enzo Chemicals), and 75 μl of a 1 mg/ml solution of 1,2-dipalmitoyl-*sn*-glycero-3-phospho-L-serine (PS(16:0/16:0) (Avanti Lipids) followed by 1,297.5 μl of 2:1 DCM-MeOH. All standards are dissolved in 2:1 mixture of DCM-MeOH prior to mixing and stored at –20 °C.

13. LC-MS internal standard: Create 100 ml of a 5 μg/ml working concentration solution of LC-MS internal standard

1,2-diheptadecanoyl-*sn*-glycero-3-[phospho-rac-(1-glycerol)] (PG 17:0/17:0) by adding 200 μl of a 2.5 mg/ml stock solution to 100 ml of a 63:30:5 ACN-IPA-H_2O mixture.

2.3 LC-MS Analysis

1. LC-MS grade acetonitrile with 0.1 % formic acid.

2. Ascentis Express C_{18} 2.1×150 mm 2.7 μm column (Sigma-Aldrich, St. Louis, MO).

3. Thermostated HOT POCKET column heater (Thermo Fisher Scientific, San Jose, CA).

4. Thermo Fisher Scientific PAL autosampler (Thermo Fisher Scientific, San Jose, CA).

5. Accela quaternary HPLC pump (Thermo Fisher Scientific, San Jose, CA).

6. Exactive benchtop Orbitrap mass spectrometer with HCD (Thermo Fisher Scientific, San Jose, CA) and heated electrospray ionization (HESI) probe.

7. Mobile phase A was prepared by dissolving 630 mg of ammonium formate into a mixture of 600 ml water and 400 ml ACN with 0.1 % formic acid. This bottle was then sonicated for several minutes to remove any gases from mixing solvents.

8. Mobile phase B was prepared by dissolving 630 mg of ammonium formate into 1 ml of water before adding to 900 ml of IPA and 100 ml of ACN with 0.1 % formic acid. This bottle was then sonicated for several minutes to facilitate dissolution and remove any gases from mixing solvents.

9. SIEVE v 1.3 differential analysis software (Thermo Fisher Scientific, San Jose CA).

3 Methods

3.1 Liver Mitochondria Isolation

1. All isolation steps have to be done on ice, and all operations have to be carried out at 0–4 °C as described previously [22] (*see* **Note 7**). Alternative mitochondrial isolation approaches from liver and/or other tissues are also expected to provide suitable starting material, although the user should be aware of the characteristics of any given preparation with regard, for example, to purity. The use of other preparations would replace **steps 1–12** of this protocol.

2. Take liver tissue (8–10 g) from the animal (*see* **Note 8**) and place it immediately in Petri dish containing ice-cold 20–30 ml of the isolation buffer #1 (IB1). Wash out blood.

3. Transfer liver tissue in a 50 ml plastic disposable beaker containing 20–30 ml of IB1, and mince the tissue with scissors, or use a press.

4. Pour the suspension in Potter-Elvehjem glass-Teflon homogenizer and add IB1 to a total volume of 40 ml.

5. Homogenize the suspension using 7–8 up-and-down strokes with the pestle rotating at 500–600 rpm (*see* **Notes 9** and **10**). Suspension should look homogeneous.

6. Transfer the obtained homogenate into 50 ml round-bottom centrifuge tubes. Centrifuge the tubes at $1,000 \times g \times 10$ min (*see* **Note 11**).

7. Collect the supernatant and centrifuge it at $11,000 \times g \times 10$ min (*see* **Note 12**).

8. Collect the pellet and resuspend it in 15 ml of IB1 in the Potter-Elvehjem glass-Teflon homogenizer using 4–5 up-and-down strokes by hand. Suspension should look homogeneous. Adjust volume to 40 ml, mix, and centrifuge the suspension at $11,000 \times g \times 10$ min.

9. Resuspend the pellet in 15 ml of isolation buffer #2 (IB2) in the Potter-Elvehjem glass-Teflon homogenizer with 4–5 up-and-down strokes by hand. Suspension should look homogeneous. Adjust volume to 40 ml, mix, and centrifuge the suspension at $11,000 \times g \times 10$ min.

10. Resuspend final pellet in 0.5 ml of IB2 and store on ice (*see* **Note 13**).

11. Add 10 μl of mitochondrial suspension into 990 μl of IB2 to measure protein concentration by bicinchoninic acid-based method using BSA as a standard (*see* www.sigmaaldrich.com for details).

12. Aliquot mitochondrial sample by 1 mg of protein, resuspend it in 125 mM KCl, 10 mM HEPES, pH 7.4 buffer, centrifuge at $11,000 \times g \times 10$ min, remove supernatant, and freeze as dry pellets at −80 °C for further analysis.

3.2 Mitochondrial Lipid Extraction

1. Add 40 μl DMSO to each aliquot of mitochondria (containing 1 mg of protein) (*see* **Note 14**).

2. Place each mitochondria/DMSO sample into the floating microcentrifuge rack and sonicate for 1 h to disrupt the membranes.

3. Pipette 10 μl from each sample into a 10 ml Falcon tube to create a mitochondrial pool sample. The pool samples will be processed for quality control (QC) and lipid identification studies.

4. Add 5 μl of lipid internal standard to each 30 μl mitochondria sample; this includes samples from the study and pool samples created in **step 3**.

5. Add 190 μl of MeOH to each sample and vortex for 10 s.

6. Next, add 380 µl of DCM and vortex for 20 s (*see* **Notes 15** and **16**).

7. Finally, add 120 µl of water to induce phase separation.

8. Vortex samples for 10 s and allow to equilibrate at room temperature for 10 min.

9. Centrifuge each sample at $8,000 \times g$ for 10 min at 10 °C (*see* **Note 17**).

10. Using a 500 µl pipette, transfer 320 µl of the lower lipid-rich DCM layer to a clean microcentrifuge tube. Be careful to push the pipette tip along the side of the container when piercing the protein disk as not to disrupt the phase separation or transfer any protein with the lipid-rich DCM layer.

11. Evaporate the samples to dryness under vacuum using a SpeedVac (*see* **Note 18**).

12. Reconstitute samples in 300 µl of LC-MS internal standard (PG (17:0/17:0)) before LC-MS analysis (*see* **Note 19**).

3.3 LC-MS Conditions

1. Separate 10 µl lipid extracts on an Ascentis Express C_{18} 2.1×150 mm 2.7 µm column connected to a Thermo Fisher Scientific PAL autosampler, Accela quaternary HPLC pump, and an Exactive benchtop Orbitrap mass spectrometer equipped with a heated electrospray ionization (HESI) probe (*see* **Notes 20** and **21**).

2. Run separations for 30 min with mobile phases A and B consisting of 60:40 water-ACN in 10 mM ammonium formate and 0.1 % formic acid and 90:10 IPA-ACN also with 10 mM ammonium formate and 0.1 % formic acid, respectively.

3. The gradient starts at 32 % B for 1.5 min; from 1.5 to 4 min increases to 45 % B, from 4 to 5 min to 52 % B, from 5 to 8 min to 58 % B, from 8 to 11 min to 66 % B, from 11 to 14 min to 70 % B, from 14 to 18 min to 75 % B, and from 18 to 21 min to 97 % B; during 21 to 25 min 97 % B is maintained; from 25 to 30 min solvent B is decreased to 32 % and then maintained. The flow rate is 260 µl/min.

4. Maintain the column oven temperature at 45 °C and set the temperature of the autosampler to 4 °C. The same LC conditions and buffers are used for all MS experiments.

5. Set the spray voltage to 3.5 kV, whereas the heated capillary and the HESI probe are held at 250 °C and 350 °C, respectively.

6. Set the sheath gas flow to 25 units and the auxiliary gas to 15 units.

7. Hold these conditions constant for both positive and negative ionization mode acquisitions.

8. The instrument is tuned by direct infusion of PG (17:0/17:0) in both positive and negative modes, and external mass calibration is performed using the standard calibration mixture and protocol from Thermo Fisher approximately every 5 days.

3.4 Full Scan Profiling Experiments

1. Operate the MS in high-resolution mode, corresponding to a resolution of 60k and a 2 Hz scan speed, and hold the scan range between m/z 120 and 2,000.

2. Mitochondrial lipid extracts are profiled by injecting each sample once, in randomized order, with pool samples, blanks, and IS mixture injections spread throughout the analysis.

3.5 Lipid Identification Studies

1. Run HCD experiments on the pool, lipid IS mixture, and blank samples only.

2. Perform these experiments by alternating between full scan acquisitions and HCD scans, both run at 2 Hz. Three different HCD energies, 30, 60, and 100 eV, are used in separate experiments in both positive and negative modes.

3.6 Data Analysis

All LC-MS profiling samples are analyzed using the MS label-free differential analysis software package SIEVE v 1.3 (Thermo Fisher Scientific and Vast Scientific, Cambridge, MA) (*see* **Notes 22** and **23**).

The frame m/z values were used to do batch searches on the METLIN database [23], the Human Metabolome Database (HMDB) [24], and the LIPID MAPS Structure Database [25], and those matches were confirmed using the intact molecule's exact mass observed during the analysis; RT regions based on the lipid IS mixture elution times run throughout the sequence and HCD fragmentation patterns that chromatographically align with the intact exact mass extracted ion chromatograms for the parent compound (*see* **Note 24**).

4   Notes

1. Use deionized water (Milli-Q, "Millipore") for mitochondrial buffer preparation.

2. Airtight glass containers or Thermo Scientific Nalgene disposable bottles are strongly recommended for storage of mitochondrial buffers.

3. Sterilize all buffers by filtering through 0.22 μm membrane (we recommend to use Thermo Scientific Nalgene disposable filter units that come with bottles). The buffers are stable for at least 3 months if kept sterile.

4. Do not freeze isolation and incubation buffers.

5. The buffer composition for mitochondria isolation and incubation may vary. For example, examination of the literature

will show changes in relative concentrations of sucrose and/or mannitol, in the choice of buffering agent and counter ion, the inclusion/exclusion of divalent cation chelators, osmolarity, and ionic strength. The buffers described have been used by our lab for upstream respiration and/or calcium overload experiments and subsequent downstream lipidomics studies.

6. Alternative protein concentration approaches can be used, but linear range and potential interference must be examined.

7. Quality and purity of isolated mitochondria are very important because contamination of mitochondrial preparation with cytosolic structures can interfere/mask lipid profiling results.

8. Animals have to be sacrificed, and liver tissue has to be removed and placed in ice-cold buffer as quickly as possible.

9. During tissue homogenization steps, do not draw a vacuum and avoid bubble creation.

10. Avoid very fast pestle rotation during homogenization.

11. Do not use detergents to clean centrifuge tubes.

12. Carefully remove fat from centrifuge tube walls with a delicate wiper after second centrifugation.

13. Make final mitochondria suspension rather concentrated than diluted (we recommend a concentration of 60–80 mg/ml of mitochondria).

14. Minimize contamination in mass spectrometry:
Do not store organic solvents used for the sample preparation in plastic tubes. PEG contaminants and plasticizers such as phthalates can leach out of plastic into the sample and make their way to the LC-MS system. Always use glass bottles.

If you have to use detergent to wash the glassware, be sure to rinse them with large volume of deionized water before use. Detergents contain PEG materials which can contaminate reversed-phase columns and cause ion suppression in mass spectrometry.

15. Dichloromethane is a harmful organic solvent with some evidence of carcinogenic effects. It must be handled wearing safety gloves and in a fume hood at all times. Be sure to read dichloromethane MSDS before use.

16. Dichloromethane has a high vapor pressure at room temperature that makes it difficult to be accurately measured and transferred using pipettes. To minimize loss of DCM through evaporation during pipetting, pipette up and down several times in DCM to ensure that the inside of the pipette tip is saturated with CDM vapors.

17. Centrifugation of the sample at 4 °C helps keep the DCM layer cool. Transfer the DCM layer while it's still cool. Cold DCM is easier to transfer using pipette tips and less prone to loss via evaporation.

18. Before drying the lipid extracts in a vacuum centrifuge, cool the extracts down in a –20 °C freezer. This will help minimize exposure to DCM vapors when opening and placing sample tubes in the vacuum centrifuge, in case the instrument is not placed in a fume hood.

19. Reconstitute the samples with the appropriate solvent right before LC-MS analysis. Avoid keeping the reconstituted samples for prolonged time before LC-MS analysis to minimize potential lipid degradation. Store the extra samples in a –80 °C freezer.

20. MS preparation for lipidomics profiling, mass calibration: Mass calibration should be done to the instrument vendor's specifications. On the Exactive in our laboratory, mass calibrations are done weekly. Calibrations are done separately in positive and negative ion modes.

    MS tuning: Tuning of the MS is done to optimize ion transmission through the ion optics to the analyzer. Tuning is therefore performed at the operating LC flow rate, and the compound chosen as the tuning compound should have similar ESI characteristics as your molecules of interest and should be at the same concentration as a majority of molecules being analyzed. As an example, for our LC-MS analysis, we tee-in (using a splitter) 2.5 µg/ml of PG 17:0/17:0 flowing at 10 µl/min and 50:50 of mobile phase A-mobile phase B at a flow of 250 µl/min (total flow into MS is 260 µl/ml). The final concentration of PG 17:0/17:0 into the MS is ~100 ng/ ml.

    LC solvents: Always use LC-MS grade solvents to avoid contamination of mass spectrometer with impurities. To assist with solubilization of ammonium formate in mobile phase B, first dissolve it in ~1 ml of deionized water.

21. Analysis of lipid samples by reversed-phase LC-MS: Lipids (especially triglycerides) are very hydrophobic, and thus they tend to stick to the injector port which leads to carryover from sample to sample. To avoid carryover, it is important to flush the injector port thoroughly. We usually flush the injector port with 50:50 acetonitrile-isopropanol and 50:50 isopropanol-dichloromethane (10× with each). The isopropanol in dichloromethane wets the needle and prevents drying (which can ruin the needle plunger). It is also important to ensure that the solvent used for flushing the area is miscible.

22. The framing parameters in these experiments are set at 0.01 daltons for the m/z window and 1.00 min for the RT window; 1,000 is used at the intensity threshold. Peaks under this intensity will normally be background or too low to quantify robustly. These parameters can be adjusted based on your own data. A pool from the middle of the sample sequence should be used as a qualitative reference and for relative quantitation, and frames built off the reference are then applied to

all samples in the experiment. If a different reference is used, the intensity values may change slightly; however, the overall lipid ratios should be the same.

23. Chromatographic alignment and framing using Sieve: For Sieve analysis on the LC-MS lipidomics data, the small molecule, chromatographic alignment and framing, and the nondifferential single class analysis options are used. Positive and negative ion data are analyzed separately, and only the full scan raw data is used for alignment and framing. The framing parameters used are determined by the user. The m/z window chosen depends on the mass accuracy of the instrument used. The retention time window chosen should be large enough to enable correct chromatographic alignment to account for retention time drifts over time, but not too large so as to minimize false positives. We therefore typically use 1 min for the retention time window and an intensity threshold of 1,000 (to ensure that as many features as possible are identified but without picking out the noise).

24. Lipid identification: To assist with lipid identification of the lipids obtained from the framing data, HCD analysis (done separately in the positive and negative modes) is performed on a pooled sample. HCD fragmentation is done at three energies: 30 eV (low energy), 60 eV (medium energy), and 100 eV (high energy).

    Lipid identification is done by matching the retention time of extracted ion chromatogram of a particular exact mass and its fragments obtained using the all-ion fragmentation data. This alignment helps in distinguishing lipids with the same exact mass but different fatty acyl chains.

Acknowledgments

These studies were funded by U01-ES16048 (BSK, PI) and P30-DK040561 (W. Allan Walker, PI). The authors thank Thermo Fisher for the loan of an Exactive benchtop Orbitrap for demonstration testing (later purchased) and financial support for scientific meeting attendance.

Financial Disclosures

SSB currently works for Thermo Fisher.

IGS, RMG, and FT have no financial disclosures.

BSK is the inventor on general metabolomics-related IP that has been licensed to Metabolon via Weill Medical College of Cornell University and for which he receives royalty payments via

Weill Medical College of Cornell University. He also consults for and has a small equity interest in the company. Metabolon offers biochemical profiling services and is developing molecular diagnostic assays detecting and monitoring disease. Metabolon has no rights or proprietary access to the research results presented and/or new IP generated under these grants/studies. BSK interests were reviewed by the Brigham and Women's Hospital and Partners Healthcare in accordance with their institutional policy. Accordingly, upon review, the institution determined that BSK financial interest in Metabolon does not create a significant financial conflict of interest (FCOI) with this research. The addition of this statement where appropriate was explicitly requested and approved by BWH.

References

1. Kroemer G, Galluzzi L, Brenner C (2007) Mitochondrial membrane permeabilization in cell death. Physiol Rev 87(1):99–163

2. Rasola A, Bernardi P (2007) The mitochondrial permeability transition pore and its involvement in cell death and in disease pathogenesis. Apoptosis 12:815–833

3. Osman C, Voelker D, Langer T (2011) Making heads or tails of phospholipids in mitochondria. J Cell Biol 192(1):7–16

4. Horvath SE, Daum G (2013) Lipids of mitochondria. Prog Lipid Res 52(4):590–614

5. Claypool SM, Koehler CM (2012) The complexity of cardiolipin in health and disease. Trends Biochem Sci 37:32–41

6. Pfeiffer K, Gohil V, Stuart RA, Hunte C, Brandt U, Greenberg ML, Schagger H (2003) Cardiolipin stabilizes respiratory chain supercomplexes. J Biol Chem 278:52873–52880

7. Klingenberg M (2009) Cardiolipin and mitochondrial carriers. Biochim Biophys Acta 1788:2048–2058

8. Kagan VE, Tyurin VA, Jiang J, Tyurina YY, Ritov VB, Amoscato AA, Osipov AN, Belikova NA, Kapralov AA, Kini V, Vlasova II, Zhao Q, Zou M, Di P, Svistunenko DA, Kurnikov IV, Borisenko GG (2005) Cytochrome c acts as a cardiolipin oxygenase required for release of proapoptotic factors. Nat Chem Biol 1(4):223–232

9. Lesnefsky EJ, Hoppel CH (2008) Cardiolipin as an oxidative target in cardiac mitochondria in the aged rat. Biochim Biophys Acta 1777:1020–1027

10. Esterbauer H, Schaur RJ, Zollner H (1991) Chemistry and biochemistry of 4-hydroxynonenal, malonaldehyde and related aldehydes. Free Radic Biol Med 11:81–128

11. Stavrovskaya IG, Baranov SV, Guo X, Davies SS, Roberts LJ 2nd, Kristal BS (2010) Reactive gamma-ketoaldehydes formed via the isoprostane pathway disrupt mitochondrial respiration and calcium homeostasis. Free Radic Biol Med 49(4):567–579

12. Keller JN, Mattson MP (1998) Roles of lipid peroxidation in modulation of cellular signaling pathways, cell dysfunction, and death in the nervous system. Rev Neurosci 9:105–116

13. Kristal BS, Park BK, Yu BP (1996) 4-Hydroxynonenal is a potent inducer of the mitochondrial permeability transition. J Biol Chem 271:6033–6038

14. Ejsing CS, Moehring T, Bahr U, Duchoslav E, Karas M, Simons K, Shevchenko A (2006) Collision-induced dissociation pathways of yeast sphingolipids and their molecular profiling in total lipid extracts: a study by quadrupole TOF and linear ion trap-orbitrap mass spectrometry. J Mass Spectrom 41:372–389

15. Wikoff WR, Anfora AT, Liu J, Schultz PG, Lesley SA, Peters EC, Siuzdak G (2009) Metabolomics analysis reveals large effects of gut microflora on mammalian blood metabolites. Proc Natl Acad Sci U S A 106(10):3698–3703

16. Picotti P, Clément-Ziza M, Lam H, Campbell DS, Schmidt A, Deutsch EW, Röst H, Sun Z, Rinner O, Reiter L, Shen Q, Michaelson JJ, Frei A, Alberti S, Kusebauch U, Wollscheid B, Moritz RL, Beyer A, Aebersold RA (2013) A Complete mass-spectrometric map of the yeast proteome applied to quantitative trait analysis. Nature 494(7436):266–270

17 Fahy E, Subramaniam S, Brown HA, Glass CK, Merrill AH Jr, Murphy RC, Raetz CR, Russell DW, Seyama Y, Shaw W, Shimizu T, Spener F, van Meer G, VanNieuwenhze MS, White SH,

Witztum JL, Dennis EA (2005) A comprehensive classification system for lipids. J Lipid Res 46: 839–861

18. Bird SS, Marur VR, Sniatynski MJ, Greenberg HK, Kristal BS (2011) Lipidomics profiling by high-resolution LC-MS and high-energy collisional dissociation fragmentation: focus on characterization of mitochondrial cardiolipins and monolysocardiolipins. Anal Chem 83: 940–949

19. Bird SS, Marur VR, Sniatynski MJ, Greenberg HK, Kristal BS (2011) Serum lipidomics profiling using LC-MS and high-energy collisional dissociation fragmentation: focus on triglyceride detection and characterization. Anal Chem 83:6648–6657

20. Gregory KE, Bird SS, Gross VS, Marur VR, Lazarev AV, Walker WA, Kristal BS (2013) Method development for fecal lipidomics profiling. Anal Chem 85(2):1114–1123

21. Stavrovskaya IG, Bird SS, Marur VR, Sniatynski MJ, Baranov SV, Greenberg HK, Porter CL, Kristal BS (2013) Dietary macronutrients modulate the fatty acyl composition of rat liver mitochondrial cardiolipins. J Lipid Res 54(10): 2623–2635

22. Stavrovskaya IG, Narayanan MV, Zhang W, Krasnikov BF, Heemskerk J, Young SS, Blass JP, Brown AM, Beal MF, Friedlander RM, Kristal BS (2004) Clinically approved heterocyclics act on a mitochondrial target and reduce stroke-induced pathology. J Exp Med 200:211–222

23. Smith CA, O'Maille G, Want EJ, Qin C, Trauger SA, Brandon TR, Custodio DE, Abagyan R, Siuzdak G (2005) METLIN: a metabolite mass spectral database. Ther Drug Monit 27:747–751

24. Wishart DS, Knox C, Guo AC, Eisner R, Young N, Gautam B, Hau DD, Psychogios N, Dong E, Bouatra S, Mandal R, Sinelnikov I, Xia J, Jia L, Cruz JA, Lim E, Sobsey CA, Shrivastava S, Huang P, Liu P, Fang L, Peng J, Fradette R, Cheng D, Tzur D, Clements M, Lewis A, De Souza A, Zuniga A, Dawe M, Xiong Y, Clive D, Greiner R, Nazyrova A, Shaykhutdinov R, Li L, Vogel HJ, Forsythe I (2009) HMDB: a knowledgebase for the human metabolome. Nucleic Acids Res 37:D603–D610

25. Fahy E, Sud M, Cotter D, Subramaniam S (2007) LIPID MAPS online tools for lipid research. Nucleic Acids Res 35:W606–W612

Chapter 37

Characterization of Mitochondrial Populations During Stem Cell Differentiation

Petra Kerscher, Blakely S. Bussie, Katherine M. DeSimone, David A. Dunn, and Elizabeth A. Lipke

Abstract

Mitochondrial dynamics play an important role in numerous physiological and pathophysiological phenomena in the developing and adult human heart. Alterations in structural aspects of cellular mitochondrial composition as a function of changes in physiology can easily be visualized using fluorescence microscopy. Commonly, mitochondrial location, number, and morphology are reported qualitatively due to the lack of automated and user-friendly computer-based analysis tools. Mitochondrial Quantification using MATLAB (MQM) is a computer-based tool to quantitatively assess these parameters by analyzing fluorescently labeled mitochondria within the cell; in particular, MQM provides numerical information on the number, area, and location of mitochondria within a cell in a time-efficient, automated, and unbiased way. This chapter describes the use of MQM's capabilities to quantify mitochondrial changes during human pluripotent stem cell (hPSC) differentiation into spontaneously contracting cardiomyocytes (SC-CMs), which follows physiological pathways of human heart development.

Key words Mitochondria, Cardiac development, Cardiomyocyte, MATLAB, Quantitative analysis, Fluorescence

1 Introduction

Mitochondrial function is crucial to a myriad of cellular, physiological, and pathological processes throughout eukaryotic life. Cellular localization of mitochondria and their morphology are correlated with the regulation of mitochondrial function. Dysregulation of mitochondrial dynamics is implicated in a wide range of pathologies including, but not limited to, neurodegenerative diseases [1–6], metabolic disorders [7, 8], cancer metastasis [9], and heart disease [10–13]. Changes in mitochondrial morphology are also associated with developmental processes, including cardiogenesis [14], and, analogously, differentiation of pluripotent stem cells into cardiomyocytes [15]. These morphologic changes often occur in response to changing cellular energetic demands.

Volkmar Weissig and Marvin Edeas (eds.), *Mitochondrial Medicine: Volume I, Probing Mitochondrial Function*,
Methods in Molecular Biology, vol. 1264, DOI 10.1007/978-1-4939-2257-4_37, © Springer Science+Business Media New York 2015

In the case of cardiac differentiation of human pluripotent stem cells (hPSCs), this morphologic change is the result of a switch from primarily glycolytic [16] to primarily oxidative metabolism [17] and is accompanied by changes in subcellular mitochondrial localization in addition to the morphology [16] of the organelle itself. Quantifying these mitochondrial changes in hPSCs can provide insight into the process of cellular differentiation and specification.

hPSCs have great potential for use in cell therapy and for enhancing our understanding of organ development, normal biological processes, and disease mechanisms, due to their ability to differentiate into the multitude of cell types found within the human body. Through culture and differentiation of hPSCs into specialized cell types, normal and abnormal changes throughout differentiation can be studied, and results used to provide insight into analogous processes that occur during human development. Mitochondria, the "power house" of cells and producer of the primary energy source (ATP), play an essential role in stem cell proliferation, differentiation, function, and death. Dynamic changes in mitochondria number, location, and morphology throughout hPSC differentiation define cell fate, function, and maturation and are particularly important in dynamic cell types like neurons and cardiomyocytes. However, quantitative information about these mitochondrial changes during different stages of stem cell differentiation is limited.

Changes in mitochondria are recognized as an indicator of differentiation, including the use of mitochondrial staining for stem cell-derived cardiomyocytes (SC-CMs) purification from mixed cell populations [18]. Despite this utility, quantitative correlations between mitochondrial number, location, and stage of SC-CM differentiation have not been reported. Current methods for assessing mitochondrial dynamics are subjective and rely on qualitative descriptions. Establishing time-efficient methods for accurately and quantitatively describing mitochondrial dynamics enables the use of these metrics as a biomarker for a number of cellular and physiological processes. Thus, there was a need for creating a platform capable of quantifying mitochondrial dynamics [19]. Here, we describe the implementation of a MATLAB script for quantifying mitochondrial dynamics via automated image analysis.

This chapter provides details on tracking mitochondrial location and number during cardiac differentiation of human-induced pluripotent stem cells (hiPSCs). To visualize mitochondria using fluorescence, cells with mitochondria marked chemically (e.g., staining) or with green fluorescent protein (GFP) targeted to mitochondria are imaged using confocal microscopy. These images are then exported to MATLAB and analyzed quantitatively using Mitochondrial Quantification using MATLAB (MQM). Analysis of the dynamics of mitochondrial number and location using these methods are broadly applicable to other experimental systems.

2   Materials

Store all reagents at 4 °C and pre-warm to room temperature before use (unless indicated otherwise). Perform all cell culture and staining procedures under sterile conditions using a laminar flow hood.

2.1 hIPSC Culture and Maintenance

1. 6-well plates (tissue culture polystyrene treated).

2. CO_2 incubator: 37 °C, 5 % CO_2, and 85 % relative humidity.

3. Human-induced pluripotent stem cells (e.g., IMR-90 Clone 1, WiCell).

4. Human embryonic stem cell (hESC)-qualified Matrigel (BD Bioscience) (*see* **Note 1**).

5. DMEM-F12 media.

6. mTeSR-1 media (Stemcell Technologies).

7. ROCK inhibitor (Y-27632, R&D Systems): Combine 6.24 mL PBS with 10 mg ROCK inhibitor. Aliquot and store at -20 °C (*see* **Note 2**).

8. Versene (Gibco).

2.2 Differentiation of hIPSCs to CMs

1. 12-well or 6-well plate.

2. CO_2 incubator: 37 °C, 5 % CO_2, and 85 % relative humidity.

3. Human embryonic stem cell (hESC)-qualified Matrigel (BD Bioscience) (*see* **Note 1**).

4. DMEM-F12 media.

5. Accutase (Innovative Cell Technologies).

6. 0.25 % trypsin (EDTA, Mediatech).

7. mTeSR-1 media (Stemcell Technologies).

8. ROCK inhibitor (Y-27632, R&D Systems): Combine 6.24 mL PBS with 10 mg ROCK inhibitor. Aliquot and store at -20 °C (*see* **Note 2**).

9. RPMI 1640 media.

10. B-27 supplement without insulin (*see* **Note 3**).

11. B-27 supplement (*see* **Note 4**).

12. CHIR99021 (Selleckchem): Add 1.49 mL DMSO to 25 mg CHIR99021. Aliquot and store at –20 °C.

13. IWP-2 (Tocris): Add 4.28 mL DMSO to 10 mg IWP-2. Incubate at 37 °C for 10 min or until IWP-2 is in solution. Aliquot and store at -20 °C.

14. Fetal bovine serum (FBS, Atlanta Biologicals) (*see* **Note 5**).

2.3 Preparation of Glass Coverslips for Cell Seeding

1. Glass coverslips (21 mm).

2. Polydimethylsiloxane (PDMS) precursor: Mix SYLGARD 184 silicone elastomer curing agent and SYLGARD 184 elastomer base at a ratio of 1:10. Add PDMS precursor onto glass coverslips, and evenly coat glass by spin coating using a WS-400-6NPP spin coater at 106 RCF (3,000 RPM) for 10 s. PDMS-coated glass coverslips are dried at 60 °C for several hours (*see* **Note 6**) and sterilized using 70 % ethanol and UV light.

3. Fibronectin (40×): Add 25 μL fibronectin to 975 μL ice cold, sterile ultrapure water.

2.4 Fluorescent Labeling of Mitochondria

1. MitoTracker Red working solution (1 μM, Molecular Probes): Dilute 1 μL MitoTracker Red stock solution (1 mM) in 1 mL mTeSR-1 (hiPSCs) or RPMI/B27 (CMs).

2. Phosphate-buffered saline without calcium and magnesium (PBS, 10×): Add 10 mL 10× PBS into 990 mL ultrapure water. Filter sterilize.

3. Filter for sterilization: polyethersulfone filter, pore size 0.22 μm.

4. Blocking buffer: Combine 3 mL FBS with 97 mL 1× PBS. Filter before use.

5. Paraformaldehyde (16 %): Add 10 mL 16 % paraformaldehyde to 30 mL 1× PBS.

6. PBS-T: Add 25 mL 10× PBS, 225 mL ultrapure water, 2,500 mg bovine serum albumin (BSA), and 0.5 mL Triton X-100. Filter sterilize.

7. 4′,6-diamidino-2-phenylindole (DAPI, Molecular Probes) (1:36,000 of primary stock in 1× PBS).

8. 50 % ethanol: Combine 25 mL DI water with 25 mL ethanol.

9. 70 % ethanol: Combine 15 mL DI water with 35 mL ethanol.

10. 95 % ethanol: Combine 2.5 mL DI water with 47.5 mL ethanol.

11. ProLong Gold antifade reagent (Life Technologies).

12. Rectangular glass coverslide: e.g., 25 × 75 mm, 1 mm thick.

13. Clear nail polish.

2.5 Image Acquisition

1. Confocal microscope (Nikon A1): 40×–100× oil objective, TRITC (Ex/Em: 561/595 nm) and DAPI (Ex/Em: 405/450 nm) filter sets.

2. Imaging acquisition software (e.g., NIS Elements).

2.6 Image Analysis Using MQM

1. Windows 7 or higher or equivalent operating system.

2. MATLAB 12 or higher with Statistics Toolbox (MathWorks, Inc.).

3. MQM script: function files *MitoMAT* and *Measurements* are available for download from MATLAB Central, the Lipke Lab website www.auburn.edu/lipkelab, or by request from the corresponding author.

4. Microsoft Excel 2007 or higher.

---

3   Methods

Current state-of-the-art stem cell culture and differentiation change rapidly, and in-depth training is available through WiCell and many major research institutions. Based on selected hiPSC cell lines, differentiation parameters might have to be adjusted [20]. Parameters influencing current state-of-the-art SC-CM differentiation have been covered in detail by Lian et al. [20].

3.1 hiPSC
Maintenance
and Expansion

1. Culture hiPSCs as colonies on Matrigel-coated 6-well plates in mTeSR-1 with daily media exchange until passage.

2. Passage cells of a confluent well using Versene. Aspirate off old mTeSR-1, rinse, and incubate cells in Versene at 37 °C for 4 min. Aspirate off Versene and resuspend cells in 1 mL mTeSR-1 + RI. Passage cells (e.g., 1:10 ratio) into a new Matrigel-coated well containing 2 mL mTeSR-1 + RI.

3.2 hiPSC
Dissociation
for Mitochondria
Labeling

1. Rinse hiPSCs of a confluent well with 1× PBS, and incubate in 1 mL Accutase at 37 °C for 8 min.

2. Combine cells with mTeSR-1 and centrifuge cells at $200 \times g$ for 5 min.

3. Aspirate off supernatant, and resuspend cells in 1 mL mTeSR-1 + RI and plate onto a Matrigel-coated PDMS glass coverslip (*see* **Note** 7).

4. Culture hiPSCs for 2 days in mTeSR-1 before fluorescently labeling cells.

3.3 CM
Differentiation
of hiPSCs

1. Rinse confluent well of hiPSCs using 1× PBS.

2. Incubate hiPSCs in Accutase at 37 °C for 8 min.

3. Combine cells with mTeSR-1; centrifuge at $200 \times g$ for 5 min.

4. Resuspend cells in 3 mL mTeSR-1 + RI (1×10^6 cells/mL), and add 1 mL into each well of a Matrigel-coated 12-well plate.

5. After 24 h, replace media with 2 mL mTeSR-1. Repeat this step twice.

6. Replace media with 2 mL RPMI/B27 without insulin +0.67 μL CHIR99021.

7. After 24 h, replace media with 2 mL RPMI/B27 without insulin.

8. After 48 h, apply combined media (1 mL old RPMI/B27 without insulin and 1 mL new RPMI/B27 without insulin +2 μL IWP2).

9. After 48 h, replace media with 2 mL RPMI/B27 without insulin.

10. After 48 h, change media to 2 mL RPMI/B27. Replace RPMI/B27 every 3 days until use.

3.4 CM Dissociation for Mitochondria Labeling

1. Rinse and incubate spontaneously contracting CMs in 0.25 % trypsin (EDTA) at 37 °C for 5 min.

2. Singularize cells using a P1000 pipette tip, combine with RPMI20, and centrifuge cells at $200 \times g$ for 5 min.

3. Aspirate off supernatant and resuspend cells in 1 mL RPMI20 + RI (*see* **Note 8**), and plate cells onto fibronectin-coated PDMS coverslips (*see* **Note 7**).

4. Culture cells for 3 days post dissociation to allow cells to adhere and reestablish their phenotypic cell morphology and function before fluorescently labeling the cells (*see* **Note 9**).

3.5 Fluorescence Labeling of Cells (See Notes 10–12)

1. Rinse cells with 1× PBS.

2. Add sufficient MitoTracker Red working solution into each well, and incubate at 37 °C for 30 min.

3. Rinse three times with 1× PBS.

4. Add 4 % paraformaldehyde; incubate cells at room temperature for 10 min.

5. Rinse with 1× PBS.

6. Incubate cells in PBS-T at room temperature for 10 min. Repeat this step twice.

7. Block cells in blocking buffer at 4 °C overnight.

8. Remove blocking buffer and add DAPI to all wells. Incubate at room temperature for 30 min.

9. Rinse off DAPI with 1× PBS. Repeat this step twice.

10. Dehydrate all samples using 50, 70, 95, and 100 % ethanol for 5 min each.

11. Air-dry all samples until completely dry.

12. Apply a small drop of ProLong Gold to each sample; invert sample onto a rectangular glass coverslide (*see* **Note 13**).

13. All samples should be dried overnight, sealed using nail polish, and stored at 4 °C.

3.6 Image Acquisition

1. Follow appropriate procedure to turn on and start up confocal microscope.

2. While using the scanning feature, adjust the scanning time and size of the image to be captured to at least 1/8 frame/s

and 1,024 pixels, respectively (*see* **Note 14**). Also adjust the confocal laser settings and pinhole to ensure clear, bright, and crisp images of the mitochondria and nuclei (*see* **Note 15**).

3. Capture and save the image as both .nd2 and separate .png, .jpg, or .tiff files in a location to be retrieved for later image analysis (*see* **Note 16**).

3.7 Image Analysis Using MQM

1. Start MATLAB by clicking on the MATLAB icon on the start menu or by opening the MQM files, *MitoMAT*, and *Measurements*.

2. Copy and import all image files to be analyzed to the same folder containing the MQM files.

3. Prepare for analysis by running the *Measurements* function file by clicking the green play button at the top of the home screen in MATLAB. Switch between files by clicking on the different function file tabs within MATLAB.

4. Press the green play button at the top of the home screen when the *MitoMAT* file is on the editor screen to begin the analysis (Fig. 1).

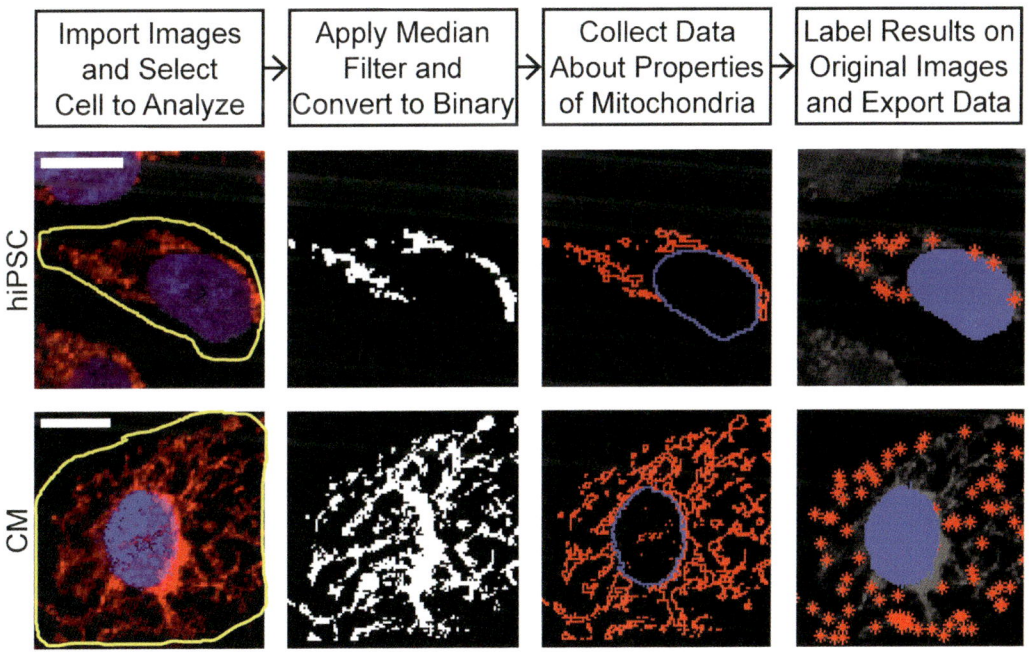

| Import Images and Select Cell to Analyze | → | Apply Median Filter and Convert to Binary | → | Collect Data About Properties of Mitochondria | → | Label Results on Original Images and Export Data |

Fig. 1 Application of MQM to fluorescently labeled mitochondria of hiPSCs and SC-CMs. In MATLAB, fluorescent images of mitochondria and nuclei are imported and overlaid, and a cell of interest is manually outlined. MQM converts fluorescent images to grayscale, applies filter to remove background, and converts images to binary. Finally, MQM can detect mitochondria and cell nuclei of hiPSCs and SC-CMs. Numerical data can be collected, exported to Excel, and analyzed. Scale bar, 10 μm

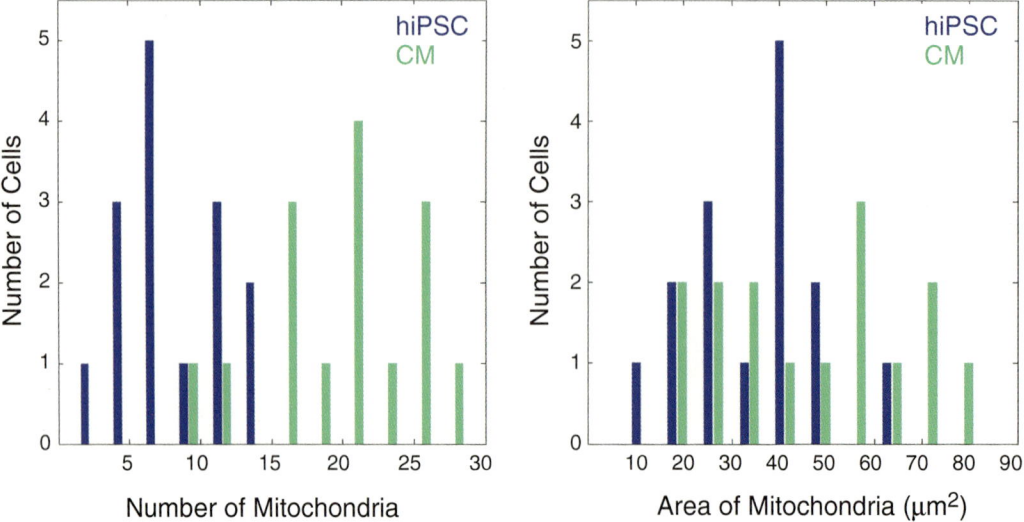

Fig. 2 Number and total area of mitochondria per cell increases during hiPSC differentiation. Mitochondria play an essential role in actively contracting SC-CMs and need to be analyzed at different stages of development. Mitochondria of hiPSCs and SC-CMs are labeled with MitoTracker Red and DAPI (nuclei). Cells are processed using MQM and analyzed in Excel to detect number and area of mitochondria per cell

5. When prompted, enter information about the images to be analyzed, i.e., image name, image magnification (in μm/pixel), and cell type for a label, if desired.

6. In the user interface, select the types of results to be collected, and specify a name for the Excel results file. After verifying this information, press continue (*see* **Note 17**).

7. An image of the fluorescently labeled mitochondria will appear. Click around the edges of the cell to be analyzed (*see* **Notes 18–21**).

8. Data can be analyzed in Excel (or similar software, e.g., Minitab, Origin) (Fig. 2).

4   Notes

1. Matrigel-coated 6-well and 12-well plates: Remove Matrigel aliquot from −80 °C freezer and transfer into a sterile tissue culture hood. Resuspend Matrigel aliquot in 1 mL cold DMEM-F12 and transfer to a 15-mL centrifuge tube. Add additional 11 mL cold DMEM-F12, mix, and add 1 mL Matrigel solution into each well of a 6-well plate (0.5 mL into each well of a 12-well plate). Incubate Matrigel solution for 30 min at room temperature. Add an additional 1 mL DMEM-F12 to each well and store at 37 °C.

2. mTeSR-1 + RI: Add 50 μL 5 mM ROCK inhibitor to 50 mL mTeSR-1 media and mix well.

3. RPMI/B27 without insulin: Combine 500 mL RPMI1640 media with 10 mL B-27 supplement without insulin, mix well, and store at 4 °C.

4. RPMI/B27: Combine 500 mL RPMI1640 media with 10 mL B-27 supplement and mix well.

5. RPMI20: Combine 40 mL RPMI1640 media with 10 mL FBS. Filter sterilize.

6. PDMS-coated coverslips can also be dried at room temperature overnight or until dry.

7. Seeding density of cells used for imaging experiments must be uniform across all groups to eliminate cell density effects as a source of variation in mitochondrial dynamics.

8. RPMI20 + RI: Add 50 µL 5 mM ROCK inhibitor to 50 mL RPMI20 and mix well.

9. CMs can be multinucleated; for the MQM module needed to analyze cells with more than one nucleus, email the corresponding author.

10. If necessary, adjust immunostaining protocol to ensure bright, clear images with no unspecific antibody binding or background fluorescence—it is impossible to analyze the mitochondria if the staining is not clear.

11. After MitoTracker Red working solution is applied, cells have to be handled in the dark to avoid photobleaching.

12. All staining procedures should be done under sterile conditions.

13. Any glass coverslide that fits onto an available confocal microscope stage holder can be chosen.

14. Images can be acquired at a slower scan speed (e.g., 1/16 frame/s) and higher resolutions (e.g., 2,048 pixels), if desired.

15. Making meaningful comparisons between samples requires that all images must be acquired at the same magnification and using identical image capture methods. Data from confocal images will not be comparable to those from non-confocal fluorescence images due to unequal contributions from nonfocal planes. Similarly, images from different magnifications will have unequal amounts of pixel bleeding from adjacent areas of the sample.

16. In order to be able to access the image files of interest, files must be stored in the same folder as the MQM files.

17. MATLAB is unable to write to an Excel document while it is open. Therefore, it is important to make sure that all Excel result files are closed during analysis.

18. Currently, MQM can only analyze one cell at a time.

19. Analysis time is between 2 and 4 min per cell, with time decreasing with use.

20. If the experimenter decides to reject the data from a particular cell, if there was inconsistency with labeling or a general error in use, the user must delete the data written in the Excel file associated with that run.

21. Label cell membrane to make it easier to identify which mitochondria belongs to which cell.

Acknowledgment

This work was supported by AHA 13PRE1470078 (P.K.), NSF-CBET-1150854 (E.A.L.), NSF-EPS-1158862 (D.A.D.), and NSF-EEC-1063107 (K.M.D.). D.A.D. is now located at the Department of Biological Sciences, State University of New York at Oswego, Oswego, NY.

References

1. Park SJ, Shin JH, Jeong JI et al (2014) Downregulation of mortalin exacerbates Abeta-mediated mitochondrial fragmentation and dysfunction. J Biol Chem 289:2195–2204

2. Rana A, Rera M, Walker DW (2013) Parkin overexpression during aging reduces proteotoxicity, alters mitochondrial dynamics, and extends lifespan. Proc Natl Acad Sci U S A 110:8638–8643

3. Abou-Sleiman PM, Muqit MM, Wood NW (2006) Expanding insights of mitochondrial dysfunction in Parkinson's disease. Nat Rev Neurosci 7:207–219

4. Hashimoto M, Rockenstein E, Crews L et al (2003) Role of protein aggregation in mitochondrial dysfunction and neurodegeneration in Alzheimer's and Parkinson's diseases. Neuromolecular Med 4:21–36

5. Castellani R, Hirai K, Aliev G et al (2002) Role of mitochondrial dysfunction in Alzheimer's disease. J Neurosci Res 70:357–360

6. Reddy PH, Beal MF (2008) Amyloid beta, mitochondrial dysfunction and synaptic damage: implications for cognitive decline in aging and Alzheimer's disease. Trends Mol Med 14:45–53

7. Yoon Y, Galloway CA, Jhun BS et al (2011) Mitochondrial dynamics in diabetes. Antioxid Redox Signal 14:439–457

8. Lowell BB, Shulman GI (2005) Mitochondrial dysfunction and type 2 diabetes. Science 307:384–387

9. Desai SP, Bhatia SN, Toner M et al (2013) Mitochondrial localization and the persistent migration of epithelial cancer cells. Biophys J 104:2077–2088

10. Chen L, Knowlton AA (2011) Mitochondrial dynamics in heart failure. Congest Heart Fail 17:257–261

11. Ong SB, Hausenloy DJ (2010) Mitochondrial morphology and cardiovascular disease. Cardiovasc Res 88:16–29

12. Liu S, Bai Y, Huang J et al (2013) Do mitochondria contribute to left ventricular non-compaction cardiomyopathy? New findings from myocardium of patients with left ventricular non-compaction cardiomyopathy. Mol Genet Metab 109:100–106

13. Dhalla NS, Rangi S, Zieroth S et al (2012) Alterations in sarcoplasmic reticulum and mitochondrial functions in diabetic cardiomyopathy. Exp Clin Cardiol 17:115–120

14. Lopaschuk GD, Jaswal JS (2010) Energy metabolic phenotype of the cardiomyocyte during development, differentiation, and postnatal maturation. J Cardiovasc Pharmacol 56:130–140

15. Folmes CD, Dzeja PP, Nelson TJ et al (2012) Metabolic plasticity in stem cell homeostasis and differentiation. Cell Stem Cell 11:596–606

16. Mitra K, Lippincott-Schwartz J (2010) Analysis of mitochondrial dynamics and functions using imaging approaches. Curr Protoc Cell Biol Chapter 4:Unit 4.25.1–21

17. Chung S, Dzeja PP, Faustino RS et al (2007) Mitochondrial oxidative metabolism is required for the cardiac differentiation of stem cells. Nat Clin Pract Cardiovasc Med 4(Suppl 1):S60–S67

18. Hattori F, Chen H, Yamashita H et al (2010) Nongenetic method for purifying stem cell-derived cardiomyocytes. Nat Methods 7:61–66

19. Rafelski SM (2013) Mitochondrial network morphology: building an integrative, geometrical view. BMC Biol 11:71

20. Lian X, Zhang J, Azarin SM et al (2013) Directed cardiomyocyte differentiation from human pluripotent stem cells by modulating Wnt/beta-catenin signaling under fully defined conditions. Nat Protoc 8:162–175

Chapter 38

An Ex Vivo Model for Studying Mitochondrial Trafficking in Neurons

Helena Bros, Raluca Niesner, and Carmen Infante-Duarte

Abstract

Distribution of mitochondria throughout the cytoplasm is necessary for cellular function and health. Due to their unique, highly polarized morphology, neurons are particularly vulnerable to defects of mitochondrial transport, and its disruption can contribute to neuropathology. In this chapter, we present an *ex vivo* method for monitoring mitochondrial transport within myelinated sensory and motor axons from spinal nerve roots. This approach can be used to investigate mitochondrial behavior under a number of experimental conditions, e.g., by applying ion channel modulators, ionophores, or toxins, as well as for testing the therapeutic potential of new strategies targeting axonal mitochondrial dynamics.

Key words Mitochondrial transport, Mitochondrial trafficking, Axonal transport, Mitochondria live imaging, Spinal nerve roots, *Ex vivo* explants

1 Introduction

Mitochondrial transport is required in most cells to ensure a correct distribution of functional mitochondria throughout the cytoplasm. This is particularly important in the case of neurons, because due to their elongated and asymmetric morphology, sites of mitochondrial biogenesis can be distal to sites of mitochondrial function. Mitochondria are typically recruited to the areas of high energy consumption, such as synapses [1], nodes of Ranvier [2], and active growth cones in growing axons [3]. To recruit mitochondria where they are most needed, neurons have specialized adaptor and motor proteins that attach mitochondria to the cytoskeleton and transport them along microtubules and actin filaments. There are two types of long-run mitochondrial movements in the axon: anterograde and retrograde. Anterograde transport, which uses kinesin motors, powers the transport of mitochondria toward synaptic terminals; retrograde transport, which involves dynein, translocates mitochondria from distal areas back to the cell body [4]. Mitochondria are able to quickly change direction, and

Volkmar Weissig and Marvin Edeas (eds.), *Mitochondrial Medicine: Volume I, Probing Mitochondrial Function*,
Methods in Molecular Biology, vol. 1264, DOI 10.1007/978-1-4939-2257-4_38, © Springer Science+Business Media New York 2015

periods of movement are frequently interspersed with pauses. In cultured neurons most of the mitochondria are stationary, and only about 20–30 % of axonal mitochondria are in motion [5–8]. This saltatory and bidirectional nature of mitochondrial movement is highly specific and differs from the transport of other organelles and axonal vesicles [4].

In view of the crucial role of mitochondrial transport for neuronal function, it is no surprise to find abnormal distributions of mitochondria in models of degenerative disorders of both the central [9–12] and peripheral [13] nervous system. Previous studies have shown that exogenous application of nitrosative [14] and oxidative stress [15] in neurons interfered with mitochondrial transport and caused cellular damage. Thus, many factors can influence mitochondrial trafficking and promote further pathology.

Mitochondrial transport in neurons has long been monitored in simplified in vitro systems either by using mitochondrial dyes [5–7] or by transfecting the cells with constructs encoding mitochondrially targeted fluorescent proteins [8, 14, 15]. More recently, the development of transgenic animals expressing mitochondria-targeted fluorescent proteins has allowed the study of mitochondrial transport in vivo [16, 17]. In this chapter, we present an intermediate approach that combines the simplicity of in vitro preparations with a preserved tissue cytoarchitecture and cellular interactions. This method makes use of acute explants of spinal nerve roots from adult mice, which contain peripheral myelinated axons. The somata, dendrites, and synaptic terminals are not included in the preparation; therefore, it will be suitable for investigating the movement of mitochondria selectively in the axon. Axonal mitochondria are easy to identify and can be tracked for several micrometers by using cationic, membrane-permeant dyes. The experimentator can decide whether to dissect dorsal or ventral roots, which will contain sensory or motor neurons, respectively. Because the explanted roots are maintained in solution, the extracellular environment can be modified to investigate how regulating ion homeostasis, intracellular signaling, or pathological cellular conditions can alter mitochondrial behavior [18]. We also encourage the use of this *ex vivo* model system as a platform for screening therapeutic molecules targeting mitochondrial dynamics, without the problems derived from poor penetration and low bioavailability.

2   Materials

2.1 Tools for Explanting Peripheral Roots

1. Straight surgical scissors.

2. Micro scissors for cutting bone.

3. Micro scissors for cutting nervous tissue.

4. Dissecting forceps.

5. Tweezers.

6. Strainer spoon.

7. Petri dishes.

8. pH meter.

9. Carbogen cylinder (95 % O_2 and 5 % CO_2).

10. Dissecting microscope.

11. Common labware such as pipettors and tips.

2.2 Reagents

1. Artificial cerebrospinal fluid (ACSF): 124 mM NaCl, 1.25 mM $NaH_2PO_4 \times H_2O$, 10 mM glucose $\times H_2O$, 1.8 mM $MgSO_4$, 1.6 mM $CaCl_2 \times 2H_2O$, 3 mM KCl, 26 mM $NaHCO_3$; pH 7.4 (adjusted with carbogen; *see* **Note 1**).

2. 1 mM MitoTracker Orange stock solution (Life Technologies, Darmstadt, Germany) in dimethyl sulfoxide, stored at –20 °C.

3. 100–500 nM MitoTracker Orange working solution in ACSF.

4. 70 % ethanol in spray bottle.

5. Anesthesia.

2.3 Equipment for Live Imaging

1. Inverted laser-scanning confocal microscope adapted for live cell imaging. The microscope software must have a time-lapse imaging function. We used the LSM 710 and ZEN imaging software (Carl Zeiss, Jena, Germany).

2. High magnification objective (e.g., 100×/1.46 oil immersion objective Plan-Apochromat from Carl Zeiss).

3. Microscope incubator for an atmosphere of 5 % CO_2. We used the XL-3 incubator from Carl Zeiss (*see* **Note 2**).

4. Live cell imaging microscope chamber, such as the open bath chambers from the RC-40 series (Warner Instruments, Hamden, USA).

5. Glass coverslips for the microscope chamber.

6. Custom-made net (or equivalent) to immobilize spinal root explants (*see* **Note 3**; Fig. 1).

3 Methods

All procedures are carried out at room temperature.

3.1 Explantation of the Spinal Cord

1. Sacrifice the mouse by a method approved at your institution.

2. Place the animal with the dorsal side up and immobilize the extremities in an extended position.

3. Spray back fur with 70 % ethanol.

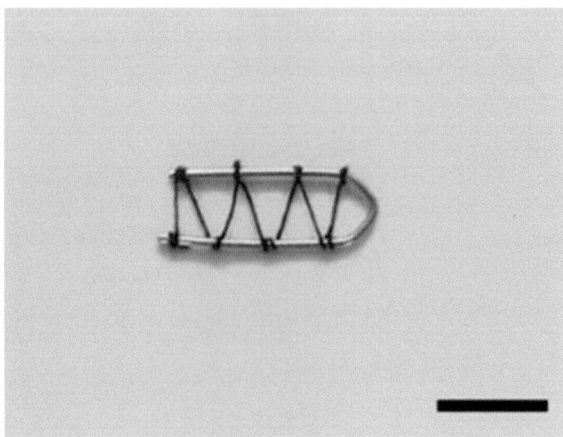

Fig. 1 Custom-made net used to immobilize the spinal nerve roots during time-lapse imaging. Scale bar: 1 cm

4. With the straight surgical scissors, make a long longitudinal cut in the middle of the back skin. Using the tweezers, pinch the skin from both sides of the cut and pull in opposite directions. Push the back muscles aside to expose the cervical, thoracic, and lumbar spinal regions. Manipulate the least tissue possible in order to minimize bleeding (*see* **Note 4**).

5. Make a large transversal cut to the vertebrae and spinal cord at the cervical level.

6. From this point, cut the dorsal side of the vertebral column longitudinally, until the lower lumbar level, to expose the spinal cord (Fig. 2a).

7. Pinch the spinal cord from the area of the transverse cut and gently pull to make the attached spinal nerve roots visible on both sides.

8. Using the micro scissors, cut the spinal nerve roots as distal as possible from the spinal cord.

9. Proceed likewise until reaching the lower lumbar level. Explant the spinal cord by making a transverse cut.

10. Transfer the explanted portion of the spinal cord with attached spinal nerve roots into a petri dish containing artificial cerebro-spinal fluid.

11. By pinching the spinal cord from one end with the forceps, move the tissue within the bathing solution to eliminate any rest of blood.

3.2 Separation of the Spinal Roots

Work with a dissecting microscope.

1. Identify the ventral and dorsal sides of the spinal cord. The roots that exit the spinal cord from the ventral side will contain motor axons; the roots leaving from the dorsal side have sensory axons.

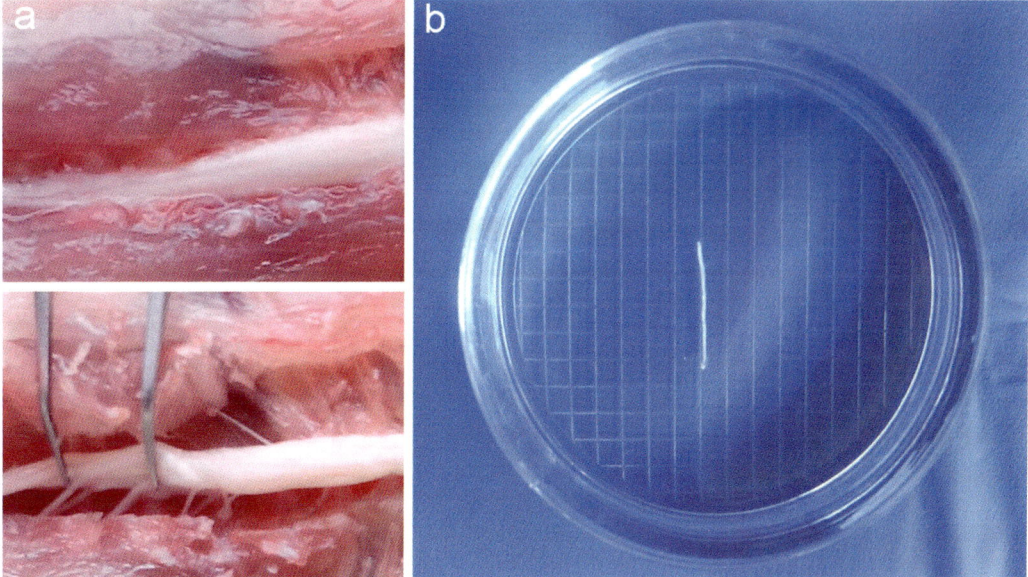

Fig. 2 (**a**) *Top*: dorsal view of the spinal cord after removing the overlying muscles and vertebral column. For clarity, the mouse has been perfused transcardially with PBS and paraformaldehyde prior to the dissection. *Bottom*: the spinal cord is being pulled aside to reveal the roots of the spinal nerves. (**b**) Explanted peripheral root. The squares measure 2×2 mm

2. On the desired area, identify the roots that are not damaged and approximately 1 cm long, and detach them from the spinal cord by cutting with micro scissors (Fig. 2b). For experiments where a control is needed, we suggest to select those roots that are preserved on both sides and work on a paired basis.

3. Preserve the orientation (i.e., proximal vs. distal) of the roots during the entire procedure. This will permit the differentiation between anterograde and retrograde transport of mitochondria (*see* **Note 5**).

3.3 Labeling of Mitochondria

1. Using the strainer spoon, transfer the spinal nerve roots into a new petri dish containing 100–500 nM of MitoTracker Orange in fresh ACSF (*see* **Note 6**).

2. Incubate in the dark for 15–30 min.

3. Transfer the roots into fresh ACSF.

3.4 Time-Lapse Imaging

1. Turn on the microscope incubator and set at 5 % CO_2.

2. Transfer a spinal root onto the live cell imaging microscope chamber, preserving the orientation of the tissue.

3. Immobilize the root by placing a net on the top of it (*see* **Note 3**).

4. Immediately after placement, fill in the microscope chamber with ACSF.

5. Place the sample under the confocal microscope. Position the root horizontally to facilitate the identification of anterograde and retrograde mitochondrial movement during the subsequent image analysis.

6. Excite MitoTracker Orange at 561 nm.

7. With a low magnification objective, screen the spinal nerve root for obvious axonal damage that might have been caused during the preparation, and discard the sample if necessary.

8. Using high magnification (*see* Subheading 2.3, **item 2**), select an area of interest in the middle of the root. Minimize the laser power and scanning time to avoid photobleaching and phototoxicity. MitoTracker Orange labels both myelin and mitochondria; fluorescent mitochondria should be visible within most axons (Fig. 3).

9. Turn on the time-lapse function on your software, and determine the frequency and total duration of the recordings according to your experimental question. For our experiments, one frame every 2 s over an imaging period of 1–5 min was sufficient to characterize mitochondrial transport.

3.5 Image Analysis   The open source ImageJ software, in combination with the *Difference Tracker* plug-in [19], can be used for analyzing mitochondrial transport and are both available online (http://rsbweb.nih.gov/ij/).

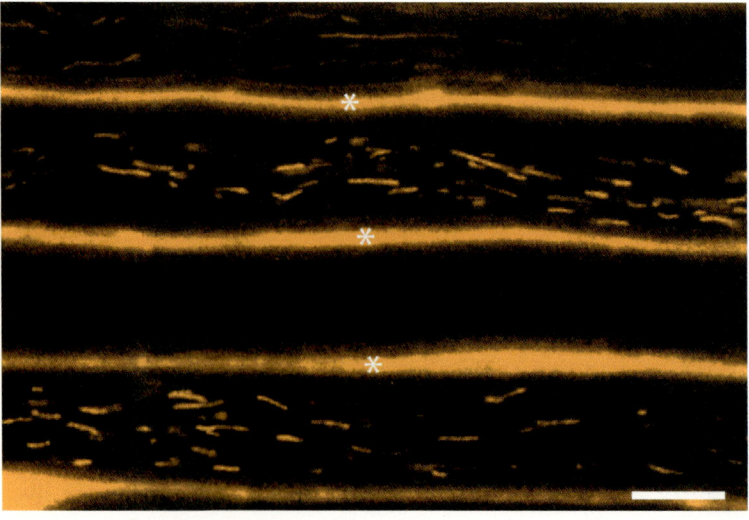

Fig. 3 Fluorescence picture of an explanted ventral root labeled with MitoTracker Orange. MitoTracker Orange labels both the myelin surrounding the axons (marked with an *asterisk*) and the mitochondria. Scale bar: 10 μm

Other software packages such as Volocity (PerkinElmer, Rodgau, Germany) and Imaris (Bitplane, Zurich, Switzerland) are commercially available and have a function of automatic particle tracking. However, for the highest precision, manual tracking of mitochondria is preferred.

4   Notes

1. To ensure a physiologic pH throughout the preparation, the different ACSF solutions must be continuously gassed with carbogen (a mixture of 95 % O_2 and 5 % CO_2), and their pH must be periodically checked. Increasing the amount of CO_2 bubbled into the ACSF will progressively lower the pH.

2. During the live cell imaging, the pH of the ACSF in the microscope chamber should be checked at various time points to confirm that it remains within the physiological range. The CO_2 levels in the microscope incubator should be adjusted accordingly.

3. It is essential to avoid the x–y drift during the time-lapse imaging. To immobilize the tissue onto the glass coverslip that will fit the microscope chamber, we placed a custom-made net on the top of the spinal nerve root. The net was made by bending a piece of a 40×1 mm aluminum wire into a U-shape and rolling a thread around it. A picture of the net is shown in Fig. 1.

4. If excessive bleeding complicates the visibility of the tissue, pipette ACSF onto the spinal cord and absorb the excess fluid with a cotton swab. Alternatively, transcardial perfusion with PBS prior to the dissection can be used to remove blood and facilitate the procedure.

5. For some experiments it may be important to preserve the orientation of the spinal nerve roots (i.e., proximal vs. distal) during the entire procedure. If this proves difficult to achieve, especially in the first attempts, one end of the root may be marked by applying 1 µL of MitoTracker Orange dye directly onto the area. This will immediately color that end of the root, will last for several minutes, and will not affect the mitochondria in the middle of the root, which should be used in the experiment.

6. MitoTracker Orange is a cationic membrane-permeant dye that accumulates within mitochondria because of their highly negative electrical charge. However, it may also label myelin. If the concentration of MitoTracker is too high, it will mainly colocalize with the myelin and will accumulate less in the mitochondria. To obtain a brighter and more selective mitochondria staining, use a lower concentration of the dye.

Acknowledgments

This work was supported by a fellowship from La Caixa and the Deutscher Akademischer Austauschdienst to H. Bros. We thank the JIMI network for infrastructural imaging support and J. Millward for reading the manuscript.

References

1. Gotow T, Miyaguchi K, Hashimoto PH (1991) Cytoplasmic architecture of the axon terminal: filamentous strands specifically associated with synaptic vesicles. Neuroscience 40:587–598

2. Fabricius C, Berthold CH, Rydmark M (1993) Axoplasmic organelles at nodes of Ranvier. II. Occurrence and distribution in large myelinated spinal cord axons of the adult cat. J Neurocytol 22:941–954

3. Morris RL, Hollenbeck PJ (1993) The regulation of bidirectional mitochondrial transport is coordinated with axonal outgrowth. J Cell Sci 104(Pt 3):917–927

4. Hollenbeck PJ, Saxton WM (2005) The axonal transport of mitochondria. J Cell Sci 118: 5411–5419

5. Overly CC, Rieff HI, Hollenbeck PJ (1996) Organelle motility and metabolism in axons vs. dendrites of cultured hippocampal neurons. J Cell Sci 109(Pt 5):971–980

6. Miller KE, Sheetz MP (2004) Axonal mitochondrial transport and potential are correlated. J Cell Sci 117:2791–2804

7. Ligon LA, Steward O (2000) Movement of mitochondria in the axons and dendrites of cultured hippocampal neurons. J Comp Neurol 427:340–350

8. Wang X, Schwarz TL (2009) The mechanism of Ca^{2+}-dependent regulation of kinesin-mediated mitochondrial motility. Cell 136: 163–174

9. De Vos KJ, Chapman AL, Tennant ME et al (2007) Familial amyotrophic lateral sclerosis-linked SOD1 mutants perturb fast axonal transport to reduce axonal mitochondria content. Hum Mol Genet 16:2720–2728

10. Pigino G, Morfini G, Pelsman A et al (2003) Alzheimer's presenilin 1 mutations impair kinesin-based axonal transport. J Neurosci 23: 4499–4508

11. Rui Y, Tiwari P, Xie Z et al (2006) Acute impairment of mitochondrial trafficking by beta-amyloid peptides in hippocampal neurons. J Neurosci 26:10480–10487

12. Trushina E, Dyer RB, Badger JD II et al (2004) Mutant huntingtin impairs axonal trafficking in mammalian neurons in vivo and in vitro. Mol Cell Biol 24:8195–8209

13. Baloh RH, Schmidt RE, Pestronk A et al (2007) Altered axonal mitochondrial transport in the pathogenesis of Charcot-Marie-Tooth disease from mitofusin 2 mutations. J Neurosci 27:422–430

14. Rintoul GL, Bennett VJ, Papaconstandinou NA et al (2006) Nitric oxide inhibits mitochondrial movement in forebrain neurons associated with disruption of mitochondrial membrane potential. J Neurochem 97:800–806

15. Fang C, Bourdette D, Banker G (2012) Oxidative stress inhibits axonal transport: implications for neurodegenerative diseases. Mol Neurodegener 7:29

16. Misgeld T, Kerschensteiner M, Bareyre FM et al (2007) Imaging axonal transport of mitochondria in vivo. Nat Methods 4:559–561

17. Plucinska G, Paquet D, Hruscha A et al (2012) In vivo imaging of disease-related mitochondrial dynamics in a vertebrate model system. J Neurosci 32:16203–16212

18. Bros H, Millward JM, Paul F et al (2014) Oxidative damage to mitochondria at the nodes of Ranvier precedes axon degeneration in ex vivo transected axons. Exp Neurol 261:127–35

19. Andrews S, Gilley J, Coleman MP (2010) Difference Tracker: ImageJ plugins for fully automated analysis of multiple axonal transport parameters. J Neurosci Methods 193:281–287

INDEX

Volkmar Weissig and Marvin Edeas (eds.), *Mitochondrial Medicine: Volume I, Probing Mitochondrial Function*,
Methods in Molecular Biology, vol. 1264, DOI 10.1007/978-1-4939-2257-4, © Springer Science+Business Media New York 2015

Printed by Printforce, the Netherlands